Sulfur Assimilation and Abiotic Stress in Plants

Nafees A. Khan • Sarvajeet Singh • Shahid Umar
Editors

Sulfur Assimilation and Abiotic Stress in Plants

Springer

Dr. Nafees A. Khan
Department of Botany
Aligarh Muslim University
Aligarh 202002
India
naf9@lycos.com

Dr. Sarvajeet Singh
Department of Botany
Aligarh Muslim University
Aligarh 202002
India
ssgill14@yahoo.co.in

Dr. Shahid Umar
Department of Botany
Jamia Hamdard
New Delhi 110062
India
s_umar9@hotmail.com

ISBN 978-3-642-09518-4 e-ISBN 978-3-540-76326-0

Softcover reprint of the hardcover 1st edition 2008

Cover Design: WMX Design GmbH, Heidelberg, Germany

Printed on acid-free paper

5 4 3 2 1 0

springer.com

Dedicated to Professor Samiullah, who nurtured the study of plant physiology and instilled the values of perseverance

Preface

Sulfur is one of the four major essential elements in the plant life cycle. Its assimilation in higher plants and its reduction in metabolically important sulfur compounds are crucial factors determining plant growth and vigor and resistance to stresses. The range of biological compounds that contain sulfur is wide. Sulfur serves important structural, regulatory, and catalytic functions in the context of proteins and as a cellular redox buffer in the form of tripeptide glutathione and certain proteins such as thioredoxin, glutaredoxin, and protein disulfide. In a cascade of enzymatic steps inorganic sulfur is converted to the nutritionally important sulfur-containing amino acid cysteine. Cysteine is the essential precursor of all organic molecules containing reduced sulfur; these range from the amino acid methionine to peptides such as glutathione, or phytochelatins, protein, vitamins, cofactors such as s-adenosyl methionine (SAM), and hormones. Cysteine and derived metabolites have the ability to regulate and repair abiotic stress-induced reactive oxygen species. They regulate the expression of many gene-encoding antioxidants, defense proteins, and signaling proteins. The information on sulfur assimilation can be exploited in tailoring transgenics for efficient sulfur utilization and in applied approaches for the sustenance of agricultural productivity through nutritional improvement and increased stress tolerance. The chapters in this book deal with the importance of sulfur in sustainable crop production, the role of sulfur-derived compounds in abiotic stress tolerance, and the enzymology of sulfur assimilation and its importance in stress tolerance. The physiology of sulfur assimilation in lower plants has also been discussed. Chapters 1 to 4 include the physiological aspects of sustainable crop production with sulfur. In addition, Chapter 4 deals with sulfur deficiency in agricultural soils and its impact on crop yield loss. Chapters 5, 6, and 7 describe the physiology of sulfur-metabolizing enzymes in abiotic stress management. Chapter 8 deals with stress-induced redox signals generated in chloroplast and modulation with sulfur metabolism. Chapters 9 and 10 are concerned with the role of cysteine and glutathione, respectively, in abiotic stress tolerance. The aspects of metal tolerance and its relationship with sulfur assimilation are described in Chapters 11, 12, and 13. Chapters 14 and 15 describe the physiology of sulfur assimilation in lower plants. Chapter 16 addresses the key problem of xenobiotic detoxification, as well as the potential role of the apoplast and possible links with sulfur metabolism. Chapter 17 is concerned with the interaction of sulfur and nitrogen.

Sincere thanks are expressed to the team at Springer-Verlag for their efficient and friendly cooperation in bringing out this volume. Thanks are also due to well-wishers, research students, and family for contributing productively in the completion of this task.

Nafees A. Khan
Sarvajeet Singh
Shahid Umar

Contents

Contributors

N.A. Anjum
Department of Botany, Aligarh Muslim University, Aligarh 202002, India

Sylvie Aubert
Laboratory for Environmental Biotechnology (LBE), Swiss Federal Institute of Technology Lausanne (EPFL), Station 6, 1015 Lausanne, Switzerland

Avtar Singh Bimbraw
Department of Agronomy, Agro Meteorology and Forestry Punjab Agricultural University, Ludhiana-141 004, Punjab, India

Basanti Biswal
School of Life Sciences, Sambalpur University, Jyotivihar-768019, Orissa, India

Udaya C. Biswal
Biology Enclave, Basant Vihar, Jyotivihar-768019, Orissa, India

Dimitris L. Bouranis
Plant Physiology Laboratory, Plant Biology Department, Faculty of Agricultural Biotechnology, Agricultural University of Athens, 75 Iera Odos, 11855 Athens, Greece

Stéphanie Braillard
Laboratory for Environmental Biotechnology (LBE), Swiss Federal Institute of Technology Lausanne (EPFL), Station 6, 1015 Lausanne, Switzerland

Peter Buchner
Plant Sciences Department, Rothamsted Research, Harpenden, Hertfordshire, AL5 2JQ, UK

David J. Burritt
Department of Botany, University of Otago, P.O. Box 56, Dunedin, New Zealand

Grazna Cacak-pietrzak
Department of Food Technology, Agricultural University, Nowoursynowska 159c, 02-787 Warszawa, Poland

Styliani N. Chorianopoulou
Plant Physiology Laboratory, Plant Biology Department, Faculty of Agricultural Biotechnology, Agricultural University of Athens, 75 Iera Odos, 11855 Athens Greece

Julie A. Francois
Plant Genetics Research Unit, USDA-ARS, Donald Danforth Plant Science Center, 975 N. Warson Rd., St. Louis, MO 63132 USA

Mario Giordano
Laboratorio di Fisiologia delle Alghe, Dipartimento di Scienze del Mare, Università Politecnica delle Marche, Via Brecce Bianche, 60131 Ancona, Italy

Malcolm J. Hawkesford
Plant Sciences Department, Rothamsted Research, Harpenden, Hertfordshire, AL5 2JQ, UK

Laura Hopkins
Plant Sciences Department, Rothamsted Research, Harpenden, Hertfordshire, AL5 2JQ, UK

Joseph M. Jez
Danforth Plant Science Center, 975 N. Warson Rd., St. Louis, MO 63132, USA

Padmanabha Joshi
Department of Physics, Anchal College, Padmapur, Orissa, India

N.A. Khan
Department of Botany, Aligarh Muslim University, Aligarh 202002, India

Renu Khanna-Chopra
Stress Physiology & Biochemistry Laboratory, Water Technology Centre, Indian Agricultural Research Institute (IARI), New Delhi - 110012, India

Marion Klein
Institute for Botany, University of Hannover, Herrenhäuserstr. 2, D-30419 Hannover, Germany

Hari B. Krishnan
Plant Genetics Research Unit, USDA-ARS, University of Missouri, Columbia, MO 65211, USA

Sangaralingam Kumaran
Danforth Plant Science Center, 975 N. Warson Rd., St. Louis, MO 63132, USA

Mario Malagoli
Department of Agricultural Biotechnology, University of Padua, Agripolis, 35020 Legnaro PD, Italy

S.S. Malhi
Agriculture and Agri-Food Canada, P.O. Box 1240, Melfort, Saskatchewan, Canada S0E 1A0

M. Müller
University of Graz, Institute of Plant Sciences, Schubertstrasse 51,
8010 Graz, Austria

R. Nazar
Department of Botany, Aligarh Muslim University, Aligarh 202002, India

Valérie Page
Laboratory for Environmental Biotechnology (LBE), Swiss Federal Institute of Technology Lausanne (EPFL), Station 6, CH-1015 Lausanne, Switzerland

Jutta Papenbrock
Institute for Botany, University of Hannover, Herrenhäuserstr. 2, D-30419 Hannover, Germany

Anna Podlesna
Plant Nutrition and Fertilisation Department, Institute of Soil Science and Plant Cultivation – National Research Institute, Czartoryskich 8, 24-100 Pulawy, Poland

Vassilis E. Protonotarios
Plant Physiology Laboratory, Plant Biology Department, Faculty of Agricultural Biotechnology, Agricultural University of Athens, 75 Iera Odos, 11855 Athens Greece

A. S. V. Chalapathi Rao
Department of Plant Sciences, School of Life Sciences, University of Hyderabad, Hyderabad 500 046, India

Simona Ratti
Laboratorio di Fisiologia delle Alghe, Dipartimento di Scienze del Mare, Università Politecnica delle Marche, Via Brecce Bianche, 60131 Ancona, Italy

Mukesh K. Raval
Department of Chemistry, Rajendra College, Bolangir, Orissa, India

Attipalli R. Reddy
Department of Plant Sciences, School of Life Sciences, University of Hyderabad, Hyderabad 500 046, India

Marjana Regvar
Department of Biology, Biotechnical Faculty, University of Ljubljana, Večna pot 111, SI-1000 Ljubljana, Slovenia

Heinrich W. Scherer
INRES - Plant Nutrition, University of Bonn, Karlrobert-Kreiten-Straße 13 D-53115 Bonn, Germany

Michela Schiavon
Department of Agricultural Biotechnology, University of Padua, Agripolis, 35020 Legnaro PD, Italy

J.J. Schoenau
Department of Soil Science, University of Saskatchewan, 51 Campus Drive, Saskatoon, Saskatchewan, Canada S7N 5A8

Jean-paul Schwitzguébel
Laboratory for Environmental Biotechnology (LBE), Swiss Federal Institute of Technology Lausanne (EPFL), Station 6, CH-1015 Lausanne, Switzerland

S. Singh
Department of Botany, Aligarh Muslim University, Aligarh 202002, India

Vassilis F. Siyiannis
Plant Physiology Laboratory, Plant Biology Department, Faculty of Agricultural Biotechnology, Agricultural University of Athens, 75 Iera Odos, 11855 Athens Greece

S. Srivalli
Stress Physiology & Biochemistry Laboratory, Water Technology Centre, Indian Agricultural Research Institute (IARI), New Delhi - 110012, India

S. Umar
Department of Botany, Faculty of Science, Hamdard University, New Delhi, 110062, India

C.L. Vera
Agriculture and Agri-Food Canada, P.O. Box 1240, Melfort, Saskatchewan, Canada S0E 1A0

Katarina Vogel-Mikuš
Department of Biology, Biotechnical Faculty, University of Ljubljana, Večna pot 111, SI-1000 Ljubljana, Slovenia

B. Zechmann
University of Graz, Institute of Plant Sciences, Schubertstrasse 51, 8010 Graz, Austria

G. Zellnig
University of Graz, Institute of Plant Sciences, Schubertstrasse 51, 8010 Graz, Austria

Chapter 1
Responses to Sulfur Limitation in Maize

Dimitris L. Bouranis, Peter Buchner, Styliani N. Chorianopoulou, Laura Hopkins, Vassilis E. Protonotarios, Vassilis F. Siyiannis and Malcolm J. Hawkesford(✉)

Abstract Maize (*Zea mays* L.) is a widely cultivated major cereal crop, and a model for a monocotyledonous C_4 plant with a substantial physiological and anatomical information base. Studies on sulfur uptake and metabolism indicate that uptake is comparable with other species, while metabolism is characterized by a segregation of components of both carbon and sulfur assimilatory pathways between different cell types. These patterns for distribution and subsequent assimilation are unique and require further elucidation. Ten distinct members of the maize sulfate transporter gene family are reported here; however specific expression and characterization data only exist for two of these. Varietal variation in uptake characteristics has been reported and may represent a potential for breeding improved sulfur use efficiency. Responses to sulfur-limitation which occur at several levels in overlapping succession are described. These include changes in gene expression focussed on cellular processes such as uptake through to wholesale changes in root: shoot biomass allocation and influences on cell death programming and the formation of aerenchyma. These provide mechanisms to maximise uptake, enhance utilization efficiency and moderate, although ultimately cannot prevent, an enhanced susceptibility to abiotic and biotic stresses.

1 Introduction

Sulfur (S) fertilization has become an issue due to reduced industrial emissions of S to the atmosphere and the consequent decreased deposition of S onto agricultural land in many areas of the world (McGrath et al. 1996). Sulfur nutrition plays an important role in the growth and development of higher plants, and sulfur limitation results in decreased yields and quality parameters of crops (Hawkesford 2000). Adequate sulfur nutrition is also required for plant health and resistance to pathogens (Rausch and Wachter 2005).

Malcolm J. Hawkesford
Plant Sciences Department, Rothamsted Research, Harpenden, Hertfordshire, AL5 2JQ, UK
malcolm.hawkesford@bbsrc.ac.uk

N.A. Khan et al. (eds.), *Sulfur Assimilation and Abiotic Stress in Plants.*

In all plant species studied to date, a series of specific responses aimed at optimizing acquisition and utilization are induced by sulfur limitation (Hawkesford 2000, Hawkesford and De Kok 2006). *Arabidopsis* has proved a useful model for basic molecular studies including the elucidation of the genes involved in these responses. The molecular knowledge from this model has been applied to several crop species, notably cereals (wheat, barley, rice) and Brassicas in relation to sulfur use efficiency in a physiological context. This review focuses specifically on sulfur nutrition in maize (*Zea mays* L.), a monocotyledonous species of the Poaceae family and a typical C_4 plant, and on responses of maize to limiting sulfur availability.

2 Characteristics of Maize

2.1 *Architecture*

The root system of maize comprises embryonic and postembryonic components (Abbe and Stein 1954). The embryonic root system consists of a single primary root and a variable number of seminal roots, while the postembryonic root system is made up of shoot-borne roots: the crown roots formed at consecutive underground nodes and the brace roots formed at consecutive aboveground nodes of the shoot. Lateral roots which emerge from all major root types also belong to the postembryonic root system. Later in development the postembryonic shoot-borne root system becomes dominant and, together with its lateral roots, is responsible for water and nutrient uptake. Although the anatomical structures of the different root types are very similar, they are initiated from different tissues during embryonic and postembryonic development (Hochholdinger et al. 2004).

The maize shoot consists of a superposition of elementary units, the phytomers. Each of these consists of a leaf, the internode below it, and the node with the axillary branch at the base of the internode. Phyllotaxy is opposite: each leaf includes the blade and the sheath. The blade unrolls progressively, while the sheath remains rolled, forming the sheath tube. Each phytomer develops within the cylinder formed by the rolled leaf of the preceding phytomer (Morrison et al. 1994).

2.2 *Functional Anatomy*

The maize leaf is characterized by Kranz anatomy, with a prominent bundle sheath cell (BSC) layer comprising concentric layers of cells having an intensely green color and, immediately surrounding, more loosely packed mesophyll cells (MC). CO_2 is initially fixed into malate in the MC and then transported into the BSC, where the formation of glycerate 3-phosphate is localized (Black 1973).

The spatial separation of phosphoenolpyruvate carboxylase and ribulose-1, 5-bisphosphate carboxylase/oxygenase (rubisco) is achieved by the anatomical differentiation of MC and BSC and cell-specific localization of the enzymes. The enzymes involved in the primary CO_2 fixation and malate and/or aspartate synthesis, such as cytosolic carbonic anhydrase, phosphoenolpyruvate carboxylase, pyruvate phosphate dikinase, and NADP-malate dehydrogenase, are localized predominantly in the MC, whereas NAD(P)-dependent malic enzyme, rubisco, rubisco activase, and some enzymes of the Calvin cycle are found exclusively in BSC (Sheen 1999, Edwards et al. 2001). BSC chloroplasts lack photosystem II and therefore exhibit very little oxygen evolution (Hatch and Osmond 1976). Consequently, noncyclic electron flow and the capacity for NADPH formation are restricted in BSC chloroplasts. The reduction of nitrate occurs exclusively in the MC (Moore and Black 1979). This combination of anatomy and physiology and the consequent division of labor is a primary factor contributing to high rates of carbon assimilation (Black 1973) and nitrogen use efficiency (Brown 1978) in C_4 plants. Evolved in the tropics in conditions of high temperatures, high light intensity, and low availability of water, maize utilizes CO_2 more efficiently than C_3 plants, and it can maintain a photosynthetic rate comparable to that of C_3 plants with reduction in water loss (Press 1999).

Maize root anatomy is typical of a monocotyledonous plant. Mature primary and seminal roots as well as shoot-borne roots exhibit a central cylinder (protostele) with many ribs of xylem. The pericycle forms the outermost layer of the central cylinder. The ground tissue consists of one layer of endodermal tissue with the suberized and often lignified casparian band and several layers of parenchymatous cortex tissue. The outermost cell layer is formed by the epidermis (rhizodermis), which consists of root-hair-forming trichoblasts and non-root-hair-forming atrichoblasts. In older roots the short-lived epidermis is replaced by a lignified and/or suberized exodermis, which develops from the outermost cells of the cortex and forms an additional casparian band. In above-ground-formed brace roots the epidermis persists and forms a protective cuticula. Maize roots do not show secondary growth of the root (Hochholdinger et al. 2004). The root apical meristem has a closed organization with three distinct tiers or layers of initials. The longitudinal structure of the maize root includes various partially overlapping specialized zones of development including the root cap, the root apical meristem, the distal elongation zone, the elongation zone and the maturation zone (Ishikawa and Evans 1995).

3 Sulfur Metabolism in Maize

3.1 Sulfate Uptake and Transport

Higher plants use inorganic sulfate as their major source of sulfur. Sulfate is actively taken up from the external environment into the symplast of the root by high-affinity sulfate transporters, and is reduced and assimilated into cysteine by the reductive sulfate assimilation pathway (Hell 1997, Leustek and Saito 1999, Hawkesford and Wray 2000).

The plant sulfate transporter gene family is divided into five distinct groups, and although not all of the respective gene products have confirmed sulfate transport activity based upon localization, functional and expression data, many may have distinct roles in S-assimilation and transport within the plant (Hawkesford 2003). Much of the data on individual isoforms has come from studies on *Arabidopsis* and *Brassica* (Buchner et al. 2004b, Buchner et al. 2004c, Hawkesford 2003, Takahashi et al. 1997), with studies on cereal sulfate transporters focused on uptake into the plant by Group 1 transporters (Buchner et al. 2004a, Smith et al. 1997, Vidmar et al. 1999). Group 1 includes the high-affinity sulfate transporters, which are responsible primarily, but not exclusively, for the transport of sulfate from the external environment into the root cells. One Group 1 isoform appears to be phloem-specific (Yoshimoto et al. 2003). Group 2 sulfate transporters have a lower affinity for sulfate, and are apparently involved in the movement of sulfate around the plant toward and between sink tissues (Hawkesford and Wray 2000). Group 3 is more enigmatic, and one isoform has been reported to be involved in a heterodimer structure, facilitating increased activity (Kataoka et al. 2004a). The Group 4 transporters are tonoplast-located and appear to be involved in efflux of sulfate from the vacuole (Kataoka et al. 2004b). Little information exists on the Group 5 transporters, which based on sequence alone are the most divergent isoforms, except that they are tonoplast-located (Buchner, Takahashi and Hawkesford, unpublished). Long-distance, inter-, and intracellular transport of sulfate around the plant depend on the coordinated expression of many of these sulfate transporters (Buchner et al. 2004b, Clarkson et al. 1993, Hawkesford and De Kok 2006).

A full-length cDNA encoding a Group 1 sulfate transporter (ZmST1;1) was isolated from maize roots (accession number AF355602, Hopkins et al. 2004), which shared 99.7% homology with the 701-bp partial sequence (accession number AF016306) reported by Bolchi et al (1999). ZmST1;1 is a 658 amino acid polypeptide (Mr 72209). This is the only maize sulfate transporter to be extensively characterized.

An extensive analysis of the databases reveals 10 maize sulfate transporter sequences, and these are shown in relation to the rice homologous gene family in Figure 1.1 and Table 1.1. Additional maize genes will most likely be identified to give a comparable number to that found for rice.

In addition to tissue specificity, there are local cellular patterns of specific expression (Hawkesford 2003). Maize represents an excellent model for the study of localization given the clear patterns of cellular organization. ZmST1;1 was expressed in epidermal cells and in the cell layer surrounding the central vascular bundle, in common with other homologous transporters of this group such as LeST1 (Howarth et al. 2003). Strongest expression away from the root tip was apparent in the epidermal and endodermal layers in common with the sites of highest expression of ZmST1;1 (Hopkins and Hawkesford 2003, Hopkins et al. 2004).

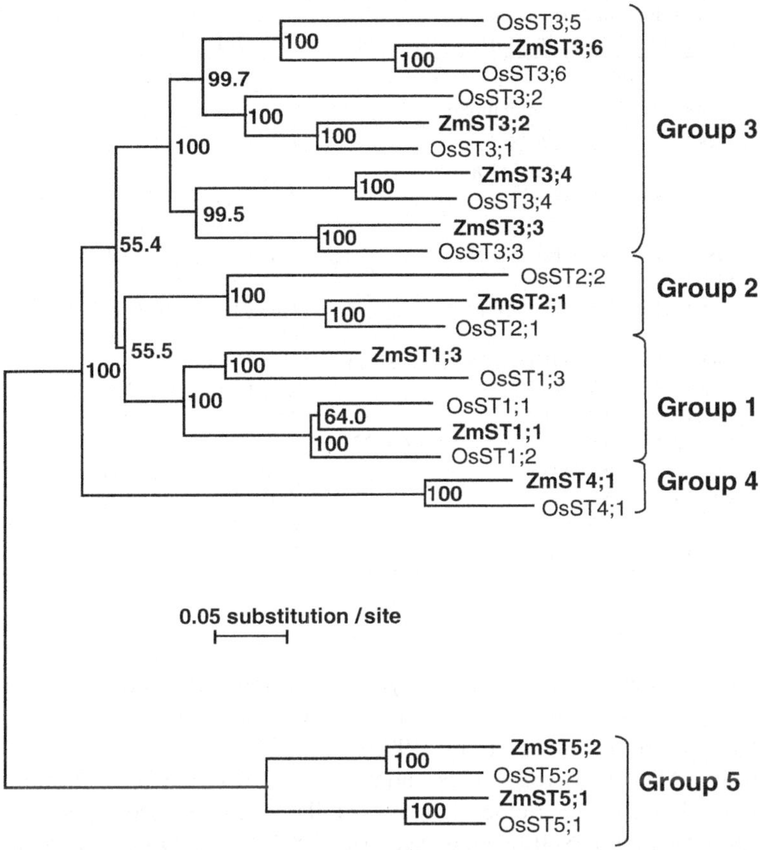

Fig. 1.1 Phylogenetic analysis of available sulfate transporter gene/mRNA sequences from maize compared to the rice sulfate transporter gene family. Neighbor joining tree (NJPLOT; Perrière and Gouy 1996) from the multiple alignment (Clustal V. 1.81; Thompson et al. 1997) of coding cDNA sequences of rice and maize sulfate transporters (see Table 1.1). The bootstrap values, expressed as a percentage, were obtained from 1,000 replicate trees

3.2 Sulfur Assimilation

Reductive sulfate assimilation is a multistep pathway. Sulfate is activated, reduced to sulfide, and incorporated into the amino acid cysteine, which is then used for the synthesis of other sulfur-containing compounds (Hawkesford and Wray 2000, Leustek and Saito 1999, Leustek et al. 2000). The sulfate assimilatory pathway includes ATP-sulfurylase (ATPS) and APS-reductase (APR). The fixation of the formed hydrogen sulfide is catalyzed by the serine acetyltransferase (SAT, EC 2.3.1.30)/O-acetylserine (thiol) lyase (OASTL, EC 4.2.99.8) bi-enzyme complex

Table 1.1 Accession numbers for rice and maize sulfate transporter mRNA and gene sequences. (DFCI = Dana Farber Cancer Institute; http://compbio.dfci.harvard.edu/tgi/)

	Rice		Maize	
Group	Name	Accession Number	Name	Accession Number
1	OsST1;1	AF493792	ZmST1;1	AF355602
	OsST1;2	NM_001055796		
	OsST1;3	AP004691	ZmST1;3	TC341973 DFCI *Zea mays* Gene Index
2	OsST2;1	NM_001055793	ZmST2;1	AY106086
	OsST2;2	AK067353		
3	OsST3;1	NM_001055577	ZmST3;1	TC318119 DFCI *Zea mays* Gene Index
	OsST3;2	NM_001071158 / AK107671		
	OsST3;3	NM_001063313	ZmST3;3	BT018869
	OsST3;4	AK104831	ZmST3;4	AY105934
	OsST3;5	AP003253 gene 30249-33228		
	OsST3;6	AK121195	ZmST3;6	CG222112 / CG117703
4	OsST4;1	AF493793	ZmST4;1	AM711891
5	OsST5;1	AK100928	ZmST5;1	BM336167
	OsST5;2	AK106547	ZmST5;2	CG145534/ CG367893

(Bogdanova and Hell 1997). SAT acetylates L-serine using acetyl-CoA to form O-acetylserine (OAS), which is then combined with sulfide in a reaction catalyzed by OASTL to form L-cysteine (Saito 1999).

In maize, ATPS and APR are essentially restricted to the BSC (Schmutz and Brunold 1984), hence restricting sulfate assimilation to these cell types (Kopriva and Koprivova 2005). Sulfite reductase (EC 1.8.7.1) and OASTL activities are found in both cell types at comparable levels (Passera and Ghisi 1982, Burnell 1984, Schmutz and Brunold 1984, 1985). The localization of ATPS and APR in BSC of C_4 plants implies a transport system for reduced sulfur compounds from BSC to MC. Cysteine synthesis seems to be located in the BSC and spatially separated from glutathione synthesis in the mesophyll cells (Burgener et al. 1998).

3.3 Glutathione Synthesis

Glutathione (GSH) is an important store of reduced sulfur, is a major form of transported reduced sulfur, and is involved in resistance to many biotic and abiotic stresses. GSH has a specific role in maintaining a cellular redox status (Kopriva and Koprivova 2005). Glutathione synthetase activity is greater in MC than in BSC, thus leading to GSH synthesis predominantly in the MC (Burgener et al. 1998) and higher GSH levels in this cell type (Doulis et al. 1997, Burgener et al. 1998, Kopriva et al. 2001). Cysteine is the suggested transport metabolite between the BSC and the MC, although the mechanism for this is unknown; this would represent a unique extracellular transport of this molecule in plants (Burgener et al. 1998). GSH content is affected by sulfur nutrition (Blake-Kalff et al. 2000).

The significance of compartmentation of sulfate assimilation in maize remains an open question. When maize plants were subjected to chilling stress, resistance to which is linked to GSH, APR activity and mRNA level were greatly increased in BSC, however, only mRNA but not enzyme activity was also detectable in MC, showing that posttranscriptional mechanisms also participate in the compartmentalization of sulfate assimilation in maize (Kopriva et al. 2001, Kopriva and Koprivova 2005).

4 Responses to Sulfur Limitation

4.1 Growth

In common with all other species examined, sulfur deprivation results in a shift of the biomass allocation program toward the root. When ten-day-old maize plants were deprived of a sulfur source, a small increase in shoot biomass was observed for the first four days, accompanied by a progressive increase in root biomass for the first six days, followed by a decline in both cases afterward (Louka and Bouranis, unpublished data; Fig. 1.2a). With regard to dry biomass, sulfur deprivation significantly reduced shoot growth and enhanced root proliferation (Bouranis et al. 2006). By day 6, the accumulation rate of dry mass in the shoot was reduced, resulting in a 44% decrease of the sulfur-deprived shoot by day 18. In contrast, dry mass accumulation in the root system was enhanced by day 18, resulting in a 63% increase of the sulfur-deprived root compared with the control. As a consequence, the root:shoot ratio of sulfur-deprived plants increased progressively from 0.53 at day 6 to 0.84 at day 18.

Growth of the crown roots of the sulfur-sufficient plants remained stable at 2.2 cm d^{-1} for 18 d. By contrast, sulfur deprivation resulted in an initial decrease of growth rate to 2 cm d^{-1} in the first 6 d followed by an increase to 3.5 cm d^{-1} up to day 12 and to 4.1 cm d^{-1} up to day 18. At day 12, sulfur-deprived root length was increased by 22% compared with the control. From day 12 to day 18, sulfur-sufficient and sulfur-deprived crown roots enlarged in length by 51.3%, and 74.3%, respectively. At day 18, sulfur-deprived root length was 40.2% longer than the control (Bouranis et al. 2006).

4.2 Leaf Anatomy

Sulfur deprivation affected leaf lignification. The lamina of the fully expanded second leaf of sulfur-deprived plants presented a more developed lower sclerenchyma

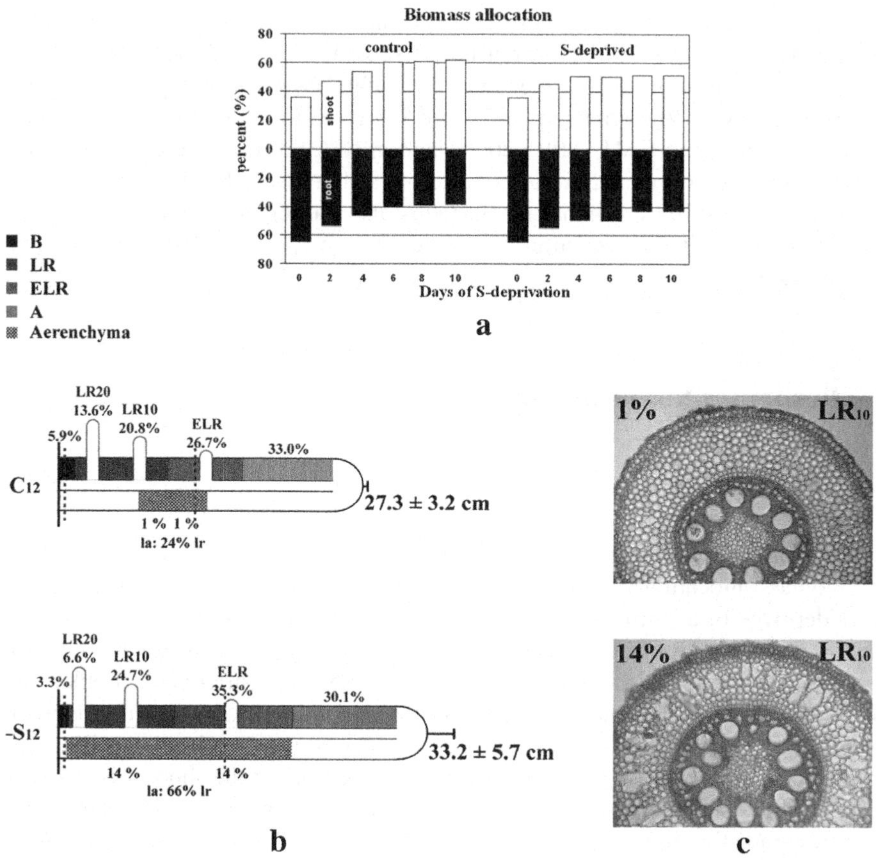

Fig. 1.2 Impacts of sulfur deprivation on root/shoot growth and morphology. (**a**) Progress of fresh biomass allocation (day 0: 10-day-old maize plants) between root and aerial part, and the corresponding effect of sulfur deprivation resulting in a shift of allocation toward the root after day 6 (Louka and Bouranis, unpublished data). (**b**) The shift in biomass allocation is accompanied by multiple progressively developed internal alterations, some of which are depicted in the schematic of the development of a crown root belonging to the second whorl in a complete (C) or sulfur-depleted (–S) nutrient solution 6 days later on. In the upper part of the cartoon (in gray scale), the various root sectors and the percentage of the total root length they occupy are given. LRx, root sector carrying lateral roots with mean length of x mm; ELR, root sector with emerging lateral roots; B and A, the basal and apical root sectors, respectively. In the lower part, the beginning and the end of aerenchyma formation are given, and within this range, the percentage of aerenchyma measured in the cortex of each root section included are provided. (**c**) Cross-sections of control (upper panel) and 12 d sulfur-deprived (lower panel) of LR10 (scale bar = 250 mm) with the percentage of aerenchyma measured in the cortex of this sector indicated (modified and used with permission from Bouranis et al. 2006)

and an intense lignification compared to sulfur-sufficient plants at day 6, mainly in the epidermal cells above the lower sclerenchyma as well as in the vascular bundles. In the lamina of the expanding fourth leaf of sulfur-deprived plants, vascular bundles were more developed, with more and larger xylem vessels compared with

S-sufficient plants (Bouranis et al. 2004). The functional significance of S-deprivation of leaf lignification is not known. Sulfur contributes in at least two key transformations of the lignification process: CoASH is required for the transformation of p-coumaric acid to p-coumaryl-SCoA, while S-adenosyl-methionine is used for the methylation process during biosynthesis of ferulic and sinapic acids.

4.3 Root Anatomy

Under well-oxygenated conditions, sulfate deprivation induced the formation of aerenchyma in maize roots (Fig. 1.2c), similar to nitrogen or phosphorus deprivation. When the beginning of sulfur deprivation coincided with the emergence of a crown root, aerenchyma started to form by day 6 of sulfur deprivation produced by lysigeny in the cortex of the root, and the first aerenchymatous spaces were created in the middle of the cortex of sulfur-deprived roots (Bouranis et al. 2003b). Hypodermis and endodermis were not affected at all by the lytic process. In a fully developed aerenchyma, chains of cells bridge hypodermis to endodermis and stele of roots. After 12 days of sulfur deprivation, aerenchyma covered the entire sector that carries the emerging lateral roots and a part of the nearby sector, with the expanded lateral roots being the 66% of root length and carrying aerenchyma in 14% of the cortex area. Developmentally, aerenchyma was disseminated toward the basal part of the S-sufficient root and toward the apical part in sulfur-deprived root. The basal and apical sectors had no aerenchyma at all (Fig. 1.2b, Bouranis et al. 2006). The functional significance of aerenchyma formation under S-deprivation is unknown and it may include redirection of scarce resources to the maintenance of essential sinks. Furthermore, it may be part of an adaptive program, which includes activation of adaptive pathways and disinvestment in nonessential sinks and pathways.

In general, programmed cell death involves fragmentation of nuclear DNA, involvement of Ca^{2+}, changes in protein phosphorylation, increases in nuclear heterochromatin, and involvement of reactive oxygen species (ROS) (Jones and Dangle 1996). The early stages of the lytic process that lead to aerenchymatous spaces were detected by assessing the loss of nuclear staining with acridine orange during S-deprivation. This revealed that in the apical sector at day 6 and in sections 2 mm from the root tip, the nuclei of the cortex of sulfur-deficient roots were shrunken and near to the cell wall, while by day 12 there was no fluorescence due to the nuclei in the cortex of sulfur-deficient roots (Bouranis et al. 2003b). Formation of ROS was detected in whole cells of the root cortex of sulfur-deprived plants by day 6. ROS appeared in groups of intact midcortex cells, and formation of superoxide anion and hydrogen peroxide was found in degenerating cells of the mid-cortex. Inhibition of superoxide dismutase activity (treatment with N, N-diethyldithiocarbamate; DDC) increased the presence of superoxide anions in the same locations and less hydrogen peroxide was apparent. Treatment of roots with ascorbate or ascorbate + DDC resulted in an almost complete inhibition of hydrogen peroxide production. By day 12, ROS were detected in the cell walls of endodermal, hypodermal, and epidermal cells of sulfur-deprived plants and were not detected in the cortical cells. The presence of hydrogen peroxide was located where superoxide anions appeared (Bouranis et al. 2003b). In the non-aerenchymatous basal sector under S-deprivation, when stained for ROS, plasma membranes of intact cortex cells fluoresced with increased intensity from day 6 onward (Bouranis et al. 2006). The alterations of calcium levels and pH in aerenchymatous sectors under sulfur deprivation were compared with the basal non-aerenchymatous region. There was a higher calcium concentration in the cell walls of the endodermis and epidermis than in the rest of the sulfur-deprived root tissues, and a higher pH was observed, mainly in the cell walls of the hypodermis and to a lesser extent in the cell walls of the endodermis of the sulfur-deprived roots (Bouranis et al. 2003b). The higher apparent Ca^{2+} concentrations may be linked to the elevated hydrogen peroxide levels: the plasma membrane NADPH oxidase, required for the controlled generation of hydrogen peroxide, is directly modulated by calcium fluxes. In addition to NADPH oxidase, pH-dependent cell wall peroxidases have been proposed as sources of hydrogen peroxide in the apoplast, activated by alkaline pH (Neill et al 2002).

The hydrogen peroxide may be directly utilized by wall-bound peroxidases in lignification and cell wall strengthening. Sulfur deprivation induced thickening of the cell walls of the endodermis, and after 12 days, thickness of the cell walls of endodermis of sulfur-deprived roots increased by 68%, estimated to be 2.7 μm. Furthermore, sulfate deprivation induced the lignification process in maize roots (Bouranis et al. 2003b). Lignified epidermal layers were located at the basal sectors, with a limited extension of the lignified layers toward the nearby lateral root carrying sector (Bouranis et al. 2006). Cell wall thickening may enhance mechanical strengthening of roots suffering sulfur deprivation.

4.4 Root Morphology

As found for other nutrients, root system architecture is affected by sulfur nutrition (see Fig. 1.2b, and Bouranis et al. 2006, Hopkins et al. 2004, Kutz et al. 2002). A proliferation of lateral roots has being reported for *Arabidopsis* in response to sulfur limitation (Kutz et al. 2002). In aeroponically-grown maize, both increased lateral root length and increased abundance of laterals near the tip of the main root occurred upon sulfur deprivation (Hopkins et al. 2004). In addition, in a hydroponic study, sulfur deprivation demonstrated shorter lateral roots in the sectors proximal to the root base. The lateral root proliferation is also linked to aerenchyma formation: aerenchyma found in the cortex along the root length was located particularly in the region of emerging or developing lateral roots (as discussed above, Section 4.3). The basal and apical sectors had no aerenchyma, and no aerenchyma connection was found with the shoot (Bouranis et al. 2006).

4.5 Sulfate Uptake and Transport

Maize responds to a limited sulfur supply by increasing influx capacity for sulfate in roots along with increased expression of genes encoding components of the uptake and assimilation pathway, a phenomenon observed in many other plant species. Removal of the sulfur source from the medium of maize seedlings led to a 3.8-fold increased capacity for sulfate uptake over a 10-min period (Hopkins et al. 2004, Quaggiotti et al. 2003). Influx was approximately equal in all sections of sulfur-sufficient and sulfur-deprived roots. In sections distal from the tip, the uptake may have been attributable to de novo influx from the external solution or from upward translocation; however, less than 0.5% of sulfate was recovered in the shoot tissues, indicating that translocation was minimal during the 10-min influx period and suggesting that all regions of the root in these young seedlings had a similar capacity for uptake (Hopkins et al. 2004). Expression analysis of abundance of sulfate transporter transcripts in the different root sections and in response to the removal of sulfur supply indicated an increased abundance of the ZmST1;1 mRNA in all root sections. There was a slightly higher apparent abundance in the sections 0-10 cm from the root base compared to the sections nearer the tip, which was most apparent 1 d after sulfur removal (Hopkins et al. 2004). Data on expression patterns of other members of the sulfate transporter gene family in maize are limited to ZmST2;1. (Hopkins et al. 2004) and a similar pattern of expression in roots (as observed by northern blotting) occurred for this transporter (Hopkins et al. 2004).

Sulfate uptake and transporter gene expression have been examined in two maize hybrids, chosen on the basis of their productivity at low nutritional inputs (Quaggiotti et al. 2003). Kinetic measurements of sulfate influx on S-deprived seedlings indicated contrasting adaptive strategies with either high affinity or high V_{max} in intact roots. Both varieties showed substantial increased capacity for uptake when S-deprived, although the response was more rapid and greater in magnitude in the variety with the higher V_{max}. Using a probe for ZmST1, a similar induction of mRNA abundance was observed for both varieties. As no other sulfate transporters were examined, the basis of the variation is not clear and deserves further attention. It may be that other members of the transporter family are involved or that variation exists in characteristics of the respective transporters. These alleles have not been isolated and sequenced; however, such variation is indicative of a potential for selection of genotypes with improved sulfur use efficiency.

The increased expression of the transcripts of the transporters involved in uptake may be due to the root proliferation but primarily represents an increase in density of the transporters in the root tissues. These responses together maximize the capacity for uptake from the soil under sulfur-limiting conditions (Hopkins et al. 2004).

Studies in maize, particularly using cell cultures, have contributed to the development of a model linking nutritional status with changes in gene expression which includes a role for OAS (Clarkson et al. 1999). The model proposes that increased expression of sulfate transporters' response to sulfur deprivation is not simply a de-repression model of regulation of sulfate uptake and assimilation involving a decrease in a downstream product of assimilation which relieves repression, but acts together with an increase in abundance of an inducer, the cysteine precursor, OAS (Smith et al. 1997, Hawkesford and Wray 2000), although not all data in other plant species are consistent with OAS as the key nutritional signal (Hopkins et al. 2005).

The sulfate transporters, ZmST1;1 and 2;1 were induced in leaf tissues (Hopkins et al. 2004). One day after the removal of the sulfur supply, a strong increase in abundance of the ZmST2;1 was found specifically in the first leaf, while this increased abundance was a transient phenomenon for the ZmST1;1. A similar increase in transcript abundance of the two sulfate transporters in the other leaves in response to the removal of sulfur supply was also observed (Hopkins et al. 2004). The occurrence of the Group 1 sulfate transporters in shoot tissues is not normally observed, but this clearly indicates that this group is not root specific. ZmST2;1 falls into Group 2 of the sulfate transporters.

A systematic analysis of 12 isoforms identified in *Brassica oleracea* (Groups 1-4) indicated a complex tissue distribution and tissue-specific responses to sulfur availability (Buchner et al. 2004b). Generally Group 1 transporters were root specific except under sulfur deprivation. The 2;1 isoforms occurred in roots, stems, and leaves, and mRNA abundance increased under S-deprivation. These data are not inconsistent with that seen for maize (Hopkins et al. 2004); however, without a more complete picture of isoforms in maize, direct comparisons are difficult.

4.6 Sulfur Assimilation

Patterns of expression of ATPS, APR, and OASTL, components of the sulfur assimilation pathway, have been studied in both roots and shoots of young aeroponically-grown maize seedlings (Hopkins et al. 2004). Increased abundances of both ATPS and APR mRNA pools were seen in both roots and leaves in response to S-limitation. In young seedlings these responses could be seen within 24 h of the removal of the external S-supply, although responses may be expected to be slower if substantial S-pools occur within more mature tissues. Levels of expression were substantial in the roots, indicating a potential for substantial S-assimilation in these tissues under these conditions. Some spatial regulation of expression was apparent, with higher expression levels occurring away from the root tip region and in the youngest leaves. Transcript abundance of OASTL did not vary in response to sulfur deprivation in the leaves or the roots in any significant pattern. In situ analysis of OASTL transcripts showed a unique spatial pattern of expression, with strongest expression throughout the cortex and noticeably absent in the root cap/quiescent zone and vascular tissues (Hopkins et al. 2004), although comparable data for other genes of the assimilatory pathway are not available. The most significant induction was of ATPS in the leaves, in contrast to *Arabidopsis*, where APR showed the greatest induction (Takahashi et al. 1997, Vauclare et al. 2002), indicating that regulation may vary between species. This contrast in regulation may be related to the spatial distribution of pathway components between different cell types, and distribution studies will be required to compare S-replete and S-deficient plant materials. Maize also differs from *Arabidopsis* in that regulation of gene expression of components of the assimilatory pathway appears to be mediated via cysteine (Bolchi et al. 1999) rather than glutathione as demonstrated for *Arabidopsis* and *Brassica* (Lappartient et al. 1999). The increased expression of enzymes of the assimilatory pathway, particularly ATPS and APR, represents an adaptation for the optimization of the assimilatory pathway under sulfur-limiting conditions and will maximize the flux of available sulfur from sulfate toward cysteine.

4.7 Glutathione Synthesis

Glutathione content of plant tissues is an indicator of sulfur nutritional status of the plant (Blake-Kalff et al. 2000), and indeed glutathione content of young maize seedlings is drastically reduced when plants are sulfur-deprived (Bolchi et al. 1999, Petrucco et al. 1996, Quaggiotti et al. 2003). As GSH is a transient store and a major transported form of reduced sulfur, sulfur-deficient conditions which limit its synthesis might be expected to cause depletion, either by dilution through growth (a minor influence if growth rate is retarded) or by consumption, as it is utilized for protein synthesis and other biosynthetic requirements for reduced sulfur. It has been suggested that glutathione plays an essential role as a signal for sulfur nutritional status; however, its importance relative to OAS has been questioned (Smith et al. 1997). As discussed above, in maize specifically, drastically reducing glutathione levels using a glutathione inhibitor (buthionine sulfoximine) did not induce a sulfate transporter or ATPS (Bolchi et al. 1999). A consequence of reduced glutathione would be impaired protection against stresses normally dependent upon its presence, including redox stress, metal exposure, chilling sensitivity, or pathogen infection. Under these conditions, other protective mechanisms may assume important roles. For example, using glutathione content as a marker, Petrucco et al. (1996) identified an isoflavone reductase by differential display, which was suggested to provide a redox protection role, compensating for decreased glutathione levels.

5 Combined Effects

5.1 Sulfur Deprivation and Nitrogen

Proteins contain both nitrogen and sulfur, and a deficiency of either will severely restrict protein synthesis and plant growth. An imbalance of supply leads to perturbations of pools of nitrate and sulfate ions and of intermediary metabolites. Plants respond at the level of gene expression and possibly enzyme activity to moderate these metabolite imbalances, and this implies interaction and coordination of the nitrogen and sulfur uptake and assimilatory pathways.

Induction (or de-repression) of the sulfate uptake and assimilatory pathway in response to sulfur deficiency will only occur in the presence of nitrogen (see, for example, Brunold 1993, Clarkson et al. 1989, Koprivova et al. 2000, Reuveny et al. 1980, Yamaguchi et al. 1999), Similarly, under sulfur-limiting conditions decreased expression and activity are seen for many enzymes of the nitrate assimilatory pathway (Friedrich and Schrader 1978, Prosser et al. 2001). In spite of the cross-coordination, accumulations of either nitrate or sulfate are seen in the vacuoles and perturbations in many amino acid pools, particularly the basic amino acids (Amancio et al. 1997, Migge et al. 2000, Prosser et al. 2001). An elegant study using ^{15}N to NMR monitor incorporation into amino acids indicated that the elevated amino acid pools were a consequence of de novo synthesis, not of protein breakdown (Amancio et al. 1997).

OAS is a key metabolite linking the nitrogen and sulfur pathways (Clarkson et al. 1999). It is the immediate precursor for cysteine combining with sulfide in a reaction catalyzed by OASTL. OAS pools will accumulate if nitrogen assimilation is greater than sulfate reduction, and the OAS has been suggested to be a positive regulator of sulfate transporter and other gene expression (for example, Smith et al. 1997). In addition, a sophisticated mechanism of control has been suggested in which SAT, which synthesizes OAS, is only active when in a complex with OASTL, and the complex is disrupted by excess OAS or cysteine and stabilized by sulfide. Hence cysteine synthesis is facilitated when sulfate reduction is active and when there are adequate sinks for the cysteine produced but also excess build-up of OAS is prevented. The complex, therefore, acts as a sensor of N/S balance (Hawkesford et al. 2006 and references therein).

5.2 *Sulfur Deprivation and Iron*

Maize has been a useful model for the examination of the interactions of iron and sulfur nutrition (Astolfi et al. 2003, 2004a, 2006a, Bouranis et al. 2003a). Sulfur metabolism is dependent upon adequate iron nutrition, for example, the upregulation of ATPS and OASTL in response to sulfur depletion (Astolfi et al. 2003, 2004a). In contrast, sulfate uptake capacity was increased by iron deficiency (Astolfi et al. 2004a, 2006b), while root cysteine content was elevated, apparently due to shoot to root translocation (Astolfi et al. 2006b). In studies on barley, sulfur deficiency has been shown to decrease phytosiderophore release and iron uptake (Astolfi et al. 2006a). Iron acquisition in graminaceous species is dependent upon phytosiderophore production, which in turn is dependent upon the sulfur assimilation pathway and methionine biosynthesis specifically. An early consequence of sulfur deficiency is therefore a decrease in iron content (Astolfi et al. 2003, Bouranis et al. 2003a).

5.3 Sulfur and Resistance to Stresses

Plant health as well as growth is dependent upon sulfur. Many sulfur-containing compounds are essential components of resistance mechanisms to abiotic and biotic stresses. Limitations in sulfur supply will influence partitioning of available sulfur. Allocation to sulfur-containing compounds of secondary metabolism may, for example, be secondary to partitioning to roots to aid proliferation. Pools of glutathione are known to be reduced under sulfur-limiting conditions (Blake-Kalff et al. 2000). Manipulation of GSH biosynthesis increases resistance to oxidative stress (Blaszczyk et al. 1999, May and Leaver 1993, May et al. 1998, Sirko et al. 2004, Youssefian et al. 2001). As discussed in section 4.3, ROS increase upon sulfur deprivation in maize. Resistance to metal stress (Astolfi et al. 2004b) or to high irradiance (Astolfi 2001) in maize has been shown to be dependent upon sulfur nutrition. Studies in other species have demonstrated clear requirements for adequate or "more than adequate" sulfur fertilization for resistance to pathogens (Cooper and Williams 2004, Rausch and Wachter 2005, Williams et al 2002).

6 Future Prospects

Maize is a unique model for studying the molecular physiology of sulfur nutrition in a monocotyledonous C_4 plant. It is an important crop, and such knowledge will be valuable for the development of low-input fertilizer strategies. The physiological information base, including detailed knowledge of anatomy and architecture, facilitates the understanding of interactions between cells and organs for the optimization of nutritional use efficiency, including sulfur. A next step will be to localize expression in relation to this physiological knowledge. As a model for the study of source sink interactions, the mature maize crops offer many opportunities, both for improving sulfur use efficiency and crop quality by optimizing sulfur nutritional content.

The ongoing genome sequencing project will give access to the required genes and will facilitate the analysis of their expression as has been determined for *Arabidopsis* and *Brassica*. No work has been undertaken examining genetic variability, which will be a vital resource for the development of low-input sustainable agriculture.

Acknowledgements Rothamsted Research receives grant-aided support from the Biotechnology and Biological Science Research Council (BBSRC) of the UK.

References

Abbe EC, Stein OL (1954) The origin of the shoot apex in maize: embryogeny. Am J Bot 41:285-293

Amancio S, Clarkson DT, Diogo E, Lewis M, Santos H (1997) Assimilation of nitrate and ammonium by sulphur deficient *Zea mays* cells. Plant Physiol Biochem 35:41-48

Astolfi S, Cesco S, Zuchi S, Neumann G, Roemheld V (2006a) Sulfur starvation reduces phytosiderophores release by iron-deficient barley plants. Soil Sci Plant Nutr 52:43-48

Astolfi S, De Biasi MG, Passera C (2001) Light-sulphur interactions on enzymes of carbon, nitrogen, and sulphur metabolism in maize plants. Photosynthetica 39:177-181

Astolfi S, Zuchi S, Cesco S, di Toppi LS, Pirazzi D, Badiani M, Varanini Z, Pinton R (2006b) Iron deficiency induces sulfate uptake and modulates redistribution of reduced sulfur pool in barley plants. Funct Plant Biol 33:1055-1061

Astolfi S, Zuchi S, Cesco S, Varanini Z, Pinton R (2004a) Influence of iron nutrition on sulphur uptake and metabolism in maize (*Zea mays* L.) roots. Soil Sci Plant Nutr 50:1079-83

Astolfi S, Zuchi S, Passera C (2004b) Effects of cadmium on the metabolic activity of *Avena sativa* plants grown in soil or hydroponic culture. Biol Plant 48:413-18

Astolfi S, Zuchi S, Passera C, Cesco S (2003) Does the sulphur assimilation pathway play a role in the response to Fe deficiency in maize (*Zea mays* L.) plants? J Plant Nutr 26:2111-21

Black CC (1973) Photosynthetic carbon fixation in relation to net CO_2 uptake. Annu Rev Plant Physiol 24:253-86

Blake-Kalff MMA, Hawkesford MJ, Zhao FJ, McGrath SP (2000) Diagnosing sulfur deficiency in field-grown oilseed rape (*Brassica napus* L.) and wheat (*Triticum aestivum* L.). Plant Soil 225:95-107

Blaszczyk A, Brodzik R, Sirko A (1999) Increased resistance to oxidative stress in transgenic tobacco plants overexpressing bacterial serine acetyltransferase. Plant J 20:237-43

Bogdanova N, Hell R (1997) Cysteine synthesis in plants. Protein-protein interactions of serine acetyltransferase from *Arabidopsis thaliana*. Plant J 11:251-62

Bolchi A, Petrucco S, Tenca PL, Foroni C, Ottonello S (1999) Coordinate modulation of maize sulfate permease and ATP sulfurylase mRNAs in response to variations in sulfur nutritional status: stereospecific down-regulation by L-cysteine. Plant Mol Biol 39:527-37

Bouranis DL, Chorianopoulou SN, Protonotarios VE, Siyiannis VF, Hopkins L, Hawkesford MJ (2003a) Leaf responses of young iron-inefficient maize plants to sulfur deprivation. J Plant Nutr 26:1189-1202

Bouranis DL, Chorianopoulou SN, Siyiannis VF, Protonotarios VE, Hawkesford MJ (2003b) Aerenchyma formation in roots of maize during sulphate starvation. Planta 217:382-91

Bouranis DL, Chorianopoulou SN, Siyiannis VF, Protonotarios VE, Hawkesford MJ (2004) Lignification of young maize plants under sulphate deprivation. Comp Biochem Physiol 137A:S239

Bouranis DL, Chorianopoulou SN, Kollias C, Maniou P, Protonotarios VE, Siyiannis VF, Hawkesford MJ (2006) Dynamics of aerenchyma distribution in the cortex of sulfate-deprived adventitious roots of maize. Annal Bot 97:695-704

Brown RH (1978) A difference in N use efficiency in C_3 and C_4 plants and its implications in adaptation and evolution. Crop Sci 18:93-8

Brunold C (1993) Regulatory interactions between sulfate and nitrate assimilation. In: de Kok LJ, Stulen K, Rennenberg H, Brunold C, Rauser WE (eds) Sulfur nutrition and sulfur assimilation in higher plants. SPB Academic Publishing, The Hague, pp. 61-75

Buchner P, Prosser IM, Hawkesford MJ (2004a) Phylogeny and expression of paralogous and orthologous sulphate transporter genes in diploid and wheats hexaploid. Genome 47:526-34

Buchner P, Stuiver CEE, Westerman S, Wirtz M, Hell R, Hawkesford MJ, De Kok LJ (2004b) Regulation of sulfate uptake and expression of sulfate transporter genes in *Brassica oleracea* as affected by atmospheric H_2S and pedospheric sulfate nutrition. Plant Physiol 136:3396-3408

Buchner P, Takahashi H, Hawkesford MJ (2004c) Plant sulphate transporters: co-ordination of uptake, intracellular and long-distance transport. J Exp Bot 55:1765-73

Burgener M, Suter M, Jones S, Brunold C (1998) Cyst(e)ine is the transport metabolite of assimilated sulfur from bundle-sheath to mesophyll cells in maize leaves. Plant Physiol 116:1315-22

Burnell JN (1984) Sulfate assimilation in C4 plants. Plant Physiol 75:873-5

Clarkson DT, Diogob E, Amâncio S (1999) Uptake and assimilation of sulphate by sulphur deficient *Zea mays* cells: The role of *O*-acetyl-L-serine in the interaction between nitrogen and sulphur assimilatory pathways. Plant Physiol Biochem 37:283-90

Clarkson DT, Hawkesford MJ, Davidian J-C (1993) Membrane and long-distance transport of sulfate. In: de Kok LJ, Stulen K, Rennenberg H, Brunold C, Rauser WE (eds) Sulfur nutrition and sulfur assimilation in higher plants. SPB Academic Publishing, The Hague, pp. 3-19

Clarkson DT, Saker LR, Purves JV (1989) Depression of nitrate and ammonium transport in barley plants with diminished sulfate status - evidence of co-regulation of nitrogen and sulfate intake. J Exp Bot 40:953-63

Cooper RM, Williams JS (2004) Elemental sulphur as an induced antifungal substance in plant defence. J Exp Bot 55:1947-53

Doulis AG, Debian N, Kingston-Smith AH, Foyer CH (1997) Differential localization of antioxidants in maize leaves. Plant Physiol 114:1031-7

Edwards GE, Franceschi VR, Ku MS, Vosnesenskaya EV, Pyankov VI, Andreo CS (2001) Compartmentation of photosynthesis in cells and tissues of C_4 plants. J Exp Bot 52:577-90

Friedrich JW, Schrader LE (1978) Sulfur deprivation and nitrogen metabolism in maize seedlings. Plant Physiol 61:900-3

Hatch MD, Osmond CB (1976) Compartmentation and transport in C4 photosynthesis. In: Stocking CR, Heber U (eds) Encyclopedia of plant physiology. New Series, Springer-Verlag, Berlin, pp. 144-84

Hawkesford MJ (2000) Plant responses to sulphur deficiency and the genetic manipulation of sulphate transporters to improve S-utilization efficiency. J Exp Bot 51:131-8

Hawkesford MJ (2003) Transporter gene families in plants: the sulphate transporter gene family - redundancy or specialization? Physiol Plant 117:155-63

Hawkesford MJ, De Kok LJ (2006) Managing sulphur metabolism in plants. Plant Cell Environ 29:382-95

Hawkesford MJ, Howarth JR, Buchner P (2006) Control of sulfur uptake, assimilation and metabolism. In: Plaxton WC, McManus MT (eds) Control of primary metabolism in plants, Annual Plant Reviews, Vol. 22, Blackwell Publishing, Oxford, pp. 348-72

Hawkesford MJ, Wray JL (2000) Molecular genetics of sulfate assimilation. Adv Bot Res 33:160-208

Hell R (1997) Molecular physiology of plant sulfur metabolism. Planta 202:138-48

Hochholdinger F, Woll K, Sauer M, Dembinsky D (2004) Genetic dissection of root formation in maize (*Zea mays*) reveals root-type specific developmental programmes. Ann Bot 93:359-68

Hopkins L, Hawkesford MJ (2003) Tissue and cell specific localization of a sulfate transporter in maize. In: Davidian J-C, Grill D, de Kok LJ, Stulen K, Hawkesford MJ, Schnug E, Rennenberg H (eds) Sulfur transport and assimilation in plants, Backhuys Publishers, Leiden, pp. 231-3

Hopkins L, Parmar S, Baszczyk A, Hesse H, Hoefgen R, Hawkesford MJ (2005) O-acetylserine and the regulation of expression of genes encoding components for sulfate uptake and assimilation in potato. Plant Physiol 138:433-40

Hopkins L, Parmar S, Bouranis DL, Howarth JR, Hawkesford MJ (2004) Coordinated expression of sulfate uptake and components of the sulfate assimilatory pathway in maize. Plant Biol 6:408-14

Howarth JR, Fourcroy P, Davidian J-C, Smith FW, Hawkesford MJ (2003) Cloning of two contrasting sulfate transporters from tomato induced by low sulfate and infection by the vascular pathogen *Verticillium dahliae*. Planta 218:58-64

Ishikawa H, Evans ML (1995) Specialized zones of development in roots. Plant Physiol 109:725-7

Jones AM, Dangl JL (1996) Logjam at the Styx: programmed cell death in plants. Trends Plant Sci 4:114-19

Kataoka T, Hayashi N, Takahashi-Watanabe A, Inoue E, Yamaya T, Takahashi H (2004a) Functional analysis of sulfate transporters SULTR2;1 and SULTR3;5 in *Arabidopsis*. Plant Cell Physiol 45:S174

Kataoka T, Hayashi N, Yamaya T, Takahashi H (2004b) Root-to-shoot transport of sulfate in Arabidopsis. Evidence for the role of SULTR3;5 as a component of low-affinity sulfate transport system in the root vasculature. Plant Physiol 136:4198-4204

Kopriva S, Jones S, Koprivova A, Suter M, von Ballmoos P, Brander K, Flückiger J, Brunold C (2001) Influence of chilling stress on the intercellular distribution of assimilatory sulfate reduction and thiols in *Zea mays*. Plant Biol 3:24-31

Kopriva S, Koprivova A (2005) Sulfate assimilation and glutathione synthesis in C_4 plants. Photosyn Res 86:363-72

Koprivova A, Suter M, Opden Camp R, Brunold C, Kopriva S (2000) Regulation of sulfate assimilation by nitrogen in *Arabidopsis*. Plant Physiol 122:737-46

Kutz A, Muller A, Hennig P, Kaiser WM, Piotrowski M, Weiler EW (2002) A role for nitrilase 3 in the regulation of root morphology in sulphur-starving *Arabidopsis thaliana*. Plant J 30:95-106

Lappartient AG, Vidmar JJ, Leustek T, Glass ADM, Touraine B (1999) Inter-organ signaling in plants: regulation of ATP sulfurylase and sulfate transporter genes expression in roots mediated by phloem-transolocated compound. Plant J 18:89-95

Leustek T, Saito K (1999) Sulfate transport and assimilation in plants. Plant Physiol 120:637-43

Leustek T, Martin MN, Bick J-A, Davies JP (2000) Pathways and regulation of sulfur metabolism revealed through molecular and genetic studies. Annu Rev Plant Physiol Plant Mol Biol 51:141-65

May MJ, Leaver CJ (1993) Oxidative stimulation of glutathione synthesis in *Arabidopsis thaliana* suspension cultures. Plant Physiol 103:621-7

May MJ, Vernoux T, Sanchez-Fernandez R, Van Montagu M, Inze D (1998) Evidence for post-transcriptional activation of gamma-glutamylcysteine synthetase during plant stress responses. Proc Natl Acad Sci USA 95:12049-54

McGrath SP, Zhao FJ, Withers PJA (1996). Development of sulphur deficiency in crops and its treatment. Proceedings of the Fertilizer Society, No. 379. The Fertilizer Society, Peterborough

Migge A, Bork C, Hell R, Becker TW (2000) Negative regulation of nitrate reductase gene expression by glutamine or asparagines accumulating in leaves of sulfur-deprived tobacco. Planta 211:587-95

Moore RC, Black CC (1979) Nitrogen assimilation pathways in leaf mesophyll and bundle sheath cells of C_4 photosynthesis plants formulated from comparative studies with *Digitaria sanguinalis* (L.) Scop. Plant Physiol 64:309-13

Morrison TA, Kessler JR, Buxton DR (1994) Maize internode elongation patterns. Crop Science 34:1055-60

Neill S, Desikan R, Hancock J (2002) Hydrogen peroxide signalling. Curr Opin Plant Biol 5:388-95

Passera C, Ghisi R. (1982) ATP sulphurylase activity and O-acetylserine sulphydrylase in isolated mesophyll protoplasts and bundle sheath strands of deprived maize leaves. J Exp Bot 83:432-8

Perrière G, Gouy M (1996) WWW-Query: An on-line retrieval system for biological sequence banks. Biochimie 78:364-9

Petrucco S, Bolchi A, Foroni C, Percudani R, Rossi GL, Ottonello S (1996) A maize gene encoding an NADPH binding enzyme highly homologous to isoflavone reductases is activated in response to sulphur starvation. Plant Cell 8:69-80

Press MC (1999) The functional significance of leaf structure: a search for generalizations. New Phytol 143:213-19

Prosser IM, Purves JV, Saker LR, Clarkson DT (2001) Rapid disruption of nitrogen metabolism and nitrate transport in spinach plants deprived of sulphate. J Exp Bot 52:113-21

Quaggiotti S, Abrahamshon C, Malagoli M, Ferrari G (2003) Physiological and molecular aspects of sulphate uptake in two maize hybrids in response to S-deprivation. J Plant Physiol 160:167-73

Rausch T, Wachter A (2005) Sulfur metabolism: a versatile platform for launching defence operations. Trends Plant Sci 10:503-9

Reuveny Z, Dougall DK, Trinity PM (1980) Regulatory coupling of nitrate and sulfate assimilation pathways in cultured tobacco cells. Proc Natl Acad Sci USA 77:6670-2

Saito K (1999) Regulation of sulfate transport and synthesis of sulfur-containing amino acids. Curr Opin Plant Biol 3:188-95

Schmutz D, Brunold C (1984) Intercellular localization of assimilatory sulfate reduction in leaves of *Zea mays* and *Triticum aestivum*. Plant Physiol 74:866-70

Schmutz D, Brunold C (1985) Localization of nitrite and sulfite reductase in bundle sheath and mesophyll cells of maize leaves. Physiol Plant 64:523-8

Sheen J (1999) C_4 gene expression. Annu Rev Plant Physiol Plant Mol Biol 50:187-217

Sirko A, Blaszczyk A, Liszewska F (2004) Overproduction of SAT and/or OASTL in transgenic plants: a survey of effects. J Exp Bot 55:1881-8

Smith FW, Hawkesford MJ, Ealing PM, Clarkson DT, Vanden Berg PJ, Belcher AR, Warrilow AGS (1997) Regulation of expression of a cDNA from barley roots encoding a high affinity sulphate transporter. Plant J 12:875-84

Takahashi H, Yamazaki M, Sasakura N, Watanabe A, Leustek T, Engler JD, van Montagu M, Saito K (1997) Regulation of sulfur assimilation in higher plants: A sulfate transporter induced in sulfate-starved roots plays a central role in *Arabidopsis thaliana*. Proc Natl Acad Sci USA 94:11102-7

Thompson JD, Gibson TJ, Plewniak F, Jeanmougin F, Higgins DG (1997) The ClustalX windows interface: flexible strategies for multiple sequence alignment aided by quality analysis tools. Nucl Acids Res 24:4876-82

Vauclare P, Kopriva S, Fell D, Suter M, Sticher L, von Ballmoos P, Krähenbühl U, Op den Camp R, Brunold C (2002) Flux control of sulphate assimilation in *Arabidopsis thaliana*: adenosine 5'-phosphosulphate reductase is more susceptible than ATP sulphurylase to negative control by thiols. Plant J 31:729-40

Vidmar JJ, Schjoerring JK, Touraine B, Glass ADM (1999) Regulation of the hvst1 gene encoding a high-affinity sulfate transporter from *Hordeum vulgare*. Plant Mol Biol 40:883-92

Williams JS, Hall SA, Hawkesford MJ, Beale MH, Cooper RM (2002) Elemental sulfur and thiol accumulation in tomato and defense against a fungal vascular pathogen. Plant Physiol 128:150-9

Yamaguchi Y, Nakamura T, Harada E, Koizumi N, Sano H (1999) Differential accumulation of transcripts encoding sulfur assimilation enzymes upon sulfur and or nitrogen deprivation in *Arabidopsis thaliana*. Biosci Biotechnol Biochem 63:762-6

Yoshimoto N, Inoue E, Saito K, Yamaya T, Takahashi H (2003) Phloem-localizing sulfate transporter, Sultr1;3, mediates re-distribution of sulfur from source to sink organs in Arabidopsis. Plant Physiol 131:1511-17

Youssefian S, Nakamura M, Orudgev E, Kondo N (2001) Increased cysteine biosynthesis capacity of transgenic tobacco overexpressing an O-acetylserine(thiol) lyase modifies plant responses to oxidative stress. Plant Physiol 126:1001-11

Chapter 2
Feasibility of Elemental S Fertilizers for Optimum Seed Yield and Quality of Canola in the Parkland Region of the Canadian Great Plains

S.S. Malhi(✉), J.J. Schoenau and C.L. Vera

Abstract The rate of sulfur (S) oxidation in Canadian prairie soils under incubation is enhanced with increasing temperature (with a maximum close to 40 °C) and moisture until lack of aeration above field capacity inhibits S oxidation. Oxidation of elemental S to sulfate-S is positively related to pH, organic matter content, nutrient-supplying power, and microbial activity, while it is inversely related to clay content of soil. Techniques that increase the oxidation rate of elemental S in soil include reducing particle size of the fertilizer product, using tillage to incorporate elemental S fertilizer into the soil to increase contact area between elemental S particles and soil microorganisms, and application in advance of crop demand to allow for oxidation to occur before plants require this element. In a 6-year (1999 to 2004) case study on a S-deficient loam soil in northeastern Saskatchewan, elemental S fertilizers were not effective in increasing seed yield and S uptake of canola in 1999, particularly when applied in spring. From 2000 onward, elemental S fertilizers (ES-90 and ES-95) increased seed yield and S uptake significantly over the zero-S control, but usually not at the same level as with sulfate-S containing fertilizers (ammonium sulfate and Agrium Plus). Autumn-applied elemental S usually had greater seed yield and S uptake than the spring-applied elemental S, and the opposite was true for ammonium sulfate in 3 of 6 years. Oil concentration in canola seed showed trends similar to seed yield, though the effects were smaller and less frequent. The need for greater dispersion of elemental S particles from granular elemental S fertilizers, to enhance microbial oxidation to sulfate-S in soil, was considered to be the main requirement for increasing the short-term availability of S in elemental S fertilizers. Canola seed yield, and N and S uptake with surface application of suspension and powder formulations of elemental S fertilizers were similar to sulfate-S fertilizer. Residual benefit of many successive annual elemental S fertilizer applications on soil S fertility was not cumulative. Residual nitrate-N in soil was much higher in the zero-S control than with applied S, and it also tended to be higher in some elemental S treatments than in spring-applied ammonium sulfate. In conclusion, the amount of elemental S fertilizer, needed to adequately meet crop S requirements, should be adjusted based on environmental conditions, soil properties, microbial activity, dispersion of elemental

S.S. Malhi
Agriculture and Agri-Food Canada, P.O. Box 1240, Melfort, Saskatchewan, S0E 1A0 Canada
malhis@agr.gc.ca

N.A. Khan et al. (eds.), *Sulfur Assimilation and Abiotic Stress in Plants.*

S particles from fertilizer granules, and balance between oxidation and immobilization, leaching, or other S losses in the Parkland region of Canadian prairies.

1 Introduction

Canola is one of the major cash crops in the Prairie Provinces of Canada (Statistics Canada 2002). The majority of canola is grown in the Parkland region, where many agricultural soils are deficient or potentially deficient in plant-available sulfur (S) for high seed yield of canola (Bettany et al. 1983, Doyle and Cowell 1993). Because of canola's high requirement for S (Grant and Bailey 1993), there is faster depletion of plant-available S from soil and increased instances and severity of S deficiency during peak growing periods, especially from the use of high-yielding cultivars and application of high rates of N and P fertilizers. Plants use only sulfate-S, and application of sulfate-S fertilizers has been successfully used to eliminate S deficiency in canola (Malhi and Gill 2002). There is now a wide variety of commercial elemental S fertilizers, which normally cost less per unit of S than sulfate-S fertilizers. However, the oxidation of elemental S to sulfate-S is dependent on soil properties, environmental conditions, and dispersion of elemental S particles from fertilizer granules (Solberg 1986, Solberg et al. 2003). In the Canadian prairie soils, granular elemental S fertilizers have been generally found to be less effective than sulfate-S fertilizers to prevent/eliminate S deficiency in canola (Ukrainetz 1982, Karamanos and Janzen 1991, Nuttall et al. 1993, Grant et al. 2001, Malhi 2005, Malhi et al. 2005). As in most of the earlier field studies, elemental S fertilizers were not applied to the same plots for a number of years, and there is limited information on the effects of long-term annual applications and residual effects of elemental S fertilizers on canola seed yield, quality, and S uptake in comparison to sulfate-S fertilizers. The objective of this paper is to discuss factors affecting the oxidation and availability of elemental S in soil and the feasibility of elemental S fertilizers for optimum seed yield and quality of canola in the Parkland region of western Canada. Selected characteristics of elemental S products are presented in Table 2.1.

2 Factors Affecting Oxidation of Elemental S

In soil, elemental S is oxidized to sulfate-S mainly by autotrophic and heterotrophic microorganisms (Schoenau and Germida 1992). Microbial activity and, consequently, the rate of elemental S oxidation are affected by environmental conditions, soil characteristics, and soil and fertilizer management practices that increase the dispersion of elemental S particles from S fertilizer granules.

2.1 Environmental Conditions (Temperature, Moisture, Aeration)

Incubation studies have shown increased rate of S oxidation with increased temperature, with an optimum close to 40 °C (Janzen and Bettany 1987b, Solberg

Table 2.1 Selected characteristics of S sources and their S particles

Name	S source: Characteristics	S source: Sulfur (%)[1]	S particles: Size Range (μm)	S particles: Weight ((%)
ES-99	Granules	ES: 99	<74	100
ES-95	Granules	ES: 95	74-44	2
	Bentonite-based		<44	98
ES-90 or Bentonitic S (A)	10% bentonite	ES: 90	Not done	
	Granules, 2-4 mm			
	1979 product			
ES-90 or Bentonitic S (B)	10% bentonite	ES: 90	2 000-1 180	17
	Dome shaped		1 180-600	36
	2-4 mm long		600-250	20
	1-3 mm wide		250-150	13
	1985 product		150-75	11
			75-45	3
			<45	1
ES-90 or Bentonitic S (C)	10% bentonite	ES: 90	>2 000	25
	Granules, 2-4 mm		2000-1180	15
			1180-600	15
			600-250	16
			250-150	13
			150-75	12
			75-45	4
			<45	1
Urea S (A)	Urea pallets	ES: 10.4	25-50	50
	1-3 mm diameter		<25	50
	Fine S impregnated			
Urea S (B)	Urea pallets	ES: 22.4	25-50	50
	1-3 mm diameter		<25	50
	Fine S impregnated			
Urea S (C)	Urea pallets	ES: 20.0	25-50	50
	1-3 mm diameter		<25	50
	Fine S impregnated			
Biosul-90	Granules	ES: 90	<10	100
Biosul-50	Suspension	ES: 50	<10	100
Sodium sulfate	Reagent grade, crystals	Sulfate-S: 22.5	Not done	
Potassium sulfate	Fertilizer grade, granules	Sulfate-S: 17	Not available	
Ammonium sulfate	Fertilizer grade, granules	Sulfate-S: 24	Not available	
Agrium Plus	Granules	ES: 21.7	90-50	15
		Sulfate-S: 18.7	<50	85
Sulgro-68	Granules	ES: 60	<74	100
	Bentonite-based	Sulfate-S: 8		
Lab fine ES-99.5 or	Reagent grade	ES: 99.5	250-150	22.9
Fine S or Fine sublimed S	Powder		150-75	58.2
			75-53	7.7
			53-45	7.2
			<45	4.2
ES SPB 571-85.8	Powder	ES: 85.8	<10	100
ES settle-47	Powder	ES: 47	<10	100
Flowable S	Suspension	ES: 52.0	1-2	100

[1]ES refers to elemental S.

et al. 2005a). Reported Q_{10} values of 3.0 to 4.7 for the relationship between temperature and oxidation indicate that elemental sulfur oxidation is more sensitive to changing temperature than many biological reactions in soil (Wen et al. 2001). Consequently, a larger proportion of applied elemental S is anticipated to be converted to sulfate-S in warmer than in colder regions.

Janzen and Bettany (1987b) reported that the oxidation rate of elemental S is related to soil moisture potential according to a parabolic relationship, and also stated that this relationship is more pronounced with increase in temperature from 3 °C to 30 °C. In their study, optimum water potential ranged from −270 kPa to > −10 kPa when clay content in soil was decreased from 60% to 7%, respectively. In another incubation study, oxidation of elemental S to sulfate-S was enhanced with increasing soil moisture until lack of aeration began to reduce S oxidation rate at soil moisture content above field capacity, reflecting interactive effects of aeration and water availability in soil (Solberg et al. 2005a). In this study, there was reduced oxidation with increase in moisture from 90% to 120% of field capacity in one soil but not in other soils, reflecting differences due to texture.

2.2 Soil Properties (pH, Texture, Microbial Activity)

Research has shown a positive relationship between oxidation of elemental S to sulfate-S and soil pH, organic C concentration, extractable P, or extractable S, while it was negatively related to clay content (Janzen 1984). Rate of oxidation is also affected by the size and activity of microbial populations carrying out the oxidation process. In incubation studies, soils that had received previous elemental S applications usually showed faster S oxidation than soils which had not received any previous elemental S (Bertrand 1973, Janzen and Bettany 1987a, Solberg et al. 2005a) due to increase in the number of S oxidizers and more effective microbial populations (Bertrand 1973). The implications of these observations are that not only soil properties and climatic conditions but also antecedent S application could influence the performance of elemental S fertilizers in a given soil.

2.3 Dispersion of Elemental S Particles from Fertilizer Granules

Being a biological process, oxidation of elemental S in soil is affected by the effective surface area of the S particles exposed to microbial activity. Techniques that increase the oxidation rate of elemental S include reduced particle size of the fertilizer product, tillage to incorporate elemental S fertilizer into the soil and increase contact area between elemental S particles and microbes, and application in advance to allow for oxidation before plants require this element (Janzen 1990). In addition, the effectiveness of these techniques varies with the characteristics of elemental S source and application method (Janzen and Bettany 1987c).

Dispersion of granular ES-95 in water prior to application greatly increased oxidation rate compared to placement of intact granules into the soil (Wen et al. 2001). In an incubation study on Alberta and Saskatchewan soils, there was greater recovery as sulfate-S from Flowable S and Fine S than the Bentonitic S (A, B and C) and Urea S (A and C) products without cultivation (Solberg et al. 2005b). In this study, the low recovery from Urea S (A and C), which had small elemental S particles (<50 μm), was due to a decrease in the exposed surface area of the elemental S particles because of impregnation in urea pellets of 1-3 mm diameter (Janzen 1984, Janzen and Bettany 1986). In soil, when the urea from Urea S is dissolved, it leaves behind the fine elemental S particles aggregated in approximately the same volume as the original granule, which results in about 40-fold reduction in the exposed surface area, as only the outermost S particles become exposed to microbial action.

Bentonitic elemental S granules are expected to break apart with time, thereby exposing much greater surface area of the finer elemental S particles to soil microbes. However, if the Bentonitic elemental S granules are not disrupted and the elemental S particles are not mixed thoroughly into the soil, the exposed surface area and oxidation potential will be low due to delayed, inadequate, or lack of dispersion of elemental S particles in soil (Noellemeyer et al. 1981, Ukrainetz 1982, Janzen 1990). In incubation studies on Alberta and Saskatchewan soils, the reduction in size of Bentonitic S (C) from 2000-4000 μm to 1500-2000 μm had no effect on recovery as sulfate-S (Solberg et al. 2005b). However, there was high sulfate-S recovery from the incorporation of crushed (<150 μm) Bentonitic S, and this suggested that if the aggregates of fractured elemental S are thoroughly dispersed into soil, the sulfate-S recovery from this material could be considerably increased.

Earlier research has shown higher oxidation rates when elemental S was mixed into soil rather than banded (Swan et al. 1986). Similarly, in an incubation study by Solberg et al. (2005b), placement of Flowable S and Fine S fertilizers in bands or in nests (point placement) showed much less sulfate-S recovery than their incorporation after broadcasting. Generally, lower sulfate-S recovery from nested than banded S could also be explained by further reduction in exposure of surface in the case of nested S. Also, more mixing by cultivation after incorporation and banding of Urea S did not increase the oxidation. This is supported by field observations, where Urea S pellets could be found four months after incorporation and a subsequent high-speed intensive rotary cultivation.

Under field conditions, the sulfate-S recovery results showed that only up to 44% of the spring-applied elemental S could be recovered during the same growing season (Solberg et al. 2003). Among the elemental S fertilizers, Fine S gave more recovery as sulfate-S compared to Urea S (A), Urea S (B), and Bentonitic S (A). Tillage increased sulfate-S recovery from applied S in one experiment but had inconsistent results in another soil. The substantial differences in the sulfate-S recovery from different soils indicated the influence of soil type on oxidation rate and recovery of applied S at a given time. These results suggest that only a portion of the applied elemental S could be recovered as sulfate-S at any of the sampling times. Partial oxidation of elemental S in the year of application was also observed in other studies on Saskatchewan soils (Janzen and Bettany 1987b, Janzen and

Karamanos 1991, Wen et al. 2001, 2003). Solberg et al. (2003) showed more sulfate-S recovery in soil from October sampling than July sampling. The sulfate-S recovered in October could be considered as the amount of sulfate-S available from an elemental S fertilizer that may be subject to losses during winter and in early spring or become available for crop use in subsequent year.

The recovery of sulfate-S in the following year tended to be lower from elemental S fertilizers (Fine S, Bentonitic S [A]) compared to sodium sulfate (Solberg et al. 2003). Also, Karamanos and Janzen (1991) did not observe any advantage of elemental S over sulfate-S fertilizers during the second and third year after their applications. They attributed the absence of greater residual benefit of elemental S products relative to the sulfate-S fertilizers to the short duration of S uptake period by crops and susceptibility of S oxidized after that period to leaching.

2.4 Balance between S Oxidation and Immobilization, Leaching, or Other Losses

The recovery of sulfate-S from elemental S in soil is affected by a balance between the amount of S oxidized into sulfate-S and the amount of this sulfate-S that is immobilized into organic S pool or lost through leaching, especially in early spring after snow melt. When soil temperature and moisture conditions favor growth of S-oxidizing microbes, they also favor the growth of other microbes that decompose organic matter residues, which may result in immobilization of sulfate-S. Since elemental S is passed through microbial biomass during oxidation, it is possible that a major portion of oxidized S may be immobilized into microbial biomass or organic S pool. At a given time, the net balance of these release-immobilization reactions is expressed as the amount of sulfate-S present in soil and is thus very difficult to predict. Large losses of sulfate-S from the inorganic S pool by immobilization have been observed in incubated soils (Freney and Spencer 1960) and under field conditions (Janzen and Bettany 1987b, Janzen and Karamanos 1991). In addition, previous field research with autumn-applied ammonium sulfate in northeastern Saskatchewan showed downward leaching of sulfate-S in the soil profile in early spring after snow melt (Malhi 2005).

3 Feasibility of Granular Elemental S on Canola: A Case Study

3.1 Materials and Methods

A field experiment was established in the autumn of 1998 on an S-deficient Gray Luvisol (Typic Cryoboralf) loam soil (1.8 mg sulfate-S kg^{-1} in 0-15 cm) at Porcupine Plain, Saskatchewan, Canada, with annual applications of various S fertilizers for 6 years (1999 to 2004) and residual effects monitored in 2005 and 2006. Growing season (May to August) precipitation was below the long-term average (244 mm) in 2001 (184 mm) and in 2003 (87 mm). In 2002, the canola crop did not mature prior to severe early frost and no seed yield was recorded. In 2004, seed yield was very low due to frost damage in late August, and the crop was completely destroyed by deer grazing in 2006. Therefore, seed yields for 2002, 2004, and 2006 are not reported.

Each treatment was replicated four times in a randomized complete block design. Individual plots were 1.8 m × 7.5 m. The treatments included two application times (preceding autumn or before seeding in spring) of four S fertilizers (at 15 kg S ha^{-1}) and a zero-S control. The four S fertilizers included ES-90 and ES-95, two bentonite-based elemental S fertilizers containing 90% and 95% elemental sulfur, respectively, Agrium Plus (16-0-0-40), a fertilizer containing both elemental S (21.7%) and sulfate-S (18.7%), and ammonium sulfate (20.5-0-0-24), a sulfate-S fertilizer. Two brands of ES-90 were used, namely, Tiger90™ in the autumn of 1999, and Tiger90CR™ in the remaining years (Tiger Industries, Calgary, Alberta, Canada). The ES-95 was Sulfer95™ brand in all years (Fernz SulFer Works, Crossfield, Alberta, Canada). The S fertilizers were surface-broadcast. Each plot received a blanket annual application of 120 kg N ha^{-1} (as ammonium nitrate), 30 kg P ha^{-1} (as triple superphosphate), and 20 kg K ha^{-1} (as KCl – muriate of potash). In treatments that received ammonium sulfate, the blanket amount of N was adjusted. All the plots were tilled before sowing in May. A double-disc press drill was used to seed canola (*Brassica rapa* L. cv. Fairview) at 17.8 cm row spacing and seed rate of 9 kg ha^{-1}.

Seed yield was determined by harvesting 1.25 m wide and 7.0 m long strips with a plot combine, and straw yield was calculated from hand-harvested samples collected from two 1 m long rows in each plot. The oven dry (60 °C) samples were analyzed for oil, total N and total S in seed, and total S in straw. Oil concentration in canola seed was determined using crude fat method (AOAC 1990). Total S in seed and straw was determined by digestion of samples in nitric acid-hydrogen peroxide and measuring its concentration in the digest by ICP-AES (Huang and Schulte 1985). Total N in seed samples was determined by sample digestion and detection of N by thermal conductivity using a CNS combustion analyzer (AOAC 1995). Protein concentration was calculated by multiplying the total N by 6.25.

Soil samples in each plot were taken from the 0-15, 15-30, and 30-60 cm depths in the spring of 1999 to 2007. Each sample was a composite of four cores (4 cm diameter) per plot. The soil samples were air dried at room temperature, ground to pass through a 2-mm sieve, and then analyzed for sulfate-S. Sulfate-S in soil was determined by extraction with $CaCl_2$ and measuring its concentration in the extract by ICP-AES (Combs et al. 1998).

The data were subjected to analysis of variance (ANOVA) using GLM procedure (SAS Institute 1993). Selected contrasts and least significant difference ($LSD_{0.05}$) were used to determine differences between treatment means.

3.2 Results

3.2.1 Yield

Seed yield was very low (4 to 29 kg ha^{-1}) in the zero-S control in all years (data not shown). Annual applications of sulfate-S containing fertilizers (ammonium sulfate and Agrium Plus) increased seed yield significantly over the zero-S control in all years. For elemental S fertilizers, spring application had no effect on seed yield in 1999, but seed yield increased significantly with elemental S fertilizers over the zero-S control in the following years (except most cases in 2001).

Averaged over time of S application, ammonium sulfate had higher seed yield than ES-90 and ES-95 in 1999, 2000, 2001, and 2003 (Table 2.2). Agrium Plus produced lower seed yield than ammonium sulfate in 1999, while seed yields were similar for both sulfate-S sources in the following years. Seed yield was lower with ES-95 than ES-90 in 3 of 4 years (2000, 2001, and 2003). Averaged over S sources, seed yields were higher with autumn application than spring application of S in 1999, 2000, and 2003.

Seed yield with spring-applied elemental S fertilizers was lower than spring-applied ammonium sulfate in all years (Table 2.2). Autumn-applied elemental S fertilizers significantly increased seed yield over the zero-S control in all years (except ES-95 in 2001), but these seed yields were significantly lower than spring-applied ammonium sulfate in 1999, 2001, and 2003 (except ES-90 in 2000 and 2001, when this elemental S fertilizer produced seed yield close to ammonium sulfate). Autumn application of elemental S was more effective in increasing seed yield of canola than spring application, but the differences were significant only in 1999 and 2000 for ES-90 and in 1999 and 2003 for ES-95. Similarly, Agrium Plus also tended to produce more seed yield with autumn than spring application in 3 of 4 years. However, ammonium sulfate tended to produce lower seed yield with autumn than spring application in 3 of 4 years.

Straw yields were recorded in all 6 years from 1999 to 2004 and were moderate (2414 to 6561 kg ha^{-1}) in the zero-S control (data not shown). Averaged over timing of S application, straw yield was significantly higher in 1999 and tended to be higher in 2003 and 2004 with ammonium sulfate than ES-90 and ES-95 (data not shown). Agrium Plus tended to produce lower straw yield than ammonium sulfate in 1999 and 2003. Straw yield tended to be lower with ES-95 than ES-90 in 4 years (2000, 2001, 2002, and 2004), and the opposite was true in one year (2003). Averaged over S sources, straw yields were generally similar for autumn and spring application in most years, with one exception in 2000, when spring-applied S had higher straw yield than autumn-applied S.

In 1999, both autumn- and spring-applied Agrium Plus tended to produce less straw yield than spring-applied ammonium sulfate (data not shown). The elemental S fertilizers increased straw yield when applied in autumn, but the increase was lower or tended to be lower than spring-applied ammonium sulfate. Spring-applied elemental S fertilizers produced significantly lower straw yield than ammonium sulfate. Spring-

Table 2.2 Seed yield of canola with application of 15 kg S ha^{-1} from various S fertilizers in preceding autumn or in spring of 1999 to 2007 near Porcupine Plain in north-eastern Saskatchewan

Fertilizer Treatment		Seed Yield (kg ha^{-1})[y]				
S source[z]	Time	1999	2000	2001	2003	2005
Fertilizer S Source x Time of Application						
ES-90	Autumn	623	1432	502	1143	1421
	Spring	27	704	278	920	1501
ES-95	Autumn	864	892	208	903	1169
	Spring	33	655	89	510	773
AP	Autumn	1644	1344	681	1396	1251
	Spring	1388	1441	565	1144	1442
AS	Autumn	1928	1449	679	1115	1363
	Spring	2108	1218	807	1473	1157
	$LSD_{0.05}$	548	458	ns	279	ns
	SEM[w]	$186^{\bullet}$	$156^{\bullet}$	123^{ns}	95^{**}	222^{ns}
Fertilizer S Source						
ES-90		325	1068	390	1031	1461
ES-95		448	774	149	707	942
AP		1526	1392	623	1270	1347
AS		2018	1333	743	1294	1275
$LSD_{0.05}$		387	324	255	197	ns
SEM		132^{***}	110^{**}	87^{***}	67^{***}	158^{ns}
Time of Application						
	Autumn	1270	1279	517	1139	1310
	Spring	889	1005	435	1012	1222
	$LSD_{0.05}$	274	229	ns	140	ns
	SEM	93^{**}	78^{*}	61^{ns}	$47^{\bullet}$	111^{ns}
[x]Contrast (autumn vs spring)						
ES-90		*	**	ns	ns	ns
ES-95		**	ns	ns	***	ns
AP		ns	ns	ns	•	ns
AS		ns	ns	ns	*	ns

[z]ES refers to elemental S; AP refers to Agrium Plus containing both elemental S and sulfate-S; and AS refers to ammonium sulfate.

[y]In the zero-S control, in 1999, 2000, 2001, 2003, 2005, 2006 and 2007, respectively, seed yield was 21, 27, 4, 29 and 4 kg ha^{-1}.

[x]Contrast refer to comparison of autumn and spring application means for the given year and S fertilizer.

[w]SEM refers to standard error of the mean. $^{\bullet}$, *, **, and *** indicate treatment effect being significant at $P \leq 0.01$, $P \leq 0.05$, $P \leq 0.01$, and $P \leq 0.001$, respectively; and ^{ns} indicate treatment effect not significant.

applied ammonium sulfate produced significantly more straw yield in 1999 and tended to produce more straw yield in 3 years than autumn-applied ammonium sulfate, but the opposite was true in 2001. Autumn-applied ES-90 had higher straw yield than spring-applied ES-90 in 1999, but the opposite occurred in 2000. For ES-95, autumn application produced higher straw yield than spring application in 1999 and 2002. Application time did not have a consistent effect on straw yield with Agrium Plus.

The residual effects of previously applied S fertilizers from 1999 to 2004 were recorded in 2005 and 2007. In 2005, there was a significant increase in seed and straw yield from previously applied S fertilizers over the zero-S control (data not shown). There were no significant differences in seed and straw yield between the elemental S and sulfate-S fertilizers, and also between autumn and spring application. In 2007, there was a significant increase in seed yield from previously applied elemental S fertilizers but little increase from sulfate-S fertilizers over the zero-S control.

3.2.2 Seed Quality

Oil concentration in canola seed was determined in 4 years, and it increased with application of S fertilizers in all years (Table 2.3). Sulfate-S-containing fertilizers had higher oil concentration in seed than the elemental S fertilizers in 1999, 2000, and 2001. Average oil concentration in seed was higher in 1999 and tended to be lower in 2003 with ammonium sulfate than Agrium Plus. Oil concentration in seed was higher in 2001 and tended to be higher in 1999, 2000, and 2003 with ES-90 than ES-95. Average oil concentration in seed was higher in 2001 and tended to be higher in 1999 with autumn than spring application. The contrasts showed that oil concentration in seed was higher when the elemental S fertilizer was applied in autumn rather than in spring for ES-90 in 1999, 2000, and 2001 and for ES-95 in 1999 and 2001. For ammonium sulfate, autumn application had higher oil concentration in seed than spring application. Application time of Agrium Plus did not have any significant effect on the oil concentration in seed.

Protein concentration in seed ranged between 192 to 304 g kg^{-1} in different years, and was decreased with S fertilization in 1999, while it increased in 2000 and 2003, and showed no significant effect in 2001 (data not shown). Sulfate-S-containing fertilizers had lower protein concentration in seed than the elemental S fertilizers in 1999, most likely due to a dilution effect of increased seed yield from sulfate-S fertilizers, but there were no differences in seed protein concentration among the various S fertilizers in other years. Autumn application had higher protein concentration in seed than spring application in 2003 for ES-90 and in 2000 and 2003 for ES-95.

In 2005, residual effects of previous 6 annual applications of S fertilizers showed significant increase in oil concentration in seed over the zero-S control for all S fertilizer treatments (data not shown). There was no residual effect of applied S on protein concentration in seed.

3.2.3 Sulfur Uptake

Like seed yield, uptake of S in canola seed was very low in the zero-S control treatment (0.01 to 0.06 kg S ha^{-1}) and the sulfate-S containing fertilizers, particularly ammonium sulfate, increased S uptake most times in all years (Table 2.4). Elemental

Table 2.3 Oil concentration in seed of canola with application of 15 kg S ha^{-1} from various S fertilizers in preceding autumn or in spring of 1999 to 2007 near Porcupine Plain in north-eastern Saskatchewan

Fertilizer treatment		Oil concentration (g kg^{-1})^{y}				
S source^{z}	Time	1999	2000	2001	2003	2005
Fertilizer S Source x Time of Application						
ES-90	Autumn	367	413	358	374	432
	Spring	339	383	338	355	432
ES-95	Autumn	362	388	320	366	387
	Spring	329	390	293	335	392
AP	Autumn	396	407	373	374	421
	Spring	393	417	367	378	433
AS	Autumn	398	417	376	333	413
	Spring	425	411	379	369	392
	$LSD_{0.05}$	29	15	ns	ns	ns
	SEM^{w}	10.0^{*}	5.1^{**}	6.5^{ns}	14.0^{ns}	8.3^{ns}
Fertilizer S Source						
ES-90		353	398	348	364	432
ES-95		345	389	307	351	390
AP		395	412	370	376	427
AS		412	414	377	351	403
$LSD_{0.05}$		20	11	13	ns	17
SEM		6.9^{***}	3.6^{***}	0.46^{***}	10.0^{ns}	6.0^{***}
Time of Application						
	Autumn	381	406	357	362	413
	Spring	371	400	344	359	412
	$LSD_{0.05}$	ns	ns	10	ns	ns
	SEM	4.9^{ns}	2.5^{ns}	3.3^{*}	7.0^{ns}	4.1^{ns}
^{x}Contrast (autumn vs spring)						
ES-90		•	***	*	ns	ns
ES-95		*	ns	**	ns	ns
AP		ns	ns	ns	ns	ns
AS		•	ns	ns	•	•

^{z}ES refers to elemental S; AP refers to Agrium Plus containing both elemental S and sulfate-S; and AS refers to ammonium sulfate.

^{y}In the zero-S control, in 1999, 2000, 2001, 2003, 2005, 2006 and 2007, respectively, oil concentration in seed was 303, 373, 252, 260 and 317 g kg^{-1}.

^{x}Contrast refer to comparison of autumn and spring application means for the given year and S fertilizer.

^{w}SEM refers to standard error of the mean. •, *, **, and *** indicate treatment effect being significant at $P \leq 0.01$, $P \leq 0.05$, $P \leq 0.01$, and $P \leq 0.001$, respectively; and ^{ns} indicate treatment effect not significant

S fertilizers increased S uptake in seed with autumn application in all years, except ES-95 in 2001, while the effect of their spring application was significant in 2000 and 2003. Average S uptake in seed was lower with elemental S fertilizers than sulfate-S-containing fertilizers in almost all cases, and it was usually lower with ES-95

Table 2.4 Total S uptake in seed of canola with application of 15 kg S ha^{-1} from various S fertilizers in preceding autumn or in spring of 1999 to 2007 near Porcupine Plain in north-eastern Saskatchewan

Fertilizer treatment		Total S uptake in seed (kg S ha^{-1})[y]				
S source[z]	Time	1999	2000	2001	2003	2005
Fertilizer S Source x Time of Application						
ES-90	Autumn	1.44	4.27	1.66	4.46	5.04
	Spring	0.06	1.83	0.79	2.75	5.04
ES-95	Autumn	1.97	2.42	0.54	2.76	2.95
	Spring	0.07	1.64	0.23	1.34	1.93
AP	Autumn	3.87	4.31	2.45	5.30	3.32
	Spring	3.19	4.80	2.09	4.52	4.44
AS	Autumn	4.87	4.87	2.64	4.67	4.20
	Spring	5.32	4.03	3.17	5.97	3.05
	$LSD_{0.05}$	1.35	1.53	1.27 ^{ns}	0.85	ns
	SEM[w]	0.46$^{•}$	0.52$^{•}$	0.43 ^{ns}	0.29***	0.74 ^{ns}
Fertilizer S Source						
ES-90		0.75	3.05	1.22	3.60	5.04
ES-95		1.02	2.03	0.38	2.05	2.37
AP		3.53	4.56	2.27	4.91	3.88
AS		5.09	4.45	2.90	5.32	3.71
$LSD_{0.05}$		0.95	1.08	0.90	0.60	1.54
SEM		0.32***	0.37***	0.31***	0.20***	0.52*
Time of Application						
	Autumn	3.04	3.97	1.82	4.29	3.94
	Spring	2.16	3.07	1.57	3.64	3.65
	$LSD_{0.05}$	0.67	0.76	Ns	0.42	ns
	SEM	0.23*	0.26*	0.22 ^{ns}	0.14***	0.37 ^{ns}
[x]Contrast (autumn vs spring)						
ES-90		*	**	Ns	***	ns
ES-95		**	ns	Ns	**	ns
AP		ns	ns	Ns	•	ns
AS		ns	ns	Ns	**	ns

[z] ES refers to elemental S; AP refers to Agrium Plus containing both elemental S and sulfate-S; and AS refers to ammonium sulfate.

[y]In the zero-S control, in 1999, 2000, 2001, 2003, 2005, 2006 and 2007, respectively, S uptake in seed was 0.05, 0.05, 0.01, 0.06 and 0.01 kg S ha^{-1}.

[x]Contrast refer to comparison of autumn and spring application means for the given year and S fertilizer.

[w]SEM refers to standard error of the mean. $^{•}$, *, **, and *** indicate treatment effect being significant at $P \leq 0.01$, $P \leq 0.05$, $P \leq 0.01$, and $P \leq 0.001$, respectively; and ^{ns} indicate treatment effect not significant

than ES-90 in 3 of 4 years. Autumn-applied ES-90 had S uptake in seed equal to ammonium sulfate or Agrium Plus in 2000 and 2003. Agrium Plus had lower S uptake in seed than ammonium sulfate only in 1999, otherwise the two sulfate-S-containing fertilizers had similar S uptake in seed. On average, autumn application had higher S uptake in seed than spring application.

Elemental S fertilizers had greater S uptake in seed when autumn-applied compared to their application in spring, with significant differences in 1999, 2000, and 2003 for ES-90 and in 1999 and 2003 for ES-95. For Agrium Plus and ammonium sulfate, application time had significant influence on S uptake in seed only in 2003, when spring-applied ammonium sulfate had greater and spring-applied Agrium Plus had lower S uptake in seed than their corresponding applications in autumn.

There was a significant increase in uptake of S in straw with application of Agrium Plus and ammonium sulfate over zero-S control in all years (data not shown). Elemental S also tended to increase S uptake in straw, with significant effect for autumn and spring applications of ES-90 in 2001, 2002, and 2004, and autumn application in 2003. ES-95 also showed significant effect for autumn and spring applications in 2003. Average S uptake in straw was higher with sulfate-S-containing fertilizers than ES-95 in all 6 years and ES-90 in only 2 of 6 years (2000 and 2003). Sulfur uptake in straw was lower with ES-95 than ES-90 in 3 of 4 years (2001, 2002, and 2004). Uptake of S in straw was lower with Agrium Plus than ammonium sulfate in 1999 and 2003, and tended to be lower in 2000, 2001, and 2002, but the opposite tended to be true in 2004. Ammonium sulfate applied in spring had higher uptake of S in straw than its autumn application in 1999, while the effect of application time of both sulfate-S-containing fertilizers was not consistent otherwise. For the elemental S fertilizers, on the other hand, much higher S uptake in straw occurred from autumn than spring application in 1999 and for ES-90 in 2001, 2002, and 2003. The elemental S fertilizers had less uptake of S in straw than Agrium Plus and ammonium sulfate, except for autumn application of ES-90 in 2001, 2002, and 2003, when there were no differences among the S sources. The differences between elemental S and sulfate-S fertilizers were relatively greater for spring than autumn application.

In 2005, there was a significant residual effect of previously applied S on S uptake in seed and straw (data not shown). A year after the last S application, uptake of S in seed and straw tended to be greatest for ES-90 and lowest for ES-95, with the exception of Agrium Plus applied in spring. Average S uptake in seed was higher with ES-90 and Agrium Plus than ES-95 and ammonium sulfate, while S uptake in straw was lower with ES-95 than the other three S sources. On average, residual effect of spring S application tended to have higher S uptake in straw than autumn application, except for ES-95 and Agrium Plus, where uptake of S in straw was higher or tended to be higher with autumn than spring application.

3.3 *Discussion*

3.3.1 Canola Yield, Seed Quality, and S Uptake

Severe S deficiency symptoms were observed on canola plants at early growth stages in the absence of S fertilizer application in all years. Under this treatment, seed yields of canola ranged from 4 to 29 kg ha^{-1}, which indicated that soil was very deficient in plant-available S during the experimental period from 1999 to 2004 and that deficiency of S in soil during the growing season can cause a major reduction in seed yield.

Averaged across 1999 to 2004, seed and straw yield of canola in plots treated with ammonium sulfate were 66.5 and 1.8 times greater than the zero-S plots, respectively. This indicated that response of canola to applied sulfate-S was much greater for seed yield compared to straw yield and that S fertilization was essential to attain optimum canola seed yield. Substantial increase in seed yield from ammonium sulfate suggests that this fertilizer provided S in the form which plants can absorb. These results are supported by earlier observations that S deficiencies in crops can be prevented or eliminated by applying sulfate-S (Malhi and Gill 2002).

Elemental S must be converted to sulfate-S before it can be utilized by a crop (Bettany and Janzen 1984). In our study, there was little or no increase in seed yield from elemental S fertilizers compared to zero-S control in the first year, especially with spring application. Similarly, in earlier studies the availability of sulfate-S from elemental S fertilizers in the year of application was considered minimal (Noellemeyer et al. 1981), and the elemental S fertilizers were inferior to sulfate-S fertilizers on canola in the year of application (Karamanos and Janzen 1991, Malhi 2005).

The amount of available S in soil from elemental S fertilizers is expected to increase with time and to produce seed yield benefit (Bettany and Janzen 1984). Therefore, cumulative effect was assessed in canola seeded with annual applications of elemental S fertilizers for 6 years on the same plots. From the second year onward, elemental S fertilizers significantly increased seed yield and S uptake over the zero-S control, except spring application of ES-95 in 2001. This indicates further dispersion of the S particles to enhance oxidation rate, and increased time for oxidation improves effectiveness of elemental S fertilizers. However, seed yield and S uptake of canola with elemental S fertilizers were lower than ammonium sulfate in many cases. This indicated that elemental S fertilizers were unable to consistently supply sufficient amounts of sulfate-S to canola plants in the growing season for optimum growth and seed yield, even after many consecutive annual applications on the same area.

Agrium Plus (which contained both sulfate-S and elemental S) was effective in correcting S deficiency in canola starting from the first year, but produced less seed yield and S uptake than ammonium sulfate in some years. This indicates that the elemental portion of S in Agrium Plus probably behaved similar to the other elemental S fertilizers and was not as effective as the sulfate-S portion in increasing seed yield of canola.

The granular elemental S fertilizers are probably less effective than granular sulate-S fertilizers because granules remain intact, and elemental S particles do not disperse from granules resulting in a low rate of microbial oxidation. The conversion of granular elemental S fertilizers to sulfate-S can be improved by increasing the contact area between elemental S particles and the soil by allowing granules to disintegrate on the soil surface prior to their incorporation at seeding (Solberg 1986). Leaving granules of elemental S fertilizers on the surface of soil exposed to frost or rain speeds up the physical breakdown of the granules into fine particles, allowing faster oxidation to sulfate (Solberg et al. 2003). In order to get maximum conversion of elemental S to sulfate-S, the autumn-applied granulated elemental S fertilizers in the present study were broadcast and left on soil surface until sowing of canola in the following spring. Autumn-applied elemental S fertilizers were more effective than their spring application in correcting S deficiency and improving seed and straw yield and S uptake of canola. This may reflect a greater conversion of elemental S to sulfate-S due to a longer period of exposure to dispersion over the autumn to spring period and then to microbial oxidation during the growing season with autumn application, but little or no oxidation with spring application, particularly in the first year of application. The canola seed yield and S uptake with both autumn and spring applications of elemental S were still less than both application times of sulfate-S-containing fertilizers, particularly ammonium sulfate, except ES-90 autumn application in 2000 and 2003. This suggests that granulated (pelletized) elemental S fertilizers cannot be recommended to correct S deficiency in S-sensitive annual crops (such as canola), because the oxidation rate is not rapid enough to satisfy the S requirements of the crop, even after six consecutive applications on the same area.

Elemental S fertilizers that disperse readily into small particles should improve the effectiveness of elemental S (Janzen 1990). The elemental S fertilizers used in this study contained bentonite or other dispersing agents to absorb water, swell, break down, and disperse into many small particles very quickly, but were still inferior to sulfate-S fertilizer. In another field experiment at same site, water suspension containing very fine elemental S particles sprayed on the soil surface after seeding of canola produced seed yield close to potassium sulfate in the year of application (Malhi et al. 2005). This suggests that physical dispersion of S particles from the elemental S granules for exposure to oxidation to sulfate-S was a major limitation under climatic conditions in the Parkland Region, as also observed by Wen et al. (2001).

Autumn application of ammonium sulfate was somewhat less effective in increasing seed and straw yield and S uptake than spring application in many cases. This was most likely due to overwinter loss of S from the soil sulfate-S pool. Sulfate-S is subject to leaching and possibly gaseous losses under wet conditions. In the Parkland region, soils are normally wet for 7-10 days in early spring after snow melting.

Increased oil concentration in seed above zero-S control was observed with Agrium Plus and ammonium sulfate in all years and with elemental S fertilizers in many cases. This indicates that when S deficiency exists, fertilization with sulfate-S can increase oil concentration in seed. Similar patterns of seed oil concentration response to S fertilizer on S-deficient soils have been obtained in Saskatchewan (Malhi and Gill 2002) and Manitoba (Grant et al. 2003a).

Application of ES-90 and ES-95 (especially when applied in spring) had little and inconsistent effect on oil concentration and S concentration in seed. This lack of consistent increase in seed S concentration and oil concentration reflects the inconsistent effect of elemental S fertilizers in increasing the quantity of sulfate-S available for crop uptake (Grant et al. 2003a, b). Sulfur fertilization decreased or tended to decrease protein concentration in 1999, whereas it increased or tended to increase protein concentration in 2000 and 2003, and had very little effect in 2001, indicating an inconsistent effect. The increased oil concentration associated with S application could have led to a proportional decrease in seed protein concentration in 1999, as seed protein and oil concentrations tend to be inversely related (Ridley et al. 1972).

In field studies related to crop response of canola to elemental S fertilizers, the performance of elemental S fertilizers was usually inferior to sulfate-S fertilizers in the initial year (Solberg et al. 2007). Exposing the elemental S fertilizers and their particles to microbes for oxidation appears to enhance their availability for plants. Incorporation or broadcast methods are better than banding for the elemental S fertilizers. Application of elemental S fertilizers much in advance of crop use time is another technique to improve their effectiveness. Overall, correction of a severe S deficiency by elemental S alone may be risky in the short term, and thus addition of some sulfate-S in the initial 1 or 2 years to supplement the sulfate-S from elemental S is advisable. Then, annual applications of elemental S alone may be sufficient to supply the S requirements of crops, depending on crop species, soil type, and climatic zone. Increased exposure of elemental S fertilizers to oxidation, either by incorporation and broadcast methods or by application ahead of plant use time, tends to improve crop response.

The residual yield benefit to future crops from S fertilizer application may also occur due to carryover of sulfate-S in soil, from the release of sulfate-S from decomposing high S crop residues, and oxidation of previously applied elemental S fertilizers that did not become available to the crop in the years of application. In a field study on canola near Star City in northeastern Saskatchewan, Grant et al. (2001, 2003a) reported that the residual seed yield benefit from broadcast-applied elemental S was similar to sulfate-S fertilizer. In Alberta, Janzen and Karamanos (1991) also found similar residual response from elemental S suspension, granular elemental S, and sulfate-S on canola, but the residual benefits of these three S sources were small. In a field study on oilseed-cereal-legume rotation in Saskatchewan, Wen et al. (2003) found residual effects of sulfate-S and elemental S fertilizers on uptake of S in the second and third years after application, and S uptake in three years tended to be greater with elemental S than sulfate-S fertilizers. Karamanos and Poisson (2004) found that both ammonium sulfate and a number of elemental S sources had the same residual effect after 3 years in a canola-barley-canola rotation. Similarly in this case study, there was a significant residual effect on canola seed yield and S uptake in 2005 from both elemental S and sulfate-S sources. In 2006, although the crop was completely destroyed by deer grazing, we visually observed much better canola growth in plots that received S fertilizers previously compared to the zero-S control. The canola growth tended to be inferior in ammonium sulfate than elemental S fertilizers.

3.3.2 Residual Sulfate-S in Soil

Compared to zero-S control, preceding autumn application tended to increase the amount of sulfate-S at spring soil sampling of 1999 in the 0-15 cm for ES-90, and in the 15-30 and 30-60 cm depths for Agrium Plus and ammonium sulfate (data not shown). In the later years, from 2000 to 2004, the residual sulfate-S amount extracted from soil was generally lower with elemental S than Agrium Plus and ammonium sulfate. The increased amount of sulfate-S in the 15-30 and 30-60 cm soil depths in some cases suggested some leaching of sulfate-S from the surface layer to subsoil layers. Autumn application generally tended to show more sulfate-S in soil than spring application, with some exceptions. However, the differences between autumn S application treatments and the zero-S control in terms of sulfate-S amount at spring sampling were always lower than the applied 15 kg S ha^{-1}. This loss from autumn application of sulfate-S probably resulted from immobilization and/or deep leaching. In the case of elemental S, this lesser recovery of S in the sulfate form was probably the result of partial oxidation of the amount applied as well as some loss after oxidation to sulfate-S. Also, larger concentration of sulfate-S was found in the deeper soil layers with autumn over spring applications and with S fertilization over the zero-S control, especially for Agrium Plus and ammonium sulfate, As samples were taken only to 60 cm soil depth, it is probable that some of the sulfate-S may have leached below the 60 cm depth.

The amount of sulfate-S in soil was less with elemental S than sulfate-S-containing fertilizers, especially with spring applications. Also, the increase in sulfate-S over the zero-S control was relatively smaller with elemental S compared to sulfate-S application, indicating more residual effect of consecutive S fertilization from sulfate-S than elemental S. The increase in sulfate-S from S fertilizers was not cumulative with time, but rather fluctuated from year to year, which suggested that S application is needed each year. This was contrary to the expectation of S fertility buildup from residual benefits of elemental S fertilizers. Earlier studies also observed no extra residual effect of elemental S compared to sulfate-S application (Janzen and Karamanos 1991, Grant et al. 2003a, b).

Elemental S sources applied annually for several years were expected to build long-term S fertility, which did not happen in this study. There was not sufficient available S from the elemental S fertilizers, as evidenced by low amounts of sulfate-S in soil at spring sampling, to produce optimum canola yield. The canola production and soil sulfate-S data clearly show that under the conditions of this study physical dispersion of the elemental S granules and their oxidation to sulfate-S were not rapid enough to supply canola plants sufficient amount of S early in the growth stage for optimum canola yield and S uptake when soils were deficient in plant-available S.

3.3.3 Effect of Balanced N and S Fertilization on Residual Nitrate-N in Soil

Because of the critical balance that exists between S and N in the plant, growth and seed production of canola has been shown to be affected by nutrient imbalances in soil, and S must be balanced with N and other nutrients for optimum yield of high-quality canola seed (Janzen and Bettany 1984, Malhi and Gill 2002). The N and S imbalance with too much N and too little S in the plant can impair protein and seed production. In the zero-S treatments, low seed yield indicates N:S imbalance, and maximum seed yield with application of both N and S fertilizers indicates a proper balance of S with N to optimize seed yield along with high utilization efficiency of nutrients, water, and energy. Nutrient balance is thus essential, as too much N with insufficient amount of S will lead to S deficiency in plants, poor seed yield, and residual N in the soil. Similarly, in the present study, residual nitrate-N in the 0-60 cm soil was considerably higher in the zero-S control (i.e., plots receiving N fertilizer without S) than with applied S (i.e., plots receiving both N and S fertilizers together) (Table 2.5). Residual nitrate-N in soil also tended to be higher in some elemental S treatments than spring-applied ammonium sulfate. The higher amount of nitrate-N in soil in the N only treatment was most likely due to poor growth and N-use efficiency of applied nitrogen alone as compared to N + S fertilizer treatment. This suggests that imbalanced fertilization can result in accumulation of nitrate-N in soil at the end of the harvest season, which is subjected to leaching and gaseous N losses over the winter, especially in early spring after snow melt (Nyborg et al. 1990, Heaney et al. 1992). In other research in Saskatchewan, Malhi et al. (2002) also noticed greater accumulation of nitrate-N in soil under organic input (i.e., no input of fertilizer) in some cases than in soil receiving adequate amounts of both N and P fertilizers due to relatively low plant-available P for optimum crop growth in soil under an organic regime.

Table 2.5 Estimated amounts of nitrate-N in the 0-60 cm soil after growing canola without and with applied sulfate-S and elemental S fertilizers on S-deficient soils in north-eastern Saskatchewan (adapted from Malhi et al. 2005)

Location	Date of initiation	Date of soil sampling	S fertilizer treatment	Nitrate-N (kg N ha^{-1}) in soil
Porcupine Plain	1999	Spring 2002	Control	194
			ES-90 – autumn	40
			ES90 – spring	50
			ES-95 – autumn	94
			ES95 – spring	73
			Agrium Plus – autumn	24
			Agrium Plus – spring	43
			Ammonium sulfate – autumn	50
			Ammonium sulfate - spring	19
			$LSD_{0.05}$	76
			SEM^y	26.1**
Star City	2004	Spring 2005	Control	149**
			Ammonium sulfate	106
Tisdale	1999	Autumn 2002	Control	179$^{\bullet}$
			Ammonium sulfate	108

^{Y}SEM refers to standard error of the mean. $^{\bullet}$, *, **, and *** indicate to treatment effect being significant at $P \leq 0.01$, $P \leq 0.05$, $P \leq 0.01$, and $P \leq 0.001$, respectively; and ^{ns} indicate to treatment effect not significant

4 Conclusions

Rate of S oxidation in Canadian prairie soils under incubation is enhanced with increasing temperature with a maximum close to 40 °C and moisture until lack of aeration above field capacity inhibited S oxidation. Oxidation of elemental S to sulfate-S is positively related to pH, organic matter content, nutrient-supplying power, and microbial activity, while it is inversely related to clay content of soil. Techniques that increase the oxidation rate of elemental S in soil include reducing particle size of the fertilizer product, using tillage to incorporate elemental S fertilizer into the soil to increase contact area between elemental S particles and soil microorganisms, and application in advance of crop demand to allow oxidation to occur before plants require this element. Elemental S fertilizers were not effective in increasing seed yield and S uptake of canola in the first year of application, particularly when applied in spring. With multiyear annual applications, elemental S fertilizers increased seed yield and S uptake significantly over the zero-S control, but usually not at the same level as with sulfate-S-containing fertilizers. Autumn-applied elemental S usually had greater seed yield and S uptake than the spring-applied elemental S, and the opposite was true for ammonium sulfate in 3 of 6 years. Oil concentration in canola seed showed trends similar to seed yield, though the effects were smaller and less frequent. The need for greater dispersion of elemental S particles from granular elemental S fertilizers, to enhance microbial oxidation to sulfate-S in soil, was considered to be the main requirement for increasing the short-term availability of S in elemental S fertilizers. Canola seed yield and N and S uptake with surface application of suspension and powder formulations of elemental S fertilizers were similar to sulfate-S fertilizer. The residual benefit of many successive annual elemental S fertilizer applications on soil S fertility was not cumulative. Residual nitrate-N in soil was much higher in the zero-S control than with applied S, and it also tended to be higher in some elemental S treatments than in spring-applied ammonium sulfate. In summary, the amount of elemental S fertilizer needed to adequately meet crop S requirements should be adjusted based on environmental conditions, soil properties, microbial activity, dispersion of elemental S particles from fertilizer granules, and balance between oxidation and immobilization, leaching, or other S losses in the Parkland region of Canadian Prairies.

Acknowledgements The authors thank Western Co-operative Fertilizers Limited, Agrium and Fernz SulFer Works for financial assistance; D. Leach, K. Fidyk and K. Hemstad-Falk for technical help and for the internal review of the manuscript.

References

Assocation of Official Analytical Chemists (AOAC) (1990) Fat (crude) or ether extract in animal feed (920.39). Official methods of analysis, 15th ed. AOAC Washington DC Association of Official Analytical Chemists (AOAC) (1995) Protein (crude) in animal feed. Combustion method (990.03). Official methods of analysis, 16th ed. AOAC Washington DC

Bertrand RA (1973) Reclaiming of soils made barren by sulfur from gas processing plants. M.Sc. Thesis, University of Alberta, Edmonton, Alberta, Canada

Bettany JR, Janzen HH (1984) Transformations of sulphur fertilizers in prairie soils. In: Terry JW (ed) Proc. international sulphur '84 conference, Sulphur Development Institute of Canada Calgary, Alberta, Canada, pp. 817-22

Bettany JR, Janzen HH, Stewart JWB (1983) Sulphur deficiency in the prairie provinces of Canada. Proc. Int. Sulphur '82 Conf., vol. 1 November 1982, London, UK, pp. 787-800

Combs SM., Denning JL, Frank KD (1998) Sulphate-sulfur. In: Recommended chemical soil test procedures for the north central region. Missouri Agric. Expt. Sta. Publication No. 221 (revised). Extension and Agricultural Information, I-98 Agricultural Building, University of Missouri, Columbia, MO 65211, USA., pp. 35-39

Doyle PJ, Cowell LE (1993) Sulphur. In: Rennie DA, Campbell CA, Roberts TL (eds) Impact of micronutrients on crop responses and environmental sustainability on the Canadian prairies. pp. 202-50

Freney JR, Spencer K (1960) Soil sulphate changes in the presence and absence of growing plants. Aust J Agric Res 11:339-45

Grant CA, Bailey LD (1993) Fertility management in canola production. Can J Plant Sci 73:651-70

Grant CA, Johnston AM, Clayton GW (2001) Sulphur fertilizer forms and placements for canola. Proc. Manitoba Agronomists Conference 2000, 12-13 December 2000, Winnipeg, Manitoba, Canada. pp. 51-59

Grant CA, Johnston AM, Clayton GW (2003a) Sulphur fertilizer and tillage effects on canola seed quality in the Black soil zone of western Canada. Can J Plant Sci 83:745-58

Grant CA, Johnston AM, Clayton GW (2003b) Sulphur fertilizer and tillage effects on early season sulphur availability and N:S ratio in canola in western Canada. Can J Soil Sci 83:451-63

Heaney DJ, Nyborg M, Solberg ED, Malhi SS, Ashworth J (1992) Overwinter nitrate loss and denitrification potential of cultivated soils in Alberta. Soil Biol Biochem 24:877-84

Huang CL, Schulte EE (1985) Digestion of plant tissue for analysis by ICP-AES. Commun. Soil Sci Plant Anal 16:943-58

Janzen HH (1984) Suphur nutrition of rapeseed. Ph.D. Thesis, University of Saskatchewan, Saskatoon, Saskatchewan, Canada

Janzen HH (1990) Elemental sulphur oxidation as influenced by plant growth and degree of dispersion within soil. Can J Soil Sci 70:499-502

Janzen HH, Bettany JR (1984a) Sulfur nutrition of rapeseed. I. Influence of fertilizer nitrogen and sulfur rates. Soil Sci Soc Am J 48:100-7

Janzen HH, Bettany JR (1984b) Sulfur nutrition of rapeseed. II. Effect of time of sulfur application. Soil Sci Soc Am J 48:107-12

Janzen HH, Bettany JR (1986) Release of available sulfur from fertilizers. Can J Soil Sci 66:91-103

Janzen HH, Bettany JR (1987a) Oxidation of elemental sulphur under field conditions in Central Saskatchewan. Can J Soil Sci 67:609-18

Janzen HH, Bettany JR (1987b) The effect of temperature and water potential on sulfur oxidation in soils. Soil Sci 144:81-9

Janzen HH, Bettany JR (1987c) Measurement of sulphur oxidation in soils. Soil Sci 143:444-52

Janzen HH, Karamanos RE (1991) Short-term and residual contribution of selected elemental S fertilizers to the S fertility of two Luvisolic soils. Can J Soil Sci 71:203-11

Karamanos RE, Janzen HH (1991) Crop response to elemental sulfur fertilizers in Alberta. Can J Soil Sci 71:213-25

Karamanos RE, Poisson DP (2004) Short and long term effectiveness of various S products in prairie soils. Commun Soil Sci Plant Anal 35:2049-66

Malhi SS (2005) Influence of four successive annual applications of elemental S fertilizers on yield, S uptake and seed quality of canola. Can J Plant Sci 85:777-92

Malhi SS, Gill KS (2002) Effectiveness of sulphate-S fertilization at different growth stages for yield, seed quality and S uptake of canola. Can J Plant Sci 82:665-74

Malhi SS, Solberg ED, Nyborg M (2005) Influence of formulation of elemental S fertilizer on yield, quality and S uptake of canola seed. Can J Plant Sci 85:793-802

Noellemeyer EJ, Bettany JR, Henry JL (1981) Sources of sulphur for rapeseed. Can J Soil Sci 61:465-7

Nuttall WF, Boswell CC, Sinclair AG, Moulin AP, Townley-Smith LJ, Galloway GL (1993) The effect of time of application and placement of sulphur fertilizer sources on yield of wheat, canola, and barley. Commun Soil Sci Plant Anal 24:2193-2202

Nyborg M, Malhi SS, Solberg ED (1990) Effect of date of application on the fate of ^{15}N-labelled urea and potassium nitrate. Can J Soil Sci 70:21-31

Ridley AO (1972) Effect of nitrogen and sulfur fertilizers on yield and quality of rapeseed. Proc. 17th Annual Manitoba Soil Science Meeting, January 1972, University of Manitoba, Winnipeg, Manitoba, Canada. pp. 182-7

SAS Institute Inc (1993) SAS/STAT user's guide. ver 6, 4th ed, vol. 2. Cary, NC, USA

Schoenau JJ, Germida JJ (1992) Sulphur cycling in upland agricultural systems. In: Howarth RW, Stewart JWB, Ivanov MV (eds.), Sulphur cycling on the continents, John Wiley and Sons, New York, pp. 261-77

Solberg ED (1986) Oxidation of elemental S fertilizers in agricultural soils of northern Alberta and Saskatchewan. MSc. thesis. University of Alberta, Edmonton, Alberta, Canada

Solberg ED, Laverty DH, Nyborg M (1987) Effect of rainfall, wet-dry, and freeze-thaw cycles on the oxidation of elemental sulfur fertilizers. Proc Alberta Soil Science Workshop, February 1987, Edmonton, Alberta, Canada. pp. 120-6

Solberg ED, Malhi SS, Nyborg M, Gill KS (2003) Fertiilzer type, tillage, and application time effects on recovery of sulfate-S from elemental sulfur fertilizers in fallow field soils. Commun Soil Sci Plant Anal 34:815-30

Solberg ED, Malhi SS, Nyborg M, Gill KS (2005a) Source, application method, and cultivation effects on recovery of elemental sulfur as SO4-S in incubated soils. Commun Soil Sci Plant Anal 36:847-62

Solberg ED, Malhi SS, Nyborg M, Gill KS (2005b) Temperature, soil moisture, and antecedent S application effects on recovery of elemental sulfur as SO4-S in incubated soils. Commun Soil Sci Plant Anal 36:863-74

Solberg ED, Malhi SS, Nyborg M, Henriquez B, Gill KS (2007) Crop response to elemental S and sulfate-S sources on S-deficient Soils in the Parkland region of Alberta and Saskatchewan. J Plant Nutr 30:321–333

Statistics Canada (2002) Agriculture 2001 Census – Data Tables: http://www.statcan.ca/english/freepub/95F0301XIE/tables.htm.

Swan M, Soper RJ, Morden G (1986) The effect of elemental sulphur, gypsum and ammonium thiosulphate as sulphur sources on yield of rapeseed. Commun Soil Sci Plant Anal 17:1383-90

Wen G, Schoenau JJ, Yamamoto T, Inoue M (2001) A model of oxidation of an elemental sulfur fertilizer in soils. Soil Sci 166:607-13

Wen G, Schoenau JJ, Mooleki SP, Inanaga S, Yamamoto T, Hamamura K, Inoue M, An P (2003) Effectiveness of an elemental sulfur fertilizer in an oilseed-cereal-legume rotation on the Canadian prairies. J Plant Nutr Soil Sci 166:54-60

Ukrainetz H (1982) Oxidation of elemental sulphur fertilizers and response of rapeseed to sulphur on Gray Wooded soils. Proc. 19th Annual Alberta Soil Science Workshop, 23-24 February, 1982. Edmonton, Alberta, Canada. pp. 278-307

Chapter 3
Impact of Sulfur on N_2 Fixation of Legumes

Heinrich W. Scherer

Abstract The fact that crop deficiencies of sulfur (S) have been reported with increasing frequency over the past years has focused on the importance of this element in plant nutrition. With legumes, limitation of S can reduce N_2 fixation by affecting nodule development and function. The present report deals with the influence of S on yield formation and N_2 fixation of legumes and tries to explain the lower yield formation and reduced N_2 fixation of S-starved legumes.

1 Introduction

It has long been known that in regions where sulfur (S) deficient soils occur, legumes are particularly responsive to S-containing fertilizers and that elementary S or sulfates increase the percentage nitrogen as well as yield on such deficient soils (Anderson and Spencer 1950). However, until 15 years ago there was little concern for S deficiency, even though the ability of the soils to retain and to release S to crops is small and the input of high-analysis low-S-containing fertilizers increased and the SO_2 emissions from industrial sources were reduced. Now, areas of S deficiency are becoming widespread throughout the world (Scherer 2001).

Understanding the role of S in legumes growth is important from the point of view that deficiency of the S-containing amino acids cysteine and methionine may limit the nutritional value of food and feed (Sexton et al. 1998). Studies with *Medicago sativa* indicate that with suboptimum S supply the mole percent of both amino acids significantly decreased (DeBoer and Duke 1982), resulting in lower protein concentrations (Rendig et al. 1976), while nonprotein N is accumulated (Bolton et al. 1976). Also in *Pisum sativum* S deficiency resulted in a decreased synthesis of the S-containing storage proteins albumin and legumin (Chandler et al. 1984). However, according to Sexton et al. (1998) the protein quality of *Glycine max* can be enhanced by increasing the concentration of S-containing amino acids.

Heinrich W. Scherer
INRES - Plant Nutrition, University of Bonn, Karlrobert-Kreiten-Straße 13,
D-53115 Bonn, Germany
h.scherer@uni-bonn.de

N.A. Khan et al. (eds.), *Sulfur Assimilation and Abiotic Stress in Plants.*

S application may not only increase quality in grain legumes (Chandler et al. 1983), but also the percentage of N as well as the yield of legumes on S-deficient soils (Eppendorfer and Bille 1992). The information concerning the requirement of S for symbiotic nitrogen fixation of legume crops, however, is still scarce. Scherer and Lange (1996) observed yield reduction and a lower N accumulation of grain and fodder legumes under S deficiency conditions. For this reason Lange (1998) stated that growth of leguminous plant species may be affected by S through its effect on N_2 fixation by *Rhizobium* bacteria.

2 Influence of S on Yield Formation

It is well established that yield formation of different plant species is reduced under S deficiency (McGrath and Zhao, 1996, Zhao et al. 1996, Andersen et al. 2007), which may be caused by an inhibited protein synthesis (Scherer et al. 2006). In pot experiments Scherer and Lange (1996) also observed a yield decrease with grain and fodder legumes under S deficiency conditions. In the treatment without S (0 S) the shortage of sulfur was visible early in the growing season with all plant species, which showed yellowish leaves and weak growth. Later in the growing season the growth of plants with low S supply (25 mg S pot^{-1}) was also reduced. Total dry matter yield of both treatments was not significantly different, while optimum S supply (200 mg S pot^{-1}) resulted in a significantly higher total dry matter yield. Fig. 3.1 demonstrates the influence of S on dry matter production of *Trifolium pratense*. However, the influence of S on yield formation differed between the legumes, being higher with *Vicia faba* and *Trifolium pratense* than with *Medicago sativa* and *Pisum sativum* (results not shown). Generally, higher plants accumulate N and S in amounts proportional to that incorporated into protein. However, under S deficiency conditions protein synthesis is inhibited and nonprotein is accumulated, resulting in yield depression.

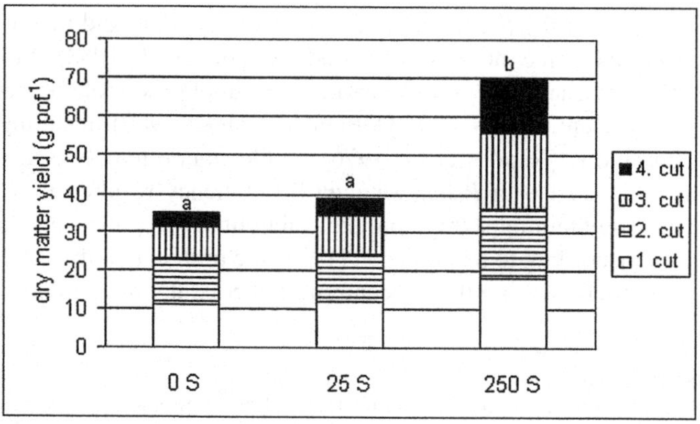

Fig. 3.1 Influence of S supply on dry matter yield of *T. pratense*. Different subscripts denote significant difference at the 5% level (ANOVA)

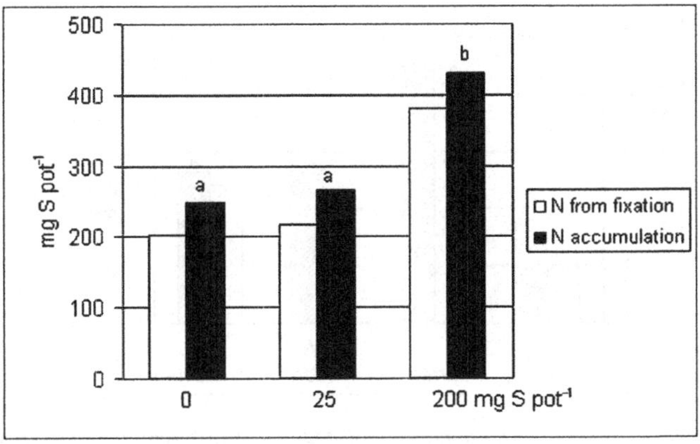

Fig. 3.2 Influence of S supply on N accumulation and N_2 fixation of *T. pratense*. Different subscripts denote significant difference at the 5% level (ANOVA)

3 Influence of S on N_2 Fixation

As shown for the first harvest of *Trifolium pratense* (Fig. 3.2) N accumulation in the plants was significantly higher with the high S application rate (Scherer and Lange 1996), which is the result of a higher N_2 fixation rate. Assuming that all N which was applied at the start of the experiment was taken up completely by the plants, about 380 mg N pot^{-1} were fixed in the treatment with optimum S supply, corresponding to 88% of the total N accumulation, while in the treatments with no or insufficient S supply only 203 and 217 mg N pot^{-1}, respectively, were fixed, corresponding to 81% and 82% of the total N accumulation. These results suggest that symbiotic N_2 fixation was impaired under S deficiency conditions and that alleviation of S deficiency increased N_2 fixation are in accordance with results of Shock et al. (1984), who found that applied S increased the percentage of symbiotically fixed nitrogen in subclover. Janssent and Vitosh (1974) state that free amino acids and other N forms, which accumulate due to lack of being synthesized into proteins with S deficiency, may have a kind of feedback repression on nitrogen fixation.

4 Reasons for a Reduced N_2 Fixation under S Deficiency Conditions

4.1 Root Nodules Number and Size

Confirming results of Collins et al. (1986) Lange (1998) found a positive correlation between nitrogen fixation of legumes and the number of nodules. Therefore, it may be speculated that under S deficiency conditions N demand of legumes is reduced

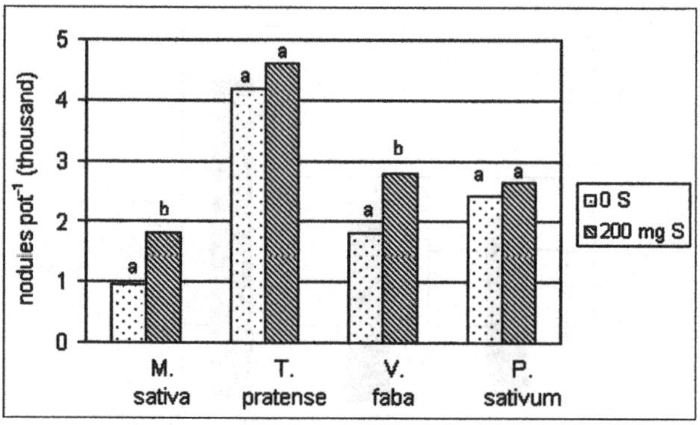

Fig. 3.3 Influence of S supply on the number of nodules of different legumes per pot. Different subscripts denote significant difference at the 5% level (ANOVA)

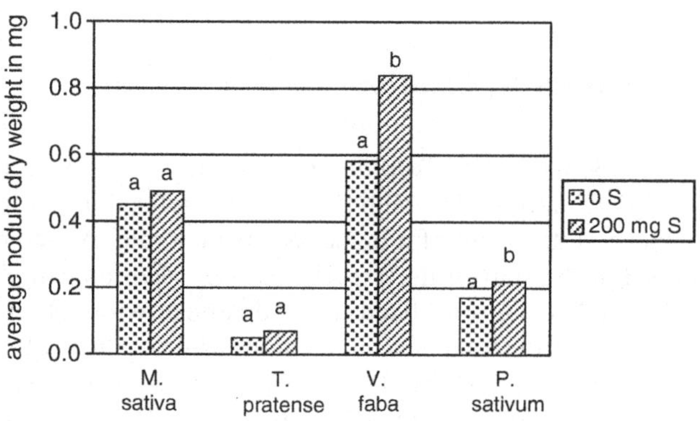

Fig. 3.4 Influence of S supply on the average nodule dry weight of different legumes (Different subscripts denote significant difference at the 5% level (ANOVA)

and for this reason the number of root nodules is decreased. According to Anderson and Spencer (1950) treatment with S increased the total weight of nodules, the total number of nodules, the number per unit root length, and the number of large nodules. Actually, in investigations of Scherer and Lange (1996) with different legumes, as compared with optimum S supply, S deficiency resulted in a significant lower number of nodules pot^{-1} with *Medicago sativa* and *Vicia faba* (Fig. 3.3), while the nodule weight was significantly decreased with *Vicia faba* and *Pisum sativum* (Fig. 3.4). However, confirming results of Gilbert and Robson (1984), who have shown that S deficiency resulted in reduced root growth, the higher number of root nodules per pot with optimum S supply in the investigations of Scherer and Lange (1996) was not caused by a higher number of root nodules per unit root length but by a better root growth.

Sulfur applied shortly before harvest markedly increased symbiotic N_2 fixation without increasing the number of nodules, confirming that N_2 fixation is not limited at a low S level by an inadequate number of nodules (Anderson and Spencer 1950).

As put forward by Parsons et al. (1993), a nitrogen feedback mechanism could be the main trigger for regulating both nodulation and N_2 fixation, both being reduced under S deficiency conditions.

4.2 *Nitrogenase*

The lower N_2 fixation of S-starved legumes may not only be caused by a reduced number of root nodules but also by lower activity of nitrogenase. Nitrogenase is a key enzyme complex, consisting of two oxygen-sensitive nonheme iron proteins (Marschner 1995), which catalyzes the ATP-dependent reduction of dinitrogen to NH_3 during the process of biological N_2 fixation. The smaller of the two proteins contains a single Fe_4S_4 unit (Jeong and Jang 2006). During the early stage of S deficiency, nitrogenase activity in the root nodules of legumes is more strongly depressed than photosynthesis (DeBoer and Duke 1982). Scherer and Lange (1996) found a close relationship between the S supply of *Medicago sativa* and the nitrogenase acitivity (Fig. 3.5) using the acetylene reduction assay as an indirect estimate of N_2 fixation. It increased with increasing S supply.

4.3 *ATP*

The nitrogenase reaction is an endergonic process. Energy, which is required in the form of ATP and reducing equivalents, is supplied by respiration and electron carriers,

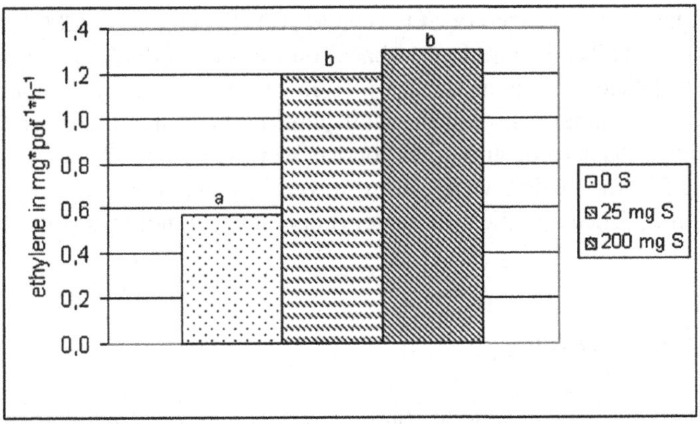

Fig. 3.5 Influence of S supply on the nitrogenase activity of *M. sativa* nodules. Different subscripts denote significant difference at the 5% level (ANOVA)

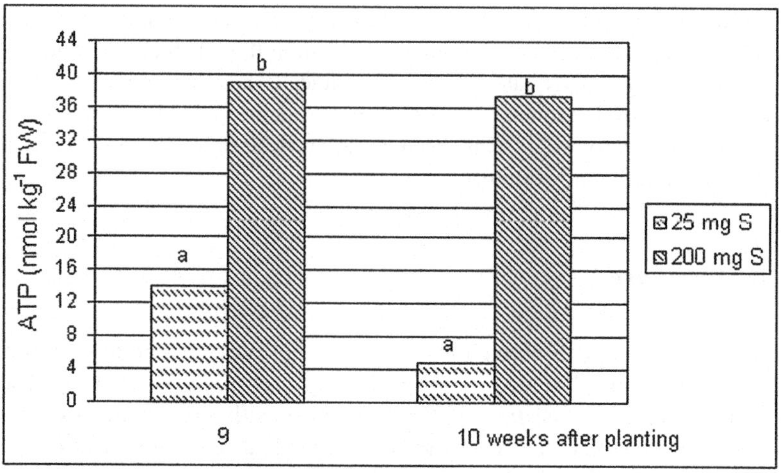

Fig. 3.6 Influence of S supply on the ATP content (nmol kg^{-1} fresh weight) of bacteroids of root nodules of *P. sativum*. Different subscripts denote significant difference at the 5% level (ANOVA)

usually ferrdoxin (Marschner 1995). The energy status could act as an indicator for an efficient symbiotic N_2 fixation (Pacyna et al. 2006), because it depends on a high-energy supply for N_2 reduction and assimilation. Mengel et al. (1974) postulated that a better carbohydrate supply of nodules of *V. faba* well supplied with potassium results in a higher carbohydrate turnover in nodules and thus the better provision of ATP required by nitrogenase. According to Bergersen (1991) the actual ATP requirement for N_2 fixation is 25 and 30 moles ATP per mole N_2. According to Pacyna et al. (2006) S deficiency resulted in significantly lower glucose concentrations in shoots of *V. faba minor*. Therefore it may be assumed that a lower provision of ATP under S deficiency conditions may be caused by a lower provision of glucose, which has to be oxidized to CO_2 to yield ATP. Scherer et al. (2007) found significantly lower ATP concentrations in bacteroids of root nodules of *P. sativum* (Fig. 3.6) under S deficiency conditions. To differentiate the impact of S supply on ATP concentrations of bacteria in nodules and plant cells, respectively, root nodules of *P. sativum* in bacteroids and mitochondria were separated. While the ATP concentrations of mitochondria were per se lower as compared with bacteroids, the concentration declined significantly under S deficiency conditions (Fig. 3.7).

4.4 Available Glucose

Forage legumes have a considerable biological N_2 fixation capacity under optimum environmental conditions. Under S deficiency conditions the chlorophyll content of leaves is drastically decreased (Friedrich and Schrader 1978) and the rate of

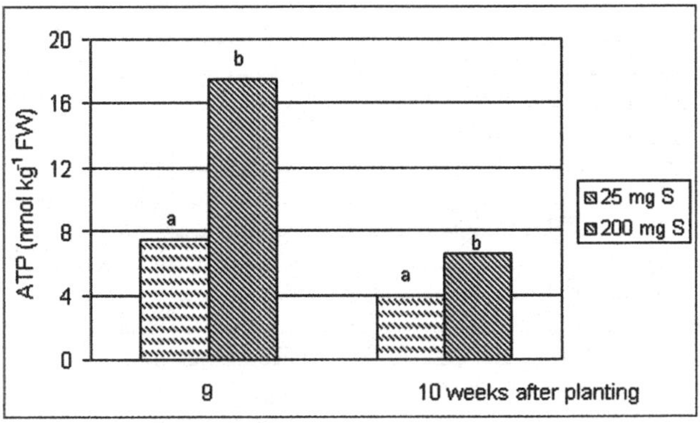

Fig. 3.7 Influence of S supply on the ATP content (nmol kg^{-1} fresh weight) of mitochondria of root nodules of *P. sativum*. Different subscripts denote significant difference at the 5% level (ANOVA)

photosynthesis is lower. However, the symbiotic nature of the plant-*Rhizobium* relationship requires a carbohydrate supply from the shoot; the main functions of carbon metabolism are to provide reducing power for the host cell cytosol and the bacteroids for N_2 fixation and subsequent amide and amino acid production in the plant. Furthermore, besides oxygen, carbohydrates must be available to sustain respiration in the nodule and thus provide energy. Therefore it may be speculated that the supply of carbohydrates could primarily limit N_2 fixation. In nodules, consumption of carbohydrates strongly varies and depends on bacterial species, plant species, and plant development (Streeter 1985, Vance and Heichel 1991). N_2 fixation costs about 6 mg C per milligram of reduced nitrogen (Ryle et al. 1986). According to Pate and Herridge (1978) nodules may consume up to 50% of the photosynthates produced by legume plants. However, as grain legumes mature, the reproductive sinks compete with the nodules for available photosynthates.

In pot experiments Pacyna et al. (2006) investigated the influence of S supply on the glucose content in shoots, roots, and nodules of *Vicia faba minor* in a time-course. The amounts of glucose were presented on per pot base, to illustrate the actual amounts of available glucose. In the treatment with optimum S supply the glucose content of the shoots ranged between 212 and 439 mg pot^{-1} during the experimental period of time with the highest content at the beginning of flowering (Fig. 3.8). The decline until 10 weeks after sowing is assumed to be an expression of sink competition for photosynthates between the developing pods and the root nodules. Therefore it may be speculated that in plants with optimum S supply the availability of glucose is influenced by the developing stage. When S was limiting the total amount of glucose in the shoots ranged between 10 and 45 mg pot^{-1}, indicating that available carbohydrates become limiting as C skeletons for ammonia assimilation and energy production (DeBoer and Duke, 1982). These results prompt to question whether a limited S supply of legumes penalizes N_2 fixation and therefore legume productivity.

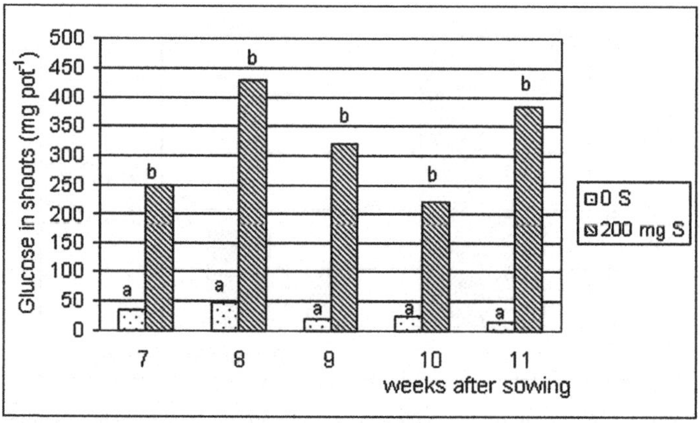

Fig. 3.8 Influence of S supply on the available amounts of glucose in shoots of *V. faba minor*. Different subscripts denote significant difference at the 5% level (ANOVA)

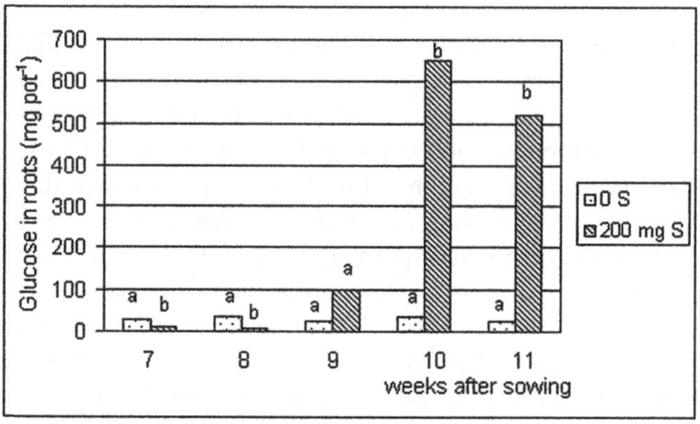

Fig. 3.9 Influence of S supply on the available amounts of glucose in roots of *V. faba minor*. Different subscripts denote significant difference at the 5% level (ANOVA)

In the roots of S-starved beans the total amount of glucose stayed on a low level during the time-course (Fig. 3.9). However, before flower setting they were significantly higher as compared to optimum supplied plants, suggesting a high metabolic activity maintained by higher ATP/ADP ratios (Dietz and Heilos 1990). The increase of the glucose content in roots of plants with optimum S supply 9 weeks after sowing was accompanied by a decrease in shoots, indicating a stronger sink of the roots.

Carbohydrates are translocated into the root nodules and metabolized to organic acids, which provide energy for the activity of nitrogenase (Vance et al. 1998). As compared to S-deficient plants the total amount of glucose in the root nodules of plants with optimum S supply was significantly higher during the whole experimental

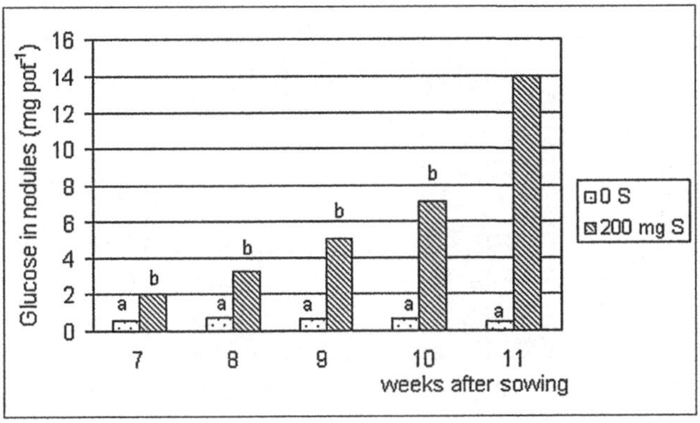

Fig. 3.10 Influence of S supply on the available amounts of glucose in root nodules of *V. faba minor* Different subscripts denote significant difference at the 5% level (ANOVA)

period of time and increased continuously (Fig. 3.10). Under S deficiency conditions the amounts of glucose stayed on the same level, indicating a preferential C allocation to growth rather than to N_2 fixation. These observations suggest that available photosynthates may limit the provision (1) of energy in the form of ATP and (2) of C skeletons for ammonia assimilation, resulting in a lower N_2 fixation and a reduced yield formation.

In infected cells, glucose is converted to acetyl-CoA, a precursor for the assimilation of ammonia by GS/GOGAT by the Embden-Meyerhof pathway. In the nodules of the plants well supplied with S the higher N_2 fixation rate, accompanied by a higher energy demand, is reflected by the higher amounts of glucose in the nodules.

4.5 Ferredoxin

Ferredoxins are acidic, low molecular weight, soluble iron-sulfur proteins, which are defined as proteins carrying iron-sulfur cluster(s) in which iron is at least partially coordinated by sulfur (Lill and Mühlenhoff 2006). The chief role of the iron-sulfur cluster is to facilitate electron transfer. In an investigation of Scherer et al. (2007) S deficiency conditions resulted in significantly reduced ferredoxin concentration of the bacteroids (Fig. 3.11). To demonstrate the total available ferredoxin, its actual amount was also shown on a per pot base (Fig. 3.12). During the time-course the actual available amounts of ferredoxin increased in the bacteroids of *M. sativa*, indicating a direct relationship between N_2 fixation and ferredoxin. Nitrogen-fixing organisms use NADPH as a predominant source of reducing power. Since resulting $NADP^+$ is reduced by ferredoxin-$NADP^+$ reductase, a low ferredoxin content influences the availability of reduced flavodoxin serving as reductant of nitrogenase

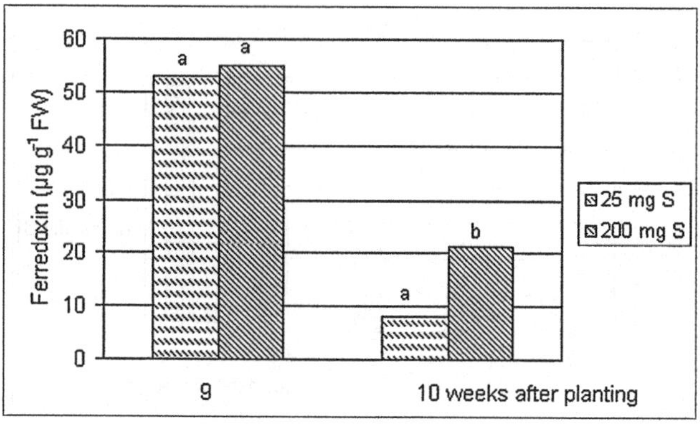

Fig. 3.11 Influence of S supply on the Ferredoxin content (µg g^{-1} fresh weight) of bacteroids of root nodules of *M. sativa*. Different subscripts denote significant difference at the 5% level (ANOVA)

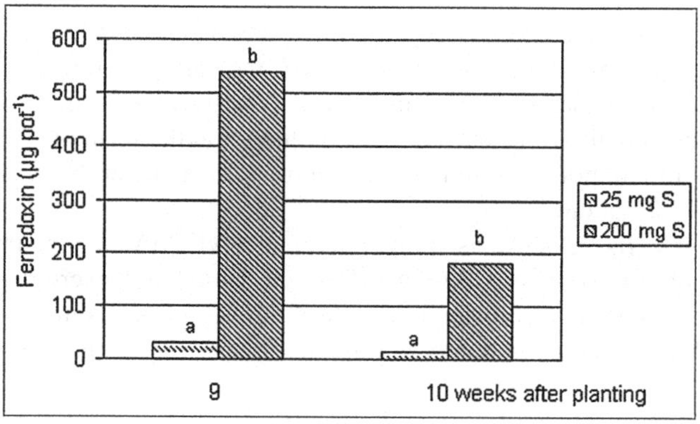

Fig. 3.12 Influence of S supply on the available amounts of ferredoxin (µg g^{-1} pot) of bacteroids of root nodules of *M. sativa*. (Different subscripts denote significant difference at the 5% level (ANOVA)

(Yoch 1973). With *V. faba* the relationship between N_2 fixation and ferredoxin was less pronounced. However, it should be pointed out that the decrease of the ferredoxin content occurred at pot filling.

5 Conclusions

It is apparent that N_2 fixation and consequently yield formation of legumes is influenced by sulfur. Therefore S fertilization for legumes should be recommended at least on soils, if the S delivery from soil organic matter is normally not sufficient to cover their N demand.

References

Andersen MK, Hauggaard-Nielsen H, Hogh-Jensen H, Jensen ES (2007) Competition for and utilisation of sulfur in sole and intercrops of pea and barley. Nutr Cycling Agroecosystems 77:143-53

Anderson AJ, Spencer D (1950) Sulphur in nitrogen metabolism of legumes and non-legumes. Aust J Sci Res (series B) 3:414-30

Bergersen FJ 1991 Physiological control of nitrogenase and uptake hydrogenase. In: Dilworth MJ, Glenn AR (eds) Biology and biochemistry of nitrogen fixation. Elsevier, Amsterdam, The Netherlands, pp. 76-102

Bolton J, Nowakowski T, Lazarus TT (1976) Sulphur-nitrogen interaction effects yield and composition of protein-N, non-protein-N, and soluble carbohydrates in perennial ryegrass. J Sci Food Agric 27:553-60

Chandler PM, Higgins TJV, Randall PJ, Spencer D (1983) Regulation of legumin levels in developing pea seeds under conditions of sulfur deficiency. Plant Physiol 71:47-54

Chandler PM, Spencer D, Randall PJ, Higgins TJV (1984) Influence of sulfur nutrition on development of some major pea seed proteins and their mRNAs. Plant Physiol 75:651-7

Collins M, Lang DJ, Kelling KA (1986) Effects of phosphorus, and sulfur on alfalfa nitrogen-fixation under field conditions. Agron J 78:959-63

DeBoer DL, Duke SH (1982) Effects of sulphur nutrition on nitrogen and carbon metabolism in lucerne (*Medicago sativa* L.). Physiol Plant 54:343-50

Dietz K-J, Heilos L (1990) Carbon metabolism in spinach leaves as affected by leaf age and phosphorus and sulphur nutrition. Plant Physiol 93:1219-25

Eppendorfer WH, Bille SW (1992) Development of S-deficiency of faba bean plants as reflected in total-S, so_4^{2-} -s and aspartic acid concentrations at various stages of growth. Proc. 2nd ESA Congress, Warwick, England, pp. 242-3

Friedrich JW, Schrader LE (1978) Sulfur depriviation and nitrogen metabolism in maize seedlings. Plant Physiol 61:900-7

Gilbert MA, Robson AD (1984) The effect of sulfur supply on the root characteristics of subterranean lover and annual ryegrass. Plant Soil 77:377-80

Janssent KA, Vitosh ML (1974) Effect of lime, sulfur, and molybdenum on N_2 fixation and yield of dark red kidney beans. Agron J 56:736-40

Jeong MS, Jang SB (2006) Electron transfer and nano-scale motions in nitrogenase Fe-protein. Curr Nanoscience 2:33-41

Lange A (1998) Einfluß der Schwefel-Versorgung auf die biologische Stickstoff-Fixierung von Leguminosen. PhD thesis, University of Bonn, Germany

Lill R, Mühlenhoff U (2006) Iron-sulfur protein biogenesis in eukaryotes: Components and mechanisms. Ann Rev Cell and Dev Biol 22:457-86

Marschner H (1995) Mineral Nutrition of Higher plants. 2nd ed, Academic Press

McGrath SP, Zhao FJ (1996) Sulphur uptake, yield responses and interactions between nitrogen and sulphur in winter oilseed rape (*Brassica napus*). J Agric Sci 126:53-62

Pacyna S, Schulz M, Scherer HW (2006) Influence of sulphur supply on glucose and ATP concentrations of inoculated broad beans (*Vicia faba minor* L.). Biol Fert Soils 42:324-9

Parsons R, Stanford A, Raven AJ, Sprent JI (1993) Nodule growth and activity may be regulated by a feed back mechanism involving phloem nitrogen. Plant Cell Environ 16:125-36

Pate JL, Herridge DF (1978) Partitioning and utilization of net photosynthate in nodulated annual legumes. J Exp Bot 29:401-12

Rendig VV, Oputa C, McComb EA (1976) Effects of sulfur deficiency on non-protein nitrogen, soluble sugars and N/S ratios in young corn (Zea mays L.) plants. Plant Soil 44:423-37

Ryle GJA, Powell CE, Gordon AJ (1986) Defoliation in white clover: nodule metabolism, nodule growth and maintenance, and nitrogenase functioning during growth and regrowth. Ann Bot 57:263-71

Scherer HW (2001) Sulphur in crop production - invited paper. Eur J Agron 14:81-111

Scherer HW, Lange A. (1996) N_2 fixation and growth of legumes as affected by sulphur fertilization. Biol Fert Soils 23:449-53

Scherer, HW, Pacyna S, Manthey N, Schulz M (2006) Sulphur supply to peas (*Pisum sativum* L.) influences symbiotic N_2 fixation. Plant Soil Environ 52:72-7

Scherer HW, Pacyna S, Spoth K, Schulz M. (2007) Low levels of ferredoxin, ATP and leghemoglobin contribute to limited N_2 fixation of peas (*Pisum sativum* L.) and alfalfa (M*edicago sativa* L.) under S deficiency conditions. Biol Fertil Soils, (submitted)

Sexton PJ, Batchelor WD, Shibles R (1998) Soybean sulfur and nitrogen balance under varying levels of available sulfur. Crop Sci 38:975-82

Shock CC, Williams WA, Jones MB, Center DM, Phillips DA (1984) Nitrogen fixation by subclover associations fertilized with sulfur. Plant Soil 81:323-32

Vance CP, Miller SS, Driscoll BT, Robinson DL, Trepp G, Gantt JS, Samas DA (1998) Nodule carbon metabolism: organic acids for N_2 fixation. In: Elmerich C (ed) Biological nitrogen fixation for the 21st century. Kluwer, The Netherlands, pp. 443-8

Yoch D (1973) Purification and characterization of ferredoxin-nicotinamide adenine dinucleotide phosphate reductase from a nitrogen-fixing bacterium. J Bacteriol 118:384-91

Zhao FJ, Hawkesford MJ, Warrilow AGS, McGrath SP, Clarkson, DT (1996) Responses of two wheat cultivars to sulphur addition and diagnosis of sulphur deficiency. Plant Soil 181:317-327

Chapter 4
Sulfur Nutrition and Assimilation in Crop Plants

Avtar Singh Bimbraw

Abstract Sulfur (S) deficiency has become common in agricultural soils and has resulted in crop yield loss. Agricultural soils are further at risk from S-deficiency in cereals and oilseed crops. The present paper discusses the effects of sulfur on plant growth and crop quality in various crops associated with S-deficiency and increase in the environment (Air pollution). Plant responses to S-supply are examined. The importance of sulfur assimilation pathways, which have pivotal role in influencing plant S-metabolism, is discussed.

1 Introduction

In recent years sulfur (S) deficiency has become an increasing problem for agriculture, resulting in decreased crop quality parameters and yields (McGrath et al 1996).). Sulfur is increasingly being recognized as the fourth major plant nutrient; its deficiency in light-textured soils that are poor in sulfur reserves, is due to low organic matter content. Its deficiency is increasing as a result of intensive cropping without the addition of sulfur-containing, high-analysis fertilizers like urea, diammonium phosphate, and muriate of potash (Kothari and Jethra 1994). Use of high-analysis fertilizer under high-intensity cropping has depleted the soil. This continuous mining has led to the deficiency of multinutrients. Sulfur has occupied an important place – after nitrogen, phosphorus, and potassium – in balanced fertilization programs due to use of sulfur-free fertilizers, mostly diammonium phosphate (Mukherjee and Singh 2002). Sulfur is known for its role in the formation of sulfur-containing amino acids, synthesis of proteins, vitamins, chlorophyll, and oil. In oilseed, sulfur promotes nodulation, while in legumes, it produces bold seeds. In oilseeds sulfur deficiency affects the quality of food for human beings (Singh and Bansal 1999). Balanced fertilization is very crucial in order to achieve potential yield with improved quality of low-demand and high-demand

Avtar Singh Bimbraw
Department of Agronomy, Punjab Agricultural University, Ludhiana-141 004, Punjab, India
avtar_bimbraw@yahoo.com

N.A. Khan et al. (eds.), *Sulfur Assimilation and Abiotic Stress in Plants.*

sulfur crops with the application of sulfur with sulfur-free fertilizers. Appropriate application of fertilizers can remedy 'e persistence of the S in the soil and the availability to the plant.

In recent decades considerable progress has been made in reducing emissions of S to the atmosphere, which has resulted in a consequent decrease in atmospheric depositions of S onto agricultural land (McGrath et al. 1996). While these depositions were once sufficient to support crop requirements for S, they now fall well below the recommended requirements for cereal and oilseed crops. Predictive modeling has shown that the occurrence of agricultural land at risk from S-deficiency will increase. Deficiencies are predicted for cereals and are more likely for oilseed rape crops, which have a higher requirement for S (McGrath et al 1996). In this paper the author discussed the effects on plant growth and crop quality in various crops associated with S-deficiency and its increase in the environment. Plant responses to S-supply will be examined and the importance of sulfur assimilation pathways, which have a pivotal role in plant S-metabolism, will be reviewed.

2 Sulfur Nutrition

2.1 Effects of S-Deficiency

The agronomic consequences of insufficient S are well documented by decreased yields and substantial impact on S-content under extreme deficiency. In many cases of mild S-deficiency stress there may be little impact on yield but important consequences for quality, with substantially modified N:S ratios (Zhao et al. 1996). A shift to higher N:S ratios has been observed in the years 1981 to 1983 in British wheat grain (Zhao et al. 1995), closely mirroring the decrease in atmospheric outputs and subsequent deposition of S (McGrath et al. 1996). Limiting S availability has been shown to favor the synthesis and accumulation of S-poor or low-S storage proteins such as gliadin and high molecular weight subunits of glutenin at the expense of S-rich proteins in wheat (Moss et al. 1981, Wrigley et al. 1984, Fullington et al. 1987). These changes in protein composition are associated with alterations of dough rheology and bread quality.

Sulfate deficiency in young wheat plants has an early effect on CO_2 assimilation rates and on Rubisco enzyme activity and protein abundance (Gilbert et al. 1997). This is a result of decreased synthesis of new protein under S-limiting conditions and, additionally, some degradation was observed in response to S-limitation in the older leaves. The lack of synthesis of Rubisco and the chlorosis of the young leaves due to decreased chlorophyll content (Burke et al. 1986) reflect a general inhibition of *de novo* synthesis of the photosynthetic apparatus.

Another metabolic effect of S-stress is a depression of the root hydraulic conductivity (Karmokar et al. 1991), an early response which may have a role in signaling nutrient starvation from root to shoot. It is proposed that stomatal closure restricts CO_2 uptake, limiting carbon assimilation and thus restricting the metabolic need for sulfur.

An obvious indication of S-deficiency is the reduction in the internal S pools, but additionally there are rises in soluble nitrogen pools, including nitrate and amides as a consequences of the N:S imbalance (Karmokar et al. 1991, Zhao et al. 1996, Prossor et al. 1997, Warrilow and Hawkesford 1998). These metabolite fluctuations have been proposed as possible diagnostic indications of S-deficiency (Zhao et al. 1996).

Several distinct pools of S occur in plant tissues, with the most occurring as sulfate or in the protein fraction. The relative abundance of these two fractions depends upon the specific tissue and the previous nutritional history of the plant (Blake-Kalff et al. 1998). Other smaller pools include free amino acids, cysteine and methionine, the tripeptide glutathione, sulfolipids, and other secondary compounds such as the glucosinolates found in the Brassicaceae. If present, the most significant and readily mobilized form is sulfate.While cytoplasmic concentrations of sulfate are kept relatively constant, sulfate taken up by the plant, which is surplus to immediate requirements for growth, is stored in the vacuole. Reports on the effectiveness of mobilization of this vacuolar sulfate pool vary, and may reflect species differences or the ability of the remobilization processes to keep pace with growth rates. The mobilization of this vacuolar pool has been reported to be a slow process in roots (Bell et al. 1994, 1995a) in mature leaves (Bell et al. 1995b), and particularly so in oilseed rape (Blake-Kalff et al. 1998).

The patterns of S-accumulation and redistribution in *Brassica napus* have been described recently in detail (Blake-Kalff et al. 1998). When supplied with adequate S, the concentrations of glutathione and glucosinolates accounted for 2% and 6% of the S-content in the youngest leaves, respectively. In the older leaves these compounds accounted for an even smaller proportion of the total S, and it was concluded that they are not major reserves of S during deficiency. The concentration of insoluble (protein S) was similar for all leaves (around 50%). In the mature leaves 70% to 90% of the total S could be accumulated as sulfate. If S-supply was withdrawn, these pools all decreased, although the decrease in concentration could be accounted for mainly by growth. There was little evidence of the large reserves of sulfate being redistributed to the younger growing parts of the plant. This inefficiency in managing S-reserves is suggested to be part of the reason for the high S requirement for oilseed rape crops.

Pulse-labeling experiments have investigated fluxes of sulfate in barley (Adiputra and Anderson 1992, 1995). These studies demonstrated redistribution of sulfate, but showed no evidence for enhanced redistribution stimulated under S-limiting conditions. In soybean, the greatest redistribution of S occurs when N-limitation induces proteolysis (Sunarpi and Anderson 1997). Studies on the remobilization of S in the flag leaf and delivery to the developing grain of wheat indicate that when there are adequate reserves of sulfate, this can be remobilized to the grain (MJ Hawkesford, unpublished results). Plants grown on an adequate S-supply (1.0 mM sulfate in the nutrient solution, applied on alternate days) until anthesis, when the supply of S was terminated, were able to maintain the S-content of the grain at near control levels (S-supply maintained after anthesis), at the expense of S-pools in the flag leaf. This was in contrast to plants grown with a suboptimal S-supply, with for example, 0.1 mM sulfate in the nutrient solution, where little S accumulated in the flag leaf, and grain contained substantially less S than the control plants.

3 Sulfur Requirement

3.1 Maize

Winter maize cultivation is now attaining greater importance in India. It has been reported that the yield level of winter maize is 30 to 40% higher with higher economic return than those obtained in *Kharif* season (Singh et al. 1998). Balanced fertilization is very crucial in order to achieve its potential yield with improved quality. Nitrogen and sulfur are known to improve the yield and quality (Tandon 1991). Based on an average of two years the magnitude of increase in grain yield of maize with 45 kg S/ha was 9.39, 5.23, and 2.43% as compared with 0, 15 and 30 kg S/ha, respectively (Maurya et al. 2005). The maximum value of harvest index (33.73%) was recorded with 45 kg S/ha, which was significantly superior to other levels of sulfur.

Application of sulfur at 45 kg S/ha significantly increased the yield attributes, number of cobs/plant, length of cob, number of grains/cob, and 1000-grain weight of maize over its lower levels of sulfur (Maurya et al. 2005).

3.2 Pearl Millet

Pearl millet is an important food and fodder crop of the arid and semi-arid tropics. The quality of the fodder, however, depends on the nourishment of the crop and virtually influences the nutritive value of the animal product feeding on pearl millet fodder (Dadhich and Gupta 2005). They reported that pooled green fodder yield increased significantly with increase in the level of sulfur up to 60 kg S/ha. The increase in pooled green fodder yields with 60 kg S/ha was 36.1, 13.5, and 2.2% as compared to 0, 20, and 40 kg S/ha. But application of 40 kg S/ha significantly increased the pooled plant height, tillers/plant, leaves/plant, stem girth and leaf area, and the magnitude of respective increase was 21.8, 30.4, 17.8, 20.1 and 12.5% over control. This might due to the fact that the soil of the experiment field had a low level of available sulfur. Application of sulfur to a deficient soil is known to have also improved the availability of other nutrients which are considered vitally important for the plant (Dadhich and Gupta 2005).

3.3 Groundnut

The yield components like pods/plant, shelling (%), 100-kernel weight along with pod yield, kernel yield, and oil content were significantly superior with gypsum and singlesuperphosphate over pyrites (Bandopadhayay and Samui 2000). The performance of gypsum and singlesuperphosphate was at par. Tandon (1986) and Patra et al. (1995) reported responses of sulfur on yield and yield attributing characters. Sulfur levels improved upon all the attributes and yield of groundnut except the haulm yield. Interaction effects between source and level of S for both pod and kernel yields of groundnut indicated that increased level of S addition through gypsum and singlesuperphosphate more yield obtained over use of pyrites.

3.4 Soybean

Sulfur is considered the fourth major nutrient and is known for its role in the formation of sulfur-containing amino acids, synthesis of proteins, vitamins, chlorophyll, and oil. In oilseed, sulfur promotes nodulation, while in legumes, it produces bold seeds. In oilseeds sulfur deficiency affects the quality of food for human beings (Singh and Bansal 1999). Soybean is emerging as an important oilseed crop of Meghalaya. Soils of the state are acidic and deficient in available P and S. The P and S are known to be involved in major metabolic processes such as protein and oil synthesis in plants (Kumar and Singh 1980). Soybean being a leguminous oilseed requires a higher and balanced dose of both P and S. The effect of applied P on crop growth depends on level of S in soil and vice versa, as P and S are both absorbed by plants in anionic form from the soil. Both synergistic and antagonistic interaction has been reported between P and S (Ali 1991). The grain and straw yields of soybean increased significantly with increasing levels of both P and S (Majumdar et al. 2001). The soil of the experimental study was deficient in both available P and S, and therefore the significant response of soybean to the applied nutrients is quite understandable. The interaction between P and S was significant for seed and straw yield of soybean. All the levels of S (Elemental Sulphur was applied 15 days before sowing to facilitate oxidation of S to sulfate form) gave significantly higher yield at each P level and the highest seed yield (2.5 tons/ha) was recorded with a combination of 60 kg P_2O_5 and 40 kg S/ha. This positive signi$_f$icant interaction might be owing to increased uptake of nutrients like N, P, and S with the combined application of P and S, which helped in better nutrition of soybean for optimum growth and development. However, Ramamoorthy et al. (1996) reported that the highest grain yield was obtained with the application of 40 kg S/ha, which was significantly superior to the control. However, Singh and Singh (1995) found that application of 30 kg S/ha significantly increased the seed yield over the control during both years. It might be due to the fact that sulfur is a constituent of amino acids and thus vital for protein production. Similar results were reported by Dhillon and Dev (1980).

The pods/plant and 100-seed weight increased by 7.6 and 2.6%, respectively, with 40 kg S/ha, and the same were increased by 37 and 8.4% with 60 kg P_2O_5 over the control (Majumdar et al. 2001). The relative contribution of P was more than S in this regard. The interaction effect of P and S was significant, thereby indicating a more beneficial effect of the two in combination. All the levels of S increased the 100-seed weight as well as pods/plant significantly at each level of P. The results indicated that the utilization of S for plant growth was associated with a concomitant supply of phosphorus. The pods/plant and 100-seed weight were maximum under a treatment combination of 60 kg P_2O_5 and 40 kg S/ha. In another study, Singh and Singh (1995) reported that application of 30 kg S/ha significantly increased leaf area/plant, dry matter/plant, number of pods/plant, number of seeds/pod, and 1000-seed weight and seed yield during both the years. The results are in conformity with the findings of Singh and Singh (1995).

Among the sources, S applied to the soybean in the form of gypsum (Ca SO_4) was found superior to S applied in the form of pyrite. Sharma and Bradford (1973) and Bansal et al. (1983) reported increased yield of soybean by sulfur fertilization.

3.5 *Pigeonpea*

Sulfur is an essential plant nutrient, and its deficiency at any stage of crop growth may lead to considerable losses in grain yield. Sulfur deficiency may cause 12% to 15% reduction in yield of soybean (Chandel et al. 1989). The deficiency of sulfur has been reported in intensively cultivated areas; it was mainly due to the higher sulfur requirement of high-yielding crops as well as due to leaching losses of sulfate (Singh et al. 1997). They reported that different sources of sulfur did not influence the grain yield. Biswas and Tewatia (1991) also reported the nonsignificant differences in grain yield due to different sources of sulfur application in chickpea, blackgram, and pigeonpea.

The number of pods/plant was significantly higher with pyrite and gypsum sources than with singlesuperphosphate; however, nonsignificant differences were observed between pyrite and gypsum sources (Singh et al. 1997).

Sulfur application at 20 and 40 kg /ha was at par but significantly increased the grain yield over the control. The magnitude of increase was 30.3 and 16.7% & 19.9 and 17.5% in grain yields over the control during 1993 and 1994, respectively. These results confirm the previous findings of Jat and Rathore (1994) and Singh and Singh (1995).

Sulfur application of 20 and 40 kg/ha increased the number of pods significantly over the control (Singh et al. 1997). This might be due to the fact that sulfur played an important role in carbohydrate metabolism, which ultimately was reflected in the production of significantly more pods/plant than in the control. These results are in conformity with findings of Jat and Rathore (1994). Nonsignificant differences were, however, observed in the number of pods recorded under 20 and 40 kg S/ha.

3.6 *Greengram*

Deficiency of phosphorus is usually one of the major constituents responsible for poor yield of greengram in all types of soil in Rajasthan, as it is an indispensable constituent of protein and nucleic acids (Yadav 2004). Sulfur also affects the growth and yield of legumes in many ways. Legumes usually require almost equal amount of sulfur and phosphorus (Singh and Agarwal 2001). Application of sulfur also significantly affected the number of branches/plant, dry matter accumulation/meter row length, number of pods/plant, seed haulm, and biological yields during both years and plant height during 1998, which was significantly higher at 40 kg S/ha. Application of 40 and 60 kg S/ha were found statistically at par with respect to these characters (Yadav 2004). Pooled analysis showed that 40 kg S/ha recorded 25.37 and 9.74% higher seed and 29.47 and 9.92% higher haulm yield of greengram over control and 20 kg/ha, respectively. Further increase in the level of sulfur to 60 kg/ha could not bring the difference in these characteristics up to levels of significance. The highest benefit:cost ratio was obtained at 40 kg S/ha. Increase in seed and haulm yield of gram might be ascribed to the cumulative effect of growth- and yield-attributing characters of the crop. Such favorable influences of sulfur on yield of greengram have also been reported earlier by Kumar (1993) & Jat and Rathore (1994).

3.7 Blackgram

In the past two decades, continuous use of major plant nutrients (N, P, and K) through mineral fertilizer under an irrigated cropping system has resulted in the depletion of secondary nutrients and micronutrients. Among the secondary nutrient elements, the deficiency of S is being observed from different parts of the country (Singh 2004). About 130 districts are considered to be suffering from S deficiency to varying degree. All indications are that S deficiencies will become even more important in coming years, and in such areas the balanced fertilizer use will includes S along with NPK application. Over the years, sulfur has become increasingly important for producing high crop yields and better quality produce. The deficiency of S which was noticed many years ago in localized areas has now engulfed much larger areas. A few years ago sulfur was considered as a nutrient of academic interest, but today it is of much importance to Indian agriculture. Generally, soils testing less than 10 ppm available S are expected to respond to S application. At present levels of fertilizer use and product pattern dominated by S-free fertilizers (Urea, DAP, MOP, NPK) more S is being removed from the soil than is being added, which is one of the reasons deficiency is being reported from more and more areas (Singh 2004). Leguminous crops in general require as much sulfur as they require phosphorus. The total sulfur requirement of a crop depends on the type of the crop, its variety, yield level, and the amount of S available to it. To produce high yields year after year it is necessary for the S and P requirement of crops to be adequately met. The role of S and P is important for maintaining the quality of grain legumes.. Sulfur plays an important role in the formation of S-containing amino acids (methionine and cystein), synthesis of proteins, and promotion of nodulation and production of bolder grains (Singh 2004). He reported that increasing levels of sulfur increased significantly the grain yield of blackgram from 8.47 to

10.39 q/ha up to 30 kg S/ha. Straw yield followed a similar trend where application of S increased the average straw yield from 30.0 to 34.55 q/ha up to 30 kg S/ha level. The average grain yield and straw yield response of 30 kg S/ha was 2. and 4.6 q/ha, respectively.

Blackgram requires higher amounts of P and S as compared to N (Trivedi et al. 1997). They reported that the seed and stover yields were increased significantly with 60 kg S/ha of blackgram.

3.8 Cowpea

Cowpea is grown for grain, vegetable, and green fodder purposes in the semi-arid eastern plains of Rajasthan. Being a crop of short duration, it fits in drought-prone areas of Rajasthan. However, it is grown on marginal lands in the *Kharif* season with no fertilizer and hence the total biomass production and nutrient uptake is adversely affected (Sharma and Jat 2002). They found that each successive increment in sulfur dose from control to 80 kg S/ha (DAP and gypsum as sources of sulfur) registered a significant increase in dry matter accumulation, which might be attributed to the beneficial effect of sulfur on balanced nutrition and thus the higher growth rate of plants.

3.9 Clusterbean

Clusterbean is an ancient multipurpose legume of arid and semi-arid regions of tropical India. The soils of arid regions are sandy in nature and are poor in native fertility. Thus, balanced fertilization along with sound crop husbandry offers a great scope for increasing productivity of clusterbean (Sharma and Singh 2005). Widespread S deficiency and profitable yield increases from S application have been recorded under field conditions (Tondon and Messick 2002). As the sources of sulfur are finite and nonrenewable, their judicious exploitation becomes imperative for sustained use. Since oxidation of elemental sulfur to sulfate in soil depends on its particle size, development of modified form of elemental S has been initiated. Beneficial effects of various substances in enhancing crop yield and quality improvement are reported (Sharma and Singh 2005). Barley is a potential winter crop of this region, requires low inputs, and can be successfully raised on residual fertility where the previous crop was a legume receiving S.

The increase in grain yield of clusterbean was 24.5 and 32.0% with 25 and 50 kg S/ha over the control, respectively. The difference in yield between 25 and 50 kg S/ha was statistically nonsignificant. The significant improvement in yield obtained under sulfur fertilization seems to have resulted in from increased concentration of sulfur in various parts of clusterbean that helped to maintain the critical balance of other essential nutrients and resulted in increased metabolic processes in plants (Sharma and Singh 2005). However, Pareek (1995) also reported increased yield of clusterbean with application of sulfur. Sulfur plays a key role in improving vegetative structure for nutrient absorption, strong sink strength through development of reproductive structures, and production of assimilates to fill economically important sinks (Sacchidanand et al. 1980).

SulFer 95 as a source of S was found superior to gypsum in improving yield attributes and produced significantly higher grain and straw yields over gypsum. The corresponding grain yield under SulFer 95 was 11.0% higher than that under gypsum (9.68 q/ha), maybe owing to constant and enhanced availability of SO_4-S to growing plants under SulFer 95 treatment (Sharma and Singh 2005, Dev 1999).

Application of 25 and 50 kg S/ha increased the pods/plant by 19.7 and 31.4%, grains/plant by 8.9 and 11.6%, and test weight by 14.9 and 17.1% over control respectively(Sharma and Singh 2005).

3.10 *Sunflower*

Krishnamurthi and Mathan (1996) recorded that seed yield increase at 45 kg S/ha was the highest in summer and *Kharif*. In summer, seed yield from 45 kg S/ha treatment was at par with 30 kg S/ha, while in *Kharif*, 45 kg S/ha produced significantly more than did the rest of treatments. This was attributed to increased seed set, head diameter, and 100-seed weight. However, Awasthi et al. (2001) reported that application of sulfur at 40 kg/ha to sunflower produced significantly higher seed and stalk yield over control and it was at par with 20 kg S/ha.

In sulfur-applied plots significantly higher growth of sunflower was observed in summer and *Kharif*. There was no significant variation among the doses. The number of days for to 50% flowering was significantly reduced by sulfur application in both the seasons (Krishnamurthi and Mathan 1996). Similar seasonal variation was observed by Singh (1975). Both in summer and Kharif, sulfur application significantly increased the filled number of seeds/head and 100-seed weight, the highest being at the 45 kg S/ha level (Krishnamurthi and Mathan 1996). Similar results were reported by Dubey et al. (1993). The disk weight was significantly increased by sulfur application in both the seasons (Krishnamurthi and Mathan 1996). Sulfur application of 45 kg S/ha recorded the highest value, being significantly higher than all other treatments. The head diameter was also significantly increased by S application, while the variation due to doses was not significant. Increase in leaf area was significant only under 45 kg S/ha treatment in both the seasons. The response to sulfur might be due to more synthesis of chlorophyll by S application (Chatterjee et al. 1985). Kernal:husk ratio increased significantly with increasing level of sulfur application, the highest being under 45 kg S/ha treatment (Krishnamurthi and Mathan 1996). This corroborates the results of Mandal and Singh (1993).

3.11 Niger

The principal niger-growing states are Madhya Pradesh, Chhatisgarh, Orissa, Maharashtra, Karnatka, Andhra Pradesh, Jharkhand, and parts of Rajasthan (Singh et al. 2003). The states of Madhya Pradesh and Orissa are the most important, together accounting for 70% of area and 67.4% of production. Niger (*Guizotia abyssinica* Cass) is a late *Kharif* oilseed crop and is known as *Surguja, Gunja, Ramtil, Kalati*l, *Khurasani, Jagni*, and other names. in various regional Indian languages. It is extensively grown as pure, mixed, or a second crop after the harvest of upland rice, *gundli* or *marua*. Niger is an important popular oilseed crop among the tribal farmers who occupy an area of 37.8 thousand hectares with poor productivity in Jharkhand state (Singh et al. 2003). Its cultivation on marginal land and fertile soil without fertilizer application is the basic reasons for poor yields. The sulfur requirement of oilseed is fairly high compared to pulses, cereals, and millets. Sulfur can play a key role in augmenting the production and productivity of this low-yield oilseed crop. The soils of Jharkhand have been grouped under the order Alfisol/Ultisol. This state has three agroclimatic subzones, viz., Central and Northeastern plateau subzone (Zone IV), Western plateau subzone (Zone V), and Southeastern plateau subzone (Zone VI). The soil samples for the red and lateritic soils of three subzones indicated 50% deficiency of plant-available sulfur, primarily due to low content of organic carbon, coarse texture, and increased loss of SO_4-S on account of leaching and erosion (Singh et al. 2003). They reported that grain yield of niger increased significantly with successive increase in the levels of S up to 45 kg S/ha and reduced at higher levels of applied S (60 kg S/ha). Application of 45 kg S/ha produced an additional mean grain yield of 159 kg/ha, with an increase of 43.5% over control.

3.12 Wheat

Wheat is one of the most important cereal crops of India and has diverse uses. Intensive cultivation has resulted in depletion of soil nutrients to great extent; thus the nutrient requirement of crops has increased considerably during the last few years. Sulfur performs many physiological functions, such as synthesis of sulfur-containing amino acids, which have a positive role in improving quality of grains (Dewal and Pareek 2004). About 38% of the soils of the semi-arid eastern plain zone of Rajasthan are deficient in sulfur (Kothari and Jethra 1994).

Wheat is considered to the backbone of the nation's food security. Fertilizer is a major input, accounting for nearly 50% increase in food grain production. Therefore, the balanced use of fertilizer is essential for further increasing productivity. Despite the application of recommended quantities of nutrients, the increase in yield is not encouraging. This indicates that in addition to major plant nutrients, there is a need to supply secondary nutrients (Sharma and

Manohar 2002). Sulfur is increasingly being recognized as the fourth major plant nutrient, as the deficiency of sulfur in light-textured soil is about 38% in the semi-arid eastern plain zone of Rajasthan (Kothari and Jethra 1994). Sulfur application benefits more than one crop and shows a significant residual effect on the following crop (Biswas and Tewatia 1991). Increasing levels of S significantly increased the grain yield of wheat up to 40 kg S/ha (Sharma and Manohar 2002 and Dewal and Pareek 2004) and straw yield up to 60 kg S/ha (Sharma and Manohar 2002). The mean increase was 26.8% in grain yield due to 40 kg S/ha and 16.5% in straw yield with 60 kg S/ha over control. The increase in the yield of wheat was because of significant improvement in effective tillers/m^2, grains/spike, and test weight under 40 kg S/ha over control. These results confirm the findings of Upadhyay and Tiwari (1996) and Dewal and Pareek (2004). However, application of 40 kg S/ha significantly increased the plant height, dry matter accumulation, and total tillers/plant over control and 20 kg S/ha. Further increase in level of 60 kg S/ha could not bring significant improvement in these parameters (Dewal and Pareek 2004). The greater photosynthetic activity and chlorophyll synthesis due to fertilization seemed to have promoted vegetative growth. Similar results were reported by Singh et al. (1986) and Dewal et al. (2000).

Maximum grain yields of 54.82 and 62.59 q/ha and straw yields of 106.5 and 124.7 q/ha of wheat were recorded with the application of ammonium sulfate during 1997-98 and 1998-99, respectively, but the differences between ammonium sulfate and pyrite for grain and straw yields were not significant in 1997-98 (Singh et al. 2001). There are ample evidences of increase in grain and straw yields of wheat with the application of sulfur. Superiority of ammonium sulfate over the other sources of S for production of wheat has been reported by Arora et al. (1991).

3.13 Oat

Oat (*Avena sativa* L.) is a fast-growing winter season cereal fodder widely grown under irrigated condition in Haryana. Balanced fertilization is very crucial in order to achieve its potential yield with improved quality (Singh et al. 2002). Nitrogen and phosphorus are known to improve the yield and quality of oat fodder (Thakuria 1992, Patel 1997). The coarse-textured soils of southern Haryana are poor in sulfur reserve due to low organic matter content, and its deficiency is increasing as a result of intensive cropping without addition of sulfur-containing high-analysis fertilizers. Thus, sulfur has become increasingly important for such areas in order to produce high crop yields (Singh et al. 2002). They reported that crops responded significantly to sulfur application up to 20 kg/ha (gypsum as source of sulphur). Increase in mean forage and dry matter yield due to application of 20 kg S/ha was 7.7 and 7.0% during 1994-95 and 1995-96, respectively, over no sulfur application. Net returns increased up to 40 kg S/ha.

3.14 Rapeseed and Mustard

Fertilizer management, especially of nitrogen and phosphorus, is one of most important factors that affect the crop yield. Of late, sulfur has become an important nutrient. Rapeseed and Indian mustard (*Brassica juncea*) crops respond to sulfur remarkably, depending upon soil type and source of S used (Singh and Sahu 1986). Singh and Singh (1984) reported a 31% increase in yield with S application (through elemental sulfur) at 250 kg/ha, 21 days before sowing on clay-loam calcarious soils of south Rajasthan.

Rapeseed and mustard are the most important oilseed crops next to groundnut in the country. Use of high-analysis fertilizer under high-intensity cropping has depleted the soil. This continuous mining has led to the deficiency of multinutrients. Sulfur has occupied an important place only after nitrogen, phosphorus, and potassium in a balanced fertilization program due to use of sulur-free fertilizers, mostly Diammonium Phosphate (Mukherjee and Singh 2002). They observed that the positive effect of sulfur application on yield attributes and yield was due to the increased siliqua formation, siliqua-bearing branches, and overall increase in growth attributes. Crops responded to sulfur up to 60 kg/ha. Similar findings were reported by Trivedi et al. (1997). Increase in plant height, branches, and dry matter production may be attributed to the essential role of sulfur in cell division. The importance of sulfur in cell division, cell elongation, and setting of cell structure was also advocated by Hashimoto and Kameoka (1985). They also reported that mustard has an additional requirement for sulfur due to the presence of various glucosinolates and a number of neutral volatile sulfur and nitrogen compounds.

The increase in stover yield was due to production of taller plants with well-developed branches assimilating higher dry matter under the influence of sulfur application (Singh and Mukharjee 2004).

The application of S up to 30 kg/ha to Indian mustard increased the seed yield over the control during both the years (Dubey and Khan 1993). However, 30, 40, and 50 kg S/ha were at par. The increases in seed yield due to 30 kg S/ha over control were 30.4 and 44.1% during 1988-89 and 1989-90, respectively. Such an increase in the seed yield might have resulted from the corresponding improvement of yield attributing characters. However, Khurana et al. (2003) reported that response to application in raya and ghobhi sarson was found to be spectacular, and the optimum rate for obtaining best yields of these crops was 20 kg S/ha as Gypsum. The average yield of raya and ghobi sarson could be increased by 2.2 and 4.0 q/ha through optimum use of sulfur.

Sulfur application had no significant effect on growth and yield attributes of toria (*Brassica campestris* subsp. Oleifera var. toria) except seeds/siliqua and 1000-seed weight (Das and Das 1995). The highest seed yield was obtained with 45 kg S/ha and the magnitude of increase was 23.1, 16.2, and 11.1% over the control, 15 and 30 kg S/ha. Similar findings were reported by Singh and Gangasaran (1987) and Singh et al. (2000). However, Narang et al. (1993) reported that application of S at 50 kg /ha to toria produced seed yield that was significantly higher than control. Singh and Gangasaran (1987) and Singh et al. (1988) also received the same results in rapeseed and mustard. The sulfur application influenced the productivity of toria by improving both the basic infrastructural frame (bearing capacity) and the leaf area (the photosynthate production efficiency as well as the siliqua size). It regulates these parameters because it is part of amino acids like methionine, cystine,and cystein, lipoic acid, co-enzyme (SCOAH) and vitamins (thiamine pyrophosphate). Singh et al. (2000) reported that the improvement in yield and its attributes with increased supply of S may be due to enhanced photosynthesis in brassicas, as S is involved in the formation of chlorophyll and activation of enzymes (Upasni and Sharma 1986).

Studies conducted by Kumar et al. (2001) reported that application of sulfur in Indian mustard through gypsum, elemental sulfur, and pyrite was found equally effective for plant height, branches/plant, siliquae/plant, seeds/siliqua, 1000-seed weight, and harvest index. The variation in seed and stover yields due to various sources of sulfur was also found nonsignificant.

The plant height, branches/plant, siliquae/plant, seeds/siliqua, 1000-grain weight and harvest index (Kumar et al. 2001), and leaf area index, primary branches/plant and dry matter production (Singh and Mukharjee 2004) significantly increased with increasing levels of sulfur up to 40 kg /ha. The results are in agreement with the findings obtained by Singh and Kumar (1996) and Singh et al (2002). The increase in mean seed yield with 40 kg S/ha was 6.45 and 18.56% and in stover yield was 4.77 and 14.53% over 20 kg and no sulfur application, respectively. The soil low in S status was unable to supply the sufficient nutrient for optimum growth and yield of the crop. There was no significant difference between 40 and 60 kg S/ha. The results are in conformity with the findings of Singh et al. (2002) and Singh et al. (1997). The increase in seed yield was mainly due to enhanced rate of photosynthesis and carbohydrate metabolism as influenced by sulfur application. Sulfur resembles nitrogen in its capacity to augment cell division and cell elongation besides chlorophyll synthesis, and it also helps in increasing photosynthetic activity of the plant, resulting in increased dry matter production (Singh and Singh 1983b and Diepenbrock 2000). The optimum economic dose was worked out to be 32.1 kg S/ha and the maximum sulfur dose of 59.4 kg/ha had shown consequently lower yield of Indian mustard/ha than that produced by 40 kg S/ha (Singh and Mukharjee 2004). However, Chauhan et al. (1996) and Kachroo and Kumar (1997) reported significant increase in the seed yield of Indian mustard with sulfur application up to 50 kg S/ha.

Siliqua/plant, seeds/siliqua, test weight, and seed yield increased significantly with increasing sulfur levels up to 40 kg S/ha (Singh and Mukharjee 2004). Siliquae/plant is largely governed by the branch number/plant (Giri and Saran 1987). Higher sulfur induced translocation of photosynthates from leaves (source) toward sink site (siliquae and seeds). This resulted in large siliquae containing bold and more seeds with superior test weight (Purakayastha and Nad 1996).

Indian mustard is one of the important oilseed crops of Agra region in western Uttar Pradesh. The yield potential of different Indian mustard genotypes may differ under different agroclimatic conditions because of their inherent capacity (Bora 1997). Sulfur plays an inevitable and imperative role so far as the formation of amino acids (methionine 21% and cystine 27%), synthesis of protein, chlorophyll, and oil in the oilseed are concerned (Aulakh et al. 1980). The adequate supply of sulfur increases crop yields appreciably (Mohan and Sharma 1992).

Sulfur fertilization at 100 kg /ha to Indian mustard significantly improved the siliquae/plant, seeds/siliqua, seed yield/plant and 1000-seed weight and it could be ascribed to the overall improvement in plant growth, vigor, and production of sufficient photosynthates through increased leaf area index and chlorophyll content of leaves with S fertilization (Khanpara et al. 1993). A part of the beneficial effect of S on yield attributes seems to be due to better availability of N, K, and S and their translocation, which is reflected by increased yield attributes of the crop. Singh and Bairathi (1980) and Rathore and Manohar (1989) also reported similar results.

Khanpara et al. (1993) found that the combined application of elemental S and gypsum to Indian mustard was a more effective agrotechnique than elemental S application 21 days before sowing. Combined application of 60 kg N/ha and 100 kg S/ha significantly increased the seed yield. The seed yield recorded with this treatment combination was 17.3 q/ha, which was significantly higher (by 58.1%) than that obtained under 50 kg S/ha applied without N.

3.15 *Taramira*

Sulfur application in oilseed crops is considered to be very important from an oil yield point of view, since S is an essential constituent of oil (Kumar et al. 2004). They reported that application of 30 kg S/ha produced a significantly higher seed yield over the control, but it was at par with the application of 60 kg S/ha in this regard.

A significant increase in plant height, dry matter production/plant, branches/plant, chlorophyll content, and LAI at different stages of crop growth was observed as a result of the application of 60 kg S/ha (Kumar et al. 2004).

3.16 *Safflower*

A significantly superior harvest index of safflower was observed for 60 kg S/ha with no significant difference between 20 and 40 kg S/ha for it (Abbas et al. 1995). Biomass yield increased significantly up to 40 kg S/ha, but seed yield increased significantly up to 20 kg S/ha; however, 60, 40, and 20 kg S/ha recorded 16.35, 20.07, and 14.88% higher seed yield than the control, respectively. Similar results are reported by Rathore and Tomar (1990). The improved nutritional management as a result of increased sulfur supply might have been favorably influenced by carbohydrate metabolism, and this favorable effect led to increased transformation of photosynthates toward yield and yield-attributing characters with slight increases in test weight. The highest seed yield was obtained at 17.6 kg P_2O_2 + 40 kg S/ha. The best combinations for seed yield would be 17.6 kg P_2O_2 + 20.0 kg S/ha and 8.8 kg P_2O_2 ha + 40 kg S/ha on the basis of economics (Abbas et al. 1995).

Increasing levels of S increased plant height, branches, and capitula/plant of safflower (Abbas et al. 1995). Significantly higher plant height and capitula/plant were recorded up to 40 kg S/ha. No significant differences existed between 40 and 60 kg S/ha for plant height, number of branches, and capitula/plant.

3.17 Chickpea

Sulfur is an essential constituent of sulfur-containing amino acids and plays a key role in protein synthesis and nitrogen metabolism. Sulfur fertilization improves both yield and quality of crop (Ali and Singh 1995, Kachhave et al. 1997, Patel and Patel 1994, Tripathi et al. 1997 and Tiwari 1990). Chickpea (*Cicer arietinum* L.) is a major pulse crop, and average yield levels of the crop in many parts of the country are quite low. Increasing cropping intensity and use of high-analysis S-free fertilizers have led to sulfur deficiency in many areas, especially in light-textured soils. The productivity of chickpea, which is the premier pulse crop of India, is governed by several biotic and abiotic stresses, including the inadequate and imbalanced supply of plant nutrients (Singh et al. 1998). They reported that application of 20 kg S/ha on alluvial soils of central Uttar Pradesh significantly increased the grain yield of both the genotypes (BG 256 and KPG 59) of chickpea during 1993-94, but in 1994-95 this response was observed only in BG 256. The differential response pattern of chickpea genotypes to sulfur application could be due to variation in the sulfur requirement of different chickpea genotypes. The differential requirement of sulfur by crop varieties was earlier reported by Dev and Arora (1984). The low level of yield increase (13-14%) in the subsequent year (1994-95) as compared to the previous year (18-30%) during 1993-94 could be due to residual effect of sulfur applied to the previous crop. The increase in grain yield of chickpea on addition of sulfur was attributed to their deficiency in the soils of the experimental fields as it was reported by Tiwari (1990) that about 21 to 40% samples collected in Kanpur District were found to be sulfur deficient. Addition of sulfur significantly improved the protein of chickpea grain at 20 kg S/ha, registering 23.2 and 23.8% in genotypes BG 256 and KPG 59, respectively during 1993-94. Similarly, in the subsequent year the significant increase in protein content was observed at 20 kg S/ha. There was about a 4.5% increase in protein content due to sulfur addition. Similar results have been reported by Singh (1998). However, Tripathi et al. (1997) reported a significant increase in the protein content of chickpea at 40 kg S/ha over control in similar soil conditions. Test weight of the two genotypes indicates that BG 256 had significantly higher test weight than KPG 59. Application of 20 kg S/ha significantly increased the test weight of both genotypes. The increase in test weight was about 2.7 to 3.0% in case of BG 256, whereas in KPG 59 it was 6.4 to 7.2%. Results suggest that proper attention to sulfur nutrition must be paid to maximize chickpea productivity.

The intensive cropping system used in the recent past has led to depletion of many nutrients, including secondary nutrients like sulfur and magnesium (Singh and Ram 1990). Sulfur is involved in the synthesis of proteins, oils, and vitamins in plant systems.

3.18 *Till* (Sesamum indicum)

Sulfur application of 10 kg S/ha significantly increased the seed yield of sesame (*Sesamum indicum*) as compared to control (Nageshwar et al. 1995).

3.19 *Lucerne*

Lucerne is the major winter (*Rabi*) forage crop grown by the farmers of the Gujarat. The adequate growth of most of forage crops is observed when the N:S ratio is between 14:1 and 16:1 (Hazra and Tripathi 1998). Most forage crops respond to S application, especially in red soils with low S content. The sulfur content of Lucerne was below the critical level of 0.22%, and yield of the Lucerne crop increased with S application on sandy soils (Patel and Patel 1994). Sulfur plays a significant role in legumes, particularly where the sink material is largely sulfur-containing amino acids (Hazra and Sinha 1996). The Lucerne plant has great affinity with sulfur because 2S albumins are the major source materials for pod and seed development. Sulfur has a specific vital role in the growth, development, and quality of crops (Patel et al. 2004). They reported that application of sulfur significantly increased seed yield of Lucerne compared to no application of sulfur during 1998-99 and 1999-2000 and in pooled analysis. The increase in seed yield with sulfur application was 17.9% over control, on an average of 3 years. Increasing levels of sulfur from 20 to 40 kg S/ha significantly increased seed yield of Lucerne from 523.9 to 594.5 kg in pooled analysis. The forage yields of Lucerne did not differ significantly due to application of sulfur over control (Patel et al. 2004).

Application of S from gypsum sources recorded the highest seed yield of Lucerne as compared to elemental sulfur (Patel et al. 2004). They also reported that elemental sulfur showed slight superiority over the gypsum in increasing forage yields on a pooled basis. Thus, results indicated that either gypsum or elemental sulfur could be used as a sulfur source.

4 Economics

4.1 *Maize*

The sulfur application of 45 kg /ha to maize increased the net return more significantly than did lower levels of sulfur (Maurya et al. 2005). The maximum net return (Rs. 14,711/- per ha) and benefit:cost ratio (2.26) were obtained with 45 kg S/ha.

4.2 Soybean

Sulfur application of 30 kg S/ha to soybean significantly increased the net return in 1989 only, but in the following year the differences between 0 and 30 kg S/ha were not significant (Singh and Singh 1995). The highest net return was obtained with 40 kg S/ha in the form of gypsum (Ramamoorthy et al. 1996). This was followed by 40 kg S/ha applied as pyrite. The return per rupee invested was also more under the above treatments.

4.3 Sunflower

Sulfur applied at 20 kg/ha to sunflower resulted in significantly higher total income, net return and benefit:cost ratio over control. Lowest net returns and benefit:cost ratio were found under 40 kg S/ha (Awasthi et al. 2001).

4.4 Clusterbean

Highest net returns (Rs. 18,777) from clusterbean were recorded with 50 kg S/ha, which was significantly higher by Rs. 3,795 and 6,441 over 25 kg S/ha and control, respectively. Of the S sources, SulFer 95 recorded higher net returns than did gypsum (Sharma and Singh 2005).

4.5 Wheat

Application of 40 kg S/ha to wheat fetched significantly higher net return and benefit:cost ratio over preceding levels (Dewal and Pareek 2004).

4.6 Rapeseed and Mustard

Net return increased with increased S rate (up to 40 kg S/ha) to Indian mustard, which gave the highest net returns in both the years (Kumar et al. (2001). The net returns/rupee invested in sulfur fertilizer worked out to Re 0.81, 085 and 0.81 due to application of 20, 40, and 60 kg S/ha, respectively. The results confirm the findings of Sharma (1994). Whereas, Singh et al. (2002) also found that the highest benefit:cost of Indian mustard was registered with the application of 40 kg S/ha, which clearly shows the economic viability of the sulfur fertilization at 40 kg S/ha. However, Singh et al. (2000) recorded that application of S in brassicas also increased the mean net return and benefit:cost ratio up to 45 kg/ha.

Net returns and net returns/Re invested were maximum in the case of pyrite application and minimum in the case of elemental sulfur application (Kumar et al. (2001).

4.7 *Safflower*

Maximum additional net return of Rs 3,963/-/ha was observed with a 40kg S/ha application to safflower. However, a better cost:benefit ratio was obtained at 20kg S/ha (Abbas et al. 1995).

4.8 *Lucerne*

Maximum net returns as well as benefit:cost ratio were obtained with the application of 40kg S/ha to Lucerne as a gypsum over elemental sulfur (Patel et al. 2004).

5 Sulfur Effect on Protein Content, Oil Content, and Yield

5.1 *Pearl Millet*

Sulfur application of 60kg S/ha to pearl millet increased crude protein yield over 0, 20 and 40kg S/ha and the increase was 67.6, 21.8, and 4.2%, respectively. The improvement in crude protein due to application of sulfur could be improvement in nitrogen absorption by the plant (Dadhich and Gupta 2005).

5.2 *Soybean*

Protein and oil contents of the soybean seed increased significantly with increasing dose of S (Majumdar et al. 2001). The increase in protein and oil content due to 40kg S/ha was 11.26 and 24.17%, respectively. The increase in oil content with S application might be due to the fact that S helped in oil synthesis by enhancing the level of thioglucosides (Dwivedi and Bapat 1998). Soybean responded more to S in increasing oil and protein content in seed, as also reported by Kumar and Singh (1981). The interaction between P and S was significant. All the S levels increased both oil and protein contents significantly at every level of P. The maximum protein and oil content was recorded with a treatment combination of 60kg P_2O_5 and 40kg S/ha.

The oil and protein yields of soybean were increased by 111.4 and 95.4% with 60kg P_2O_5 /ha, while the same were increased by 42.9 and 27.2%, respectively, by 40kg S/ha over the control (Majumdar et al. 2001). The interaction between P and S was significant, indicating that the combined application of P and S would be more useful for the improvement of seed quality of soybean when both of these nutrients were deficient in the soil. All the levels of S increased the protein as well as oil yield significantly at each level of P_2O_5. Maximum yield of oil and protein (571.2 and 1108.7kg /ha respectively) was obtained by combined application of 60kg P_2O_5 and 40kg S/ha.

5.3 *Blackgram*

Increased levels of S significantly increased the protein content of grain of black gram from 22.50 to 23.12%, which accounted for 2.75% increase over control (Singh 2004). Increasing levels of S enhanced the methionin and cystine content in the grain from 2.12 to 3.14 and 2.54 to 4.12g/100g proteins. The relative increase in methionin and cystine content were 48.11 and 63.38% over control, respectively.

5.4 *Sunflower*

Increase in oil content of sunflower by S application might be attributed to involvement of S in the biosynthesis of oil. The higher oil yield was obviously because of higher seed yield and oil content (Awasthi et al. 2001).

5.5 *Niger*

The mean oil content at 60kg S/ha application to niger increased by 3.5, 2.7, and 4.9% in Ranchi, East Singhbhum, and Dumka, respectively, with mean increase of 3.7% as compared to control. Oil yield of niger in both the years increased steeply with S application up to 45kg S/ha and decreased at 60kg S/ha level. Mean oil yield of niger at three locations ranged from 80.5 to 154.6kg/ha in control to 122.7 to 234.9kg/ha at 45kg S/ha with the additional benefit of 65.9 at 45kg S/ha (Singh et al. 2003).

5.6 *Rapeseed and Mustard*

The different sources of sulfur did not influence sulfur protein and oil content of Indian mustard (Kumar et al. 2001 and Rajput et al. 1993). The magnitude of increase in oil content over control due to application of 40kg S/ha was 10.9 and 9.8% in both the years, respectively (Kumar et al. 2001). However, sulfur application significantly increased protein content, oil yield of mustard, and the maximum protein content and oil yield were obtained with 40 kg S/ha. Similar results were reported by Singh et al. (2002); however, Dubey and Khan (1993) found that application of S up to 40 kg/ha increased the oil content significantly and an appreciable improvement in oil yield of Indian mustard. The result confirms the findings of Tandon (1990). However, differences were not significant between 40 and 60 kg S/ha. Singh and Kumar (1996) also reported similar results. Oil yield is directly related to seed yield, hence with increasing seed yield, oil yield also increased with increased sulfur application (Singh and Mukharjee 2004). Increased oil content was mainly due to an increase in glucoside formulation (Singh and Singh 1977). Higher sulfur utilization occurred in crops with an adequate supply of sulfur, which enhances the protein synthesis in plant and ultimately increases the protein content in mustard. In another study, Das and Das (1995) found that sulur application had marked effect on the oil content and oil yields of toria (*Brassica campestris* subsp. Oleifera var. toria). The increase in oil content due to 45 kg S/ha was 12.5% compared to control. The maximum oil yields were obtained with 45 kg S/ha and magnitude on increase was 36.7, 21.9, and 11.8% compared to control, 15 and 30 kg S/ha, respectively. However, Narang et al. (1993) reported that sulfur at 50 kg/ha increased the seed-protein content from 24.5% to 25.1% and oil content from 41.1 and 42.2%. Singh and Gangasaran (1987) also reported the beneficial effect of sulfur. Sulfur application benefited both seed and oil yields, emphasizing the need of these nutrients in yield maximization of toria on light sandy soils of low-inherent fertility (Narang et al. 1993).

Sulfur application to toria also increased iodine index. Thus, the extent of unsaturated fatty acids was improved with S. Likewise; the pungency (as indexed by allyl-iso-thiocyanate) was also increased with S fertilization. Allyl-iso-thiocyanate increased with S application because it is a part of sinigrin glycoside, the causal agent for pungency in brassicas.

5.7 *Taramira*

Oil content and oil yield taramira significantly improved with application of sulfur up to 60 kg S/ha (Kumar et al. 2004). This might be due to the fact that sulfur enhanced cell multiplication, elongation, and expansion and led to better chlorophyll synthesis. Sulfur is also an integral part of oil and thus increased oil yield. These findings are in conformity with the results of Singh and Singh (1983a) and Rathore (1998).

5.8 *Till*

Sulfur application resulted in significantly more oil yields from sesame (*Sesamum indicum*) than control (Nageshwar et al. 1995), since sulfur is an integral part of acylcoenzyme A, which promotes synthesis of fatty acids. Similar findings were reported by Singh et al. (1987).

6 Sulfur Uptake

6.1 *Soybean*

Singh and Bansal (1999) reported that application of S at 60 kg/ha accumulated the maximum nitrogen in the soybean crop, which was at par with 30 kg S/ha but statistically superior over control during the first year. In the second year the N uptake increased up to 30 kg S/ha. The same trend was observed in phosphorus uptake. Sulfur uptake was significantly increased up to 30 kg S/ha according to the mean per cent increase of 12.3 over control. Nutrient uptake by crop is a function of nutrient concentration and dry matter production by plants. The marked improvement in accumulation of plant nutrients by crops under the influence of sulfur fertilization can easily be explained. The results obtained in this investigation are in conformity with the findings of Bansal (1991) and Abbes et al. (1992). Other studies were conducted by Majumdar et al. (2001), who reported that S uptake by soybean seed and straw increased significantly with their respective increase in doses. The S uptake was higher in seed than straw, which might be due to the fact that absorbed S was partitioned more in the seed, which led to apparent depletion in straw. The P and S interaction was significant for P and S uptake and followed the pattern of seed yield. The highest P and S uptake by seed was recorded with the combination of 60 kg P_2O_5 and 40 kg S/ha. The increase in P uptake by S application might be due to the increase in soil of available P, as both of these nutrients are absorbed by plants as anions, and if one is available in higher amount, the available pool of other will also increase. Application of P significantly increased S uptake of soybean, which could be due to greater root proliferation and increased activity of S-solubilizing bacteria, resulting in a beneficial effect on S absorption (Singh et al 1986). The N uptake of soybean seed increased significantly by all the levels of P and S (Majumdar et al. 2001). This increase in N uptake of soybean by P and S application might be due to increased nodulation as well as N fixation by nodules (Agarwal and Mishra 1994). A combination of P and S significantly increased the N uptake, showing a synergistic effect and the maximum value of N uptake recorded in the treatment combination of 60 kg P_2O_5 and 40 kg S/ha.

6.2 *Blackgram*

Application of S levels produced significant impact on the sulfur content of grain and straw blackgram. As the level of sulfur increased from 0 to and 30 to 60 kg S/ha, S content of grain and straw increased significantly. Application of S significantly increased the S uptake by blackgram grain, straw, and produce from 2.15 to 2.49, 4.13 to 5.14 and 6.28 to 7.60 kg/ha, respectively, in the first year and from 1.33 to 1.92, 3.13 to 3.78, and 4.40 to 5.65 kg/ha, respectively in second year, with average total uptake, which was enhanced from 5.37 to 6.62 kg/ha (Singh 2004).

The application of 60 kg S/ha significantly increased the N, P, and S uptake in blackgram (Trivedi et al 1997 and Singh et al 1992).

6.3 Clusterbean

Successive increase in S fertilization significantly increased the S uptake by clusterbean up to 50 kg S/ha. The crop fertilized with 25 and 50 kg S/ha recorded 34.8 and 50.2% higher total uptake by clusterbean over control (3.51 kg/ha), respectively. Similarly, SulFer 95 proved better for S uptake by succeeding barley crops over gypsum as a source of S. This increase in uptake can be ascribed to the influence of applied S on availability of S in the soil and its extraction by plants as well as concomitant increase in crop yields (Sharma and Singh 2005).

6.4 Frenchbean

Increased inorganic fertilizer use has led to poor soil health, which thus warrants organic matter addition (Shivananda et al 1998).They recorded that the ammonium sulfate-treated plants accumulated much less sulfur. These results indicated that the pods accounted for nearly 22 to 23% of total plant sulfur, and the rest remained in root, leaves, and stem frenchbean. The roots accounted for 47 to 48% of total plant sulfur in vermicompost and FYM applied pots at preflowering and flowering stages. However, at harvest this pool reduced from 29 to 34%, indicating redistribution of sulfur from roots to fruits.

6.5 Cowpea

Application of sulfur at the rate of 60 kg/ha to cowpea recorded a significant increase in nitrogen uptake. However, in the case of phosphorus and sulfur uptake, a significant response was obtained up to 80 kg S/ha. This might be attributed to higher seed and stover yield at this level (Sharma and Jat 2002).

6.6 Potato

Application of P did not affect the S content in tubers of potato significantly, whereas S uptake by potato crop increased significantly with P application. Increase in S uptake due to P addition can be attributed to increased tuber production (Singh et al. 1996). Application of FYM increased the content and uptake of S in tubers over control. Maximum values of content and uptake of S by potato tubers were recorded under FYM at 15 tons/ha. This might be attributed to the increased availability of S in soil and greater tuber production due to FYM addition (Singh et al. 1996).

6.7 *Niger*

Significant progressive increase in total S uptake by niger seed and straw with increasing levels of S up to 45 kg /ha was noticed during 1997-98 and 1998-99 (Singh et al. 2003). The total mean S uptake was two- to three-fold higher at 45 kg/ha of S application over control. Phosphogypsum contains sulfur in the form of SO_4-S. Higher S uptake by the niger crop may be attributed to the increased availability of SO_4-S in soil solution through added phosphogypsum and consequent better utilization by plant roots.

Mean S recovery of added S in niger crop ranged from 10.9% to 18.0%. The sulfur use efficiency by the crop showed an increasing trend with increasing levels of S application up to 45 kg/ha. Crop response in terms of kg grain/kg S ranged from 2.0 to 3.6 (Singh et al. 2003).

6.8 *Wheat*

Increasing doses of S up to 60 kg S/ha significantly increased N and S uptake of wheat over the preceding levels, whereas increase in uptake of P and Zn was noted up to 40 kg S/ha (Dewal and Pareek 2004). Nutrient uptake by crops is mainly a function of crop yield and nutrient concentration in grain and straw. The concentration of nutrients also increases due to S fertilization because of improved nutritional environment in rhizisphere and consequently in the plant system (Dewal and Pareek 2004). These results are in close conformity with the findings of Varavipour et al. (1999).

Maximum sulfur contents of 207 and 219 mg/100 in grain of wheat were recorded with gypsum, whereas in straw 113 and 129 mg/100 g were noted under ammonium sulfate in 1997-98 and 1998-99, respectively (Singh et al. 2001). Similar findings have been reported by Arora and Kaur (1995). The highest significant potassium content in wheat straw of 2.04 and 1.89% was recorded in 1997-98 and 1998-99 under singlesuperphosphate followed by gypsum, pyrite, and ammonium sulfate (Singh et al. 2001). The higher production with less availability of K from soil may lower K content in straw under ammonium sulfate due to the dilution effect. Similar findings have also been reported by Auti et al. (1999). Potassium content in grain and sulphur content in straw were higher under 100 kg N: 50 kg P_2O_5: 5 kg Zn/ha + ammonium sulfate, whereas maximum value for potassium content in straw and sulfur content in grain were recorded under 100 kg N: 50 kg

P_2O_5: kg Zn/ha + singlesuperphosphate and 100 kg N:50 kg P_2O_5:5 kg Zn/ha + gypsum, respectively.

6.9 Rapeseed and Mustard

Sulfur uptake by Indian mustard was not influenced by various sources of sulfur (Kumar et al. 2001 and Rajput et al. 1993). Uptake of sulfur by seeds was significantly increased with increasing levels of sulfur up to 40 kg S/ha; however, it was found at par with 60 kg S/ha. The magnitude of increase over control due to application of 40 kg S/ha was 38.2 and 34.2% during both the years. The application of 60 kg S/ha significantly increased the N, P, and S uptake in mustard (Trivedi et al. 1997 and Dubey et al. 1994), Dubey and Khan (1993) reported that application of nitrogen up to 90 kg/ha increased the S concentration significantly in plants at various growth stages and in seed and stalk in both the years. However, the increase was higher in seed than stalk, since both the nutrients are closely linked with protein metabolism and their relationship is synergistic (Aulakh and Pasricha 1983). Sulfur application up to 30 kg /ha to Indian mustard significantly increased the S content in plants at every crop growth stage during both the years and higher levels were at par (Dubey and Khan 1993). The S content in plants was found to increase with the advancement of plant growth up to 45 days after sowing and thereafter declined gradually up to 90 days after sowing. Higher values of S at early stages could be due to greater root activity and less plant growth and also due to basal application of S, resulting in more soil S concentration and its availability, which increased the absorption by plant roots. The low absorption of S in the later growth stages might be due to a diluted effect (Aulakh et al. 1977). Decrease in S content of the mature crop might be attributed to the growing parts, especially to seed (Tandon 1990, Dubey and Khan 1993). Application of S up to 40 kg/ha increased the S content in seed and stalk during both the years (Dubey and Khan 1993). The N and S uptake by mustard crops increased significantly with 90 kg N/ha and 30 kg S/ha during both the years, which may be due to the greater production of dry matter and seed yield with increasing N and S levels. Singh and Singh (1984) and Tandon (1990) reported similar results.

6.10 Lucerne

Sulfur-treated plots showed 4.1% higher N uptake by Lucerne plants in a total of two cuts compared to the control in an average of 3 years (Patel et al. 2004). This might be due to higher N concentration and higher dry matter yield in sulfur-treated plots of Lucerne compared to control (Aulakh et al. 1976). Similarly, sulfur-treated plots showed significantly higher S uptake in Lucerne seed than the control in an average of 3 years (Patel et al. 2004). This increase in S uptake may be attributed to increased

S concentration in Lucerne seed and higher seed yield due to addition of S. The increase in sulfur level from 20 to 40 kg S/ha significantly increased the S uptake by Lucerne seed, but further increase in S level from 40 to 60 kg S/ha significantly decreased the S uptake by seed (Patel et al. 2004). Similar results were reported by Bhilare et al. (2001). Thus, the results specifically indicated that 40 kg S/ha was sufficient for the better utilization of nitrogen and sulfur by Lucerne crops (Patel et al. 2004).

Application of sulfur through a gypsum source removed higher S by Lucerne seed over elemental sulfur source, but the differences did not reach the level of significance (Patel et al. 2004).

7 Sulfur Residual Effect

7.1 Pearl Millet

Sulfur fertilization of wheat recorded significant residual effects on succeeding pearl millet (Sharma and Manohar 2002). Application of 40 kg S/ha to wheat increased the grain and stover yields of succeeding pearl millet by 0.80 and 2.0 q/ha over control. Bapat et al. (1994) also observed increased yield of soybean by 22.2% when 30 kg S/ha was added through gypsum to succeeding wheat.

7.2 Barley

Application of 50 kg S/ha to clusterbean increased the grain and straw yields of barley by 21.0 and 15.6% over the control, respectively. This increase in grain yield may be attributed to enrichment of soil with sulfur, resulting in its greater uptake (Sharma and Singh 2005). The results are in close conformity with the findings of Nad and Goswami (1983). SulFer 95 also left more residual S, thereby increasing the yield of barley grains by 6.7% over that under gypsum (Sharma and Singh 2005).

7.3 Rapeseed and Mustard

Sulfur plays a key role in the quality and development of seeds in oilseed crops, need comparatively higher amounts of sulfur than do cereals and legumes (Singh and Mukherjee 2004). They found that although plants remained at par with sulfur application at 45 and 30 kg S/ha, primary branches/plant increased significantly over lower doses. LAI and dry matter increased significantly with each increment in sulfur levels (Singh and Mukherjee 2004). The higher leaf area index with increasing sulfur levels resulted in better utilization of photosynthates and total photosynthetic activity per unit area (Singh and Mukherjee 2004 and Jain et al. 1995).

Siliquae/plant and seeds/siliqua of mustard increased with increasing sulfur levels up to 45 kg/ha, although the difference between 30 and 45 kg S/ha was not significant. However, residual sulfur failed to show any significant effect on the test weight of mustard grown in rice-mustard sequence (Singh and Mukherjee 2004). Increase in branches/plant due to increasing sulfur levels (Yadav et al. 2000) may be the reason for the increased number of siliqua/plant. Once the siliqua formation started, their development was better under higher sulfur supply resulting in larger siliquae having more number of seeds. In the case of seed and stover yields, responses up to 45 kg S/ha more were also observed, but the difference between 45 and 30 kg S/ha was not significant during first year.

Residual sulfur levels significantly increased protein content and oil yield of mustard. The highest protein content, oil content, and oil yield in mustard were recorded with residual effect of 45 kg S/ha, which was significantly superior to their lower levels; however, 45 and 30 kg S/ha were statistically at par during first year of the study with respect to oil yield (Singh and Mukharjee 2004).

8 Effect on Soil Fertility

8.1 *Blackgram*

Trivedi et al. (1997) found that available sulfur content in soil was increased in soil after the harvest of blackgram. Among the different treatments of N, P, and S, the effect of 60 kg S/ha was most pronounced in increasing the available S status of soil.

8.2 *Wheat*

Sulfur fertilization of wheat brought out significant improvement on available P_2O_5 and S content of soil after harvest of both the crops (Sharma and Manohar 2002). However, Ram et al. (1998) also reported that the sulfur fertilization improved the residual fertility of soil by lowering down the soil pH appreciably.

8.3 *Rapeseed and Mustard*

Trivedi et al. (1997) reported that the available sulfur content decreased after the harvest of mustard as compared to its initial value (6.69 ppm). The effect of 60 kg S/ha was most pronounced in increasing the available S status of soil among the different treatments of N, P, and S. For the residual balance of S in soil at post harvest stage, it was not so much increased with the increasing S levels (Dubey and Khan 1993). Sulfur status was raised marginally with applied N and S.

8.4 Nitrogen: Sulfur Ratio

The application of P and S significantly reduced N:S ratio in soil (Majumdar et al. 2001). The wide N:S ratio at lower doses of P and S application indicated a deficiency of S, while a narrow ratio indicated sufficiency of S or possibly a deficiency of N (Dwivedi and Bapat 1998), as increasing levels of N to meet the S levels were not applied (Agarwal and Mishra 1994). Lowering of N:S ratio proved beneficial in increasing the yield and quality of soybean seeds. The N:S ratio of 17.7 at 60 kg P_2O_5 and 40 kg S/ha was the most optimum for soybean under soil.

9 Plant Sulfur Status

9.1 Niger

Singh et al. (2003) reported that samples of niger at flowering stage from two districts of Ranchi and Lohardaga (Jharkhand) in four blocks showed that overall plant S content ranged from 0.04 to 1.08% with a mean value of 0.38%. Out of 574 plant samples, 256 and 318 samples were found to be deficient and sufficient in plant sulfur status, respectively. Considering 0.24% as the critical concentration of S in plant, 44.6% samples as a whole may be rated as deficient in plant sulfur. These results indicated that deficient S status in niger plants may possibly be due to coarse texture and low content of organic carbon and nonapplication of fertilizers containing S. Therefore, S management strategies are paramount to maintain soil fertility levels for higher productivity of niger crops.

10 Sulfur Assimilation

Sulfur is an important element for physiological processes and growth of plants and its content varies in plants due to species, varieties, type of soil, from location to location, and climatic conditions, and it ranges from 0.1 to 6% on a dry weight basis. Elemental sulfur- or sulfate-containing fertilizers are applied to the plants, but the elemental sulfur is applied to the plants 15-21 days before sowing for the conversion in to sulfate. Sulfur in the sulfate form taken up by the roots has to be reduced to sulfide before it is further metabolized. However, sulfate reduction enzymes are present in the root plastids and reduce the sulfate in to sulfide, and its incorporation into cysteine takes place in the shoot in the chloroplast. Cysteine is the precursor (reduced sulfur donor) for the synthesis of most other organic sulfur compounds in plants. The maximum proportion of the organic sulfur is present in the protein fraction (about 70% of total sulfur) as cysteine and methionine residues. Cysteine and methionine have significance in structure, conformation, and function of proteins. Plants have a large number of other organic sulfur compounds as thiols (glutathione), sulfolipids, and secondary sulfur compounds (alliins, glucosinolates, phytochelatins), which play key roles in physiology and protection against environmental stress and pests. Sulfur compounds have great significance for food quality and for the production of phytopharmaceutics. Sulfur deficiency impairs the production of plants, health, and resistance to environmental stress and pests (Saito et al. 2005).

The uptake of sulfate by the roots and its transport to the shoot is strictly controlled, and it and it appears to be one of the primary regulatory sites of sulfur assimilation. Sulfur is actively taken up across the plasma membrane of the root cells, subsequently loaded into the xylem vessels, and transported to the shoot by transpiration stream. The uptake and transport of sulfate are energy dependent and driven by a proton gradient generated by ATPases through a proton/sulfate cotransport. In the shoot the sulfate is unloaded and transported to the chloroplasts, where it is reduced. The remaining sulfate in plant tissue is predominantly present in the vacuole, since the concentration of sulfate in the cytoplasm is kept rather constant. Distinct sulfate transporter proteins mediate the uptake, transport, and subcelluler distribution of sulfate. Cellular and subcellular gene expression and possible function of the sulfate transporter's gene family has been classified in up to five different groups. Some groups are expressed exclusively in the roots or shoots or expressed both in the roots and shoots.

Group 1 are high-affinity sulfate transporters, which are involved in the uptake of sulfate by the roots.

Group 2 are vascular transporters and are low-affinity sulfate transporters.

Group 3 is the "leaf group," however, little is known about the characteristics of this group..

Group 4 transporters might be involved in the transport of sulfate into plastids prior to its reduction.

Group 5 sulfate transporters of this group are not known yet.

Regulation and expression of the majority of sulfate transporters are controlled by the sulfur nutritional status of the plants. Upon sulfate deprivation, the rapid decrease in root sulfate is regularly accompanied by a strongly enhanced expression of most sulfate transporter genes (up to 100 fold), accompanied by a substantially enhanced sulfate uptake capacity. It is not yet solved, whether sulfate itself or metabolic products of the sulfur assimilation (O-acetyl-serine, cysteine, glutahione) act as signals in the regulation of sulfate uptake by the root and its transport to the shoot, and in the expression of the sulfate transporters involved.

Root plastids contain all sulfate reduction enzymes, but sulfate reduction predominantly takes place in the leaf chloroplasts. The reduction of sulfate to sulfide occurs in three steps.

Step 1 Sulfate needs to be activated to adenosine 5'-phosphosulphate (APS) prior to its reduction to sulfite.

Step 2 The activation of sulfate is catalyzed by ATP sulfurylase, whose affinity for sulfate is rather low (km approximately 1 mM), and the in situ sulfate concentration in the chloroplast is most likely one of the limiting/regulatory steps in sulfur reduction.

Step 3 Subsequently APS is reduced to sulfite, catalyzed by APS reductase, with glutathione as the likely reductant. This reaction is assumed to be one of the primary regulation points in sulfate reduction, since the activity of APS reductase is the lowest of the enzymes of the sulfate reduction pathway and has a fast turnover rate (Fig. 4.1). Sulfite with high affinity is reduced by sulfite reductase to sulfide with ferredoxin as a reductant. The remaining sulate in plant tissue is transferred into the vacuole. The remobilization and redistribution of the vacuolar sulfate reserves appear to be rather slow, and sulfur-deficient plants may still contain detectable levels of sulfate (Saito et al. 2005).

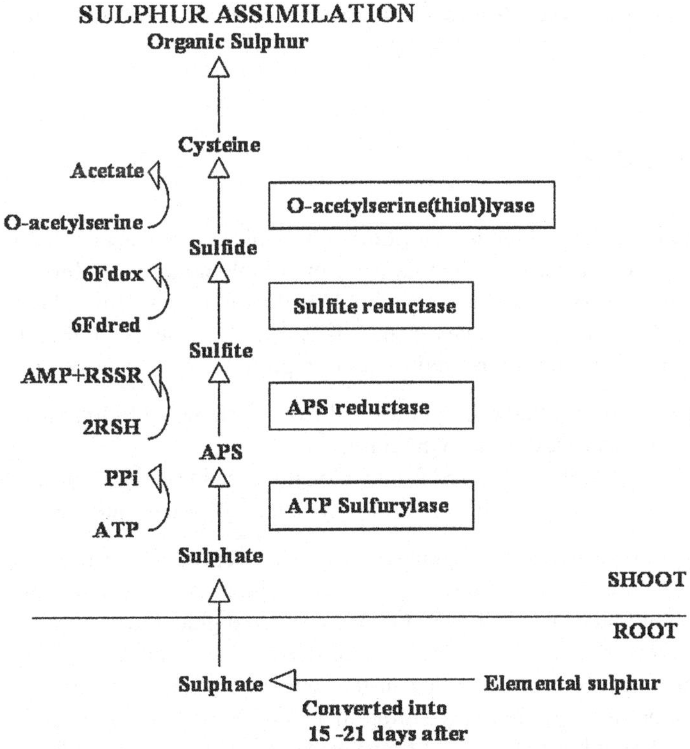

Fig. 4.1 Sulfate reduction and assimilation pathway in plants (APS, adenosine 5'-phosphosulfate; Fdred, Fdox, reduced and oxidized ferredoxin; RSH, RSSR, reduced and oxidized glutathione)

11 Synthesis and Function of Sulfur Compounds

11.1 Cysteine and Cystine

Sulfide is incorporated into cysteine, catalyzed by O-acetylserine (thiol)lyase, with O-acetylserine as substrate. The synthesis of O-acetylserine is catalyzed by serine acetyltransferase, and together with O-acetylserine (thiol)lyase it is associated as an enzyme complex named cysteine synthase. The formation of cysteine is the direct coupling step between sulfur and nitrogen assimilation in plants. Cysteine is the sulfur donor for the synthesis of methionine, the major other sulfur-containing amino acid present in plants. Both sulfur-containing amino acids are of great importance in the structure conformation and function of proteins and enzymes, but high levels of these amino acids may also be present in seed storage proteins. The thiol groups of the cysteine residues in proteins can be oxidized, resulting in disulfide bridges with other cysteine side chains, and form cystine and/or linkage of polypeptides. Disulfide bridges make an important contribution to the structure of proteins. The thiol groups are also of great significance in substrate binding of enzymes, in metal-sulfur clusters in proteins, e.g., ferredoxins and in regulatory proteins, e.g., thioredoxins (Schnug 1998, Abrol and Ahmed 2003, Saito et al. 2005).

11.2 Glutathione

Glutathione or its homologues (e.g., homoglutathione in Fabaceae; hydroxymethylglutathione in Poaceae) are the major water-soluble nonprotein thiol compounds present in plant tissue and account for 1%-2% of total sulfur. The content of glutathione in plant tissue ranges from 0.1-3 mM. Cysteine is the direct precursor for the synthesis of glutathione and its homologues.

Step 1 γ-Glutamylcysteine is synthesized from cysteine and glutamate catalyzed by γ-glutamylcysteine synthetase.

Step 2 Glutathione is synthesized from γ-glutamylcysteine and glycine in glutathione homologues, β-alanine or serine catalyzed by glutathione synthetase.

Both steps of the synthesis of glutathione are ATP-dependent reactions. Glutathione is maintained in the reduced form by an NADPH-dependent glutathione reductase and the ratio of reduced glutathione (GSH) to oxidized glutathione (GSSG) generally exceeds a value of 7. Glutathione fulfills various roles in plant functioning. In sulfur metabolism it functions as reductant in the reduction of APS to sulfite. It is also a major transport form of reduced sulfur in plants. Roots largely depend for their reduced sulfur supply on shoot/root transfer of glutathione via the phloem, since the reduction of sulfur occurs predominantly in the chloroplast. Glutathione is directly involved in the reduction and assimilation of selenite into selenocysteine. Furthermore, glutathione is of great importance in the protection of plants against oxidative and environmental stress, and it depresses/scavenges the formation of toxic reactive oxygen species, e.g., superoxide, hydrogen peroxide, and lipid hydroperoxides. Glutathione functions as reductant in the enzymatic detoxification of reactive oxygen species in the glutathione-ascorbate cycle and as thiol buffer in the protection of proteins via direct reaction with reactive oxygen species or by the formation of mixed disulfides. The potential of glutathione as protectant is related to the pool size of glutathione, its redox state (GSH/GSSG ratio), and the activity of glutathione reductase. Glutathione is the precursor for the synthesis of phytochelatins, which are synthesized enzymatically by a constitutive phytochelatin synthase. The number of γ-glutamylcysteine residues in the phytochelatins may range from 2-5, sometimes up to 11. Despite the fact that the phytochelatins form complexes which a few heavy metals, viz. cadmium, it is assumed that these compounds play a role in heavy metal homeostasis and detoxification by buffering of the cytoplasmatic concentration of essential heavy metals. Glutathione is also involved in the detoxification of xenobiotics, compounds without direct nutritional value or significance in metabolism, which at too high levels may negatively affect plant functioning. Xenobiotics may be detoxified in conjugation reactions with glutathione catalyzed by glutathione S-transferase, whose activity is constitutive; different xenobiotics may induce distinct isoforms of the enzyme. Glutathione S-transferases have great significance in herbicide detoxification and tolerance in agriculture, and their induction by herbicide antidotes (Safeners) is the decisive step for the induction of herbicide tolerance in many crop plants. Under natural conditions glutathione S-transferases are assumed to have significance in the detoxification of lipid hydroperoxides, in the conjugation of endogenous metabolites, hormones, and DNA degradation products, and in the transport of flavoniods (Grill et al. 2001).

11.3 Sulfolipids

Sulfoquinovosyl diacylglycerol is the predominant sulfur-containing lipid present in plants. In leaves its content comprises up to 3%-6% of the total sulfur present. This sulfolipid is present in plastid membranes and likely is involved in chloroplast functioning. The route of biosynthesis and physiological function of sulfoquinovosyl diacylglycerol is still under investigation. From recent studies it is evident that sulfite is the likely sulfur precursor for the formation of the sulfoquinovose group of this lipid (Schnug 1998, Abrol and Ahmed 2003, Saito et al. 2005).

11.4 Secondary Sulfur Compounds

Brassica species contain glucosinolates, which are sulfur-containing secondary compounds. Glucosinolates are composed of a β-thioglucose moiety, a sulfonated oxime, and a side chain. The synthesis of glucosinolates starts with the oxidation of the parent amino acid to an aldoxime, followed by the addition of a thiol group through conjugation with cysteine to produce thiohydroximate. The transfer of glucose and a sulfate moiety completes the formation of the glucosinolates. The physiological significance of glucosinolates is still ambiguous, though they are considered to function as sink compounds in situations of sulfur excess. Upon tissue disruption glucosinolates are enzymatically degraded by myrosinase and may yield a variety of biologically active products such as isothiocyanates, thiocyanates, nitriles, and oxazolidine-2-thiones. The glucosinolate-myrosinase system is assumed to play a role in plant-herbivore and plant-pathogen interactions. Furthermore, glucosinolates are responsible for the flavor properties of Brassicaceae and recently have received attention in view of their potential anticarcinogenic properties. Allium species contain γ-glutamylpeptides and alliins (S-alkenyl cysteine sulfoxides). The content of these sulfur-containing secondary compounds strongly depends on stage of development of the plant, temperature, water availability, and the level of nitrogen and sulfur nutrition. In onion bulbs their content may account for up to 80% of the organic sulfur fraction. Less is known about the content of secondary sulfur compounds in the seedling stage of the plant. It is assumed that alliins are predominantly synthesized in the leaves, from where they are subsequently transferred to the attached bulb scale. The biosynthetic pathways of synthesis of γ-glutamylpeptides and alliins are still ambiguous. γ-Glutamylpeptides can be formed from cysteine via γ-glutamylcysteine or glutathione and can be metabolized in to the corresponding alliins via oxidation and subsequent hydrolyzation by γ-glutamyltranspeptidases. However, other possible routes of the synthesis of γ-glutamylpeptides and alliins may not be excluded. Alliins and γ-glutamylpeptides are known to have therapeutic utility and might have potential value as phytopharmaceutics. The alliins and their breakdown products, e.g., allicin, are the flavor precursors for the odor and taste of species. Flavor is only released when plant cells are disrupted and the enzyme alliinase from the vacuole is able to degrade the alliins, yielding a wide variety of volatile and nonvolatile sulfur-containing compounds. The physiological function of γ-glutamylpeptides and alliins is rather unclear (Schnug 1998, Abrol and Ahmed 2003, Saito et al. 2005).

12 Sulfur Metabolism and Air Pollution

Rapid economic growth, industrialization, and urbanization are associated with a strong increase in energy demand and emissions of air pollutants, including sulfur dioxide and hydrogen sulfide, which may affect plant metabolism. Sulfur gases are potentially phytoxic; however, they may also metabolized and used as a sulfur source and even be beneficial if sulfur fertilization of the roots is not sufficient. Plant shoots form sinks for atmospheric sulfur gases, which can directly be taken up by the foliage (dry deposition). The foliar uptake of sulfur dioxide is generally directly dependent on the degree of opening of the stomates, since the internal

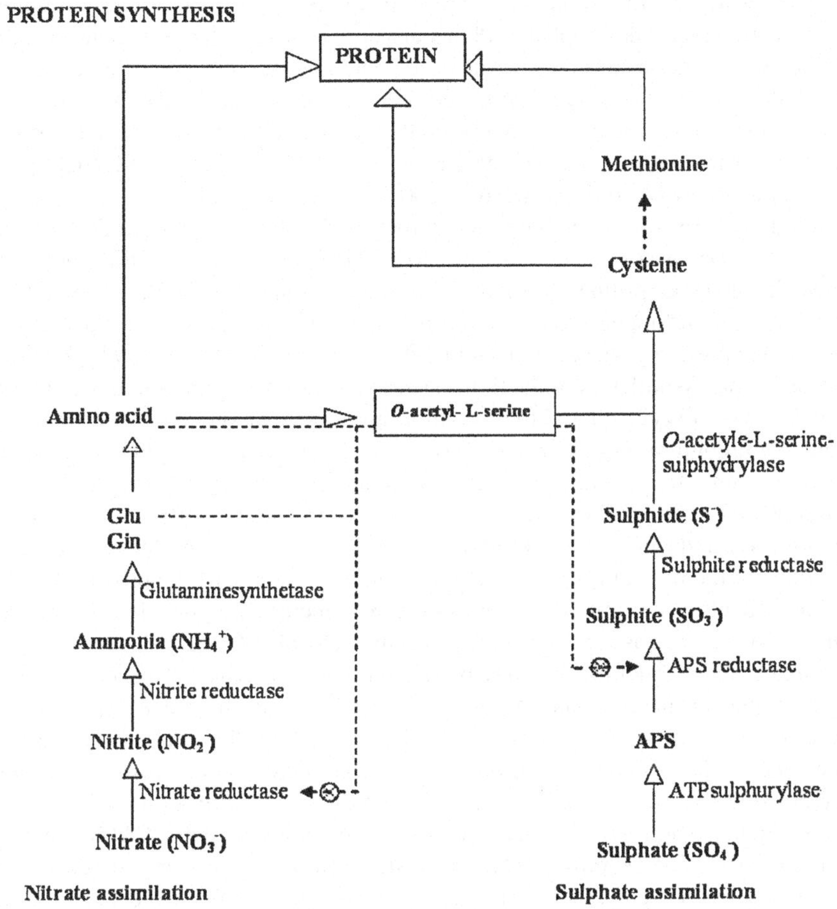

Fig. 4.2 Enzymes of the assimilatory nitrite and sulfate reduction pathways in plants

resistance to this gas is low. Sulfur is highly soluble in the apoplastic water of the mesophyll, where it dissociates under formation of bisulfite and sulfite. Sulfite may directly enter the sulfur reduction pathway and be reduced to sulfide or incorporated in to cysteine and subsequently into other sulfur compounds. Sulfite may also be oxidized to sulfate, extra and intracellularly by peroxidases, or nonenzymatically catalyzed by metal ions or superoxide radicals and subsequently reduced and assimilated again. Excessive sulfate is transferred into the vacuole; enhanced foliar sulfate levels are characteristic for exposed plants. The foliar uptake of hydrogen sulfide appears to be directly dependent on the rate of its metabolism into cysteine and subsequently into other sulfur compounds. There is strong evidence that

O-acetyl-serine (thiol)lyase is directly responsible for the active fixation of atmospheric hydrogen sulfide by plants. Plants are able to transfer from sulfate to foliar absorbed atmospheric sulfur, as sulfur source and levels of 60 ppb or higher appear to be sufficient to cover the sulfur requirement of plants. There is interaction between atmospheric and pedospheric sulfur utilization. For instance, hydrogen sulfide exposure may result in decreased activity of APS reductase and a depressed sulate uptake (Hawkesford and De Kok 2006).

Protein synthesis requires inorganic carbon and reduced N and S. The organic N:S ratio of various species ranges from 20 to 40 on a molar basis (Dijkshoorn and Van Wijk 1967). Coordination of the assimilatory reduction pathways of nitrate and sulfate is therefore necessary so that appropriate proportions of amino acids are available for protein synthesis (Brunold 1993, Stulen and De Kok 1993). The first enzyme in the assimilatory reduction pathway of nitrate to ammonium is nitrate reductase, which is considered the rate-limiting enzyme in the overall reduction of nitrate to ammonium (Fig. 4.2). The enzyme APS reductase is considered as one of the prime points for regulatory control in the sulfate reduction pathway (De Kok et al 2000). For the synthesis of S-containing amino acids, sufficient reduced N and S compounds must be available. Several metabolites might act as signal molecules in the mutual regulation of both pathways. Some of the possibilities for regulatory control of nitrate reductase and APS reductase are given in the pathway. Amino acids, amides as asparagines and arginine, and also *O*-acetyl-L-serine, which accumulate under S deficiency, may be related to the decrease in nitrate reductase activity found in these plants. Migge et al. (2000) found that gene expression of nitrate reductase in the shoot was negatively affected by glutamine and /or asparagines accumulated under conditions upon sulfate deprivation. However, a more indirect effect of the amides on nitrate reductase via an effect on the uptake of nitrate might also play a role. Other studies showed that the activity of APS reductase changed rapidly in response to S or N starvation or exposure to reduced S compounds. At present O-acetyl-L-serine, sulfide, or cysteine are considered the most likely signal molecules (Brunold 1993, Hawkesford and Wary 2000, Koprivova et al. 2000, Leustek and Saito 1999).

References

Abbas Mohd, Tomar Surendra S, Nigam KB (1995) Effect of phosphorus and sulphur fertilization in safflower (*Carthamus tinctorius*).. Indian J Agron 40(2): 243-8

Abbes C, Karan R, Isfan D, Parent IE (1992) Effect of sulphur fertilization on soybean. Canad J Plant Sci 72(2):377-82

Abrol YP, Ahmed A (2003) Sulphur in Plants. Kluwer Academic Publishers, Dordrecht, ISBN 1-4020-1247-0. http://en.wikipedia.org/wiki/Sulfur_assimilation

Adiputra IGK, Anderson JW (1992) Distribution and redistribution of sulphur taken up from nutrient solution during vegetative growth in barley. Physiologia Plantarum 85:453-60

Adiputra IGK, Anderson JW (1995) effect of sulphur nutrition on redistribution of sulphur in vegetative barley. Physiologia Plantarum 95:643-50

Agarwal HP, Mishra AK (1994) Sulphur nutrition of soybean. Commun Soil Sci Plant 25:1302-12

Ali M (1991) Consolidated reports on *Kharif* pulses,. 1990-91. Directorate of Pulses Research, Kanpur, Uttar Pradesh

Ali M, Singh KK (1995) Dalhani Phashalo Me Gandhak Ka Mahatva. Bull Indian Inst Pulses Res, Kanpur

Arora BR, Hundal HS, Sekhan GS (1991) Efficiency of different sulphur carriers to peanut, maize and wheat on an alkaline soil. J Indian Soc Soil Sci 39(3):591-92

Arora CL, Kaur NP (1995) Mineral gypsum. Fertilizers for Indian Agriculture. A Guide book. Dr. H L S Tandon, Fertilizer Development and Consultation Organization, New Delhi, pp. 35-45

Aulakh MS, Dev G, Arora BR (1976) Effect of sulphur fertilization on the nitrogen sulphur relationship in alfalfa (*Medicago sativa* L.). Plant Soil 45(1):75-80

Aulakh MS, Pasricha NS (1983) Interrelationship between sulphur, magnesium and potassium in Indian mustard and their influence on nutrient recovery, protein content and nitrogen: sulphur ratio. Indian J Agr Sci 53(3):192-4

Aulakh MS, Pasricha NS, Sahota NS (1977) Nitrogen sulphur relationship in brown sarson and Indian mustard. Indian J Agr Sci 45(5):249-53

Aulakh MS, Pasricha NS, Sahota NS (1980) Yield, nutrient concentration and quality of mustard crop as influenced by nitrogen and sulphur fertilization. J Agr Sci Cambridge 84:545-9

Auti AK, Wadile SC, Pawar VS (1999) Yield, quality and nutrient removal of wheat (*Triticum aestivum*) as influenced by levels and sources of fertilizer. Indian J Agron 44(1):119-22

Bansal KN (1991) Effect of levels of sulphur on the yield and composition of soybean (*Glysine max* L.) Merrill, green gram (*Vigna radiate*), black gram (*Vigna munga*) and cowpea (*Vigna sinensis*). Madras Agr J 98 (5-8):188-90

Bansal KN, Motiramani DP, Pal AR (1983) Studies on sulphur in vertisol. 1. Soil and plant tests for diagnosing sulphur deficiency in soybean (*Glycine max* L Merr.). Plant Soil 70:133-40

Bandopadhyay P, Samui RC (2000) Response of groundnut (*Arachis hypogaea*) to levels and sources of sulphur in West Bengal. Indian J Agron 45(4):761-4

Bapat PN, Rathore GS, Tomar VS (1994) Proceedings of National Seminar on Nutrient Management in Wheat-Soybean Cropping System in M.P., held at Indian Institute of Soil Science, Bhopal, 22-23 September 1994

Bell CI, Clarkson DT, Cram WJ (1994) Compartmental analysis of ^{35}SO exchange kinetics in roots and leaves of a tropical legume *Macroptilium atropurpureum* cv. Siratro. J Exp Bot 45:879-86

Bell CI, Clarkson DT, Cram WJ (1995a) Sulphate supply and its regulation of transport in roots of a tropical legume *Macroptilium atropurpureum* cv. Siratro. J Exp Bot 46:65-71

Bell CI, Clarkson DT, Cram WJ (1995b) Partitioning and redistribution of sulphur during S-stress in *Macroptilium atropurpureum* cv. Siratro. J Exp Bot 46:73-81

Bhilare RL, Deasale JS, Pathan SH (2001) Evaluation of response of sulphur to winter forage Lucerne. Madras Agr J 88 (1-3):137-8

Biswas BC, Tewatia RK (1991) Results of FAO. Sulphur trials network in India. Fert News 36(4):11-35

Blake-Kalff MMA, Harrison KR, Hawkesford MJ, Zhao FJ, McGrath SP (1998) Allocation of sulphur with in oilseed rape (*Brassica napus* L.) leaves in response to sulphur-deficiency. Physiologia Plantarum 118:1337-44

Bora PC (1997) Effect of gypsum and lime on performance of Brassica varieties under rain fed conditions. Indian J Agron 42(1):155-8

Brunold C (1993) Regulatory interactions between sulfate and nitrate assimilation. In: Sulphur nutrition and assimilation in higher plants: regulatory, agricultural and environmental aspects. (Eds. De Kok et al.), SPB Ac. Publ., The Hague, pp. 61-75

Burke JJ, Holloway P, Dalling MJ (1986) The effect of sulphur deficiency on the organization and photosynthetic capability of wheat leaves. J Plant Physiol 125:371-5

Chandel AS, Rao GP, Saxena SC (1989) Effect of sulphur nutrition on soybean (*Glycine max* L.) Merrill Proc World Soybean Research Conference 4(1):363-6

Chauhan DR, Paroda S, Mangat R (1996) Response of Indian mustard (*Brassica juncea*) to biofertilizers, sulphur and nitrogen. Indian J Agron 41(4):620-3

Chatterjee BN, Ghosh RK, Chakraborthy PK (1985) Response of mustard to sulphur and micronutrients. Indian J Agron 30(1):75-8

Dadhich LK, Gupta AK (2005) Growth and yield of fodder pearl millet as influenced by sulphur, zinc and intercropping with cowpea. Fertilizer News 50(3):55-7

Das KN, Das K (1995) Effect of sulphur and nitrogen fertilization on growth and yield of toria (*Brassica campestris* subsp. Oleifera var. toria). Indian J Agron 40(2):329-31

De Kok LJ, Westerman SI, Stuiver CEE, Stulen I (2000) Atmospheric H_2S as plant sulfur source: interaction with pedospheric sulfur nutrition- a case study with *Brassica oleracea* L. In: Sulpfur nutrition and assimilation in higher plants: molecular, biochemical and physiological aspects (Eds. Brunold et al.), Paul Haupt, Bern, pp. 41-56

Dev G (1999 SulFer 95-Information Brochure. Fernz Sulfer World Inc. PO Box 9, Irricana, Alberta, Canada-TOMIBO.

Dev G, Arora CL (1984) Crop response to secondary nutrients in the northern region of India. In: Balanced fertilizer programme with specila reference to secondary and micronutrient nutrition of crops under intensive cropping. Proceedings of FAINR Seminar held at Jaipur during March 30-31, 1984. Fertilizer Association of India, New Delhi

Dewal GS, Pareek RG (2004) Effect of phosphorus, sulphur and zinc on growth, yield and nutrient uptake of wheat (*Triticum aestivum*). Indian J Agron 49(3):160-2

Dewal GS, Sharma HS, Pareek RG (2000)Effect of sulphur and FYM on growth and yield of barley (*Hordeum vulgare* L.). Acta Ecol 22(2):107-11

Dhillon NS, Dev G (1980) Studies on sulphur nutrition of soybean from three sulphate sources. J Indian Soc Soil Sci 28(3):361-5

Diepenbrock W (2000) Yield analysis of winter oilseed rape (*Brassica napus* L.): A review. Field Crops Res 67:35-49

Dijkshoorn D, Van Wijk AL (1967) The sulphur requirements of plants as evidenced by the sulphur-nitrogen ratio in the organic matter, a review of published data. Plant Soil 26:129-57

Dubey OP, Khan RA (1993) Effect of nitrogen and sulphur on sulphur content in Indian mustard (*Brassica juncea*) and their residual balance in soil. Indian J Agron 38(4):582-7

Dubey OP, Sahu TR, Garg DC (1994) Response and economics in relation to nitrogen and sulphur nutrition in Indian Mustard (*Brassica juncea*). Indian J Agron 39(1):49-53

Dubey OP, Sahu TR, Garg DC, Khan RA (1993) Response of mustard to sulphur and nitrogen under irrigated vertisol condition of S and N on ancillary characters, yield and quality. J Oilseed Res 10(1):11-15

Dwivedi AK, Bapat PN (1998) Sulphur phosphorus interaction on the synthesis of nitrogenous fractions and oil in soybean. J Indian Soc Soil Sci 46:254-7

Fullington JG, Miskelly DM, Wrigley CW, Kasarda DD (1987) Quality-related endosperm proteins in sulphur-deficient and normal wheat grain. J Cereal Sci 5:233-45

Gilbert S, Clarkson DT, Cambridge M, Lambers H, Hawkesford MJ (1997) Sulphate-deprivation has an early effect on the content of ribulose 1,5-bisphosphate carboxylase/oxygenase and photosynthesis in young leaves of wheat. Plant Physiol 115:1231-9

Giri G, Saran G (1987) Influence of doses of S application on mustard under rainfed conditions. Abstract: Symposium on Alternate Farming System, Feb. 21-23, 1987. Indian Soc Agron,, IARI, New Delhi

Grill, D, Tausz M, De Kok LJ (2001) Significance of glutathione to plant adaptation to the environment. Kluwer Academic Publishers, Dordrecht, ISBN 1-4020-0178-9. http://en.wikipedia.org/wiki/Sulfur_assimilation

Hashimoto S, Kameoka (1985) Sulphur and nitrogen containing volatile components of cruciferae. J Food Sci 50:847-8

Hawkesford MJ, De Kok LJ (2006) Managing sulfur metabolism in plants. Plant Cell Environ 29:382-95 http://en.wikipedia.org/wiki/Sulfur_assimilation

Hawkesford MJ, Wary JL (2000) Molecular genetics of sulphate assimilation. Adv Botan Res 33:159-223

Hazra CR, Sinha NC (1996) Forage Seed Production,. South Asian Publishers Pvt. Ltd., New Delhi, pp. 46
Hazra CR, Tripathi SB (1998) Effect of secondary and micronutrients on yield and quality of forages. Fertilizer News 43(12):77-82
Jain K, Vyas AK, Singh AK (1995) Effect of phosphorus and sulphur fertilization on growth and nutrient uptake by mustard (*Brassica juncea* L. Czern and Coss). Ann Agr Res 16(2):389-90
Jat RL, Rathore PS (1994) Effect of sulphur, molybdenum and *Rhizobium* inoculation on greengram. Indian J Agron 39(4):651-4
Kachhave KG, Gawande SD, Kohire OD, Mane SS (1997) Influence of various sources and levels of sulphur on nodulation, yield and uptake of nutrients by chickpea. J Indian Soc Soil Sci 45:590-1
Kachroo D, Kumar A (1997) Nitrogen and sulphur fertilization in relation to yield attributes and seed yield of Indian mustard (*Brassica juncea*). Indian J Agron 42(1):145-7
Karmokar JL, Clarkson DT, Saker LR, Rooney JM, Purves JV (1991) Sulphate deprivation depresses the transport of nitrogen to the xylem and the hydraulic conductivity of barley (*Hordeum vulgare* L.) roots. Planta 185:269-78
Khanpara VD, Porwal BL, Sahu MP, Patel JC (1993) Effect of sulphur and nitrogen on yield attributes and seed yield of Indian mustard (*Brassica juncea*) on vertisol. J Agron 38(4):588-92
Khurana MPS, Nayyar VK, Sidhu BS, Gill MS (2003) Response of raya and ghobi sarson to sulphur in Punjab. Fertilizer News 48(7):39-41
Koprivova A, Suter M, Op den Camp R, Brunold C, Kopriva S (2000) Regulation of sulfate assimilation by nitrogen in Arabidopsis. Plant Physiol 122:737-46
Kothari ML, Jethra JK (1994) Available sulphur status of soils of semi-arid eastern plain zone of Rajasthan. In: Diamond Jubilee National Seminar of Indian Society of Soil Science, held at New Delhi.
Krishnamurthi VV, Mathan KK (1996) Influence of sulphur and magnesium on growth and yield of sunflower (*Helianthus annus*). Indian J Agron 41(4):627-9
Kumar D (1993) Response of cowpea varieties to sulphur through gypsum in loamy sand soil. M.Sc. Ag. Thesis, Rajasthan Agricultural University, Bikaner
Kumar S,, Singh B, Rajput AL (2001) Response of Indian mustard (Brassica juncea) to source and level of sulphur. Indian J Agron 46(3):528-32
Kumar S,, Gaur BL, Sumeriya HK (2004) Effect of nitrogen, phosphorus and sulphur levels on growth and oil yield of taramira under rainfed conditions of southern Rajasthan. Haryana J Agron 20(1):4-6
Kumar V, Singh M (1980) Interaction of sulphur, phosphorus and molybdenum in relation to uptake and utilization of phosphorus by soybean. Soil Sci 130:26-31
Kumar V, Singh M (1981) Effect of sulphate, phosphate and molybdenum application on quality of soybean. Plant Soil 59:3-8
Leustek T, Saito K (1999) Sulfate transport and assimilation in plants. Plant Physiol 120:637-43
Majumdar B, Venkatesh MS, Lal B, Kumar K (2001) Response of soybean (Glycine max) to phosphorus and sulphur in acid alfisol of Meghalaya. India J Agron 46(3):500-5
Mandal S, Singh R (1993) Oil and protein contents in sunflower genotypes. J Oilseeds Res 10(1):161-2
Maurya KL, Sharma HP, Tripathi HP, Singh S (2005) Effect of nitrogen and sulphur application on yield attributes, yield and net returns of winter maize (*Zea mays* L.). Haryana J Agron 21(2):115-16
McGrath SP, Zhao FJ, Withers PJA (1996) Development of sulphur deficiency in crops and its treatment. Proceedings of the Fertilizer Society, No. 379. Peterborough, The Fertilizer Society.
Migge A, Bork C, Hell R, Becker TW (2000) Negative regulation of nitrate reductase gene expression by glutamine or asparagines accumulation inleaves of sulfur-deprived tobacco. Planta 211:587-95

Mohan K, Sharma HC (1992) Effect of nitrogen and sulphur on growth, yield attributes, seed and oil yield of Indian mustard (*Brassica juncea*). Indian J Agron 37(4):748-54

Moss HJ, Wrigley CW, MacRitchie F, Randall PJ (1981) Sulphur and nitrogen fertilizer effects on wheat. II. Influence on grain quality. Aust J Agr Res 32:213-26

Mukherjee D, Singh R K (2002) Influence of sulphur, iron and silicon nutrition on growth and yield of irrigated mustard. Haryana J Agron 18(1 & 2):50-2

Nad BK, Goswami NN (1983) Response to legume and oilseed crops to different sources of sulphur and magnesium on some alluvial soils. J Indian Soc Soil Sci 31:60-4

Nageshwar L, Sarawgi SK, Tripathi RS, Bhambri MC (1995) Effect of nitrogen, potassium and sulphur on seed yield, nutrient uptake, quality and economics of summer sesame (*Sesamum indicum*). Indian J Agron 40(2):333-5

Narang RS, Mahal SS Gill MS (1993) Effect of phosphorus and sulphur on growth and yield of toria (*Brassica campestris* subsp. Oleifera var toria). Indian J Agron 593-7

Pareek RG (1995) Response of clusterbean to sulphur levels and sources, phosphorus and plant growth regulators and the residual effect on tarameeru. Ph.D thesis, Rajasthan Agricultural University, Bikaner.

Patel JR (1997) Effect of nitrogen and phosphorus levels on growth and yield of forage oat. Forage Research 23(3 & 4):235-6

Patel PC, Patel JR (1994) Effect of phosphorus and sulphur on forage yield and nutrient uptake by Lucerne. J Indian Soc Soil Sci 42:154-6

Patel PC, Yadavendra JP, Kotecha AV (2004) Effect of source and level of sulphur on seed yield and nitrogen and sulphur uptake by Lucerne (*Medicago sativa*). Indian J Agron 49(2):128-30

Patra AK, Tripathy SK, Samui RC (1995) Response of groundnut varieties to sulphur on alluvial soils of West Bengal. Indian Agriculturist 39(2):137-42

Prosser IM, Schneider A, Hawkesford MJ, Clarkson DT (1997) Changes in nutrient composition, metabolite concentrations and enzyme activities in spinach in the early stages of S-deprivation. In: Cram W J, De Kok L J, Stulen I, Brunold C, Rennenberg, H (eds.) Sulphur metabolism in higher plants. Leiden, The Netherlands: Backhuys Publishers: 339-42

Purakayastha TJ, Nad BK (1996) ffect of sulphur magnesium and molybdenum on the utilization of sulphur by mustard and wheat. J Nuclear Agr Biol 25(3):159-63

Rajput RL, Yadav RP, Verma OP (1993) Effect of sulphur levels and sources on mustard production. Bhartiya Krishi Anusandhan Patrika 8(3-4):85-6

Ram H, Jha CK, Prasad J (1998) Effect of sulphur and rhizobium on soil pH, available N and P in green gram. Fertilizer News 43(10):51-5

Ramamoorthy K, Ramasamy M, Vairavan (1996) Effect of source and level of sulphur on production of soybean (*Glycine max*). Indian J Agron 41(4):654-5

Rathore PS (1998) Response of mustard (*Brassica juncea* L. Czern and Coss) to levels of nitrogen and sulphur and the relationship between yield and leaf analysis. Ph.D Thesis, Raj.Agril. Univ. Bikaner, Campus: Jobner.

Rathore PS, Manohar SS (1989) Response of mustard to nitrogen and sulphur. 1. Effect of nitrogen and sulphur on ancillary charaters and yield of mustard. Indian J Agron 34:333-6

Rathore DS, Tomar SS (1990) Effect of sulphur and nitrogen on seed yield and nitrogen uptake by mustard. Indian J Agron 35(4):361-3

Sacchidanand B, Sawarkar MJ, Glurayya RS, Subde DA, Sinha SB (1980) Response of soybean (*Glycine max* L.) to sulphur and phosphorus. J Indian Soc Soil Sci 28:189-92

Schnug E (1998) Sulphur in Agroecosystems. Kluwer Academic Publishers, Dordrecht, 221 pp, ISBN 0-7923-5123-1. http://en.wikipedia.org/wiki/Sulfur_assimilation

Saito K, De Kok LJ, Stulen I, Hawkesford MJ, Schnug E, Sirko A, Rennenberg H (2005) Sulphur transport and assimilation in plants in the Post Genomic Era. Backhuys Publishers, Leiden, ISBN 90-5782-166-4 http://en.wikipedia.org/wiki/Sulfur_assimilation

Sunarpi, Anderson JW (1997) Effect of nitrogen on the export of sulphur from leaves in soybean. Plant Soil 188:177-87

Sharma GC, Bradford RR (1973) Effect of sulphur on yield and amino acid content of soybean. Communication Soil Sci Plant Anal 4:77-82

Sharma JP (1994) Response of Indian mustard to different irrigation schedules, nitrogen and sulphur levels. India J Agron 39(3):421-5

Sharma OP, Singh GD (2005) Effect of sulphur in conjunction with growth substances on productivity of clusterbean (*Cyamopsis tetragonoloba*) and their residual effect on barley (*Hordeum vulgare*). Indian J Agron 50(1):6-18

Sharma PK, Manohar SS (2002) Response of wheat (*Triticum aestivum*) to nitrogen and sulphur and their residual effect on succeeding peralmillet (*Pennisetum glaucum*). Indian J Agron 47(4):473-6

Sharma SK, Jat NL (2002) Effect of phosphorus and sulphur on growth and nutrients uptake in cowpea.Haryana J Agron 18(1 & 2):168-9

Shivananda TN, Sreerangappa KG, Lalitha BS, Parama Ramakrishna VR, Siddaramappa R (1998) Dynamics of nitrogen, phosphorus, potassium and sulphur in frenchbean as influenced by source of nitrogen application. Indian J Pulses Res 11(2):56-60

Singh AK, Singh SB, Sharma RP (1998) Rabi maize excels wheat in diara land of Bihar. Indian Fmg 38(6):13-15

Singh AK, Singh T,, Singh S,, Kumar S, Tomar S (1997) Response of Indian Mustard (*Brassica juncea*) to nitrogen, phosphorus and sulphur fertilization. Indian J Agron 42(1):148-51

Singh A, Singh V, Mehra VS (1988) Effect of nitrogen and sulphur on yield and nutrients uptake by rapeseed. J Indian Soc Soil Sci 36:182-4

Singh HG and Sahu RC (1986) Response of oil seed crops to fertilizers in dryland agriculture. Proceedings of FAI–NR Seminar Varanasi, Fertilizer Association of India, NewDelhi, pp. 147-59

Singh B, Singh D, Tomar RAS, Trivedi SK (2001) Effect of fertility levels and sulphur sources on yield and quality of wheat (*Triticum aestivum*). Indian J Agron 17(1&2):148-50

Singh B, Kumar A, Yadav YP, Dhankhar RS (2000) Response of Brassicas to sulphur application. Indian J Agron 45(4):752-5

Singh KS, Bairathi RC (1980) A study in the sulphur fertilization of mustard in semi-arid tract of Rajasthan. Ann Arid Zone 19(3):197-202

Singh B, Kumar V (1996) Response of Indian mustard to nitrogen and sulphur application under rainfed condition. Indian J Agron 41(2):286-9

Singh BP, Prakash, Singh B, Singh SK 2002. Comparative performance of Indian mustard (*Brassica juncea*) genotypes in relation to sulphur fertilization. Indian J Agron 47(4):531-6

Singh BP, Singh HG (1983a) Effect of sulphur on oil production, amino acid and thio-glucoside-content of seed of mustard grown on vertisols. Crop Physiol 1:76-86

Singh BP, Singh HG (1983b) Comparative efficacy of S on production of green matter, mineral composition and uptake of mustard of vertisols. Forage Res 9(1):37-41

Singh BP, Singh HG (1984) Comparative efficacy of sulphur content at growth stages and uptake by mustard on vertisols. Indian J Agron 29(3):179-84

Singh D, Singh V (1995) Effect of potassium, zinc and sulphur on growth characters, yield attributes and yield of soybean. Indian J Agron 40(2):223-7

Singh J, Yadav JS, Sheoran RS, Kumar V (2002) Response of oat to nitrogen, phosphorus and sulphur in light textured soils. Haryana J. Agron 18(1& 2):36-8

Singh M, Singh H (1977) Effect of sulphur and selenium on sulphur combining Aminoacids and quality of oil in raj in normal and sodic soils. Indian J Physiol 20(1):56-62

Singh M, Satyavan, Tomer DPS, Singh RC, Kumar R (1997) Effect of sources and levels of sulphur on productivity of pigeonpea. Haryana J Agron 13:69-71

Singh Rajesh KR, Mukherjee D (2004) Effect of sulphur fertilization in sustaining mustard productivity in rice-mustard cropping system. Haryana J Agron 20(1):7-9

Singh SS (1975) Pick bound sunflower varieties. Intensive Agr 13(4):364-9

Singh S, Gangasaran (1987) Effect of sulphur and nitrogen on growth, yield, quality and nutrient uptake on Indian rape. Indian J Agron 32(4):474-5

Singh SB (1998) Effect of sulphur and magnesium fertilizer on yield and quality of chickpea. Indian J Pulses Res 11(2):142-3

Singh SB, Rao Ch, Srinivasa, Ali M (1998) Response of chickpea genotypes to sulphur application. Indian J Pulses Res 11(2):61-4

Singh Shashi P, Singh V, Lakhan R (1996) Effect of phosphorus and farm yard manure application on yield, content and uptake of nitrogen, phosphorus and sulphur by potao (*Solanum tuberosum*). Indian J Agron 41(4):630-2

Singh SP, Bansal KN (1999) Influence of nitrogen, its application time and sulphur on nitrogen, phosphorus and sulphur uptake by soybean (*Glysine max*). Haryana J Agron 15(1):30-4

Singh S,, Sarkar AK, Singh BP (2003) Plant status and response of niger to sulphur in acidic soils of Jharkhand. Fertilizer News 48(10):57-8, 61

Singh S,, Singh AP, and Singh B (1992) Direct and residual effect of pyrite on yield, protein content and S uptake by blackgram and lentil in Entisol J Indian Soc Soil Sci 40:584-5

Singh V, Chauhan YS, Tripathi NK (1987) Comparative efficacy of level of nitrogen and sulphur to production and biochemical value of till (*Sesamum indicum* L.) var. C-6. Res Dev Rep 4(2):213-17

Singh V, Mehta VS, Singh B (1986) Individual and interaction effect of sulphur, phosphorus and molybdenum in mustard. J Indian Soc Soil Sci 39:535-8

Singh YP (2004) Role of sulphur and phosphorus in black gram production. Fertilizer News 49(2):33-6

Singh YP, Agarwal RL (2001) Effect of sulphur and phosphorus levels on yield and quality of blackgram. Ann Pl Soil Res 3(2):298-300

Stulen I, De Kok LJ (1993) Whole plant regulation of sulphur metabolism: a theoretical approach and comparison with current ideas on regulation of nitrogen metabolism. In: Sulphur nutrition and assimilation in higher plants: regulatory, agricultural and environmental aspects. (Eds. De Kok et al.), SPB Ac. Publ., The Hague, pp. 77-91

Tandon HLS (1986) Sulphur Research and Agriculture Production in India. Fertilizer Development and Consultation Organization, New Delhi, pp. 76

Tandon HLS (1990) Fertilization Recommendation for Oilseed Crops: A Guide Book. Development and Consultation Organization, New Delhi

Tandon HLS (1991) Sulphur Research and Agriculture Production in India, 3rd ed. The sulphur Institute, Washington, DC, USA.

Tandon HLS, Messick DL (2002) Practical sulphur Guide Book pp. 1-3. The Sulphur Institute, Washington, DC

Thakuria K (1992) Effect of nitrogen and *Azotobacter* on fodder yield and quality of oat (*Avena sativa*). Indian J Agron 37(3):571-2

Tiwari KN (1990) Sulphur research and Agricultural Production in U.P. Sulphur in Agriculture Bulletin No. 14:29-34

Tripathi HC, Singh RS, Mishra VK (1997) Effect of S and Zn nutrition on yield and quality of chickpea (*Cicer arietinum* L.). J Indian Soc Soil Sci 45:123-6

Trivedi SK, Singh V, Shinde CP (1997) Effects of nitrogen, phosphorus and sulphur on productivity, nutrient uptake and soil properties in a blackgram-mustard sequence. Haryana J Agron 13(1):1-6

Upadhaya VB, Tiwari JP (1996) Influence of nitrogen, seed rate and mulch on wheat (*Triticum aestivum*) varieties under late sown conditions. Indian J Agron 41(4):562-5

Upasani RK, Sharma HC (1986) Effect of nitrogen and sulphur on some growth parameters, evapotranspiration and moisture use efficiency of mustard under dryland conditions. Indian J Agron 31(3):222-8

Varavipour M, Hassan R, Singh (1999) Effect of applied phosphorus, sulphur and zinc on yield and yield parameters of wheat (*Triticum aestivum* L.) and soybean (*Glycine max*) grown on a loamy sand. Indian J Agr Sci 67:1-4

Warrilow AGS, Hawkesford MJ (1998) Separation, subcellular location and influence of sulphur nutrition on isoforms of cysteine synthase in spinach. J Exp Bot 49:1625-36

Wrigley CW, Du Cros DL, Fullington JG, Kasarda DD (1984) Changes in polypeptide composition and grain quality due to sulphur deficiency in wheat. J Cereal Sci 2:15-24

Yadav SS (2004) Growth and yield of green gram (*Vigna radiate* L.) as influenced by phosphorus and sulphur fertilization. Haryana J Agron 20(1):10-12

Zhao FJ, Hawkesford MJ, Warrilow AGS, McGrath SP, Clarkson DT (1996) Responses of two wheat varieties to sulphur addition and diagnosis of sulphur deficiency. Plant Soil 181:317-27

Zhao FJ, McGrath SP, Crosland AR (1995) Changes in the sulphur status of British wheat grain in the last decade, and its geographical distribution. J Sci Food Agr 68:507-14

Chapter 5
Regulatory Protein-Protein Interactions in Primary Metabolism: The Case of the Cysteine Synthase Complex

Sangaralingam Kumaran, Julie A. Francois, Hari B. Krishnan and Joseph M. Jez(✉)

Abstract Sulfur is an essential nutrient for plant growth and development. In plant sulfur assimilation, cysteine biosynthesis plays a central role in fixing inorganic sulfur from the environment into the metabolic precursor for cellular thiol-containing compounds. A key regulatory feature of this process is the physical association of the two enzymes involved in cysteine biosynthesis (serine acetyltransferase, SAT, and O-acetylserine sulfhydrylase, OASS) to form the cysteine synthase complex. Physiologically, this multienzyme complex acts as a molecular sensor in a regulatory circuit that coordinates sulfur assimilation and modulates cysteine production. Here we focus on aspects of the protein-protein interactions in the plant cysteine synthase complex and how formation of the complex has been studied. In addition, we summarize the initial efforts to understand the structural, kinetic, and thermodynamic basis for association of SAT and OASS in the multienzyme assembly.

1 Introduction

In plants and bacteria, cysteine biosynthesis is the final pathway in the conversion of inorganic sulfur into a chemically stable organic compound. Two enzymes catalyze the reactions of this pathway (Rabeh and Cook 2004, Wirtz and Droux 2005, Kopriva 2006). In the first step, serine acetyltransferase (SAT, EC 2.3.1.30) generates O-acetylserine by transferring acetate from acetyl-coenzyme A to serine to form O-acetylserine. During the second step, O-acetylserine sulfhydrylase (OASS, EC 4.2.99.8) uses pyridoxal phosphate (PLP) as a cofactor to yield cysteine from O-acetylserine and sulfide. Cysteine biosynthesis is the metabolic link between sulfur assimilation and the myriad of sulfur-containing molecules in the cell. For example, this pathway provides essential metabolites for production of glutathione, a key regulatory agent of intracellular redox environment during abiotic and biotic stresses.

Cysteine biosynthesis in plants also contributes to regulating sulfur assimilation by influencing the expression of genes involved in sulfur uptake and assimilation

---

Joseph M. Jez
Danforth Plant Science Center, 975 N. Warson Rd., St. Louis, MO 63132, USA
jjez@danforthcenter.org

N.A. Khan et al. (eds.), *Sulfur Assimilation and Abiotic Stress in Plants.*

and by modulating enzymatic activities in response to sulfur demand (Hell and Hillebrand 2001, Saito 2004, Kopriva 2006, Wirtz and Hell, 2007). This mechanism involves the physical association of SAT and OASS to form the hetero-oligomeric cysteine synthase complex; however, little is known about the molecular mechanism of how these proteins associate and the three-dimensional architecture of the complex. This chapter focuses on aspects of protein-protein interactions in the plant cysteine synthase complex and how its formation has been studied.

2 Overview of the Cysteine Synthase Complex

Kredich and co-workers (1969) first described the cysteine synthase complex after isolating it from *Salmonella typhimurium*. Initially, metabolic channeling was suggested as the function of this macromolecular assembly, but later studies demonstrated this was not the case (Cook and Wedding 1977). Nearly two decades after the bacterial enzyme complex was identified, Nakamura and Tamura (1990) identified the complex in plants. Subsequently, formation of the cysteine synthase complex by interaction of SAT and OASS from plants and bacteria has been shown using a variety of experimental approaches, including size-exclusion chromatography, yeast two-hybrid analysis, surface plasmon resonance, fluorescence spectroscopy, and calorimetry. Efforts to map the protein-protein interaction regions in the complex indicate that the C-terminus of SAT plays a critical role in association with OASS (Bogdanova and Hell 1997, Mino et al. 1999, 2000, Wirtz et al. 2001, Francois et al. 2006, Kumaran and Jez 2007). Recently, determination of the x-ray crystal structures of the *Haemophilus influenzae* OASS (HiOASS) and *Arabidopsis thaliana* OASS (AtOASS) in complex with peptides corresponding to the C-termini of the SAT from each organism and protein-protein interaction studies revealed that the OASS active site forms the SAT interaction site (Bonner et al. 2005, Huang et al. 2005, Campanini et al. 2005, Zhao et al. 2006, Francois et al. 2006, Liszewska et al. 2007). Although individual components of the complex have been identified and the biochemical properties of SAT and OASS well studied, the mechanism of assembly and structure of the cysteine synthase complex remains to be defined.

2.1 Protein-Protein Interaction Modulates the Activities of SAT and OASS

The cysteine synthase complex is a macromolecular assembly that plays a regulatory role in plant sulfur assimilation with protein-protein interactions between SAT and OASS modulating each enzyme's activity. Functional studies have provided a wealth of information on the kinetic and chemical mechanisms of both SAT and OASS. The biochemical properties of the plant and bacterial SAT have been extensively examined (Noji et al. 1998, Leu and Cook 1994, Droux et al. 1998, Johnson et al. 2004a, 2004b). Likewise, the reaction mechanism of OASS has been dissected in detail (Rabeh and Cook 2004).

Interestingly, the activity of each protein changes upon association in the plant cysteine synthase complex. SAT activity is up to 20-fold higher in the complex than as an isolated protein (Droux et al. 1998, Berkowitz et al. 2002, Hell et al. 2002, Wirtz and Droux 2005). In contrast, OASS activity is almost completely abrogated when complexed with SAT (Kredich et al. 1969, Droux et al. 1998, Berkowitz et al. 2002). Furthermore, in plants, unbound SAT rapidly loses its activity in the absence of OASS with maximal SAT activity for cysteine synthesis requiring a nearly 400-fold excess of OASS activity (Droux et al. 1998). Overall, these findings led to the conclusion that association of SAT and OASS into the cysteine synthase complex is a prerequisite for optimal flux of cysteine synthesis in vivo (Hell and Hillebrand 2001).

2.2 Role of the Complex in a Regulatory Circuit Controlling Sulfur and Cysteine Metabolism

SAT and OASS associate to form the cysteine synthase complex under sulfur-sufficient conditions in plants (Droux et al. 1998, Saito 2004) (Fig. 5.1). As part of the complex, the bound form of OASS retains very little activity, but SAT activity is enhanced. The mechanism for increased SAT activity remains to be established, but structural studies have revealed how OASS activity is reduced (see Section 5.3.1.). Complex formation results in the production of the pathway intermediate O-acetylserine (OAS). Under sulfur-sufficient conditions, free OASS catalyzes cysteine formation from OAS and sulfide. If intracellular sulfur levels are low, then OAS accumulates, as free OASS is unable to produce cysteine due to a lack of sulfide. Elevated levels of OAS promotes dissociation of the cysteine synthase complex and effects sulfur metabolism, as the higher OAS concentration activates expression of genes encoding sulfate transporters, ATP sulfurylase, OASS, and SAT (Smith et al. 1997, Koprivova et al. 2000, Hopkins et al.

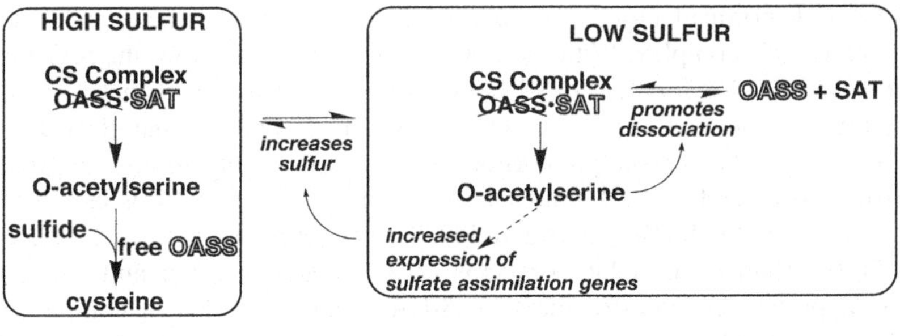

Fig. 5.1 Model for regulation of sulfur assimilation and cysteine synthesis by the plant cysteine synthase complex. Highly active forms of OASS and SAT are indicated by the outlined text. The inactivated form of OASS in the complex is indicated with an "X." Modified from Hell and Hillebrand 2001.

2005). This leads to increased sulfur uptake and assimilatory reduction. As sulfur levels elevate, free OASS begins to catalyze cysteine formation, which reduces OAS levels. Decreased concentration of OAS allows for association of SAT and OASS, which results in activation of SAT and resumption of cysteine biosynthesis (Hell and Hillebrand 2001).

2.3 *Initial Structural Investigations of the Cysteine Synthase Complex*

Although the three-dimensional structure of the cysteine synthase complex remains to be solved, x-ray crystal structures of the OASS from bacteria (Burkhard et al. 1998, Claus et al. 2005) and *Arabidopsis thaliana* (Bonner et al. 2005) and the bacterial SAT (Olsen et al. 2004, Pye et al. 2004) are available. The OASS from plants and bacteria form stable homodimers with a molecular weight of 68-73 kDa (Burkhard et al. 1998, Bonner et al. 2005, Claus et al. 2005) (Fig. 5.2a). Each OASS monomer consists of two α/β structural domains. The smaller N-terminal domain (residues 45-150 in AtOASS) contains a central four-stranded parallel β-sheet flanked by four α-helices. The larger C-terminal domain (residues 3-44 and 151-303 in AtOASS) centers on a six-stranded β-sheet surrounded by four α-helices. The active site is located at the interface of these domains and is defined by the location of the PLP cofactor. The three-dimensional structures of bacterial SAT have been solved using x-ray crystallography (Olsen et al. 2004, Pye et al. 2004). SAT functions as a hexameric protein with a molecular weight of 180-200 kDa (monomer Mr 30-35 kDa). The overall structure consists of a largely α-helical N-terminal domain and a C-terminal domain dominated by a left-handed β-helix structure consisting of 14 β-strands forming five coils of the helix (Fig. 5.2b). The N-terminal domain is involved in interaction with other monomers in the structure with the C-terminal domain containing the active site of the enzyme.

Relatively little information is available about the structural and molecular properties of the cysteine synthase complex from either plants or bacteria. Kredich et al. (1969) determined a total molecular weight of 310 kDa for the *S. typhimurium* hetero-oligomeric complex. With estimated weights of 160 kDa for the SAT homo-oligomer and 68 kDa for the OASS homodimer, it was assumed that two OASS dimers bind to one SAT hexamer; however, so far no precise analytical data are available to support this assumption. Biochemical and molecular biology approaches established that the 10 to 20 amino acid C-terminal tail of SAT is essential for association with OASS (Bogdanova and Hell 1997, Mino et al. 1999, 2000, Wirtz et al. 2001, Zhao et al. 2006, Francois et al. 2006, Kumaran and Jez 2007). Moreover, protein crystal structures of AtOASS and HiOASS in complex with peptides corresponding to the C-termini of their cognate SAT reveal that the OASS active site forms the binding site for the C-terminus of SAT (Huang et al. 2005, Francois et al. 2006). Nonetheless, how the two full-length proteins interact to form the complex remains to be elucidated.

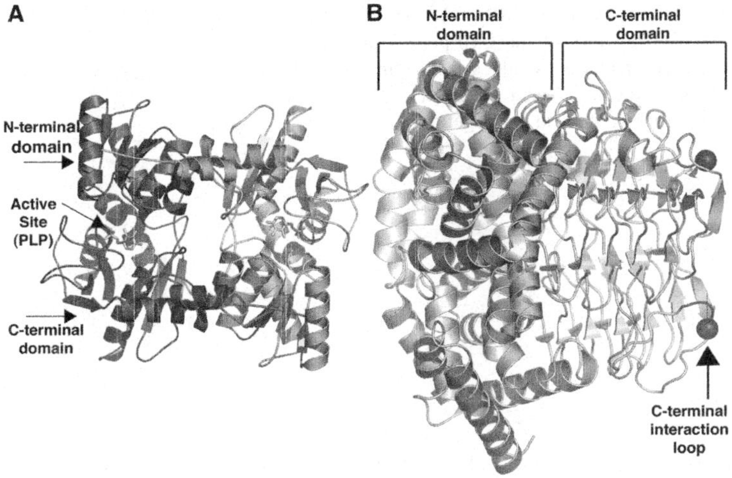

Fig. 5.2 (A) Ribbon diagram of the AtOASS dimer (Bonner et al. 2005). Each monomer is shaded in different grays. The location of the PLP cofactor (drawn as a stick model) and active site are indicated. (B) Ribbon diagram of one trimer of the *E. coli* SAT (Pye et al. 2004) with the N- and C-terminal domains indicated. The C-terminal loop believed to interact with OASS is indicated by the three spheres.

3 Methods for Studying Protein-Protein Interactions in the Cysteine Synthase Complex

Structural, kinetic, and thermodynamic information is required to completely understand how SAT and OASS assemble to form the cysteine synthase complex. While trying to characterize protein-protein interactions, experimentalists encounter two major difficulties. First, biophysical methods require the purification of sufficient amounts of stable protein to perform the necessary studies. The second challenge is to find optimal solution conditions under which the interacting proteins are stable, maintain their native structure, and do not undergo any non-native oligomerization. All of the possible methods and approaches to solve these problems will not be discussed here; however, assuming one has overcome such obstacles, complete understanding of any macromolecular interaction requires a multifaceted approach to obtain quantitative information about the system. In this section, we discuss quantitative approaches for dissecting protein-protein interactions in the cysteine synthase complex.

3.1 Crystallographic Analysis of the Interaction between AtOASS and the AtSAT C-Terminus

The ultimate goal of studying protein-protein interactions is to understand the recognition principles involved in macromolecular assembly. An essential part of this process

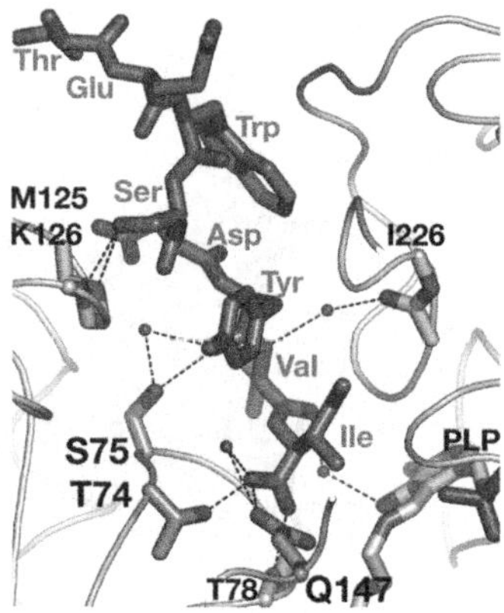

Fig. 5.3 Interactions in the AtOASS C10 peptide binding site. All modeled residues of the C10 peptide are shown and labeled. Side chains of residues interacting with the C10 peptide are indicated. Water molecules are shown as spheres. Dotted lines represent hydrogen bonds.

involved structure determination by protein crystallography, which provides a detailed view of spatial arrangement of atoms and interactions. The 2.5 Å resolution x-ray crystal structure of AtOASS in complex with a peptide corresponding to the 10 C-terminal residues of AtSAT (C10 peptide) revealed the molecular mechanism for downregulation of OASS activity in the plant cysteine synthase complex (Francois et al. 2006). The overall structure of the AtOASS C10 peptide complex is similar to that of AtOASS and clearly shows 8 of 10 amino acids in the C10 peptide bound to AtOASS, with the N-terminal two residues disordered. The C10 peptide binds at the AtOASS active site in an extended conformation with the C-terminal isoleucine of the peptide positioned near the PLP and the remaining residues occupying a cleft that forms the active site entrance (Fig. 5.3). Thus, changes in OASS activity upon interaction with SAT results from blocked access to the catalytic center of the enzyme. Although structural studies indicate that several key residues from both AtOASS and the C10 peptide contribute to binding, the relative contribution of these individual residues to complex formation and activity cannot be accessed without functional analyses. Thus, to test the specific role of these residues and to quantify their relative contribution, solution-based binding and activity studies complement crystallographic data.

3.2 *Mapping Hot-Spots That Play Dual Roles in Cysteine Biosynthesis Using Mutagenesis and Calorimetry*

Previously, crystallographic studies provided a guide for extensive site-directed mutagenesis, steady-state kinetic characterization, and ligand binding analysis of AtOASS (Bonner et al. 2005). A similar approach was adopted to dissect the role of residues in binding the C10 peptide using isothermal titration calorimetry (ITC) (Francois et al. 2006). In AtOASS, Thr74, Ser75, and Gln147 form an extensive interaction network with the C-terminal isoleucine of the C10 peptide, locking it in the active site/peptide-binding pocket. Mutation of each of these three residues to alanine decreased the OASS activity more than 1,000-fold (Bonner et al. 2005).

Similarly, site-directed mutations of these residues also effected C10 peptide binding, as determined by ITC. Compared to wild-type enzyme, the T74S, S75A, S75T, and Q147A mutants all displayed lower affinity for the C10 peptide with up to 100-fold increases in K_d values (Francois et al. 2006). As these residues form hydrogen bonds with the C10 peptide through their side chains, loss of these interactions results in decreased binding affinity. Interestingly, mutation of these residues also had a profound effect on the stoichiometry of complex formation. Each mutant enzyme bound only one C10 peptide in contrast to the 2:1 stoichiometry (i.e., two moles of C10 peptide bound to one mole of AtOASS dimer) observed for wild-type AtOASS. These mutational and binding studies suggest that protein-protein interaction in the cysteine synthase complex and the catalytic activity of OASS rely on a common set of critical or "hot-spot" residues. Additional studies are needed to completely map the hot-spot residues beyond the OASS active site that dictate interaction with SAT in the plant cysteine synthase complex. (More information about ITC and the analysis of the thermodynamics of binding interactions is given in section 5.4.)

3.3 Fluorescence-Based Assays to Monitor Complex Formation and Allostery

An alternative method to ITC for probing binding events is to employ fluorescence-based assays that depend on either intrinsic protein fluorescence or that of a chromophore. In addition to determining binding constants, equilibrium binding studies performed under very high affinity solution conditions can provide the absolute stoichiometry of an interaction.

The active site of OASS contains a natural fluorphore – the PLP cofactor (Campanini et al. 2005, Kumaran and Jez 2007). Excitation of PLP at 412 nm generates a fluorescence emission spectrum with a maximum around 505 nm. Binding of either SAT or C10 peptide at the OASS active site can be determined by monitoring changes in the intrinsic fluorescence emission signal of PLP. Upon addition of either SAT or C10 peptide, the relative fluorescence of PLP in the OASS active site increases. Titrations of AtOASS with the C10 peptide show a maximal 1.8-fold change in emission signal at saturation, indicating that a maximum binding of two C10 peptides bind per AtOASS dimer (Kumaran and Jez 2007). Further, the C10 peptide binds to AtOASS with very high affinity, i.e., K_d = 10 nM for each binding site.

Structural studies reveal that AtOASS is a symmetric dimer with two identical active sites, but analysis of C10 peptide binding to AtOASS using a fluorescence-based assay revealed that high concentrations of NaCl affects binding of the peptide (Kumaran and Jez 2007). Binding of the C10 peptide to AtOASS is only weakly sensitive to NaCl concentration below 0.3 M, suggesting that the binding energy is derived from mainly nonelectrostatic interactions. In contrast, at higher salt concentrations the two binding sites in the dimer exhibit completely different binding affinities. The binding of the first C10 peptide at the active site of one monomer decreases affinity for binding of the second peptide at the other active site through negative cooperativity between each subunit. Allosteric structural changes and inhibition of the bacterial OASS occur at less than 0.05 M NaCl (Tai et al. 2001). It is conceivable that chloride may be a potential allosteric modulator of the plant OASS; however, the biological function of this allosteric mechanism in plants is unclear.

3.4 Surface Plasmon Resonance (BIAcore) Analysis of the OASS-SAT Interaction

Unlike calorimetry and fluorescence assays that measure equilibrium binding constants, surface plasmon resonance or biomolecular interactions analysis (BIAcore) is used to determine the kinetic association and dissociation rates of a binding interaction. A BIAcore instrument monitors molecular interactions by surface plasmon resonance, which detects changes in the mass of molecules at an interface by probing changes in refractive index (Malmquist 1993). For these experiments, one interaction partner is either chemically immobilized or noncovalently tethered to a gold-film sensor chip. While the second interaction partner flows over the sensor chip and complexes form and dissociate at the metal surface, the refractive index of the interface changes. The time-dependent change in the refractive index is used to determine association and dissociation rate constants.

Berkowitz and co-workers (2002) used this method to evaluate formation of the cysteine synthase complex with the mitochondrial isoforms of AtSAT and AtOASS and dissociation of the complex by O-acetylserine. Using a 1:1 binding model, they determined a $K_d = 25\,\text{nM}$ for the complex formation. In addition, they analyzed the effect of OAS on dissociation of the complex, which showed a half-maximal rate at 77 μM with strong positive cooperativity. Importantly, this value is in the range of OAS concentrations found in the cell and is consistent with the regulatory model in which changes in OAS levels affect formation of the complex and modulates cysteine synthesis.

4 Thermodynamic Basis for Formation of the Plant Cysteine Synthase Complex

Although structural and binding experiments revealed the specificity of interaction and quantification of binding affinity in the plant cysteine synthase complex, these studies alone cannot provide information about the thermodynamic forces that define the binding affinity and drive interaction of SAT and OASS. Understanding the thermodynamic basis for formation of the cysteine synthase complex would provide insight into the different physical forces that optimize complex formation. Complete thermodynamic characterization of a binding interaction requires (1) that the affinity and stoichiometry of the interaction are known, (2) determination of the net enthalpy change upon interaction (i.e., the heat change that occurs upon breaking or forming bonds), and (3) the entropy change that results from changes in order of the system. Such an analysis depends on using calorimetry to directly measure the heat changes during a molecular recognition process and allows for the determination of all three thermodynamic parameters (ΔH, ΔS, and ΔG) in one experiment (Stites 1997). Since many texts describe the principles, design, and application of ITC, we will not discuss those details here.

4.1 AtOASS is a Symmetric Dimer: Structurally and Thermodynamically

Crystallographic and fluorescence binding experiments demonstrate that each subunit of AtOASS tightly binds a C10 peptide molecule; however, the energetics governing complex formation can only be probed using ITC. In an ITC experiment, the instrument detects the changes in heat occurring upon injection of small volumes of one ligand (C10 peptide) into a chamber containing the interaction partner (AtOASS) (Fig. 5.4a). Analysis of this data yields a binding isotherm (Fig. 5.4b) and estimates of the thermodynamic properties of the interaction.

ITC binding studies demonstrates that the two C10 peptide binding sites of the AtOASS dimer are functionally independent and bind the peptide with very similar affinity: $K_{eq} = 6.8 \times 10^7$ M^{-1} ($K_d = 14.7$ nM) (Kumaran and Jez 2007). The enthalpy change associated with binding at a single active site is ΔH = -12.7 kcal mol^{-1}. Although binding of OAS to the plant OASS can exhibit positive cooperativity (Droux et al. 1998), which suggests that structural changes during the catalytic cycle may mediate allosteric communication between the active sites in the homodimer, analysis of the AtOASS•C10 peptide interaction shows no evidence for any cooperativity between subunits at 25 °C and low (<0.3 M) NaCl concentrations. Thus, the two binding sites of AtOASS are symmetric in nature, so that C10 peptide binding at one site is independent of binding at the second site over broad temperature (10-30 °C) and salt ranges (0.02-0.3 M NaCl). These results agree with crystallographic studies of AtOASS alone and in complex with the C10 peptide that shows no global structural changes occur upon peptide binding.

4.2 Effect of Temperature on Formation of the Plant Cysteine Synthase Complex

Temperatures above 35 °C reduce the demand for cysteine synthesis (Nieto-Sotelo and Ho 1986), but increased sulfur assimilation and cysteine production help maintain glutathione levels in response to chilling stress in plants (Kocsy et al. 2000, Gomez et al. 2004, Phartiyal et al. 2006). Interestingly, ITC analysis of the temperature-dependence of interaction between AtOASS and the C10 peptide demonstrates that above 35 °C changes occur in how the subunits respond to peptide binding (Kumaran and Jez 2007). At elevated temperatures, the binding of a C10

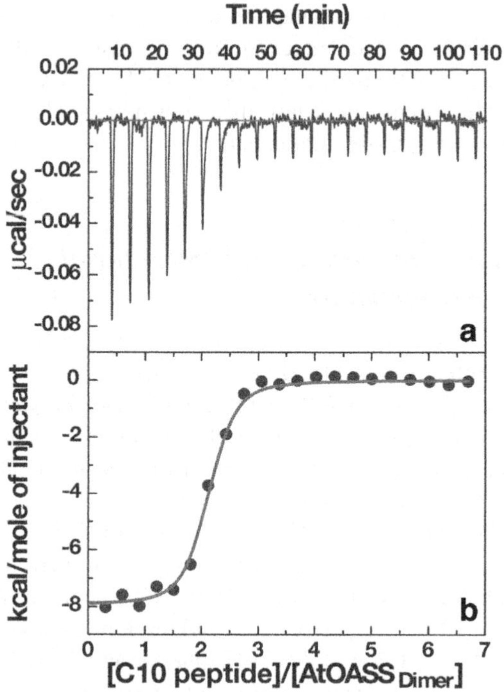

Fig. 5.4 (**a**) Titration of AtOASS with C10 peptide. ITC data is plotted as heat signal (μcal/sec) versus time (min). The experiment consisted of 20 injections of 12 μl each of C10 peptide (21.8 μM) into a solution containing AtOASS (0.65 μM) at 20 °C. (**b**) The integrated heat responses per injection from panel **a** are plotted as normalized heat per mole of injectant. The solid line represents the best fit to a two independent sites binding model.

peptide to one subunit decreases the affinity for the second subunit through an apparent negative cooperativity. The molecular origin of negative cooperativity is difficult to trace, but a reasonable explanation could be a conformation change that may have transformed the nonallosteric enzyme to an allosteric one. When the enzyme is in allosteric state, binding at one subunit may lead to closing of the active site in the other subunit. Thus, at higher temperatures, the activity of OASS in the plant cysteine synthase may be lost due to structural changes that occur as the enzyme transforms to a closed state from the open state upon complex formation. Transition between open and closed forms of the bacterial OASS occurs upon substrate binding (Rabeh and Cook 2004), but a similar transition has not been reported for the plant enzyme.

4.3 *Thermodynamic Basis for Interaction of AtOASS and the C-Terminus of AtSAT*

The temperature-dependence of thermodynamic parameters of a binding interaction also provides detailed information about the different forces that drive binding. The observed enthalpy change for binding of the C10 peptide by AtOASS increases as temperature rises (Kumaran and Jez 2007). Moreover, the binding interaction is exothermic at all temperatures examined, which indicates a favorable negative enthalpy of binding. Overall, both enthalpy and entropy contribute to the binding free energy, although the relative contributions depend on the temperature. At lower temperatures (10 and 15 °C), the entropic contribution is more favorable, and this contribution decreases with increasing temperature. Interaction at these temperatures is driven by both enthalpy and entropy, but enthalpy becomes the dominant force at higher temperatures. The crystal structure of unbound AtOASS revealed that multiple water molecules occupy the active site (Bonner et al. 2005). Presumably, binding of the C10 peptide releases nonspecifically bound water molecules from AtOASS at lower temperatures, thereby increasing the net entropy of the system and contributing to the free energy of binding.

An important thermodynamic property of molecular interactions is derived from temperature-dependence of the observed enthalpy of interaction. The first derivative of temperature-dependence of enthalpy change, heat capacity (ΔC_p) for AtOASS-C10 complex, is $\Delta C_p = -0.401 \pm 0.025$ kcal mol^{-1} deg^{-1} (Kumaran and Jez 2007). The negative value of ΔC_p indicates that the interaction is specific and is accompanied by burial of nonpolar surface area. In addition, the negative ΔCp is responsible for the shift from a predominantly entropic contribution to an enthalpic one as the temperature rises. At 22 °C (T_S), the entropic contribution to the binding process is zero and binding process becomes completely driven by enthalpy at higher temperatures. On the other hand, at −3.0 °C (T_H), the net enthalpic contribution is ~0 kcal mol^{-1}, with the binding process driven completely by entropy. Thus, both enthalpy and entropy contribute favorably to the Gibbs energy of complex formation between −3.0 °C and 22 °C.

5 Conclusion

The series of structural and functional studies discussed here provide an initial view of the plant cysteine synthase complex and the molecular basis for interaction of SAT and OASS. Although the experimental approaches discussed here extensively examine protein-peptide interactions, the general strategy for dissecting how SAT and OASS interact will be applicable with the full-length proteins.

Acknowledgements The authors acknowledge grant support from the Illinois-Missouri Biotechnology Alliance (H.B.K. and J.M.J.) and the U.S. Department of Agriculture (NRI-2005-02518 to J.M.J)

References

Berkowitz O, Wirtz M, Wolf A, Kuhlmann J, Hell R (2002) Use of biomolecular interaction analysis to elucidate the regulatory mechanism of the cysteine synthase complex from *Arabidopsis thaliana*. J Biol Chem 277:30629-34

Bogdanova N, Hell R (1997) Cysteine synthesis in plants: protein-protein interactions of serine acetyltransferase from *Arabidopsis thaliana*. Plant J 11:251-62

Bonner ER, Cahoon RE, Knapke SM, Jez JM (2005) Molecular basis of cysteine biosynthesis in plants: structural and functional analysis of O-acetylserine sulfhydrylase from *Arabidopsis thaliana*. J Biol Chem 280:38803-13

Burkhard P, Rao GS, Hohenester E, Schnackerz KD, Cook PF, Jansonius JN (1998) Three-dimensional structure of O-acetylserine sulfhydrylase from *Salmonella typhimurium*. J Mol Biol 283:121-33

Campanini B, Speroni F, Salsi E, Cook PF, Roderick SL, Huang B, Bettati S, Mozzarelli A (2005) Interaction of serine acetyltransferase with O-acetylserine sulfhydrylase active site: evidence from fluorescence spectroscopy. Protein Sci 14:2115-24

Claus MT, Zocher GE, Maier, THP, Schulz GE (2005) Structure of the O-acetylserine sulfhydrylase isoenzyme CysM from *Escherichia coli*. Biochemistry 44:8620-6

Cook PF, Wedding RT (1977) Initial kinetic characterization of the multienzyme complex, cysteine synthetase. Arch Biochem Biophys 178:293-302

Droux M, Ruffet ML, Dounce R, Job B (1998) Interactions between serine acetyltransferase and O-acetylserine(thiol)lyase in higher plants: structural and kinetic properties of the free and bound enzymes. Eur J Biochem 255:235-45

Francois JA, Kumaran S, Jez JM (2006) Structural basis for interaction of O-acetylserine sulfhydrylase and serine acetyltransferase in the *Arabidopsis* cysteine synthase complex. Plant Cell 18:3647-55

Gomez LD, Vanacker H, Buchner P, Noctor G, Foyer CH (2004) Intercellular distribution of glutathione synthesis in maize leaves and its response to short-term chilling. Plant Physiol 134:1662-71

Hell R, Hillebrand H (2001) Plant concepts for mineral acquisition and allocation. Curr Opin Biotech 12:161-8

Hell R, Jost R, Berkowitz O, Wirtz M (2002) Molecular and biochemical analysis of the enzymes of cysteine biosynthesis in the plant *Arabidopsis thaliana*. Amino Acids 22:245-57

Hopkins L, Parmar S, Blaszczyk A, Hesse H, Hoefgen R, Hawkesford MJ (2005) O-acetylserine and the regulation of expression of genes encoding components for sulfate uptake and assimilation in potato. Plant Physiol 138:433-40

Huang B, Vetting MW, Roderick SL (2005) The active site of O-acetylserine sulfhydrylase is the anchor point for bienzyme complex formation with serine acetyltransferase. J Bacteriol 187:3201-5

Johnson CM, Huang B, Roderick RL, Cook PF (2004a) Kinetic mechanism of the serine acetyltransferase from *Haemophilus influenzae*. Arch Biochem Biophys 429:115-22

Johnson CM, Huang B, Roderick RL, Cook PF (2004b) Chemical mechanism of the serine acetyltransferase from *Haemophilus influenzae*. Biochemistry 49:15534-9.

Kocsy G, Szalai G, Vagujfalvi A, Stehli L, Orosz G, Galiba G (2000) Genetic study of glutathione accumulation during cold hardening in wheat. Planta 210:295-301

Kopriva S (2006) Regulation of sulfate assimilation in *Arabidopsis* and beyond. Ann Bot 97:479-95

Koprivova A, Suter M, den Camp RO, Brunold C, Kopriva S (2000) Regulation of sulfate assimilation by nitrogen in *Arabidopsis*. Plant Physiol 122:737-46

Kredich NM, Becker MA, Tomkins GM (1969) Purification and characterization of cysteine synthetase, a bifunctional protein complex, from *Salmonella typhimurium*. J Biol Chem 244:2428-39

Kumaran S, Jez JM (2007) Thermodynamics of the interaction between O-acetylserine sulfhydrylase and the C-terminus of serine acetyltransferase. Biochemistry 46:5586-94

Leu LS, Cook PF (1994) Kinetic mechanism of serine transacetylase from *Salmonella typhimurium*. Biochemistry 33:2667-71

Liszewska F, Lewandowska M, Plochocka D, Sirko A (2007) Mutational analysis of O-acetylserine (thiol) lyase conducted in yeast two-hybrid system. Biochim Biophys Acta 1774:450-5

Malmquist M (1993) Biospecific interaction analysis using biosensor technology. Nature 361:186-7

Mino K, Yamanoue T, Sakiyama T, Eisaki N, Matsuyama A, Nakanishi K (1999) Purification and characterization of serine acetyltransferase from *Escherichia coli* partially truncated at the C-terminal region. Biosci Biotech Biochem 63:168-79

Mino K, Hiraoka K, Imamura K, Sakiyama T, Eisaki N, Matsuyama A, Nakanishi K (2000) Characteristics of serine transacetylase from *Escherichia coli* deleting different lengths of amino acid residues from the C-terminus. Biosci Biotech Biochem 64:1874-80

Nakamura K, Tamura G (1990) Isolation of serine acetyltransferase complexed with cysteine synthase from *Allium tuberosum*. Agric Biol Chem 54:649-56

Nieto-Sotelo J, Ho TH (1986) Effect of heat shock on the metabolism of glutathione in maize roots. Plant Physiol 82:1031-5

Noji M, Inoue K, Kimura N, Gouda A, Saito K (1998) Isoform-dependent differences in feedback regulation and subcellular localization of serine acetyltransferase involved in cysteine biosynthesis from *Arabidopsis thaliana*. J Biol Chem 273:32739-45

Olsen LR, Huang B, Vetting MB, Roderick SL (2004) Structure of serine acetyltransferase in complexes with CoA and its cysteine feedback inhibitor. Biochemistry 43:6013-19

Phartiyal P, Kim WS, Cahoon RE, Jez JM, Krishnan HB (2006) Soybean ATP sulfurylase, a homodimeric enzyme involved in sulfur assimilation, is abundantly expressed in roots and induced by cold treatment. Arch Biochem Biophys 450:20-9

Pye VE, Tingey AP, Robson RL, Moody PCE (2004) The structure and mechanism of serine acetyltransferase from *Escherichia coli*. J Biol Chem 279:40729-36

Rabeh WM, Cook PF (2004) Structure and mechanism of O-acetylserine sulfhydrylase. J Biol Chem 279:26803-6

Saito K (2004) Sulfur assimilatory metabolism: the long and smelling road. Plant Physiol 136:2443-50

Smith FW, Hawkesford MJ, Ealing PM, Clarkson DT, Vandenberg PS, Belcher AR, Warrilow AG (1997) Regulation of expression of a cDNA from barley roots encoding a high-affinity sulfate transporter. Plant J 12:875-8

Stites WE (1997) Protein-protein interactions: interface structure, binding thermodynamics, and mutational analysis. Chem Rev 97:1233-50

Tai CH, Burkhard P, Gani D, Jenn T, Johnson C, Cook PF (2001) Characterization of the allosteric anion-binding site of O-acetylserine sulfhydrylase. Biochemistry 40:7446-52

Wirtz M, Berkowitz O, Droux M, Hell R (2001) The cysteine synthase complex in plants: mitochondrial serine acetyltransferase from *Arabidopsis thaliana* carries a bifunctional domain for catalysis and protein-protein interaction. Eur J Biochem 268:686-93

Wirtz M, Droux M (2005) Synthesis of the sulfur amino acids: cysteine and methionine. Photosynth Res 86:345-6

Wirtz M, Hell R (2007) Dominant-negative modification reveals the regulatory function of the multimeric cysteine synthase protein complex in transgenic tobacco. Plant Cell 19:625-39

Zhao C, Moriga Y, Feng B, Kumada Y, Imanaka H, Imamura K, Nakanishi K (2006) On the interaction site of serine acetyltransferase in the cysteine synthase complex from *Escherichia coli*. Biochem Biophys Res Comm 341:911-16

Chapter 6
Glutathione Reductase: A Putative Redox Regulatory System in Plant Cells

A.S.V. Chalapathi Rao and Attipalli R. Reddy(✉)

Abstract Glutathione reductase (GR, EC 1.6.4.2) and glutathione (GSH, γ-Glu-Cys-Gly) are important components of the cell's scavenging system for reactive oxygen compounds in plants. GSH is a major reservoir of nonprotein reduced sulfur. In addition, GSH plays a crucial role in cellular defense, where it gets oxidized to glutathione disulfide (GSSG). GR mediates the reduction of GSSG to GSH by using NADPH as an electron donor, and thus a highly reduced state of GSH/GSSG and ASA/DHA ratios is maintained at the intracellular level by this reaction during oxidative stress. GR activity has been shown to increase in various plant species under different types of stresses. Studies using transgenic plants have shown that GR plays an important role in providing resistance to oxidative stress caused by paraquat, methyl viologen, ozone, moderate chilling at high light intensity, drought, heavy metals, high light, salinity, and chilling. From the existing literature, it is clear that among the enzymatic and nonenzymatic antioxidative pathways, GR is one of the key enzymes in the active oxygen scavenging system, involving superoxide dismutase (SOD, EC 1.15.1.1) and the enzymes of ascorbate–glutathione cycle in higher plants. In this chapter, we review most recent information on the structural details of GR, its conserved domains, different isoforms, its role in sulfur assimilation and also its importance in maintaining redox balance of the plant cell under environmental and biotic stresses. Further, we discuss the transgenic approach to increase GR activities for developing improved tolerance to oxidative stress in plants.

1 Oxidative Stress in Plants

Plant cells are continuously exposed to various environmental and biotic cues which lead to the increased production of reactive oxygen species (ROS). The responses of plants to these excess ROS have recently been analyzed extensively at biochemical

Attipalli R. Reddy
Department of Plant Sciences, School of Life Sciences, University of Hyderabad, Hyderabad 500 046, India
arrsl@uohyd.ernet.in

N.A. Khan et al. (eds.), *Sulfur Assimilation and Abiotic Stress in Plants.*

and molecular levels (Gachomo et al. 2003, Kotchoni and Gachomo 2006, Kouril et al. 2003, Kreps et al. 2002, Rizhsky et al. 2004). Under normal physiological conditions, ROS are continuously produced in the chloroplasts, mitochondria, and peroxisomes as byproducts of aerobic metabolic processes like photosynthesis, respiration, and photorespiration. These ROS are scavenged by both enzymatic and nonenzymatic antioxidant pathways for the maintenance of the normal plant growth (Asada 2006, Davletova et al. 2005, del Rio et al. 2002, Jimenez et al. 1998, Kotchoni and Gachomo 2006, Mittler et al. 2004, Nathawat et al. 2007). For example, chloroplasts generate H_2O_2 in the range of 150-250 pmoles mg^{-1} chlorophyll h^{-1} during photosynthesis (Asada 1994) through the Mehler reaction (Mehler 1951), which are continuously removed by the ascorbate-glutathione cycle. However, an increase in the production of ROS results when the production exceeds the scavenging capacity of the cell during unfavorable conditions like drought, salinity, ozone stress, high light, UV irradiation, heat shock, chilling, heavy metals, air pollutants, and biotic stresses (Agarwal 2007, Chaitanya et al. 2001, Desikan et al. 2001, Guo et al. 2006, Mittler 2002, Romero-Puertas et al. 2006, Sumithra et al. 2006, Zgallai et al. 2006). These excess ROS cause damage to proteins, lipids, carbohydrates, and DNA and ultimately result in cell death (Fadzilla et al. 1997, Fahmy et al. 1998, Foyer and Noctor 2005, Gueta-Dahan et al. 1997, Hernandez et al. 1995, Mittler et al. 2004, Sairam et al. 2005, Shulaev et al. 2006). Development of antioxidative system is one of the mechanisms by which plant cells can avoid the deleterious effects caused by ROS. Among the enzymatic and nonenzymatic antioxidative pathways, it has been frequently interpreted that GR is one of the key enzymes in the active oxygen scavenging system, involving SOD and the enzymes of the ascorbate–glutathione cycle in higher plants. (Asada 1994, Reddy and Raghavendra 2006).

2 Generation and Scavenging of Free Radicals in Plant Cells

Production of ROS is seen in different compartments of the plant cells during normal metabolic events and during environmental and biotic stresses (Fig. 6.1). Electron transport chains of chloroplasts and mitochondria are responsible for the generation of ROS like superoxide, hydrogen peroxide, and singlet oxygen in these organelles. In peroxisomes, H_2O_2 is produced mainly during the photorespiration and also during ß-oxidation of fatty acids, in the enzymatic reactions of flavin oxidases as well as in the dispropotionation of superoxide radicals. In glyoxysomes, acyl-coA oxidase is the primary enzyme responsible for H_2O_2 generation. Membrane-bound NADPH-oxidase and soluble xanthine oxidase activities also generate free radicals in this organelle (Jimenez et al. 1997, Slooten et al. 1998). Generation of free radicals is also seen in the apoplastic region during oxidative stress. Highly energetic reactions of photosynthesis and availability of abundant oxygen in the chloroplast makes it a rich source of ROS. High light intensity can lead to excess reduction of PSI; the Calvin cycle cannot keep pace, which results in the shortage of electron acceptor, $NADP^+$. Under these conditions, O_2 can compete for electrons from PSI, leading to the generation of superoxide through the Mehler reaction. This superoxide reacts

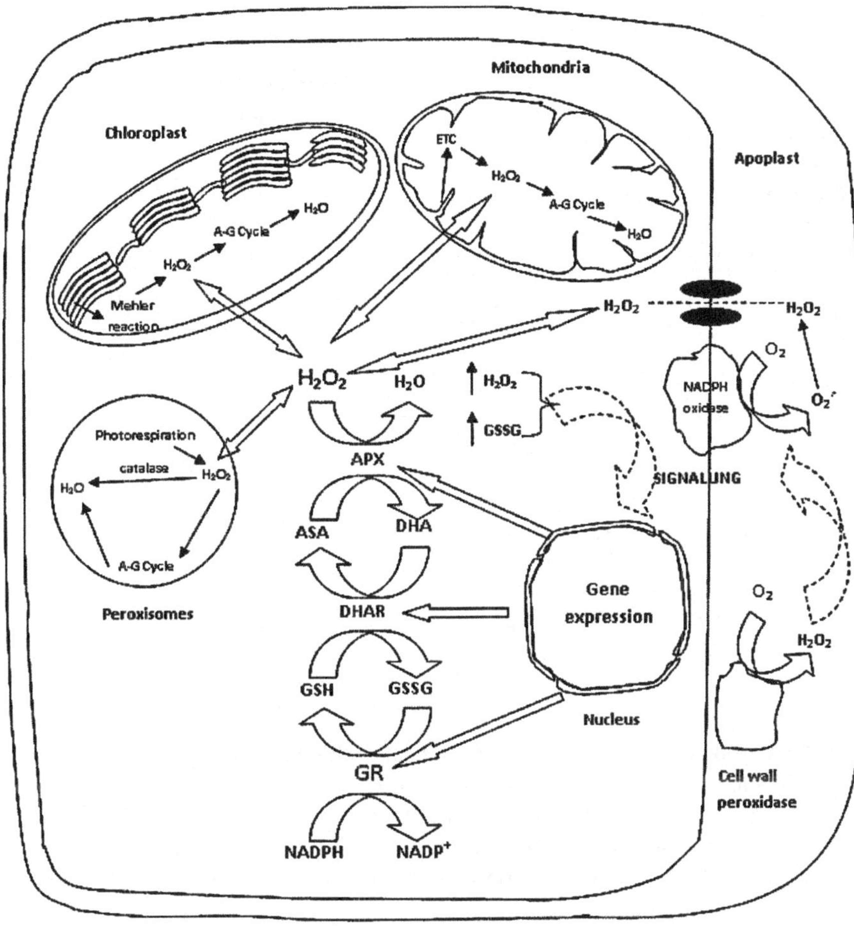

Fig. 6.1 Free radical formation and scavenging in plant cell. ETC, electron transport chain; A-G Cycle, Ascorbate-glutathione cycle

with H_2O_2 and produces hydroxyl radicals through the Haber-Weiss reaction. This reaction is thermodynamically slow, but in the presence of the catalyst iron, significant production of hydroxyl radicals are formed through Fenton's reaction. These hydroxyl radicals are highly reactive molecules, which are responsible for damaging the membranes and other essential macromolecules, including photosynthetic pigments, proteins, DNA, and lipids (Fadzilla et al. 1997, Fahmy et al. 1998, Foyer and Noctor 2005, Gueta-Dahan et al. 1997, Hernandez et al. 1995, Shulaev et al. 2006). Therefore ROS produced as a result of various abiotic and biotic stresses need to be scavenged for maintenance of normal growth. Plant cells have developed inherent capacity to scavenge these ROS, generated during normal metabolic processes and stressful conditions. Superoxide produced in the cells is less stable, and it is immediately dismutated to H_2O_2 in the presence of SOD. The excess H_2O_2 molecules should be immediately removed by the plant defense system; otherwise these

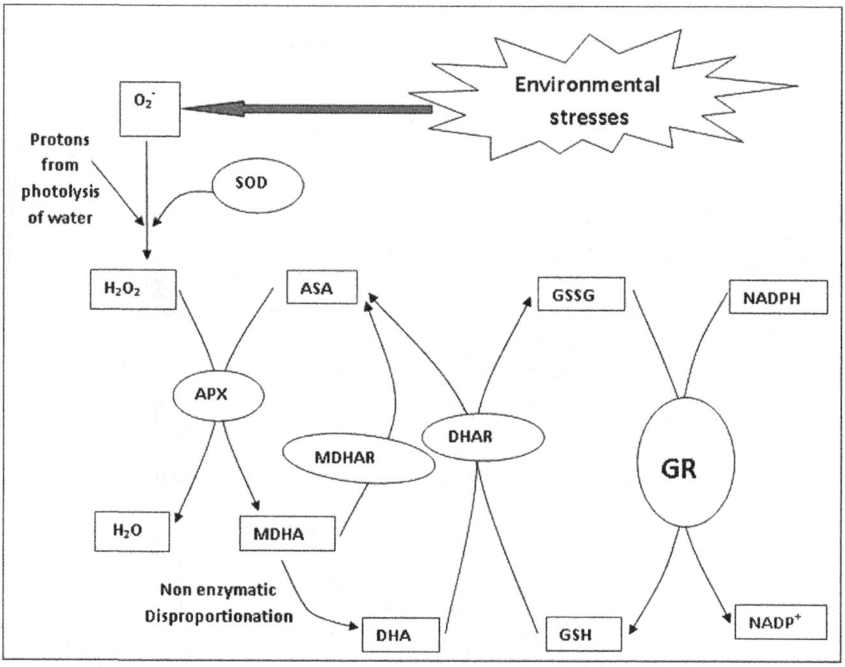

Fig. 6.2 Ascorbate-glutathione cycle in plant cell under abiotic stresses

molecules will be converted to hydroxyl radicals, which are very powerful oxidants. H_2O_2 molecules can also be efficiently scavenged by ascorbate-glutathione cycle as shown in Fig. 6.2. H_2O_2 is converted to H_2O by the activity of ascorbate peroxidase (APX) using ascorbate as the substrate. During this process, ascorbate is oxidized to either monodehydroascorbate (MDA) or dehydroascorbate (DHA). MDA will be reduced to ascorbate mediated by monodehydroascorbate reductase (MDAR) and NADPH. DHA which will then be reduced to ascorbate by the activity of dehydroascorbate reductase (DHAR), using GSH as a substrate. During this process GSH is oxidized to GSSG, which is further reduced back to GSH with the activity of GR using NADPH. Peroxisomal catalase can also scavenge H_2O_2, formed during normal and stress conditions. However, because of its low affinity for H_2O_2 (K_m over 1 M), catalase (CAT) is probably unable to scavenge H_2O_2 completely in plant cells, and hence additional scavenging mechanisms are needed, among which ascorbate-glutathione cycle plays an important role in scavenging H_2O_2 and GR is a key enzyme in this cycle (Slooten et al. 1998).

In this review, we discuss recent perspectives on GSH metabolism during stress, the importance of GR during sulfur assimilation, structure of GR and its catalytic mechanism, homology of GR among various species, GR isoforms, and regulation of GR synthesis during stress to maintain the cellular redox potential. Further, we review the biochemical and molecular evidence showing the importance of GR through transgenic approaches to understand the mechanism of oxidative stress for developing stress-tolerant crops.

3 Sulfur Assimilation in Higher Plants

The source of sulfur in plants is mainly the sulfate obtained from the soil in addition to sulfur dioxide and hydrogen sulfide from the atmosphere absorbed by the leaves through the stomata. The sulfate is usually reduced to organic sulfur compounds, although directly sulfate can be used for some of the metabolic processes. Unlike the assimilation of sulfur in bacteria and yeast, sulfate assimilation pathway in plants is a complex process (Fig. 6.3). The final product of sulfate assimilation is cysteine, which is involved in synthesis of tripeptide, γ-glutamylcycteinyl-x or GSH (x-being β-alanine, serine, or glutamate), depending on plant species (Foyer et al. 2001, Innocenti et al. 2006, Noctor et al. 2002, Xiang et al. 2001). GSH acts as a storage and transport form for cysteine; otherwise the excess cysteine present in the cell becomes toxic (Cobbett and Goldsbrough 2002, Droux 2004, Foyer et al. 2001, Noctor et al. 2002). GSH acts as a signal, controlling the interorgan regulation of sulfur nutrition, and is mainly confined to the leaves (Hartmann et al. 2004, Nocito et al. 2006). Sulfur assimilation in tobacco is a light-enhanced process, and the GR activity is modulated by phytochrome (Drumm-Herrel et al. 1989, Foyer et al. 1989, Rennenberg et al. 1979).

Sulfate taken from the soil is first activated to adenosine 5-phosphosulfate (APS) mediated by ATP sulphurylase. APS is converted to sulfite, in the presence of APS reductase and GSH. APS can also be further phosphorylated to adenosine 3- phosphate 5- phosphate (PAPS), which serves as a source for a variety of sulphotransferases. In some lower plants, PAPS can be directly reduced to sulfite by PAPS reductase, but the significance of this reaction in higher plants is uncertain. The sulfite formed is reduced to sulfide through sulfite reductase and ferredoxin. Sulfide then reacts with O-acetyl serine (OAS) to form cysteine by the activity of OAS (thiol) lyase. APS reducatse is the key enzyme in the cysteine biosynthesis pathway, where in GSH is oxidized to GSSG. The GSSG formed during this process is converted to GSH by the flavoprotein GR, which uses NADPH as a cofactor (Fig. 6.3).

4 Metabolism of GSH in Plant Cells with Special Reference to GR

GSH is a multifunctional metabolite which is present in all organisms (Fig. 6.4). It is present in millimolar concentrations in various plant tissues (Creissen et al. 1999, Meyer and Fricker 2002, Noctor et al. 2002) GSH synthesis, compartmentation within the cell, homeostasis, and degradation during various biological functions have been well established as depicted in Fig. 6.5 (Alscher 1989, Kumar et al. 2003, May et al. 1998, Mayer and Hell 2005, Meister 1988, Mullinaeux and Rausch 2005, Noctor et al. 1998, 2002). Analogs of GSH also exist in some plants where the glycine from GSH is replaced by alanine (γ-glutamyl-cysteinyl-ß-alanine) or serine (γ-glutamylcysteineserine) (Innocenti 2006, Xiang et al. 2001). Reactivity of GSH

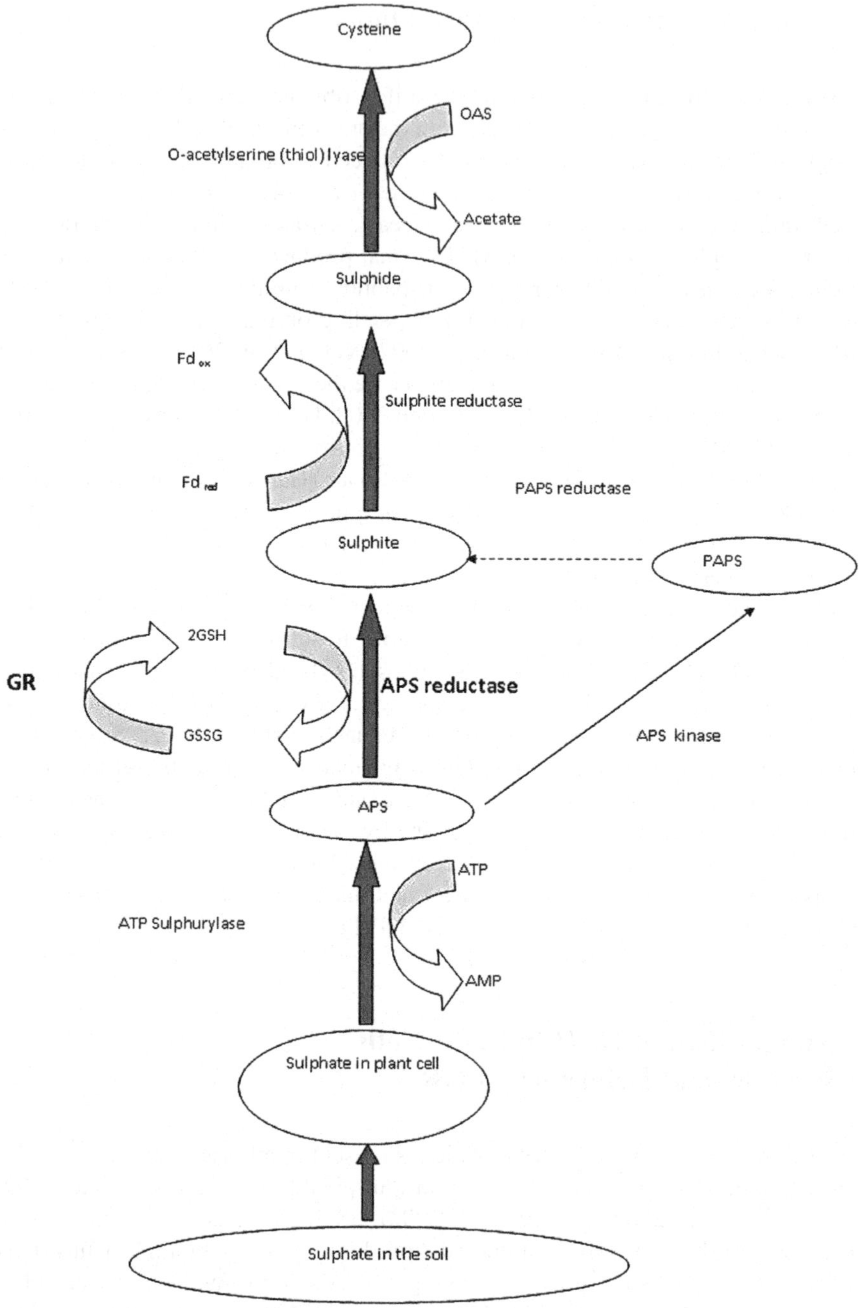

Fig. 6.3 Sulfur assimilatory pathway in higher plants. APS, adenosine 5[1]-phosphosulphate; PAPS, adenosine 3[1]-phosphate 5[1]-phosphosulphate; OAS, O-acetylserine

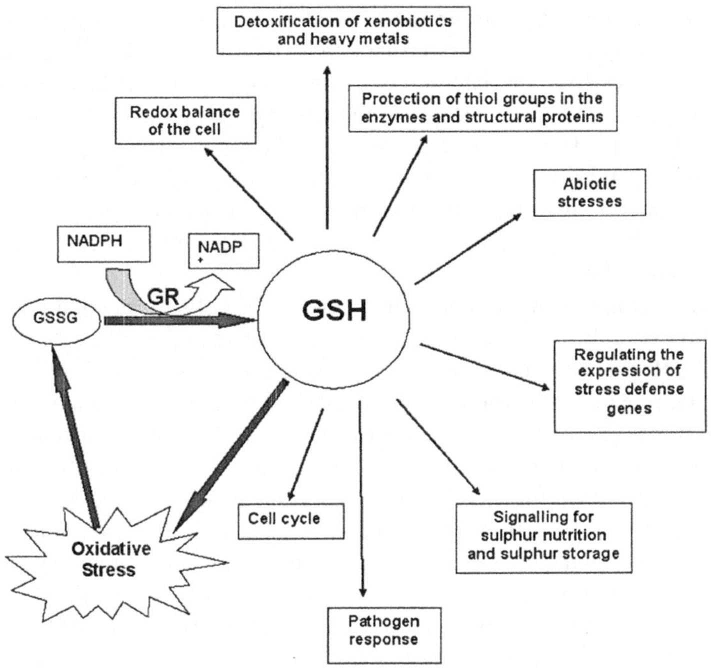

Fig. 6.4 Role of glutathione reductase (GR) and glutathione (GSH) in plant metabolism

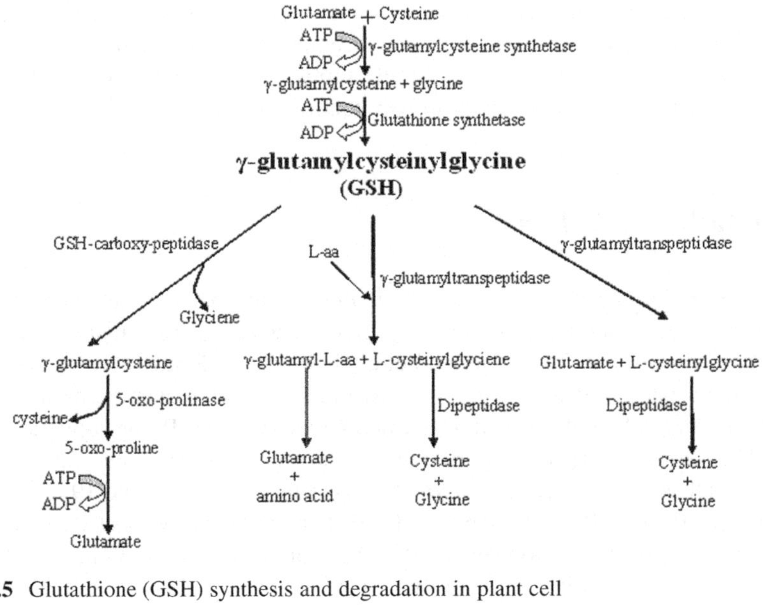

Fig. 6.5 Glutathione (GSH) synthesis and degradation in plant cell

depends on the thiol (-SH) group, and hence glutathione and homo-glutathiones may not show significant differences in their redox chemistry (Hausladen and Alscher 1993). GSH has an oxidation reduction potential of –0.23 V that allows it to act as an effective electron acceptor and donor for numerous biological reactions (Xiang et al. 2001). The nucleophilic nature of the thiol group is also important in the formation of mercaptide bonds with metals and for reacting with select electrophiles. (Xiang et al. 2001) This reactivity, along with the relative stability and high water solubility of GSH makes it an ideal biomolecule to protect plants against environmental and biotic stresses.

GR and GSH are the important components of the cells for maintaining antioxidant machinery of the cell (Ansel et al. 2006, Foyer and Halliwell 1976, Kurnet and Foyer 1993, Rennenberg and Brunold 1994, Romero-Puertas et al. 2006). GSH also plays crucial role in sulfur storage, the plant defense system, and in maintaining the redox balance of the cell (Kopriva and Koprivova 2005, Tausz et al. 2004). Other functions of GSH are detoxification of xenobiotics and heavy metals (Tausz et al. 2004), as an important cofactor for enzyme activities and DNA synthesis (Potters et al. 2004), induction of flowering (Ogawa et al. 2004), and cell cycle and plant development (Maughan and Foyer 2006). It is also known to regulate expression of certain stress defense genes during environmental stresses (Ball et al. 2004, Wingate et al. 1988), including desiccation (Kraner 2002), drought (Herbinger et al. 2002, Tausz et al. 2004), salinity (Vaidyanathan et al. 2003), chilling (Kocsy et al. 2000, 2001), and ozone (Herbinger et al. 2002, Noctor et al. 2002). GSH is also known to protect –SH groups of some enzymes and structural proteins against oxidation, either by acting as scavenger for oxidizing substances or by repairing the –SH groups through the GSH-disulphide exchange reaction (Mayer and Hell 2005). Further, GSH is an effective donor of reducing equivalents to ascorbate in the active oxygen scavenging system. In almost all of its metabolic functions, oxidation of GSH and its further reduction by GR are extremely efficient reactions to dissipate energy and to modulate the ATP: NADPH ratios, at times when CO_2 fixation is limited in plants under unfavorable growth conditions (Chen et al. 2004).

5 Glutathione Reductase

Glutathione reductase (GR, EC 1.6.4.2) maintains the balance between reduced glutathione (GSH) and ascorbate pools (Fig. 6.6), which in turn maintain cellular redox state (Ansel et al. 2006, Lascano et al. 1999, 2001, Reddy and Raghavendra 2006, Romero-puertas et al. 2006). It is a flavo-protein oxidoreductase, found in both prokaryotes and eukaryotes (Mullineaux and Creissen 1997, Romero-puertas et al. 2006). This enzyme was first reported in eukaryotes and yeast (Meldrum and Tarr 1935), as well as in plants, almost 57 years ago (Conn and Vennesland 1951, Mapson and Goddord 1951). GR has been purified and characterized from bacteria including, cyanobacteria (Libreros-Minotta et al. 1992, Rendon et al.1986, Serrano et al. 1984), *Chlamydomonas* (Serrano and Liobell 1993), yeast (Massey and Williams

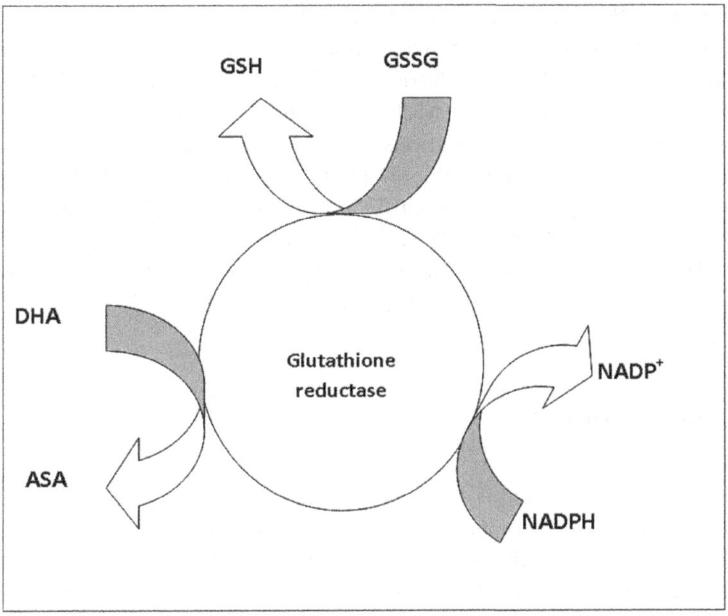

Fig. 6.6 Role of glutathione reductase (GR) in the maintenance of cellular redox status

1965), filamentous fungi (Woodin and Segal 1968), pea (Bielawski and Joy 1986, Connell and Mullet 1986, Creissen et al. 1991, Edwards et al. 1990, Kalt-Torres et al. 1984), maize (Mahan and Burke 1987), red spruce (Hausladen and Alscher 1994a, b), eastern white pine (Anderson et al. 1990), scot pine (Wingsle 1989), *Arabidopsis thaliana* (Kubo et al. 1993), rice (Kaminaka et al. 1998), wheat (Lascano et al. 2001), spinach (Schaedle and Bassham 1977), and tobacco (Creissen and Mullineaux 1995). Although GR is localized mainly in the chloroplasts (70-80%) (Connell and Mullet 1986), it is also found in cytosol (Drumm-Herrel et al. 1989, Edwards et al. 1990), mitochondria, peroxisomes (Jimenez et al. 1997, Romero-Puertas et al. 2006), and even in certain nonphotosynthetic tissues and organelles (Edwards et al. 1990, Ida and Morita 1971, Mapson and Isherwood 1963, Young and Conn 1956).

GR from different organisms showed optimum activity at different pH values (Mullineaux and Creissen 1997). Wheat chloroplastic-GR showed optimum activity at pH 8 (Lascano et al. 2001), human erythrocytes at pH 6.8 (Mullineaux and Creissen 1997) and pea chloroplastic-GR showed maximum activity at pH 7.6 (Kalt-Torres et al. 1984). Within the same plant, different isoforms showed different pH values for optimum activity (Mullineaux and Creissen 1997). GRs have high specificities for their substrates, although some glutathione conjugates and mixed glutathione disulphides can also be reduced (Gaullier et al. 1994). The specificity of GR to non–GSSG substrates is apparent among GRs from different sources. GR from mouse liver, yeast, and *E.coli* were inactivated by low concentrations of

NADPH (Lopez–Barea and Lee 1979, Mata et al. 1984, Pinto et al. 1985), whereas the activity of GR from spinach, maize, and *anabena* was not completely inhibited, even with relatively high concentrations of NADPH (0.1-0.5 mM) (Mahan and Burke 1987, Serrano et al. 1984). Mullinaeux and Creissen (1997) suggested that plant GRs modified their confirmations during the course of evolution to protect themselves from high concentrations of NADPH generated during photosynthesis. Most GRs can catalyze the reduction of GSSG by using NADH, but the efficiency is quite low (Halliwell and Foyer 1978) while most GRs from plants have a high affinity for NADPH (<10 μM). However, there is a considerable variation in the affinity of GR for GSSG (from 10-7300 μM) (Mullineaux and Creissen 1997).

5.1 Structure of GR

GR contains different domains for binding internal flavin molecules (FAD), the reducing substrate (NADPH), and the interface domain for joining the two subunits. The active site is between the two subunits where GSSG is bound, and it is at the site where the redox active disulphide of the enzyme is located (Lascano et al. 1999,

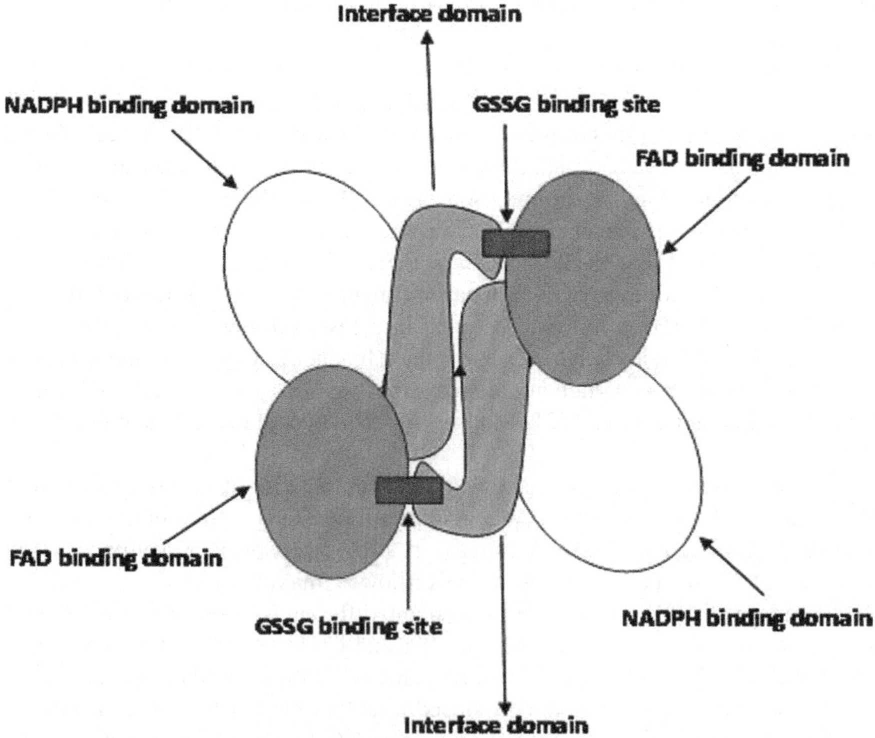

Fig. 6.7 Structure of GR homodimer (modified after Schulz et al. 1978)

Mullineaux and Creissen 1997). Mostly GR exists as homodimer (α2 configuration) as depicted in Fig. 6.7. In *Chalmydomonas reinharditii*, GR is a monomer (Takeda et al. 1993), while in pea and maize it is a heterodimer (Kalt-Torres et al. 1984, Mahan and Burke 1987). GR dimers may be further assembled into tetramers or even higher aggregative states as reported in *Spiriluna maxima* and pea. The type of higher-order state is pH- and temperature-dependent but not on their substrates or products. This may be one of the mechanisms of regulation of GR activity (Mullinaeux and Creissen 1997). GR from different sources consists of a number of domains, which in terms of their overall structure and key amino acid residues are absolutely conserved. Details of the domains given here are mostly based on *E. coli* enzyme (Mittl and Schulz 1994, Mullinaeux and Creissen 1997, Rouhier et al. 2006, Scrutton et al. 1990) (Fig. 6.7).

- FAD domain with amino acid residues of 1-140 and 265-336.
- NADPH domain covering residues of 141-264. This domain consists of altering ß sheet and α helix (ß α ß α ß arrangement), which is a dinucleotide binding fold typical for NAD linked dehydrogenases.
- GSSG binding region consists of domines from all over the GR polypeptide: from residues 11-22, 41-52, 90-103, 310-320, and 438-449. The two cysteines of the GR redox center are separated by four amino acid residues in a highly conserved motif GGTCV (I/V) RGCVPKK (I/L) LVY.
- The interface region runs from amino acid residues of 337-450, and it is the region on which each monomer that allows GSSG to be bound between the subunits by cooperation between domains of each GSSG binding region. This confirmation also brings the FAD domain of each subunit into close proximity with the opposite catalytic site of each monomer.

5.2 Catalytic Mechanism of GR

GR undergoes redox interconversions, depending upon the substrate availability. Oxidized form of the enzyme is more stable than the reduced form. The oxidized form of GR also showed tolerance to divalent metal ions like Zn^{+2}, Cu^{+2}, Fe^{+2} (Smith et al. 1989). During the conversion of GSSG to GSH, glutathione reductase splits the two electrons provided by NADPH and donates them one at a time to each of the two sulfur atoms of GSSG (Fig. 6.8). GR contains FAD and it also uses a redox active enzyme disulphide (Ghisla and Masey 1989, Jiang et al. 1995, Mullinaeux and Creissen 1997). GR has cysteine residues in the active sites which catalyze the NADPH-dependent reduction of GSSG to GSH. The reaction involves the following sequential steps:

- GR contains a highly conserved GXGXXA "fingerprint" motif in the NADPH binding domain. Arginine residues, which are present in the NADPH binding domain, are important for binding NADPH to this domain. These arginine residues are highly conserved but in *anabaena* one of the arginine residues is replaced with lysine.

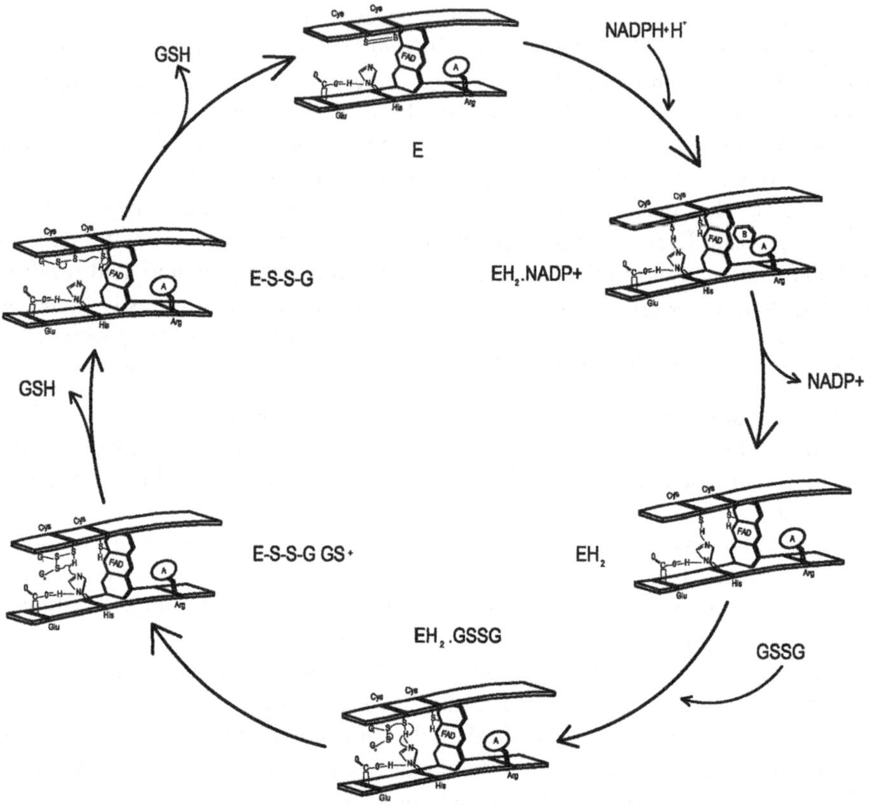

Fig. 6.8 Catalytic conversion of GSSG to GSH mediated by GR in plant cell (modified after Campbell 1995). E, oxidized form of the enzyme; EH_2.$NADP^+$, enzyme with $NADPH+H^+$; EH_2, reduced form of the enzyme; EH_2.GSSG, EH_2 with GSSG, E-S-S-G+GS^-, reaction of GSSG with FAD moiety and disulphide bond of the enzyme, E-S-S-G, enzyme after releasing one GSH molecule; A, NADPH domain; B, $NADP^+$

- The enzyme-bound FAD is converted to $FADH_2$. The reduced FAD reacts with internal disulphide of the enzyme (internal disulphide of the enzyme is formed between Cys [42] and Cys [47]) where one -SH group is stabilized by interaction with FAD (enzyme-SH-FAD) while the other is stabilized by histidine in combination with glutamate.
- GSSG reacts with the reduced form of the enzyme.
- One of the glutathione moieties forms a bond with –SH group of histidine, and a mixed disulphide bond is formed between the second glutathione and the enzyme thiol (G-S-S-enzyme).
- The bond formed between the glutathione and –SH group of histidine releases one molecule of GSH, and later the enzyme-SH-FAD complex attacks the G-S-S-enzyme to release the second GSH, which results in the regeneration of the oxidized form of the enzyme.

5.3 Genes of GR and Their Homology

In higher plants, GR is encoded by more than one gene (Edwards et al. 1990, 1994, Madhamanchi et al. 1992). Two genes were identified in *Nicotiana tobaccum* (Creissen and Mullineaux 1995), *Arabidopsis thaliana* (Kubo et al. 1993), *Pisum sativum* (Creissen et al. 1991, Stevens et al. 1997); three genes were identified in *Oryza sativa* and *Populus trichocarpa* (Rouhier et al. 2006). The native molecular mass of plant GRs ranges from 60-190 KD (Mullineaux and Creissen 1997). The molecular mass of GR from *Chlamydomonas reinherditii* is 56 KD and exists as monomer and in pea chloroplast GR exists as tetramer with a molecular mass of 220 KD (Mullineaux and Creissen 1997). Depending on the presence of N-terminal extension, Rouhier et al. (2006) have classified GR proteins into two classes: GR1- the shorter cytosolic enzymes and GR2- elongated organellar proteins. Creissen et al. (1995) showed that the N-terminal sequence in pea GOR1 can target the protein to both chloroplast and the mitochondria, but they could not identify which parts of the sequence of GOR1 transist peptide is having the cotargeting function. Peeters and Small (2001) indicated that dual targeting signals are more hydrophobic than mitochondrial or chloroplastic targeting signals. Chew et al. (2003a) have shown the hydrophobic residues, particularly at the N-terminal region, were important for targeting to both mitochondria and chloroplasts. Single mutations of the targeting signal on the positive residues affected mitochondrial import while double mutants (both on positive and hydrophobic residues) affected both mitochondrial and chloroplastic import. It was suggested that the dual targeting signal sequence of pea GOR1 had overlapping nature by changing positive and hydrophobic residues in the targeting signals. Not only GR but also other antioxidative enzymes have dual targeted nature, which was seen in *Arabidopsis* by Chew et al. (2003b). More recently Ansel et al. (2006) have cloned dual-targeted GR from *Vigna unguiculata* (cowpea).

Homology of amino acid sequence of GR1 from three model plants- *Arabidopsis thaliana, Oryza sativa*, and *Populus trichocarpa* ranges from 71% to 91%, whereas the identity of GR2 oscillates between 64% to 77%. When the two subgroups are compared, the homology was lower ranging, from 43% to 51% (Rouhier et al. 2006). GR amino acid sequence alignment from various plant sources showed key amino acids governing the catalytic function and interaction with substrate are conserved or substituted with similar amino acids (Ansel et al. 2006, Mullineaux and Creissen 1997). Stevens et al. (1997) isolated cytosolic glutathione reductase cDNA *gor2* from pea and showed that antibody to the GOR1 protein (chloroplastic GR from pea) recognized GOR2, which shows that the two proteins share sequence homology. On the basis of DNA and protein sequence homology it was concluded that cytosolic and plastidic GRs have quite different evolutionary histories as evidenced by dendrogram from mature peptide sequences of various GRs from plants.

cDNAs for GR have been isolated from cowpea (Ansel et al. 2006) *Arabidopsis thaliana* (Kubo et al. 1993) and soybean (Tang and Webb 1994), rice (Kaminaka et al. 1998), pea (Creissen et al. 1991, Mullineaux et al. 1996, Stevens et al. 1997), tobacco (Creissen and Mullineaux 1995), wheat (Lascano et al. 2001), *Brassica campestris* (Lee and Son 1998, Lee et al. 1998, 2002), *Anabaena* PCC7120 (Jiang et al. 1995),

Escherichia coli (Greer and Perham 1986), *Pseudomonas aeruginosa* (Perry et al. 1991), and yeast (Collinson and Dawes 1995). Creissen et al. (1995) showed that the pea *gor1* pre-sequence has a targeting capability for chloroplasts and mitochondria. Later, Stevens et al. (1997) reported the cloning and characterization of a putative cytosolic GR cDNA from pea, although they could not clearly show that the protein encoded by this cDNA was actually localized in the cytosol.

5.4 Isoforms of GR

Several studies have shown the increased GR activity in plants under varying environmental constraints. The increase in GR activity may be due to increased protein synthesis or by changed kinetic properties of the enzyme resulting in the formation of GR isoforms which will be involved in the detoxification of H_2O_2 produced during normal metabolic reactions as well as during stress conditions through enzymatic and nonenzymatic antioxidative pathways (Casano et al. 1999, Lascano et al. 1998, 1999). It would be of greater interest to understand changes in the level of a particular GR isoform than changes in total GR activity in stressed plants. In pea, GR activity increased up to two-fold under ozone stress, while there was no increase either in the protein content or in steady-state mRNA levels. However, the K_m of GR for GSSG was significantly decreased in all stress treatments, suggesting that the change in the composition of the isoform population may be more important than the transcriptional induction or a net increase in protein synthesis. GR isoforms have been reported in tobacco (Foyer et al. 1991), spinach (Guy and Carter 1984), red spruce (Hausladen and Alscher 1994a), eastern white pine (Anderson et al. 1990), scot pine (Wingsle and Karpinspi 1996, Wingsle 1989), mustard (Drumm–Harrel et al. 1989), pea (Edward et al. 1994), *Chlamydomonas reinharditii* (Serrano and Llobell 1993), and wheat (Lascano et al. 2001).These isoforms have been separated based on charge by using ion exchange chromatography, isoelectric focusing, and native gels. Two GR isoforms with different N-terminal amino acid sequences were isolated from red spruce subjected to freezing temperatures, providing evidence that more than one GR gene exists in this species (Haulsder and Alscher 1994a). However, not all isoforms were encoded by different genes; in pea, only two genes encoding eight isoforms have been identified (Edwards et al. 1990). Further, two genes encoding four isoforms have been identified in tobacco (Creissen and Mullineaux 1995). GR sequences from plants contain a C-terminal extension of unknown function of 22-26 amino acids, compared to the C-terminal of humans, *E. coli*, or *Pseudomonas aeroginosa* (Creissen et al. 1991, Kubo et al. 1993). It is possible that this domain is responsible for the appearance of different isoforms of GR in plants exposed to a variety of environmental stresses, and the following two possibilities have been known to induce the isoforms under varying growth regimes:

- **Posttranscriptional modifications:** alternative processing of nascent GR mRNAs leading to synthesis of different isoforms. In tobacco, isoforms have been reported resulting from the same gene by intron skipping during the splicing of nascent GR mRNAs (Creissen and Mullinaeux 1995). Changes in *gor* transcript stability during oxidative stress could also be an important factor.

- **Posttranslational modifications:** Occurrence of five isoforms in pea chloroplasts explains the possibility of posttranslational modifications (Madhamanchi et al. 1992). C-terminal regions of plant GRs are rich in lysine residues. When the enzyme is in the native state, these lysine residues are exposed and, depending, on the degree of carbamylation, different charge separable isoforms are observed (Mullineaux and Creissen 1997).

6 Significance of GR Activities during Abiotic and Biotic Stresses

GR and GSH play a crucial role in determining the tolerance of a plant during environmental and biotic stresses. In almost all the biological functions, GSH is oxidized to GSSG, which should be converted back to GSH in plant cells to perform normal physiological functions. Hence, rapid recycling of GSH is more essential than synthesis of GSH, which is a highly regulated and ATP-requiring process. During environmental cues, ABA levels are increased in plants which are used for signal transduction pathways specifically for scavenging the excess ROS produced during oxidative stress (Anderson et al. 1994, Ansel et al. 2006, Bueno et al. 1998, Gong et al. 1998, Hung and Kao 2003, Jiang and Zhang 2001, 2002). Increase in GR activity has been reported in maize seedlings and rice roots with increased ABA treatment (Jiang and Zhang 2001, 2002, 2003, Tsai and Kao 2004). H_2O_2 is known to be involved in ABA-induced GR activity in plant tissues (Anderson et al. 1994, Bueno et al. 1998, Gong et al. 1998, Jiang and Zhang 2001, 2002, Hung and Kao 2003, Tsai and Kao 2004). Kaminaka et al. (1998) have shown that cytosolic GR in rice was modulated by two ABA-responsive elements, and the expression of rice cytosolic GR gene was regulated by ABA-mediated signal transduction pathway. Most recently, Ansel et al. (2006) stated that the ABA-mediated signal transduction pathway is involved in the expression of GR genes during drought and desiccation. However, in pea there are no such ABA-responsive elements for GR cDNAs. It is now evident that GR responds differentially to different environmental and biotic stresses even within the same plant species (Lascano et al. 2001, Mullineaux and Creissen 1997, Pastori and Trippi 1992, Romero-Puertas et al. 2006). GR was partly inactive in red spruce during normal conditions and it was activated when the levels of oxidants increased in the cell (Hausladen and Alscher 1994b). In plants, the majority of the H_2O_2 generated during stress may be scavenged by the ascorbate-glutathione cycle, and GR has a central role for the H_2O_2 scavenging cycle in plants growing under different environmental and biotic stresses as discussed below.

6.1 *Drought Stress*

During drought stress, stomata are closed, and the availability of the CO_2 for carbon assimilation is decreased, which in turn results in the nonavailability of electron acceptor (NADP) leading to the generation of free radicals (Asada 2000, Reddy et al. 2004, Sairam and Saxena 2000, Smirnoff 1993). In addition to the antioxidant enzymes described already, an increase in GR activity during stomatal closure in response to water stress has been reported (Sairam et al. 1997/98). Therefore, a drought-tolerant genotype will not only be able to retain sufficient water but also possesses a highly active antioxidative system to guard against oxidative stress. As a part of this system, GR plays a key role, as evidenced by increased activities of the enzyme during stress (Ansel et al. 2006, Gamble and Burke 1984, Lascana et al. 2001, Romero-Puertas et al. 2006) Increased GR activity during drought was reported in various plant species, including barley (Smirnoff and Colombe 1988), maize (Jiang and Zhang 2002, Pastori and Trippi 1992), tobacco (Van Rensburg and Kruger 1994), wheat (Chen et al. 2004, Sairam et al. 1997/8, Selote and Chopra 2006), rice (Lin and Kao 2000a, Selote and Chopra 2004, Sharma and Dubay 2005, Srivalli et al. 2003), *Tortula ruralis* (Dhindsa 1991), and pea (Gogorcena et al. 1995).

6.2 *Salt Stress*

Salinity is one of the major abiotic stresses that adversely affect crop productivity. Salinity stress results in osmotic inhibition and ionic toxicity, which can affect the physiological and biochemical functions of the plant cell (Chinnusamy et al. 2005, Sumithra et al. 2006). Salinity stress exerts many symptoms similar to those observed under water deficit. Increased GR activities during salinity stress were reported in pea (Hernandez et al. 1993, 1995, 2000), cantaloupe (Fahmy et al. 1998), citrus (Gueta-Dahan et al. 1997), soybean (Comba et al. 1998), rice (Demiral and Turkan 2005, Dionisio-Sese and Tobita 1998, Lin and Kao 2000b, Tsai et al. 2005, Vaidyanathan et al. 2003), *Cicer arietinum* (Kukreja et al. 2005), tomato (Molina et al. 2002, Mittova et al. 2003, Shalata et al. 2001), *Arabidopsis thaliana* (Huang et al. 2005), wheat (Sairam et al. 2005), *Vigna radiate* (Sumithra et al. 2006), *Setaria italica* (Sreenivasulu et al. 2000), and *Helianthus annuus* (Davenport et al. 2003).

6.3 *Chilling Stress*

Chilling injury is a physiological disorder commonly found in crops indigenous to subtropical and temperate regions (El-Saht 1998). Chilling temperatures limit the activity of the Calvin cycle enzymes by disrupting the sulphahydryl groups, thus reducing the utilization of absorbed light energy for CO_2 assimilation (Leegood 1985, Payton et al. 2001, Wise 1995). Restricted carbon metabolism is the symptom of low temperature stress, which leads to an inadequate supply of natural electron acceptor, resulting in an overreduction of the photosystem reaction centers leading to increased ROS in the cell (Foyer et al. 1994, Hodgson and Raison 1991, Oquist and Huner 1993, Schoner and Krause 1990, Tsang et al. 1991). Acclimation to chilling temperatures generally leads to increased GSH contents and GR activities (Anderson et al. 1992, Logan et al. 1998). A positive correlation between resistance to chilling-induced photoinhibition and high GR activities has been shown in frenchbean seedlings (EL-Saht 1998), rice (Guo et al. 2006, Huang and Guo 2005, Kuk et al. 2003), maize (Hodges et al. 1996, 1997a, b), tomato (Walker and McKersie 1993), and eastern white pine (Anderson et al. 1992).

6.4 *Ozone Stress*

Ozone is a widespread phytotoxic secondary air pollutant that is formed by photochemical interactions of nitrogen oxides, hydrocarbons, and carbon monoxide in the atmosphere (Kangasjarvi et al. 1994). Current ambient ozone concentrations in certain geographical locations have been shown to cause not only significant yield losses in agricultural crops but also to be linked with forest decline. The ingress of ozone seems to depend on the number and size of stomata in different plant species (Conklin and Last 1995, Heath 1994, McLaughlin and Taylor 1981, Rich et al. 1970). Ozone effects on plants have been found dependent on leaf age, with young leaves hardly affected (Strohm et al. 1999). Once ozone enters the stomata, it reacts instantaneously with the apoplastic cell structures and generates secondary oxy-products like H_2O_2 and O_2^- which are known to alter plant metabolism by structurally modifying proteins and enhancing their susceptibility to proteolytic degradation (Pell and Dann 1991). During ozone stress, minimal protection would be at the stomatal level and is largely dependent on protective mechanisms operating at the cellular level (Rao 1992). O_3-induced alteration of antioxygenic activities and/or the gene expression of several antioxidant enzymes have been reported in various plants, indicating antioxidative systems linking to ozone tolerance (Creissen et al. 1994, Kangasjarvi et al. 1994, Willekens et al. 1994). Increase in GR activity during ozone stress was reported in pea (Edwards et al. 1994, Madamanchi et al. 1992, Mehlhorn et al. 1987), *Spinacia oleracea* (Tanaka et al. 1988), *Triticum aestivum* (Rao et al. 1995), and *Arabidopsis thaliana* (Kubo et al. 1995, Rao et al. 1996). Edwards et al. (1994) reported that O_3 exposure induced two isoforms of GR in pea plants with no significant changes either in the GR protein or in the mRNA transcripts encoding GR. Activities of GR and other enzymes of ascorbate-glutathione cycle were higher in sensitive rice cultivar (TN1) than in the tolerant cultivar (TNG67), suggesting that other physiological processes may confirm ozone tolerance in these cultivars. Pretreatment of TN1 with ABA significantly facilitated stomatal closure and prevented the ingress of ozone (Lin et al. 2001). However, the actual role of ABA in ozone detoxification in plants needs to be studied (Plochi et al. 2000). Stevens et al. (2000) showed the presence of two ABA-responsive elements in the rice cDNA sequence of GR. It is believed that increase in GSH synthesis or the GSH pool in cytosol or chloroplast may not be sufficient for ozone stress tolerance, and enhanced tolerance may be achieved by increasing antioxidant capacity in the apoplast, which is likely to be the first target of ozone stress (Polle 1998).

6.5 Herbicide Stress

Herbicides are also known to generate ROS such as superoxide radicals, hydrogen peroxide, hydroxyl radicals, and singlet oxygen within the chloroplast in light (Aono et al. 1995a, Foyer et al. 1995). The free radicals generated by the herbicides are effectively scavenged by efficient regeneration of ascorbate and glutathione (Chang and Kao 1997). These antioxidants are maintained in the cell by a rate-limiting enzyme GR. Increased GR activity to paraquat treatment was reported in pea (Donahue et al. 1997, Edwards et al. 1994) and maize (Pastori and Trippi 1992). The increased ratio of GSH to GSSG was maintained by GR during herbicide treatment, indicating the key role of GR in herbicide stress tolerance in plants (Romero-Puertas et al. 2006).

6.6 Heavy Metal Stress

Industrial mining and burning of fossil fuels have contaminated large areas of land in developed countries with high concentrations of heavy metals causing environmental pollution. Heavy metals, unlike organic pollutants, cannot be chemically degraded or biodegraded by microorganisms. Phytoremediation is an alternative cheap and efficient biological approach to deal with this problem (Peuke and Rennenberg 2005a, b, Salt et al. 1995). Some of the heavy metals are essential for the plant growth when they are present in normal levels; but when they are present in excess, they cause toxic effects on plant growth, ultimately resulting in decreased yields or death of the plant. For example, various crops are affected by Cr toxicity (Barcelo et al. 1996, Panda and Patra 1997, 1998, 2000, Poschenrieder et al. 1991), Zn toxicity (Bhattacharjee and Mukherjee 1994, Cakmak 2000, Fargasova 2001, Prasad et al. 1999), Ni toxicity (Ewais 1997, Parida et al. 2003, Vinterhatler and Vinterhaltler 2005), and certain other heavy metal toxicities (Gallego et al. 1996, Mazhoudi et al. 1997, Panda and Patra 2000, Somashekaraiah et al. 1992, Weckx and Clijster 1997). One of the mechanisms by which plants can withstand heavy metal toxicity is by maintaining high levels of phytochelatin or its precursor, GSH. Heavy metals are also known to induce free radicals in plants and, consequently, oxidative damage (Dietz et al. 1999). GR plays an important role in the detoxification of Cd-induced ROS, possibly via the ascorbate-glutathione cycle (Clijsters et al. 1999, Pereira et al. 2002). Cd inhibits the activities of several groups of enzymes involved in the Calvin cycle (Stevens et al. 1997), nitrogen metabolism (Boussama et al. 1999a, b, Kumar and Dubey 1999), sugar metabolism (Verma and Dubey 2001), and sulfate assimilation (Lee and Leustek 1999). Increased GR activity in the roots exposed to Cd was reported in plants, including *Phaseolus vulgaris* (Chaoui et al. 1997), potato (Stroinski et al. 1999), radish (Vitoria et al. 2001), soybean (Ferreira et al. 2002), sugarcane (Fornazier et al. 2002), and *Arabidopsis thaliana* (Skorzynska et al. 2003/4). Increased GR activity is also recorded in plants exposed to other heavy metals like Cu in mulberry plants (Tewari et al. 2006) and Fe stress in rice (Fang et al. 2001).

GSH is the known precursor for the formation of phytochelatins in plants (Steffens 1990). Phytochelatin synthase is constitutively expressed in plants and is mostly activated by heavy metal stress (Cobbett 2000, Cobbett and Goldsbrough 2002, Rauser 1990, Rea 2006, Rea et al. 2004, Zenk 1996). Phytochelatins belongs to the family of γ-glutamyl cysteine oligopeptides with glycine or other amino acids at the carboxy terminal end in which γ-Glu-Cys units are repeated 2-11 times. They are synthesized from GSH and its derivatives through phytochelatin synthase (Rea et al. 2004). Hence, GSH availability is the rate-limiting step for the formation of phytochelatin. These phytochelatins react with heavy metal ions, and the complex is sequestered to the vacuole for further processing or degradation, leading to less damage to cell membranes (Marrs 1996, Pilon-Smits et al. 2000). Therefore reduced GSH is required for all the reactions to sequester heavy metals, and these reduced GSH levels are to be constantly maintained by GR, which suggests the crucial role of GR in heavy metal detoxification.

6.7 Biotic Stress

Although plant-pathogen interactions are well documented, very little information is available on the involvement of ascorbate-glutathione cycle in plant defense against biotic stress (Baker and Orlandi 1995, De Gara et al. 2003, Kuzniak and Sklodowska 2005). Generation of ROS during pathogen infection occurs mainly extracellularly via membrane-bound NADPH oxidase, cell wall peroxidase, or amine/diamine/polyamine oxidase type enzymes (Bolwell et al. 2002, Wojtaszek 1997a, 1997b). Although pathogen-induced oxidative stress is generated extracellularly, organelles including chloroplast, mitochondria, and peroxisomes as well as their antioxidant machinery could also be targeted during pathogenesis. In tomato, increased GR activity is seen in almost all the major cell organelles, which are responsible for generation of ROS during stress conditions showing the importance of GR during biotic stress (Foyer and Noctor 2003, Kuzniak and Sklodowska 1999, 2001, 2004, 2005). Vanacker et al. (1998) have shown the presence of ascorbate-glutathione cycle enzymes in apoplasts of barley leaves and the role of GR in barley leaves was highly significant during biotic stress.

7 GR Transgenics

From the foregoing account, it is increasingly evident that GR is a key enzyme for scavenging the ROS and maintenance of redox potential of the cell during oxidative stress. With the availability of the GR cDNAs from various sources, it is possible to study the actual biochemical and molecular responses of GR during stress by overexpressing the gene in various heterologous systems as shown in Table 6.1. GR transgenic plants were first developed by Foyer et al. (1991) using *Escherichia coli gor* gene in *Nicotiana tobaccum* cv. samsun. The gene was targeted to the cytosolic compartment. Transgenic plants showed 2-10 times increased GR activity, but the cellular glutathione pools were not affected under both normal and methyl viologen treatments. Reduced ascorbate pools were also noticed in transgenic plants, suggesting that under natural conditions of growth, GR exists at levels above those required for maximal operation of ascorbate-glutathione pathway in tobacco. Later on, Aono et al. (1991, 1993) transformed bacterial *gor* gene and targeted to both cytosolic and chloroplastic compartments in *Nicotiana tabaccum* cv. SR1. Both the transgenic plants showed three-fold increase in GR activity, but there was no effect on the glutathione pool. The transgenic plants showed less sensitivity to paraquat in light but not to the ozone stress. Chloroplastic GR-transgenic plants showed lower susceptibility to SO_2 in light, suggesting that increasing GR activity both in chloroplasts and cytosol is effective in protecting cells from photooxidative damage. Comai et al. (1985) showed both the systems may cooperate in transporting the substrates and products across the envelope membranes of chloroplasts and suggested that increasing GR activity in one system may affect the same in the other system. Aono et al. (1995b) developed transgenic *Nicotiana tobaccum* cv. SR1 by cross-fertilization of transgenic *Nicotiana tabaccum* cv. SR1 with bacterial *gor* gene and transgenic *Nicotiana tabaccum* cv.SR1 with *oryza sativa Cu/Zn sod* gene. This is the first report in manipulating more than one gene for antioxidative enzymes to mitigate stress tolerance. Transgenic plants showed higher tolerance to paraquat upto 50 µM than either of the parent hybrids and concluded that comanipulating more than one gene could be an efficient strategy for lowering the levels of oxidative stress. Most recently Lee et al. (2007) showed the expression of three antioxidative genes Cu/Zn SOD, APX and DHAR in tobacco and showed that these plants are more tolerant to paraquat than the transgenic plants with Cu/Zn SOD or APX.

Transgenic *Nicotiana tobaccum* cv. SR1 with reduced GR activity by using spinach antisense GR cDNA, and the transgenic plants were more sensitive to paraquat than the control plants (Aono et al. 1995a). However, transgenic plants showed no significant differences from control plants under normal conditions, indicating that there might be sufficient residual GR activity in normal plants and the enzyme activity would become limiting under stress conditions. With the availability of GR cDNA from the plants, Broadbent et al. (1995) developed transgenic *Nicotiana tobaccum* cv samsun, using pea *gor1* gene, and they have targeted *gor1* to the chloroplast (GR36), cytosol (GR32), and into both chloroplast and mitochondrial compartments (GR46). All the transgenic plants showed increased GR activity, and the total glutathione pools increased upto 50%. These transgenic plants showed more tolerance to paraquat than the bacterial *gor*-transgenic plants (Aono et al. 1991, 1993). Transgenic GR46 plant with least elevation of GR activity showed decreased sensitivity to paraquat and some of the transgenic lines with greater than 4.5 times GR activity showed reduced sensitivity to ozone but not to the paraquat stress. It was suggested that there is no correlation between the degree of stress and GR activity, and the mechanisms of protection against ozone and paraquat are different, although both can be mitigated by elevated GR activity. Foyer et al. (1995) developed transgenic poplar for enhanced GR in chloroplast and cytosolic compartments. The transgenic poplar showed 1000 times higher GR activity and the one which is targeted to the cytosol showed 2-10 times higher GR activity. When compared to the foliar glutathione pools among all the transgenic plants, only chloroplastic GR transgenic plants showed increased foliar glutathione pool, suggesting that overexpression of GR in the chloroplast increases antioxidant capacity of the leaves. Strohm et al. (1999) showed transgenic poplar with enhanced GR in chloroplast and cytosolic compartments and the transgenics targeted to the chloroplast showed 150-200 times higher GR activity while those targeted to the cytosol showed five-fold higher GR activity. None of the transgenic plants showed tolerance to ozone toxicity and it was concluded that elevated foliar activities of glutathione synthetase or GR alone may not be sufficient for acute ozone stress. The sensitivity of the poplar leaves to ozone stress might be controlled by unknown factors closely related to the leaf development, rather than by foliar activities of GR or glutathione synthetase. Tyystjarvi et al. (1999) reported transgenic *Nicotiana tobaccum* cv. samsun using *Escherichia coli gor* gene, which was targeted to the cytosolic compartment. Transgenic plants showed 5-8 times increased GR activity and the transgenic plants didn't show enhanced tolerance for the repair of photoinhibitory damage. Pilon-Smith et al. (2000) transformed bacterial *gor* gene, targeted to both chloroplastic and cytosolic compartments in *Brassica juncea*. Transgenic plants showed 50-fold increase in GR activity when targeted to chloroplast and 2-fold increase when targeted to the cytosol. Only chloroplast GR transgenic plants showed enhanced Cd tolerance. Root glutathione levels were two times higher than the wild plants, and it was suggested that the difference of cadmium tolerance and accumulation may result from increased root glutathione levels. Transgenic cotton plants (*Gossypium hirsutum* cv coker 312) for enhanced chloroplastic GR have been reported using a gene from *Arabidopsis thaliana* (Kornyeyev et al. 2001, 2003, Logan et al. 2003, Payton et al. 2001). GR-Transgenics reported by Payton et al. (2001) showed enhanced tolerance to moderate chilling at high light intensity by protecting cotton photosynthesis. Kornyeyev et al. (2001, 2003) showed transgenics with decreased chilling-induced photoinhibition of PSII through enhanced photochemical light utilization and by increasing rates of photochemistry. Kouril et al. (2003) showed transgenic rice plants (*Oryza sativa* cv Daeribbyeo) with overproducing cytosolic GR using a gene from *Brassica campestris*. These transgenic plants showed increased tolerance upto 35 °C in the presence of low concentration of methyl viologen (10 µM) but failed to show tolerance at higher concentrations of methyl viologen (50 µM). However, at lower temperatures (25 °C) transgenics showed resistance to methyl viologen upto 50 µM. Transgenic tobacco plants developed by Lederer and Boger (2003) with

Table 6.1 Chronology of tran sgenic approach for modifying glutathione reductase activity in plants

S.No.	Transgenic Plant	GR Gene Source	Localization in Cell	Response in Transgenic Plants	Reference
1	*N.tobaccum* cv Samsun	*Escherichia coli* (*gor*)	Cytosol	• 2-10 times increased GR activity. • No effect on cellular glutathione pool under both normal and methyl viologen treatment. • Pool of ascorbate increased.	Foyer et al. 1991
2	*N.tobaccum* cv SR1	*Escherichia coli* (*gor*)	Cytosol	• Increased GR activity. • No effect on glutathione pool. • Lower susceptibility to paraquat. • No resistance to ozone stress.	Aono et al. 1991
3	*N.tobaccum* cv SR1	*Escherichia coli* (*gor*)	Chloroplast	• 3-fold increase in GR activity. • No effect on glutathione pool. • Lower susceptibility to paraquat and SO_2 in light. • No resistance to ozone stress.	Aono et al. 1993
4.	*N.tobaccum* cv Samsun	*Pisum sativum gor1*	pGR32 in cytosol pGR 36 in chloroplasts PGR46 both in chloroplast and mitochondria	• 4.5-fold increase in GR activity. • Increase in total glutathione pool. • More tolerance to paraquat. • Lines with greater than 4.5-fold GR activity showed less sensitivity to ozone stress.	Broadbent et al. 1995
5.	*N.tobaccum* cv SR1	*Escherichia coli* (*gor*)	Cytosol	• Enhanced tolerance to paraquat upto 50 μM.	Aono et al. 1995b
6.	*N.tobaccum* cv SR1	Spinach GR	Chloroplast	• Suppression of GR activity. • Enhanced sensitivity to paraquat	Aono et al. 1995a
7.	Poplar hybrid	*Escherichia coli* (*gor*)	Both in cytosol and chloroplast.	• Chloroplasts showed 1000 times' higher GR activity. • *Gor* in cytosol showed 2-10 times higher GR activity. • Transgenic plants with increased GR showed resistance to methyl viologen	Foyer et al. 1995
8	Poplar hybrid	*Escherichia coli* (*gor*)	Both in cytosol and chloroplast.	• GR in the cytosol showed 5-fold increase • GR in the chloroplasts showed 150-200 fold increase. • Ozone tolerance is controlled by leaf development rather than exclusively by foliar activities of GR	Strohm et al. 1999

9.	*N.tobaccum* cv Samsun	*Escherichia coli (gor)*	Cytosol	• 5-8 times increased GR activity. • Didn't show any specific enhancement of the mechanism repairing photoinhibitory damage under high light intensity.	Tyystjarvi et al. 1999
10.	*N.tobaccum*	*Pisum sativum (gor2)*	Cytosol	• Increased GR activity in the cytosol. • Change in the kinetic properties of the enzyme was observed when pea leaves were subjected to chilling, ozone and paraquat treatments.	Stevens et al. 2000
11.	*Brassica juncea*	*Escherichia coli (gor)*	Both in cytosol and chloroplast.	• GR in cytosol showed 2-fold increase and chloroplast showed 50- fold increase. • Chloroplastic GR transgenics showed enhanced tolerance to cadmium (100 µM)	Pilon Smith et al. 2000
12.	*Gossypium hirsutum* cv coker 312	*Arabidopsis thaliana* (GR)	Chloroplast	• Increased GR activity • Enhanced photochemical light utilization and decreased chilling-induced photoinhibition of PS II.	Kornyeyev et al. 2001
13.	*Gossypium hirsutum* cv coker 312	*Arabidopsis thaliana* (GR)	Chloroplast	• Increased tolerance to moderate chilling at high light intensity by protecting cotton photosynthesis.	Payton et al. 2001
14	Poplar hybrid	*Escherichia coli gor*	Both in cytosol and chloroplast.	• GR in the cytosol showed 5-fold increase • GR in the chloroplast showed 150-200 fold increase.	Strohm et al. 2002
15	*Oryza sativa* cv. Daeribbyeo	*Brassica campestris (BcGR1)*	Cytosol	• Increased tolerance upto 35 °C in the presence of low concentration of methyl viologen (10 µM) but failed to show tolerance at 50 µM concentration.	Kouril et al. 2003
16.	*Gossypium hirsutum* cv coker 312	*Arabidopsis thaliana* (GR)	Chloroplast	• Decreased chilling-induced photoinhibition by increasing rates of photochemistry.	Kornyeyev et al. 2003
17	*Nicotiana tabacum* cv BelW3	*Escherichia coli (gor)*	Cytosol	• Increased GR activity • Increased tolerance to paraquat and H_2O_2.	Lederer and Boger et al. 2003

bacterial GR suffered lipid peroxidation under moderate light intensities in contrast to the wild-type tobacco. However, these transgenic plants were more resistant than wild-type plants towards oxidative stress induced by paraquat or hydrogen peroxide (Lederer and Boger 2003).

8 A Central Role for GR in Plant Oxidative Stress Tolerance

GR and GSH play an important role in maintaining reduced state of the ascorbate during different constrained environmental regimes. GSSG is very important in increased turnover involving regeneration of the GSH rather than the synthesis of GSH, as GSH synthesis is tightly controlled in situ. Foyer et al. (1995) suggested that GSH-mediated ascorbate regeneration via ascorbate-glutathione cycle makes a significant contribution to the size and redox state of the ascorbate pool. Romero-Puertas et al. (2006) also indicated that GR is the rate-limiting enzyme in the ascorbate-glutathione cycle. The three redox couples in the cell, including GSH/GSSG, ascorbate/dehydroascorbate, and NADPH/NADP are critically influenced by the activity of GR during stress as depicted in Fig. 6.6. GR plays an important role in maintaining the cellular ascorbate and glutathione pools, which was evidenced in transgenic plants by overexpressing GR. Therefore, more rapid recycling of the reduced ascorbate and GSH pools would be beneficial in combating the sequential production of harmful free radicals (Foyer et al. 1995). Transgenic research suggests that plant-encoded GR affords greater protection than the *Escherichia coli*-derived enzyme. This may be due to the fact that bacterial-derived enzyme is rapidly inactivated by NADPH in the absence of glutathione, whereas plant GR seems to be relatively insensitive to the high NADPH concentrations that can occur in illuminated chloroplasts. It may be an adaptive characteristic during the course of evolution for the plant GR to withstand the high concentrations of NADPH produced in the chloroplasts. When the bacterial *gor* gene was expressed in cytosol and in the chloroplast of poplar hybrids and in *Brassica juncea*, the expression of GR was very high in chloroplasts, which indicates that bacterial GR is more stable in the chloroplasts compared to the cytosol (Foyer et al. 1995, Pilon-Smith et al. 2000, Strohm et al. 1999).

It was very interesting to note the contrasting observations of Strohm et al. (1999) and Foyer et al. (1995) with the same poplar hybrid which was transformed with the same construct and the transgenics showed nearly five times variation in GR activity. Foyer et al. (1995) suggest that high chloroplastic GR expressers are more robust because chloroplasts are the primary sources for the generation of free radicals, which should be instantly removed. In contrast, Creissen et al. (1996) stated that overexpression of the cytosolic isoform (*gor2*) was proved to be greater benefit for oxidative stress tolerance. On the other hand Broadbent et al. (1995) reported that there was no significant correlation between the degree of stress and GR activity. They observed that the GR46 line with low GR activity showed some tolerance to paraquat toxicity. These differences in GR activity in different lines developed with the same construct might be due to position effect, cosuppression,

T-DNA insertion mutations, gene dosage effect, or somaclonal variation. Cotton GR transgenic plants with increased GR activity have shown increased resistance to short-term chilling stress (Kornyeyev et al. 2001, 2003, Logan et al. 2003, Payton et al. 2001). GR activity in wild-type cotton doubled over the course of the chilling exposure. Even though GR transgenic plants exhibited higher GR activity, there is no difference between the transgenic and wild-type plants when exposed to long-term chilling, indicating the need for other enzymes to combat the long-term chilling stress (Logan et al. 2003). Transcript levels of plastidic violoxanthin deepoxidase and cytosolic Cu/Zn *sod* were strongly reduced in GR transgenic *Nicotiana tabacum* cv. BelW3 plants developed by Lederer and Boger et al. (2003) as compared to the wild-type plants, which were more sensitive to high light intensities. It is believed that violoxathin deepoxidase is a predominant protective enzyme in the xanthophyll-mediated energy dissipation which protects the photosynthetic machinery. However, these transgenic plants were more resistant than the wild types towards oxidative stress induced by paraquat or hydrogen peroxide.

9 Conclusions and Perspectives

Biochemical and molecular studies have shown that GR is an important enzyme for scavenging the ROS, which are continuously generated in different compartments during environmental and biotic stresses. Although plant stresses are of different origin involving different signal transduction pathways, there may be a common mechanism of antioxidative scavenging pathway existing in the cell to detoxify the ROS involving one or more critical enzymes. Increase in GR activity during stress should be achieved not only by increased GR protein synthesis but also through changes in isoform population of GR. It is also important to characterize which isoform is crucial for a particular stress and how the expression of that particular isoform is regulated. We believe that there may be a threshold value of GR in plants beyond which there should be a requirement of other enzyme(s) or antioxidants for scavenging the ROS. Transgenic plants with more than one gene for antioxidant pathway showed enhanced tolerance compared to the transgenic plants containing a single gene product. Enhanced GR activity along with other antioxidative enzymes showed better tolerance to environmental stresses, suggesting the importance of coregulation of other enzymes along with the GR in limiting the concentrations of ROS.

Acknowledgements We thank the Department of Biotechnology, Government of India, for financial support through grant #SP/SO/PS-27/05. ASVCR acknowledges the financial support from the Indian Council of Medical Research, New Delhi, India.

References

Agarwal S (2007) Increased antioxidant activity in *Cassia* seedlings under UV-B radiation. Biol Plant 51:157-60

Alscher RG (1989) Biosynthesis and antioxidant function of glutathione in plants. Physiol Plant 77:457-64

Anderson JV, Hess JL, Chevone BI (1990) Purification, characterization and immunological properties for two isoforms of glutathione reductase from eastern white pine needles. Plant Physiol 94:1402-9

Anderson JV, Chevone BI, Hess JL (1992) Seasonal variation in the antioxidant system of eastern white pine needles: evidence for thermal dependence. Plant Physiol 98:501-8

Anderson MD, Prasad TK, Martin BA, Stewart CR (1994) Differential gene expression in chilling-acclimated maize seedlings and evidence for the involvement of abscisic acid in chilling tolerance. Plant Physiol 105:331-9

Ansel DC, Franklin MLT, De Carvalho MHC, Lameta ADA, Fodil YZ (2006) Glutathione reductase in leaves of cowpea: cloning of two cDNAs, expression and enzymatic activity under progressive drought stress desiccation and abscisic acid treatment. Ann Bot 98:1279–87

Aono M, Kubo A, Saji H, Natori T, Tanaka K, Kondo N (1991) Resistance to active oxygen toxicity of transgenic *Nicotiana tabacum* that expresses the gene for glutathione reductase from *Escherichia coli*. Plant Cell Physiol 32:691-7

Aono M, Kubo A, Saji H, Natori T, Tanaka K, Kondo N (1993) Enhanced tolerance to photooxidative stress of transgenic *Nicotiana tabacum* with high chloroplastic glutathione reductase activity. Plant Cell Physiol 34:129-35

Aono M, Saji H, Fujiyama K, Sugita M, Kondo N, Tanaka K (1995a) Decrease in activity of glutathione reductase enhances paraquat sensitivity in transgenic *Nicotiana tabacum*. Plant Physiol 107:645-8

Aono M, Saji H, Sakamoto A, Tanaka K, Kondo N, Tanaka K (1995b) Paraquat tolerance of transgenic Nicotiana *tabacum* with enhanced activities of glutathione reductase and superoxide dismutase. Plant Cell Physiol 36:1687-91

Asada K (1994) Production and action of active oxygen species in photosynthetic tissues. In: Foyer CH, Mullineaux PM (eds) Causes of photooxidative stress and amelioration of defense systems in plants, CRC Press, Boca Raton, pp. 77-104

Asada K (2000) The water-water cycle as alternative photon and electron sinks. Phil Trans R Soc Lond B 355:1419-31

Asada K (2006) Production and scavenging of reactive oxygen species in chloroplasts and their functions. Plant Physiol 141:391-6

Baker CJ, Orlandi EW (1995) Active oxygen in plant pathogenesis. Annu Rev Phytopathol 33:299-321

Ball L, Accotto GP, Bechtold U, Creissen GP, Funck D, Jimenez A, Kular B, Leyland N, Mejia-Carranza J, Reynolds H, Karpinski S, Mullineaux PM (2004) Evidence for a direct link between glutathione biosynthesis and stress defense gene expression in *Arabidopsis*. Plant Cell 16:2448-62

Barcelo J, Poschenrieder C, Vazquez MD, Gunse B (1996) Aluminium phytotoxicity: a challenge for plant scientists. Fertil Res 43:217-23

Bhattacharjee S, Mukherjee AK (1994) Influence of cadmium and lead on physiological and biochemical responses of *Vigna unguiculata* (L.) Walp. seedlings. I. Germination behaviour, total protein and proline content and protease activity. Poll Res 13:269-77

Bielawski W, Joy KW (1986) Properties of glutathione reductase from chloroplasts and roots of pea. Phytochemistry 25:2261-5

Bolwell GP, Bindschedler LV, Blee KA, Butt VS, Davies DR, Gardner SL, Gerrish C, Minibayeva F (2002) The apoplastic oxidative burst in response to biotic stress in plants: a tree component system. J Exp Bot 53:1367-76

Boussama N, Ouariti O, Ghorbal MH (1999a) Changes in growth and nitrogen assimilation in barley seedlings under cadmium stress. J Plant Nutr 22:731-52

Boussama N, Ouariti O, Suzuki A, Ghorbal MH (1999b) Cd stress on nitrogen assimilation. J Plant Physiol 155:310-17

Broadbent P, Creissen GP, Kular B, Wellburn AR, Mullineaux PM (1995) Oxidative stress responses in transgenic tobacco containing altered levels of glutathione reductase activity. Plant J 8:247-55

Bueno P, Piqueras A, Kurepa J, Savoure A, Verbruggen N, Van Montagu M, Inze D (1998) Expression of antioxidant enzymes in response to abscisic acid and high osmoticum in tobacco BY-2 cell cultures. Plant Sci 138:27-34

Cakmak I (2000) Possible roles of zinc in protecting plant cells from damage by reactive oxygen species. New Phytol 146:185-205

Campbell WH (1995) Enzyme mechanism examples. Michigan Technological University. BL/CH401 Lecture 16B

Casano LM, Martin M, Zapata JM, Sabater B (1999) Leaf age and paraquat concentration-dependent effects on the levels of enzymes protecting against oxidative stress. Plant Sci 149:13-22

Chaitanya KV, Sundar D, Masilamani S, Reddy AR (2002) Variation in heat stress-induced antioxidant enzyme activities among three mulberry cultivars. Plant Growth Regul 36:175–80

Chang CJ, Kao CH (1997) Paraquat toxicity is reduced by metal chelators in rice leaves. Physiol Plant 101:471-6

Chaoui A, Mazhoudi S, Ghorbal MH, El Ferjani E (1997) Cadmium and zinc induction of lipid peroxidation and effects on antioxidant enzyme activities in bean (*Phaseolus vulgaris* L). Plant Sci 127:139-47

Chen KM, Gong HJ, Chen GC, Wang SM, Zhang CL (2004) Gradual drought under field conditions influences the glutathione metabolism, redox balance and energy supply in spring wheat. J Plant Growth Regul 23:20-8

Chew O, Rudhe C, Glaser E, Whelan J (2003a) Characterization of the targeting signal of dual-targeted pea glutathione reductase. Plant Mol Biol 53:341-56

Chew O, Whelan J, Millar H (2003b) Molecular definition of the ascorbate glutathione cycle in *Arabidopsis* mitochondria reveals dual targeting of antioxidant defenses in plants. J Biol Chem 278:46869-77

Chinnusamy V, Jagendorf A, Zhu JK (2005) Understanding and improving salt tolerance in plants. Crop Sci 45:437-48

Clijsters H, Cuypers A, Vangronsveld J (1999) Physiological responses to heavy metals in plants: defense against oxidative stress. Z Naturforsch 54:730-4

Cobbett CS (2000) Phytochelatins and their roles in heavy metal detoxification. Plant Physiol 123:825-32

Cobbett C, Goldsbrough P (2002) Phytochelatins and metallothioneins: roles in heavy metal detoxification and homeostasis. Annu Rev Plant Biol 53:159-82

Collison LP, Dawes IW (1995) Isolation, characterization and overexpression of the yeast gene, *GLR1*, encoding glutathione reductase. Gene 156:123-7

Comai I, Facciotti D, Hiatt WR, Thompson G, Rose RE, Stalker DM (1985) Expression in plants of a mutant *aroA* gene from *Salmonella typhimurium* confers tolerance to glyphosate. Nature 317:741-4

Comba ME, Benavides MP, Tomaro ML (1998) Effect of salt stress on antioxidant defense system in soybean root nodules. Aus J Plant Physiol 25:665-71

Conklin PL, Last RL (1995) Differential accumulation of antioxidant mRNAs in *Arabidopsis thaliana* exposed to ozone. Plant Physiol 109:203-12

Conn EE, Vennesland B (1951) Glutathione reductase of wheat germ. J Biol Chem 192:17-28

Connell JP, Mullet JE (1986) Pea chloroplast glutathione reductase: purification and characterization. Plant Physiol 82:351-6

Creissen GP, Mullineaux PM (1995) Cloning and characterization of glutathione reductase cDNAs and identification of two genes encoding the tobacco enzyme. Planta 197:422-5

Creissen GP, Edwards EA, Enard C, Wellburn A, Mullineaux PM (1991) Molecular characterization of glutathione reductase cDNAs from pea *(Pisum sativum* L.). Plant J 2:129-31

Creissen GP, Edwards EA, Mullineaux PM (1994) Glutathione reductase and ascorbate peroxidase. In: Foyer CH, PM Mullineaux PM (eds). Causes of photooxidative stress and amelioration of defense systems in plants, CRC Press, Boca Raton, pp. 343-64

Creissen GP, Reynolds H, Xue Y, Mullineaux PM (1995) Simultaneous targeting of pea glutathione reductase and of a bacterial fusion protein to chloroplasts and mitochondria in transgenic tobacco. Plant J 8:167-75

Creissen GP, Broadbent P, Stevens R, Wellburn AR, Mullineaux PM (1996) Manipulation of glutathione metabolism in transgenic plants. Biochem Soc Trans 24:465-9

Creissen GP, Firmin J, Fryer M, Kular B, Leyland N, Reynolds H, Pastori G, Wellburn F, Baker N, Wellburn A, Mullineaux PM (1999) Elevated glutathione biosynthetic capacity in the chloroplasts of transgenic tobacco plants paradoxically causes increased oxidative stress. Plant Cell 11:1277-91

Davenport SB, Gallego SM, Benavides MP, Tomaro ML (2003) Behaviour of antioxidant defense system in the adaptive response to salt stress in *Helianthus annuus* L. cells. Plant Growth Regul 40:81-8

Davletova S, Rizhsky L, Liang H, Shenggiang Z, Oliver DJ, Coutu J, Shulaev V, Schlauch K, Mittler R (2005) Cytosolic ascorbate peroxidase 1 is a central component of the reactive oxygen gene network of *Arabidopsis*. Plant Cell 17:268-81

De Gara L, de Pinto MC, Tommasi F (2003) The antioxidant systems vis-a-vis reactive oxygen species during plant-pathogen interaction. Plant Physiol Biochem 41:863-70

del Rio LA, Corpas FJ, Sandalio LM, Palma Jose M, Gomez M, Barroso JB (2002) Reactive oxygen species, antioxidant systems and nitric oxide in peroxisomes. J Exp Bot 53:1255-72

Demiral T, Turkan (2005) Comparative lipid peroxidation, antioxidant defense systems and proline content in roots of two rice cultivars differing in salt tolerance. Environ Exp Bot 53:247-57

Desikan R, Mackerness AHS, Hancock JT, Neill SJ (2001) Regulation of the *Arabidopsis* transcriptome by oxidative stress. Plant Physiol 127:159-72

Dhindsa RS (1991) Drought stress, enzymes of glutathione metabolism, oxidative injury and protein synthesis in *Tortula ruralis*. Plant Physiol 95:648-51

Dietz KJ, Baier M, Kramer U (1999) Free radicals and reactive oxygen species as mediators of heavy metal toxicity in plants. In: Prasad MNV, Hagemeyer J (eds): Heavy metal stress in plants: from molecules to ecosystems, Springer-Verlag, Berlin, pp. 73-9

Dionisio-Sese ML, Tobita S (1998) Antioxidant responses of rice seedlings to salinity stress. Plant Sci 135:1-9

Donahue JL, Okpodu CM, Cramer CL, Grabau EA, Alscher RG (1997) Responses of antioxidants to paraquat in pea leaves: relationships to resistance. Plant Physiol 113:249-57

Droux M (2004) Sulfur assimilation and the role of sulfur in plant metabolism: a survey. Photosynth Res 79:331-48

Drumm-Herrel H, Gerhausser U, Mohr H (1989) Differential regulation by phytochrome of the appearance of plastidic and cytoplasmic isoforms of glutathione reductase in mustard (*Sinapsis alba* L.) cotyledons. Planta 178:103-9

Edwards EA, Rawsthorne S, Mullineaux PM (1990) Subcellular distribution of multiple forms of glutathione reductase in leaves of pea (*Pisum sativum* L.). Planta 180:278-84

Edwards EA, Enard C, Creissen GP, Mullineaux PM (1994) Synthesis and properties of glutathione reductase in stressed peas. Planta 192:137-43

El-Saht HM (1998) Responses to chilling stress in French bean seedlings: antioxidant compounds. Biol Plant 41:395-402

Ewais EA (1997) Effects of cadmium, nickel and lead on growth, chlorophyll content and proteins of weeds. Biol Plant 39:403-10

Fadzilla NM, Robert P, Finch RP, Burdon RH (1997) Salinity, oxidative stress and antioxidant response in shoot cultures of rice. J Exp Bot 48:325-31

Fahmy AS, Mohamed TM, Mohamed SA, Saker MM (1998) Effect of salt stress on antioxidant activities in cell suspension cultures of cantaloupe (*Cucumis melo*). Egyptian J Physiol Sci 22:315-26

Fang WC, Wang JW, Lin CC, Kao CH (2001) Iron induction of lipid peroxidation and effect on antioxidative enzyme activities in rice leaves. Plant Growth Regul 35:75-80

Fargasova A (2001) Phytotoxic effects of Cd, Zn, Pb, Cu and Fe on *Sinapsis alba* L. seedlings and their accumulation in roots and shoots. Biol Plant 44:471-3

Ferreira RR, Fornazier RF, Vitoria AP, Lea PJ, Azevedo RA (2002) Changes in antioxidant enzyme activities in soybean under cadmium stress. J Plant Nutr 25:327-42

Fornazier RF, Ferreira RR, Vitoria AP, Molina SMG, Lea PJ, Azevedo RA (2002) Effects of cadmium on antioxidant enzyme activities in sugarcane. Biol Plant 41:91-7

Foyer CH, Halliwell (1976) The presence of glutathione and glutathione reductase in chlorophasts: A proposed role in ascorbate metabolism. Planta 133:21–5

Foyer CH, Noctor G (2003) Redox sensing and signalling associated with reactive oxygen in chloroplasts, peroxisomes and mitochondria. Physiol Plant 119:355-64

Foyer CH, Noctor G (2005) Redox homeostasis and antioxidant signalling: a metabolic interface between stress perception and physiological responses. Plant Cell 17:1866-75

Foyer CH, Dujardyn M, Lemoine Y (1989) Responses of photosynthesis and the xanthophylls and ascorbate-glutathione cycles to changes in irradiance, photoinhibition and recovery. Plant Physiol Biochem 27:751-60

Foyer CH, Lelandais M, Galap C, Kunert KJ (1991) Effects of elevated cytosolic glutathione reductase activity on the cellular glutathione pool and photosynthesis in leaves under normal and stress conditions. Plant Physiol 97:863-72

Foyer CH, Descourvieres P, Kunert KI (1994) Protection against oxygen radicals: an important defense mechanism studied in transgenic plants. Plant Cell Physiol 17:507-23

Foyer CH, Souriau N, Perret S, Lelandais M, Kunert KJ, Pruvost C, Jouanin L (1995) Overexpression of glutathione reductase but not glutathione synthetase leads to increases in antioxidant capacity and resistance to photoinhibition in poplar trees. Plant Physiol 109:1047-57

Foyer CH, Theodoulou FL, Delrot S (2001) The functions of inter and intracellular glutathione transport systems in plants. Trends Plant Sci 6:486-92

Gachomo WE, Shonukan OO, Kotchoni OS (2003) The molecular initiation and subsequent acquisition of disease resistance in plants. Afr J Biotechnol 2:26-32

Gallego SM, Benavides MP, Tomaro ML (1996) Effect of heavy metal ion excess on sunflower leaves: evidence for involvement of oxidative stress. Plant Sci 121:151-9

Gamble PE, Burke JJ (1984) Effect of water stress on the chloroplastic antioxidant system. I. Alterations in glutathione reductase activity. Plant Physiol 76:615-21

Gaullier JM, Lafontant P, Valla A, Bazin M, Giraud M, Santus R (1994) Glutathione peroxidase and glutathione reductase activities towards glutathione derived antioxidants. Biochem Biophys Res Commun 203:1668-74

Ghisla S, Massey V (1989) Mechanisms of flavoprotein-catalyzed reactions Eur J Biochem 181:1-17

Gogorcena Y, Iturbe-Ormaetxe I, Escuredo PR, Becana M (1995) Antioxidant defenses against activated oxygen in pea nodules subjected to water stress. Plant Physiol 108:753-9

Gong M, Li YJ, Chen SZ (1998) Abscisic acid-induced the tolerance in maize seedlings is mediated by calcium and associated with antioxidant systems. J Plant Physiol 153:488-96

Greer S, Perham RN (1986) Glutathione reductase from *Escherichia coli*: cloning and sequence analysis of the gene and relationship to other flavoprotein disulfide oxidoreductases. Biochemistry 25:2736-42

Gueta-Dahan Y, Yaniv Z, Zilinkas BA, Ben-Hayyim G (1997) Salt and oxidative stress: similar and specific responses and their relation to salt tolerance in citrus. Planta 203:460- 9

Guy CL, Carter JV (1984) Characterization of partially purified glutathione reductase from cold-hardened and non hardened spinach leaf tissue. Cryobiology 21:454-64

Halliwell B, Foyer CH (1978) Properties and physiological function of a glutathione reductase purified from spinach leaves by affinity chromatography. Planta 139:9-17

Hartmann T, Honicke P, Wirtz M, Hell R, Rennenberg H, Kopriva S (2004) Regulation of sulphate assimilation by glutathione in poplars (*Populus tremula* X *P. alba*) of wild type and overexpressing γ-glutamylcysteine synthetase in the cytosol. J Exp Bot 55:837-45

Hausladen A, Alscher RG (1993) Glutathione. In: Alsher R, Hess J (eds): Antioxidants in higher plants, CRC Press, Boca Raton, pp. 1-30

Hausladen A, Alscher RG (1994a) Purification and characterization of glutathione reductase isozymes specific for the state of cold hardiness of red spruce. Plant Physiol 105:205-13

Hausladen A, Alscher RG (1994b) Cold-hardiness-specific glutathione reductase isozymes in red spruce. Thermal dependence of kinetic parameters and possible regulatory mechanisms. Plant Physiol 105:215-23

Herbinger K, Tausz M, Wonisch A, Soja G, Sorger A, Grill D (2002) Complex interactive effects of drought and ozone stress on the antioxidant defense systems of two wheat cultivars. Plant Physiol Biochem 40:691-9

Hernandez JA, Corpas FJ, Gomez M, Del Rio LA, Sevilla F (1993) Salt-induced oxidative stress mediated by activated oxygen species in pea leaf mitochondria. Physiol Plant 89:103-10

Hernandez JA, Olmos E, Corpas FJ, Sevilla F, del Rio LA (1995) Salt-induced oxidative stress in chloroplasts of pea plants. Plant Sci 105:151-67

Hernandez JA, Jimenez A, Mullineaux PM, Sevilla F (2000) Tolerance of pea (*Pisum sativum* L.) to long term salt stress is associated with induction of antioxidant defenses. Plant Cell Environ 23:853-62

Hodges DM, Andrews CJ, Johnson DA, Haniton RI (1996) Antioxidant compound responses to chilling stress in differentially sensitive inbred maize lines. Physiol Plant 98:685-92

Hodges DM, Andrews CJ, Johnson DA, Hamilton RI (1997a) Antioxidant enzyme responses to chilling stress in differentially sensitive inbred maize lines. J Exp Bot 48:1105-13

Hodges DM, Andrews CJ, Johnson DA, Hamilton RI (1997b) Antioxidant enzyme and compound responses to chilling stress and their combining abilities in differentially sensitive maize hybrids. Crop Sci 37:857-63

Hodgson RAJ, Raison JK (1991) Superoxide production by thylakoids during chilling and its implication the susceptibility of plants to chilling induced photoinhibition. Planta 183:222-8

Huang M, Guo Z (2005) Responses of antioxidative system to chilling stress in two rice cultivars differing in sensitivity. Biol Plant 49:81-4

Huang C, He W, Guo J, Chang X, Su P, Zhang L (2005) Increased sensitivity to salt stress in an ascorbate-deficient *Arabidopsis* mutant J Exp Bot 56:3041-9

Hung KT, Kao CH (2003) Nitric oxide counteracts the senescence of rice leaves induced by abscisic acid. J Plant Physiol 160:871-9

Ida S, Morita Y (1971) Studies on respiratory enzymes in rice kernel: enzymatic properties and physical and chemical characterization of glutathione reductase from rice embryos. Agric Biol Chem 35:1550-7

Innocenti G, Pucciariello C, Gleuher ML, Hopkins J, Stefano MD, Delledonne M ú Puppo A, Baudouin E, Frendo P (2006) Glutathione synthesis is regulated by nitric oxide in *Medicago truncatula* roots. Planta doi 10.1007/s00425-006-0461-3 online

Jiang F, Hellman U, Sroga GE, Bergman B, Mannervik B (1995) Cloning, sequencing, and regulation of the glutathione reductase gene from the cyanobacterium *Anabaena* PCC 7120. J Biol Chem 270:22882-9

Jiang M, Zhang J (2001) Effect of abscisic acid on active oxygen species, antioxidative defense system and oxidative damage in leaves of maize seedlings. Plant Cell Physiol 42:1265-73

Jiang M, Zhang J (2002) Water stress-induced abscisic acid accumulation triggers the increased generation of reactive oxygen species and up-regulates the activities of antioxidant enzymes in maize leaves. J Exp Bot 53:2401-10

Jiang M, Zhang J (2003) Cross-talk between calcium and reactive oxygen species originated from NADPH oxidase in abscisic acid-induced antioxidant defense in leaves of maize seedlings. Plant Cell Environ 26:929-39

Jimenez A, Hernandez J, del Rio L, Sevilla F (1997) Evidence for the presence of the ascorbate-glutathione cycle in mitochondria and peroxisomes of pea leaves. Plant Physiol 114:275-84

Jimenez A, Hernandez JA, Pastori G, del Rio LA, Sevilla F (1998) Role of the ascorbate-glutathione cycle of mitochondria and peroxisomes in the senescence of pea leaves. Plant Physiol 118:1327-35

Kalt-Torres W, Burke JJ, Anderson JM (1984) Chloroplast glutathione reductase: purification and properties. Physiol Plant 61:271-8

Kaminaka H, Morita S, Nakajima M, Masumura T, Tanaka K (1998) Gene cloning and expression of cytosolic glutathione reductase in rice (*Oryza sativa* L.). Plant Cell Physiol 39:1269-80

Kangasjarvi J, Talvinen J, Utriainen M, Karjalainen R (1994) Plant defense systems induced by ozone. Plant Cell Environ 17:783-94

Kocsy G, von Ballmoons P, Suter M, Ruegsegger A, Galli U, Szalai G, Galiba G, Brunold C (2000) Inhibition of glutathione synthesis reduces chilling tolerance in maize. Planta 211:528-36

Kocsy G, von Ballmoons P, Ruegsegger A, Szalai G, Galiba G, Brunold C (2001) Increasing the glutathione content in a chilling-sensitive maize genotype using safeners increased protection against chilling-induced injury. Plant Physiol 127:1147-56

Kopriva S, Koprivova A (2005) Sulfate assimilation and glutathione synthesis in C4 plants. Photosynth Res 86:363-72

Kornyeyev D, Logan BA, Payton P, Allen RD, Holaday AS (2001) Enhanced photochemical light utilization and decreased chilling induced photoinhibition of photosystem II in cotton over expressing genes encoding chloroplast targeted antioxidant enzymes. Physiol Plant 113:323-31

Kornyeyev D, Logan BA, Payton P, Allen RD, Holaday AS (2003) Elevated chloroplast glutathione reductase activities decrease chilling induced photoinhibition by increasing rates of photochemistry, but not thermal energy dissipation in transgenic cotton. Func Plant Bio 30:101-10

Kotchoni OS, Gachomo EW (2006) The reactive oxygen species network pathways: an essential prerequisite for perception of pathogen attack and the acquired disease resistance in plants. J Biosci 31:389-404

Kouril R, Lazar D, Lee H, Jo J, Naus J (2003) Moderately elevated temperature eliminates resistance of rice plants with enhanced expression of glutathione reductase to intensive photooxidative stress. Photosynthetica 41:571-8

Kraner I (2002) Glutathione status correlates with different degrees of desiccation tolerance in three lichens. New Phytol 154:451-60

Kreps JA, Wu Y, Chang HS, Zhu T, Wang X, Harper JF (2002) Transcriptome changes for *Arabidopsis* in response to salt, osmotic, and cold stress. Plant Physiol 130:2129-41

Kubo A, Sano T, Saji H, Tanaka K, Kondo N, Tanaka K (1993) Primary structure and properties of glutathione reductase from *Arabidopsis thaliana*. Plant Cell Physiol 34:1259-66

Kubo A, Saji H, Tanaka K, Konda N (1995) Expression of *Arabidopsis* cytosolic peroxidase gene in response to ozone or sulphur dioxide. Plant Mol Biol 29:479-89

Kuk YI, Shin JS, Burgos NR, Hwang TE, Han O, Cho BH, Jung S, Guh JO (2003) Antioxidative enzymes offer protection from chilling damage in rice plants. Crop Sci 43:2109-17

Kukreja S, Nandwal AS, Kumar N, Sharma SK, Sharma SK, Unvi V, Sharma PK (2005) Plant water status, H_2O_2 scavenging enzymes, ethylene evolution and membrane integrity of *Cicer arietinum* roots as affected by salinity. Biol Plant 49:305-8

Kumar RG, Dubey RS (1999) Glutamine synthetase isoforms from rice seedlings: Effects of stress on enzyme activity and the protective roles of osmolytes. J Plant Physiol 155:118-21

Kumar S, Singla-Pareek, Reddy MK, Sopory SK (2003) Glutathione: biosynthesis, homeostasis and its role in abiotic stresses. J Plant Biol 30:179-87

Kunert KJ, Foyer CH (1993) Thiol/disulphide exchange in plants In: De Kok LJ, Stulen I, Rennenberg H, Brunhold C and Rausen W (eds): Sulfur nutrition and assimilation in higher plants. Regulatory Agricultural and environmental Aspects, SPB Acedemics, The Hague, pp. 139–51

Kuzniak E, Sklodowska M (1999) The effect of *Botrytis cinerea* infection on ascorbate–glutathione cycle in tomato leaves. Plant Sci 148:69-76

Kuzniak E, Sklodowska M (2001) Ascorbate, glutathione and related enzymes in chloroplasts of tomato leaves infected by *Botrytis cinerea*. Plant Sci 160:723-31

Kuzniak E, Sklodowska M. (2004) The effect of *Botrytis cinerea* infection on the antioxidant profile of mitochondria from tomato leaves. J Exp Bot 397:605-12

Kuzniak E, Sklodowska M (2005) Fungal pathogen-induced changes in the antioxidant systems of leaf peroxisomes from infected tomato plants. Planta 222:192-200

Lascano HR, Gomez LD, Casano LM, Trippi VS (1998) Changes in glutathione reductase activity and protein content in wheat leaves and chloroplasts exposed to photooxidative stress. Plant Physiol Biochem 36:321-9

Lascano HR, Gomez LD, Casano LM, Trippi VS (1999) Wheat chloroplastic glutathione reductase activity is regulated by the combined effect of pH, NADPH and GSSG. Plant Cell Physiol 40:683-90

Lascano HR, Casano LM, Melchiorre MN, Trippi VS (2001) Biochemical and molecular characterisation of wheat chloroplastic glutathione reductase. Biol Plant 44:509-16

Lee SM, Leustek T (1999) The effect of cadmium on sulphate assimilation enzymes in *Brassica juncea*. Plant Sci 141:201-7

Lee H, Jo J, Son D (1998) Molecular cloning and characterization of the gene encoding glutathione reductase in *Brassica campestris*. Biochim Biophys Acta 1395:309-14

Lee H, Won SH, Lee BH, Park HD, Chung WI, Jo J (2002) Genomic cloning and characterization of glutathione reductase gene from *Brassica campestris* var. Pekinensis. Mol Cells 13:245-51

Lee YP, Kim SH, Bang JW, Lee HS, Kwak SS, Kwon SY (2007) Enhanced tolerance to oxidative stress in transgenic tobacco plants expressing three antioxidant enzymes in chloroplasts: genetic transformation and hybridization Plant cell Rep 26:591-8

Lederer B, Boger P (2003) Antioxidative responses of tobacco expressing a bacterial glutathione reductase. Z Naturforsch 58:843-9

Leegood RC (1985) The intercellular compartmentation of metabolites in leaves of *Zea mays* L. Planta 164:163-71

Libreros-Minotta CA, Pardo JP, Mendoza-Hernandez G, Rendon JL (1992) Purification and characterization of glutathione reductase from *Rhodospirillum rubrum*. Arch Biochem Biophys 298:247-53

Lin JN, Kao CH (2000a) Involvement of lipid peroxidation in water stress promoted senescence of detached rice leaves. Biol Plant 43:141-5

Lin CC, Kao CH (2000b) Effect of NaCl stress on H_2O_2 metabolism in rice leaves. Plant Growth Regul 30:151-5

Lin DI, Lur HS, Chu C (2001) Effects of abscisic acid on ozone tolerance of rice (*Oryza sativa* L.) seedlings. Plant Growth Regul 35:295-300

Logan BA, Grace SC, Adams III WW, Adams DB (1998) Seasonal differences in xanthophyll cycle charecteristics and antioxidants in *Mahonia repens* growing in different light environments. Oecologia 116:9-17

Logan BA, Monteiro G, Kornyeyev D, Payton P, Allen RD, Holaday AS (2003) Transgenic overproduction of glutathione reductase does not protect cotton, *Gossypium hirsutum* (Malvaceae), from photoinhibition during growth under chilling conditions. Am J Bot 90:1400-3

Lopez-Barea J, Lee CY (1979) Mouse liver glutathione reductase. Purification, kinetics and regulation. Eur J Biochem 98:487-99

Madamanchi NR, Anderson JV, Alscher RG, Cramer CL, Hess JL (1992) Purification of multiple forms of glutathione reductase from Pea *(Pisum sativum* L.) seedlings and enzyme levels in ozone fumigated pea leaves. Plant Physiol 100:138-45

Mahan JR, Burke JJ (1987) Purification and characterization of glutathione reductase from corn mesophyll chloroplasts. Physiol Plant 71:352-8

Mapson LW, Goddard DR (1951) Reduction of glutathione by coenzyme II. Nature 167:975-7

Mapson LW, Isherwood (1963) Glutathione reductase from germinated peas. Biochem J 86:173-91

Marrs KA (1996) The functions and regulation of glutathione S transferases in plants. Annu Rev Plant Physiol Plant Mol Biol 47:127-58

Massey V, Williams CH (1965) On the reaction mechanism of yeast glutathione reductase. J Biol Chem 240:4470-80

Mata AM, Pinto MC, Lopez-Barea J (1984) Purification by affinity chromatography of glutathione reductase (EC 1.6.4.2) from *Escherichia coli* and characterization of such enzyme. Z Naturforsch 39:908-15

Maughan S, Foyer CH (2006) Engineering and genetic approaches to modulating the glutathione network in plants. Physiol Plant 126:382-97

May MJ, Vernoux T, Leaver C, Van Montagu M, Inze D (1998) Glutathione homeostasis in plants: implications for environmental sensing and plant development. J Exp Bot 49:649-67

Mayer AJ, Hell R (2005) Glutathione homeostasis and redox-regulation by sulfhydryl groups. Photosynth Res 86:435-57

Mazhoudi S, Chaoui A, Ghorsab MW, Ferjani EE (1997) Response of antioxidant enzymes to excess copper in tomato (*Lycopersicon esculentum* Mill.). Plant Sci 127:129-37

McLaughlin SB, Taylor GE (1981) Relative humidity: important modifier to pollutant uptake by plants. Science 211:167-9

Mehler AH (1951) Studies on reactions of illuminated chloroplasts. I. Mechanisms of the reduction of oxygen and other Hill reagents. Arch Biochem Biophys 33:65-77

Mehlhorn H, Cottam DA, Lucas PW, Wellburn AR (1987) Induction of ascorbate peroxidase and glutathione reductase activities by interactions of mixtures air pollutants. Free Radical Res Commun 3:193-7

Meister A (1988) Glutathione metabolism and its selective modification. J Biol Chem 263:17205-8

Meldrum NU, Tarr HLA (1935) The reduction of glutathione by the Warburg-Christian system. Biochem J 29:108-15

Meyer AJ, Fricker MD (2002) Control of demand-driven biosynthesis of glutathione in green *Arabidopsis* suspension culture cells. Plant Physiol 130:1927-37

Mittl PRE, Schulz GE (1994) Structure of glutathione reductase from *Escherichia coli* at 1.86 Å resolution: comparison with the enzyme from human erythrocytes. Protein Sci 3:799-809

Mittler R (2002) Oxidative stress, antioxidants, and stress tolerance. Trends Plant Sci 7:405-10

Mittler R, Vanderauwera S, Gollery M, Van Breusegem F (2004) The reactive oxygen gene network of plants. Trends Plant Sci 9:490-8

Mittova V, Tal M, Volokita M, Guy M (2003) Up-regulation of the leaf mitochondrial and peroxisomal antioxidative systems in response to salt-induced oxidative stress in the wild salt-tolerant tomato species *Lycopersicon pennellii*. Plant Cell Environ 26:845-56

Molina A, Bueno P, Marín MC, Rosales MPR, Belver A, Venema K, Donaire JP (2002) Involvement of endogenous salicylic acid content, lipoxygenase and antioxidant enzyme activities in the response of tomato cell suspension cultures to NaCl. New Phytol 156:409-15

Mullineaux PM, Creissen GP (1997) Glutathione reductase: regulation and role in oxidative stress. In: Scandalios JG (ed) Oxidative stress and the molecular biology of antioxidants, Cold Spring Harbor Laboratory Press, NY, pp. 667-713

Mullineaux PM, Rausch T (2005) Glutathione, photosynthesis and the redox regulation of stress-responsive gene expression. Photosynth Res 86:459-74

Mullineaux PM, Enard C, Hylton C, Hellens R, Creissen GP (1996) Characterization of a glutathione reductase gene and its genetic locus from pea (*Pisum sativum* L.). Planta 200:186-94

Nathawat NS, Nair JS, Kumawat SM, Yadava NS, Singh G, Ramaswamy NK, Sahu MP, D'Souza SF (2007) Effect of seed soaking with thiols on the antioxidant enzymes and photosystem activities in wheat subjected to water stress. Biol Plant 51:93-7

Nocito FF, Lancilli C, Crema B, Fourcroy P, Davidian JC, Sacchi GA (2006) Heavy metal stress and sulfate uptake in maize roots. Plant Physiol 141:1138-48

Noctor G, Arisi A, Jouanin L, Kunert K, Rennenberg H, Foyer C (1998) Glutathione: biosynthesis, metabolism and relationship to stress tolerance explored in transformed plants. J Exp Bot 49:623-47

Noctor G, Gomez L, Vanacker H, Foyer CH (2002) Interactions between biosynthesis, compartmentation and transport in the control of glutathione homeostasis and signalling. J Exp Bot 53:1283-1304

Ogawa Ki Hatano-Iwasaki A, Yanagida M, Iwabuchi M (2004) Level of glutathione is regulated by ATP-dependent ligation of glutamate and cysteine through photosynthesis in *Arabidopsis thaliana*: mechanism of strong interaction of light intensity with flowering. Plant Cell Physiol 45:1-8

Oquist G, Huner NPA (1993) Cold-hardening induced resistance to photoinhibition of photosynthesis in winter rye is dependent upon an increased capacity for photosynthesis. Planta 189:150-6

Panda SK, Patra HK (1997) Physiology of chromium toxicity: a review. Plant Physiol Biochem 24:10-17

Panda SK, Patra HK (1998) Alteration of nitrate reductase activity by chromium ions in excised wheat leaves. Indian J Agr Biochem 11:56-7

Panda SK, Patra HK (2000) Does chromium (III) produce oxidative damage in excised wheat leaves? J Plant Biol 27:105-10

Parida BK, Chhibba IM, Nayyar VK (2003) Influence of nickel contaminated soils on fenugreek (*Trigonella corniculata* L.) growth and mineral composition. Sci Hort 98:113-19

Pastori GM, Trippi VS (1992) Oxidative stress induces high rate of glutathione reductase synthesis in a drought resistant maize strain. Plant Cell Physiol 33:957-61

Payton P, Webb R, Kornyeyev D, Allen R, Holaday AS (2001) Protecting cotton photosynthesis during moderate chilling at high light intensity by increasing chloroplastic antioxidant enzyme activity. J Exp Bot 52:2345-54

Pell EJ, Dann MS (1991) Multiple stress-induced foliar senescence and implications for whole-plant longevity. In: Monney HA, Winner WE and Pell EJ (eds): Responses of plants to multiple stresses, Academic Press, San Diego, pp. 389-403

Pereira GJG, Molina SMG, Lea PJ, Azevedo RA (2002) Activity of antioxidant enzymes in response to cadmium in *Crotalaria juncea*. Plant Soil 239:123-32

Perry ACF, Bhriain NN, Brown NL, Rouch DA (1991) Molecular characterization of the *gor* gene encoding glutathione reductase from *Pseudomonas aeruginosa*: determinants of substrate specificity among pyridine nucleotide–disulphide oxidoreductases. Mol Microbiol 5:163-71

Peeters N, Small I (2001) Dual targeting to mitochondria and chloroplasts. Biochim Biophys Acta 1541:54-63

Peuke AD, Rennenberg H (2005a) Phytoremediation with transgenic trees. Z Naturforsch 60:199-207

Peuke AD, Rennenberg H (2005b) Phytoremediation: molecular biology, requirements for application, environmental protection, public attention and feasibility. EMBO Reports 6:497-501

Pilon-Smith E, Zhu YL, Sears T, Terry N (2000) Over-expression of glutathione reductase in *Brassica juncea*: effects on cadmium accumulation and tolerance. Physiol Plant 110:455-60

Pinto MC, Mata AM, Lopez-Barea (1985) The redox interconversion mechanism of *Saccharomyces cerevisiae* glutathione reductase. Eur J Biochem 15:275-81

Plochi M, Lyons T, Ollerenshaw J, Barnes J (2000) Simulating ozone detoxification in the leaf apoplast through the direct reaction with ascorbate. Planta 210:454-67

Polle A (1998) Photochemical oxidants: uptake and detoxification mechanisms. In: De Kok LJ and Stulen I (eds): Responses of plant metabolism to air pollution and global change. Backhuys Publishers, Leiden, pp. 95-116

Poschenrieder C, Vazquez MD, Bonet A, Barcelo J (1991) Chromium III iron interaction in iron sufficient and iron deficient bean plants. II Ultra structural aspects. J Plant Nutr 14:415-28

Potters G, Horemans N, Bellone S, Caubergs RJ, Trost P, Guisez Y, Asard H (2004) Dehydroascorbate influences the plant cell cycle through a glutathione-independent reduction mechanism. Plant Physiol 134:1479-87

Prasad KVSK, Pardhasaradhi P, Sharmila P (1999) Concerted action of antioxidant enzyme and curtailed growth under zinc toxicity in *Brassica juncea*. Environ Exp Bot 42:1-10

Rao MV (1992) Cellular detoxifying mechanisms determine age dependent injury in tropical plants exposed to SO_2. J Plant Physiol 140:733-40

Rao MV, Hale B, Ormrod DP (1995) Amelioration of ozone induced oxidative damage in wheat plants grown under high carbon dioxide. Role of antioxidant enzymes. Plant Physiol 109:421-32

Rao MV, Paliyath G, Ormrod DP (1996) Ultraviolet-B and ozone-induced biochemical changes in antioxidant enzymes of *Arabidopsis thaliana*. Plant Physiol 110:125-36

Rauser WE (1990) Phytochelatins. Annu Rev Biochem 59:61-86

Rea PA (2006) Phytochelatin synthase, papain's cousin, in stereo. Proc Natl Acad Sci USA 103:507-8

Rea PA, Vatamaniuk OK, Rigden DJ (2004) Weeds, worms, and more. Papain's long-lost cousin, phytochelatin synthase. Plant Physiol 136:2463-74

Reddy AR, Raghavendra AS (2006) Photooxidative stress. In: KV Madhava Rao, Raghavendra AS, Reddy KJ (eds): Physiology and molecular biology of stress tolerance in plants, Springer, The Netherlands, pp.157-86

Reddy AR, Chaitanya KV, Vivekanandan M (2004) Drought-induced responses of photosynthesis and antioxidant metabolism in higher plants. J Plant Physiol 161:1189-1202

Rendon JL, Calcagno M, Mendoza-Hernandez G, Ondarza RN (1986) Purification, properties, and oligomeric structure of glutathione reductase from the cyanobacterium *Spirulina maxima*. Arch Biochem Biophys 248:215-23

Rennenberg H, Brunold C (1994) Significance of glutathione metabolism in plants under stress. Prog Bot 55:144-56

Rennenberg H, Schmitz K, Bergmann L (1979) Long distance transport of sulphur in *Nicotiana tabacum*. Planta 147:57-62

Rich W, Wagoner PE, Tomlinson H (1970) Ozone uptake by bean leaves. Science 169:19-80

Rizhsky L, Davletova S, Liang H, Mittler R (2004) The zinc finger protein Zat12 is required for cytosolic ascorbate peroxidase 1 expression during oxidative stress in *Arabidopsis*. J Biol Chem 279:11736-43

Romero-Puertas MC, Corpas FJ, Sandalio LM, Leterrier M, Rodriguez-Serrano M, del Rio LA, Palma JM (2006) Glutathione reductase from pea leaves: response to abiotic stress and characterization of the peroxisomal isozyme. New Phytol 170:43-52

Rouhier N, Couturier J, Jacquot JP (2006) Genome-wide analysis of plant glutaredoxin systems. J Exp Bot 57:1685-96

Sairam PK, Saxena DC (2000) Oxidative stress and antioxidants in wheat genotypes: possible mechanism of water stress tolerance. J Agron Crop Sci 184:55-61

Sairam RK, Shukla DS, Saxena DC (1997/98) Stress-induced injury and antioxidant enzymes in relation to drought tolerance in wheat genotypes. Biol Plant 40:357-64

Sairam RK, Srivastava GC, Agarwal S, Meena RC (2005) Differences in antioxidant activity in response to salinity stress in tolerant and susceptible wheat genotypes. Biol Plant 49:85-91

Salt DE, Blaylock M, Nanda Kumar PBA, Dushenkov V, Ensley BD, Chet I, Raskin I (1995) Phytoremediation: a novel strategy for the removal of toxic metals from the environment using plants. Biotechnol 13:468-74

Schaedle M, Bassham JA (1977) Chloroplast glutathione reductase. Plant Physiol 59:1011-12

Schoner S, Krause GH (1990) Protective systems against active oxygen species in spinach: response to cold acclimation in excess light. Planta 180:383-9

Schulz GE, Schirmer RH, Sachsenheimer W, Pai EF (1978) The structure of the flavoenzyme glutathione reductase. Nature 273:120-4

Scrutton NS, Berry A, Perham RN (1990) Redesign of the coenzyme specificity of a dehydrogenase by protein engineering. Nature 343:38-43

Selote DS, Chopra RK (2004) Drought-induced spikelet sterility is associated with an inefficient antioxidant defense in rice panicles. Physiol Plant121:462-71

Selote DS, Chopra RK (2006) Drought acclimation confers oxidative stress tolerance by inducing co-ordinated antioxidant defense at cellular and subcellular level in leaves of wheat seedlings. Physiol Plant 127:494–506

Serrano A, Liobell A (1993) Occurrence of two isoforms of glutathione reductase in the green alga *Chlamydomonas reinhardtii*. Planta 190:199-205

Serrano A, Rivas J, Losada M (1984) Purification and properties of glutathione reductase from the cyanobacterium *Anabaena sp*. strain 7119. J Bacteriol 158:317-24

Shalata A, Mittova V, Volokita M, Guy M, Tal M (2001) Response of the cultivated tomato and its wild salt-tolerant relative *Lycopersicon pennellii* to salt-dependent oxidative stress: the root antioxidative system. Physiol Plant 112:487-94

Sharma P, Dubey RS (2005) Drought induces oxidative stress and enhances the activities of antioxidant enzymes in growing rice seedlings. Plant Growth Regul 46:209–21

Shulaev V, Oliver DJ (2006) Metabolic and Proteomic Markers for Oxidative Stress: New tools for reactive oxygen species research. Plant Physiol 141:367-72

Skorzynska-Polit E, Drazkiewicz, Krupa Z (2003/4) The activity of the antioxidative system in cadmium-treated *Arabidopsis thaliana*. Biol Plant 47:71-8

Slooten L, Montagu MV, Inze D (1998) Manipulation of oxidative stress tolerance in transgenic plants In: Lindsey K (ed): Transgenic plant research, Harwood Academic Publishers, Australia, pp. 241-62

Smirnoff N (1993) The role of active oxygen in the response of plants to water deficit and desiccation. New Phytol 125:27-58

Smirnoff N, Colombe SV (1988) Drought influences the activity of enzymes of the chloroplast hydrogen peroxide scavenging system. J Exp Bot 39:1097-1108

Smith IK, Vierheller TL, Thorne CA (1989) Properties and functions of glutathione reductase in plants. Physiol Plant 77:449-56

Somashekaraiah BV, Padmaja K, Prasad ARK (1992) Phytotoxicity of cadmium ions on germinating seedlings of mung bean (*Phaseolus vulgaris*): involvement of lipid peroxides in chlorophyll degradation. Physiol Plant 85:85-9

Sreenivasulu N, Grimma B, Wobusa U, Weschkea W (2000) Differential response of antioxidant compounds to salinity stress in salt-tolerant and salt-sensitive seedlings of foxtail millet (*Setaria italica*). Physiol Plant109:435-42

Srivalli B, Sharma G, Chopra RK (2003) Antioxidative defense system in an upland rice cultivar subjected to increasing intensity of water stress followed by recovery. Physiol Plant 119:503-12

Steffens JC (1990) The heavy metal-binding peptides of plants. Ann Rev Plant Physiol Plant Mol Biol 41:553-75

Stevens R, Creissen GP, Mullineaux PM (1997) Cloning and characterization of a cytosolic glutathione reductase cDNA from pea (*Pisum sativum* L.) and its expression in response to stress. Plant Mol Biol 35:641-54

Stevens R, Creissen GP, Mullineaux PM (2000) Characterization of pea cytosolic glutathione reductase expressed in transgenic tobacco. Planta 211:537-45

Stroinski A, Kubis J, Zielezinska M (1999) Effect of cadmium on glutathione reductase in potato tubers. Acta Physiol Plant 21:201-7

Strohm M, Eiblmeier M, Langebartels C, Jouanin L, Polle A, Sandermann H, Rennenberg H (1999) Responses of transgenic poplar (*Populus tremula x P.alba*) overexpressing glutathione reductase to acute ozone stress: visible injury and leaf gas exchange. J Exp Bot 50:365-74

Strohm M, Eiblmeier M, Langebartels C, Jouanin L, Polle A, Sandermann H, Rennenberg H (2002) Responses of antioxidative systems to acute ozone stress in transgenic poplar (*Populus tremula x P.alba*) over-expressing glutathione synthetase or glutathione reductase. Trees 16:262-73

Sumithra K, Jutur PP, Dalton CB, Reddy AR (2006) Salinity-induced changes in two cultivars of *Vigna radiata*: responses of antioxidative and proline metabolism. Plant Growth Regul 50:11-22

Takeda T, Ishikawa T, Shigeoka S, Hirayama O, Mitsunaga T (1993) Purification and characterization of glutathione reductase from *Chlamydamaonas reinhardtii*. J Gen Microbiol 139:2233-8

Tanaka K, Saji H, Kondo N (1988) Immunological properties of spinach glutathione reductase and inductive biosynthesis of the enzyme with ozone. Plant Cell Physiol 29:637-42

Tang X, Webb MA (1994) Soyabean root nodule cDNA encoding glutathione reductase. Plant Physiol 104:1081-2

Tausz M, Sircelj H, Grill D (2004) The glutathione system as a stress marker in plant ecophysiology: is a stress-response concept valid? J Exp Bot 55:1955-62

Tewari RK, Parma PK, Sharma N (2006) Antioxidant responses to enhanced generation of superoxide anion radical and hydrogen peroxide in the copper-stressed mulberry plants. Planta 223:1145-53

Tsai YC, Kao CH (2004) The involvement of hydrogen peroxide in abscisic acid-induced activities of ascorbate peroxidase and glutathione reductase in rice roots. Plant Growth Regul 43:207-12

Tsai YC, Hong CY, Liu LF, Kao CH (2005) Expression of ascorbate peroxidaes and glutathione reductase in roots of rice seedlings in respons to NaCl and H_2O_2. J Plant Physiol 162:291-9

Tsang EWT, Bowler C, Herouart D, van Camp W, Villarroel R, Genetello C, van Montagu M, Inze D (1991) Differential regulation of superoxide dismutases in plants exposed to environmental stress. Plant Cell 3:783-92

Tyystjarvi E, Riikonen M, Arisi ACM, Kettunen R, Jouanin L, Foyer CH (1999) Photoinhibition of photosysem II in tobacco plants overexpressing glutathione reductase and poplars overexpressing superoxide dismtase. Physiol Plant 105:409-16

Vaidyanathan H, Sivakumar P, Chakrabarty R, Thomas G (2003) Scavenging of reactive oxygen species in NaCl-stressed rice (*Oryza sativa* L.) - differential responses in salt tolerant and sensitive varieties. Plant Sci 165:1411-18

Vanacker H, Carver TLW, Foyer CH (1998) Pathogen-induced changes in the antioxidant status of the apoplast in barley leaves. Plant Physiol 117:1103-14

Van Rensburg L, Kruger GHJ (1994) Evaluation of components of oxidative stress metabolism for use in selection of drought tolerant cultivars of *Nicotiana tobacum* L. J Plant Physiol 143:730-36

Verma S, Dubey RS (2001) Effect of cadmium on soluble sugars and enzymes of their metabolism in rice. Biol Plant 44:117-23

Vinterhalter B, Vinterhalter D (2005) Nickel hyper accumulation in shoot cultures of *Alyssum markgrafii*. Biol Plant 49:121-4

Vitoria AP, Lea PJ, Azevedo RA (2001) Antioxidant enzymes responses to cadmium in radish tissues. Phytochemistry 57:701-10

Walker MA, McKersie (1993) Role of the ascorbate-glutathione antioxidant system in chilling resistance of tomato. J Plant Physiol 141:234-9

Weckx JEJ, Clijsters HMM (1997) Zn phytotoxicity induces oxidative stress in primary leaves of *Phaseolus vulgaris*. Plant Physiol Biochem 35:405-10

Willekens H, Van Camp W, Van Montagu M, Inze D, Langerbelts C, Sandermann H (1994) Ozone, sulfur dioxide and UV-B radiation have similar effects on mRNA accumulation of antioxidant genes in *Nicotiana plumbaginifolia* L. Plant Physiol 106:1007-14

Wingate VMP, Lawton MA, Lamb CJ (1988) Glutathione causes a massive and selective induction of plant defense genes. Plant Physiol 87:206-10

Wingsle G (1989) Purification and characterization of glutathione reductase from Scots pine needles. Physiol Plant 76:24-30

Wingsle G, Karpinski S. (1996) Differential redox regulation by glutathione of glutathione reductase and CuZn-superoxide dismutase gene expression in *Pinus sylvestris* L. needles. Planta 198:151-7

Wise RR (1995) Chilling-enhanced photooxidation: The production, action, and study of reactive oxygen species produced during chilling in the light. Photosynth Res 45:79-97

Wojtaszek P (1997a) Oxidative burst: an early plant response to pathogen infection. Biochem J 332:681-92

Wojtaszek P (1997b) Mechanisms for the generation of reactive oxygen species in plant defense response. Acta Physiol Plant 19:581-9

Woodwin TS, Segel IH (1968) Isolation and characterization of glutathione reductase from *Penicillium chrysogenum*. Biochem Biophys Acta 164:64-77

Xiang C, Werner BL, Christensen EM, Oliver DJ (2001) The biological functions of glutathione revisited in *Arabidopsis* transgenic plants with altered glutathione levels. Plant Physiol 126:564-74

Young LCT, Conn EE (1956) The reduction and oxidation of glutathione by plant mitochondria. Plant Physiol 31:205-11

Zenk MH (1996) Heavy metal detoxification in higher plants: a review. Gene 179:21-30

Zgallai H, Steppe K, Lemeur R (2006) Effects of different levels of water stress on leaf water potential, stomatal resistance, protein and chlorophyll content and certain antioxidative Enzymes in tomato plants. J Integr Plant Biol 48:679-85

Chapter 7
Sulfotransferases and Their Role in Glucosinolate Biosynthesis

Marion Klein and Jutta Papenbrock(✉)

Abstract All members of the sulfotransferase (SOT, EC 2.8.2.-) protein family use 3'-phosphoadenosine 5'-phosphosulfate as sulfuryl donor and transfer the sulfate group to an appropriate hydroxyl group of several classes of substrates. In plants, sulfate conjugation reactions seem to play an important role in plant growth, development, and adaptation to stress. The genome of *Arabidopsis thaliana* contains in total 21 genes that are likely to encode SOT proteins. The respective genes are differentially expressed under various biotic and abiotic stress conditions. Most of their substrates, and therefore the respective physiological roles, of plant SOT proteins are not known. In *Arabidopsis* three SOT proteins catalyze the last step of glucosinolate (Gl) biosynthesis. In vitro enzyme assays revealed preferences of the recombinant SOT proteins for chemically different types of Gls. *In planta* Gls exist side by side; therefore initial results from one-substrate measurements were verified using a mixture of ds-Gls and Gl leaf extracts from *Arabidopsis* as substrates. These studies confirmed the in vitro measurements. The putative role of SOT proteins in the manipulation of Gl biosynthesis and regulatory aspects will be discussed.

1 The Multiprotein Family of Sulfotransferases

1.1 Sulfotransferases in Organisms

Sulfotransferases (SOT) catalyze the transfer of a sulfate group from the co-substrate 3'-phosphoadenosine 5'-phosphosulfate (PAPS) to an appropriate hydroxyl group of several classes of substrates with the parallel formation of 3'-phosphoadenosine 5'-phosphate (PAP).

$$\text{PAPS} + \text{R-OH} \xrightarrow{\text{SOT}} \text{PAP} + \text{R-OSO}_3^-$$

Jutta Papenbrock
Institute for Botany, University of Hannover, Herrenhäuserstr. 2, D-30419 Hannover, Germany
Jutta.Papenbrock@botanik.uni-hannover.de

N.A. Khan et al. (eds.), *Sulfur Assimilation and Abiotic Stress in Plants.*

In most cases, the addition of a sulfate moiety to a compound increases its water solubility and alters its biological activity. Members of the SOT family have been found in most organisms investigated to date, except in archaea. The SOT proteins were classified on basis of their affinity for different classes of substrates. One group of mammalian SOT proteins, mainly membrane-associated, accepts as substrates macromolecules such as proteins and peptides, and glucosaminoglycans (Niehrs et al. 1994). The second group, in general soluble proteins, accepts as substrates small organic molecules such as flavonoids, steroids, and xenobiotics with diverse chemical structures (Weinshilboum and Otterness 1994). Structural similarities are present among SOT proteins from eubacteria, plants, and cytosolic soluble SOT from mammalian species (Yamazoe et al. 1994). All SOT proteins have conserved amino acid motives, which seem to be involved in PAPS binding (regions I and IV) (Marsolais and Varin 1995, Weinshilboum et al. 1997). In mammals, sulfation contributes to the homeostasis and regulation of numerous biologically potent endogenous compounds such as catecholamines, steroids, and iodothyronines as well as detoxification of xenobiotics (Coughtrie et al. 1998). In bacteria, sulfation is essential for the signaling of rhizobial nod factors to the plant (Roche et al. 1991). In plants, a large proportion of the known sulfated metabolites play various roles in plant defense against biotic and abiotic stress (Varin et al. 1997).

1.2 Functional Analysis of Sulfotransferases in Plants

The substrates and therefore the physiological roles of SOTs are very diverse. There are a number of sulfated metabolites in the plant organism, such as brassinosteroids, coumarins, flavonoids, gibberellic acids, glucosinolates (Gls), phenolic acids, steroids, sulfate esters such as choline-*O*-sulfate, and terpenoids, which might be sulfated by SOT proteins. Gls are responsible for taste and flavor of cruciferous vegetables (Mithen 2001). Sulfated flavonoids are present in about 32 plant families are involved in detoxification of reactive oxygen species and regulation of plant growth (Varin et al. 1997). Several other sulfated compounds were shown to directly participate in plant defense against pathogens, such as a sulfated derivate of jasmonic acid (Gidda et al. 2003) or sulfated β-1,3 glucan (Menard et al. 2004) and fucan oligosaccharides (Klarzynski et al. 2003) that induced salicylic acid defense signaling.

To date only a few plant SOTs have been characterized functionally in detail. The flavonol 3- and 4'-SOT from *Flaveria* species (Asteraceae), which catalyze the sulfation of flavonol aglycones and flavonol 3-sulfates, respectively, were the first plant SOTs for which cDNA clones were isolated (Varin et al. 1992). The plasma membrane-associated gallic acid SOT of *Mimosa pudica* L. pulvini cells may be intrinsic to signaling events that modify the seismonastic response (Varin et al. 1997). In *Brassica napus* L. four SOT isoforms have been isolated. Two of them have been functionally analyzed; they catalyze the *O*-sulfation of brassinosteroids and thereby abolish specifically the biological activity of 24-epibrassinolide

(Rouleau et al. 1999, Marsolais et al. 2000). In *Oryza sativa* L. 26 genes potentially coding for SOT proteins were identified by BLAST searches using *Arabidopsis* SOT sequences as a query. Phylogenetic analysis revealed that all rice SOT proteins were clustered together, as well as the four *Flaveria* flavonol SOT and the four *B. napus* steroid SOT sequences (data not shown). In the green algae *Chlamydomonas reinhardtii* and in the moss *Physcomitrella patens* (Hedw.) no SOT sequences were found, which indicates a late evolutionary origin of SOT proteins.

1.3 Sulfotransferases in **Arabidopsis**

The fully sequenced genome of *Arabidopsis thaliana* (L.) Heyn. contains in total 21 genes that are likely to encode SOT proteins (AtSOT) based on sequence similarities of the translated products with an average identity of 51.1%. Originally 18 SOT had been identified, and according to their sequence similarity seven groups had been formed (Klein and Papenbrock 2004). The phylogenetic tree shown in Fig.7.1 includes all AtSOT proteins known to date and is inferred from the neighbor-joining method. The tree reflects the real relationships of the AtSOT proteins in a better way. Including recent data the current knowledge about the SOT family in *Arabidopsis* is summarized in Table 7.1. For many of them the according substrate specificity and therefore the in vivo function are not known yet. So far only a few SOT from *Arabidopsis* were functionally characterized: one protein (At5g07000) was shown to catalyze the sulfation of 12-hydroxyjasmonate, thereby inactivating excess jasmonic acid in plants (Gidda et al. 2003). For another SOT of *Arabidopsis* (At3g45070) a flavonol activity was found with strict specificity for position 7 of flavonols. It is thought that flavonoids might be involved in plant growth and development (Marsolais et al. 2000, Gidda and Varin 2006).

Recently, for two further *Arabidopsis* SOTs (AtSOT10, At2g14920; AtSOT12, At2g03760) a brassinosteroid activity has been shown. However, AtSOT10 has a different enzymatic activity profile than the previously characterized *B. napus* brassinosteroid SOT BNST3 and BNST4 (Rouleau et al. 1999, Marsolais et al. 2000), and AtSOT12. AtSOT10 was specific for biologically active end products of the brassinosteroid biosynthetic pathway. Despite their high level of sequence identity with AtSOT10, recombinant AtSOT8 and AtSOT9 did not display significant catalytic activity with brassinosteroids, either suggesting that their biochemical function is different from AtSOT10 or that the recombinant proteins were not catalytically active due to improper folding (Marsolais et al. 2007). A possible function of these enzymes could be the inactivation of brassinosteroids in *Arabidopsis* due to sulfation, and therefore an involvement in plant development (Marsolais et al. 2000, 2007). Three SOT proteins, AtSOT16, 17, and 18 (nomenclature is given in Klein and Papenbrock 2004) are desulfo- (ds) Gl SOT enzymes (see chapters 2 and 3).

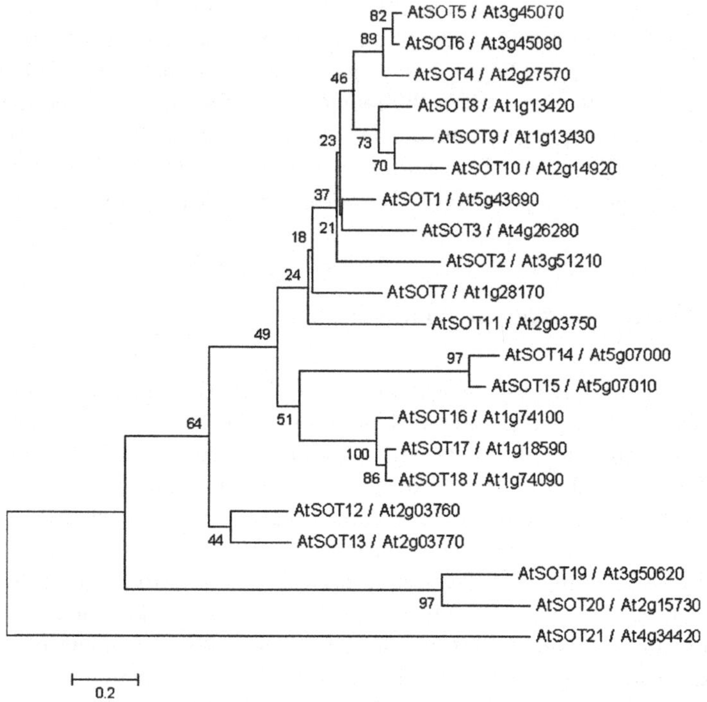

Fig. 7.1 Phylogenetic and molecular evolutionary analyses were conducted using MEGA version 3.1 (Kumar et al. 2004). The tree was constructed based on the protein sequences using the neighbor-joining method under the evolutionary model JTT (At numbers as given in Table 7.1). The bootstrap support for each node was calculated using 500 replications

The determination of the three-dimensional (3-D) structure is sometimes helpful for analyzing the binding of putative substrates. Nine crystal structures of human soluble cytosolic SOT proteins have been determined, and together with site-directed mutagenesis experiments and molecular modeling, one is beginning to understand the factors that govern distinct but overlapping substrate specificities (Gamage et al. 2006). To date the 3-D structure for the AtSOT12 (At2g03760) protein has been determined. The overall 3-D structure was nearly identical to already solved structures of soluble steroid SOT proteins from humans. The structures contain a long disordered loop not found in the electron density map, presumably because the substrates were not present. Whether substrate binding will in part order this region requires prior identification of the appropriate substrates and subsequent crystallographic investigations including the substrates (Smith et al. 2004). Thus, to date structural investigations have not been helpful in determining the putative substrates of plant SOT proteins.

Table 7.1 Summary of the members of the SOT family in Arabidopsis and their putative substrates

Gr.[a,b]	Name	Gene ID	No. aa	EST	Putative Substrate	Reference
I	AtSOT1	At5g43690	331	3[c]	not identified yet	-
	AtSOT2	At3g51210	67	-	not identified yet	-
II	AtSOT3	At4g26280	314	-	not identified yet	-
	AtSOT4	At2g27570	273	-	not identified yet	-
	AtSOT5	At3g45070	323	5	flavonoids at 7' position	Marsolais et al. 2000, Gidda and Varin 2006
	AtSOT6	At3g45080	329	1	not identified yet	-
III	AtSOT7	At1g28170	326	-	not identified yet	-
	AtSOT8	At1g13420	331	10	not identified yet	-
	AtSOT9	At1g13430	351	7	not identified yet	-
	AtSOT10	At2g14920	333	-	brassinosteroids	Marsolais et al. 2000, Marsolais et al. 2007
IV	AtSOT11	At2g03750	351	33	not identified yet	-
V	AtSOT12	At2g03760	326	16	brassinosteroids	Lacomme and Roby 1996, Marsolais et al. 2007
	AtSOT13	At2g03770	324	-	not identified yet	-
VI	AtSOT14	At5g07000	347	9	not identified yet	-
	AtSOT15	At5g07010	359	30	11-, 12-OH-jasmonate	Gidda et al. 2003
VII	AtSOT16	At1g74100	338	75	desulfo-Gls	Piotrowski et al. 2004, Klein et al. 2006
	AtSOT17	At1g18590	346	19	desulfo-Gls	Piotrowski et al. 2004, Klein et al. 2006
	AtSOT18	At1g74090	350	17	desulfo-Gls	Piotrowski et al. 2004, Klein et al. 2006
VIII	AtSOT19	At3g50620	340	1	not identified yet[d]	-
	AtSOT20	At2g15730	344	3	not identified yet[d]	-
	AtSOT21	At4g34420	403	5	not identified yet[d]	-

[a] Abbreviations used: Gr., group; No., number; aa, amino acids.
[b] The groups were formed as described (Klein and Papenbrock 2004).
[c] The number of EST clones was determined on March 26 2007 (http://arabidopsis.org).
[d] Nodulation-related protein contains weak similarity to nodulation protein H (EC 2.8.2.-). Swiss-Prot: P06237, (*Rhizobium meliloti*)

1.4 Expression Analysis of SOT *Genes Indicates General Stress Response*

The mRNA levels for most of the *Arabidopsis* SOTs characterized to date were low in plants grown at normal growth conditions (Lacomme and Roby 1996, Gidda et al. 2003, Piotrowski et al. 2004). This fact is reflected by the relatively low number or even no hits of EST clones, with the exception of *AtSOT15* and *AtSOT16*. However, in agreement with the potential role of sulfated compounds in plant defense, the mRNA levels of *AtSOT12, AtSOT15, AtSOT16*, and *AtSOT17* significantly increased upon treatment with jasmonate (Lacomme and Roby 1996, Gidda et al. 2003, Piotrowski et al. 2004). *AtSOT12* mRNA was induced also by salicylic acid and by interaction with bacterial pathogens and elicitors (Lacomme and Roby 1996), whereas *AtSOT16* mRNA level responded to coronatine, an analogue of octadecanoid signaling molecules, and to 1-aminocyclopropane-1-carboxylic acid, a precursor of ethylene (Lacomme and Roby 1996). Screening of available microarray data (https://www.genevestigator.ethz.ch) revealed that only 17 of 21 SOTs are present on the 24 k Affymetrix chip and for many of them the absolute signal intensity is very low.

Similar to *Arabidopsis*, in *B. napus* the mRNA levels of the SOTs were very low under normal conditions but were increased upon treatments with salicylic acid and ethanol, again pointing out the potential function of these enzymes in the stress response (Rouleau et al. 1999, Marsolais et al. 2000).

2 Sulfotransferases Involved in Glucosinolate Biosynthesis

2.1 Glucosinolates

Gls are a group of over 130 nitrogen- and sulfur-containing natural products found in vegetative and reproductive tissues of 16 plant families, but are most well known as the major secondary metabolites in agriculturally important crop plants of the Brassicaceae family such as oilseed rape (*B. napus*), fodder and vegetables (e.g., broccoli and cabbage) and the model plant *Arabidopsis*. Gls share a core structure containing a β-D-glucopyranose residue linked via a sulfur atom to a (*Z*)-*N*-hydroxyimino sulfate ester, and are distinguished from each other by a variable R group derived from one of several amino acids, mainly tryptophan, phenylalanine, and methionine. In *Arabidopsis* more than 30 different Gls have been described (for review see Mithen 2001). The Gl pattern of *Arabidopsis* varies within the plant as well as among *Arabidopsis* ecotypes. Gl contents on organ level of *Arabidopsis* were analyzed at different developmental stages. Significant differences could be detected in both quality and quantity of Gls (Brown et al. 2003). Due to determination of Gl content of 39 *Arabidopsis* ecotypes 34 different Gls have been identified in the genus *Arabidopsis* so far (Kliebenstein et al. 2001a).

Intact Gls are nontoxic. Damage to plant tissue results in the hydrolysis of Gls, catalyzed by thioglucosidase ("myrosinase") to produce to a variety of (volatile) hydrolysis products, such as thiocyanates, isothiocyanates, and nitriles. These breakdown products have a wide range of biological activities, which include both negative and positive effects. Gls are the best-characterized preformed defense compounds in the Brassicaceae, and the breakdown products were shown in several studies to be involved in plant defense against pathogens and herbivores. In short, Gls contribute to protection against pathogens of the generalist type (summarized in Rausch and Wachter 2005). The Gl level is reduced in the seed of rape varieties due to breeding, because of the antinutritional value of specific Gls. This allows use of protein-rich rape seed cake as animal food.

For a number of Gl breakdown products it was shown that they act anticarcinogenically by detoxification of phase II enzymes (Zhang et al. 1994). It was also demonstrated that 6-methylsulfinylhexyl isothiocyanate (MS-ITC) acts as a potential inhibitor of human platelet aggregation in vitro through extensive screening of vegetables and fruits. MS-ITC also induced glutathione S-transferase activity in RL34 cells. MS-ITC administered to rats or mice also showed both activities in vivo. MS-ITC plays an important role for antiplatelet and anticancer activities because of its high reactivity with sulfhydryl groups in biomolecules (Morimitsu et al. 2000). Also in epidemiological studies the effects of high Gl contents in the food was demonstrated. In general, it was shown that dietary supplements are not as effective as the food, presumably because of synergistic effects among different components (Weisburger and Butrum 2002). Therefore the food itself should be enriched by secondary metabolites, such as Gls. The strongest exogenous factor influencing the Gl content in *Brassica* crops is the sulfur supply, whereas the nitrogen fertilization did not significantly affect at least the content of the Gl glucotropaelin (Bloem et al. 2001). Therefore controlled enhancement of Gls in crop species might have striking effects on human and animal health.

2.2 Sulfotransferases in Glucosinolate Biosynthesis

Gl biosynthesis can be divided into three stages: (1) precursor amino acids, such as methionine and tryptophan, are elongated by one or several methylene groups; (2) precursor amino acids are converted into parent Gls; and (3) finally, parent Gls can be secondarily modified (Fig.7.2) (Wittstock and Halkier 2002). for a long time it has been suggested that the last step in the Gl biosynthesis core structure of the different aliphatic, aromatic, and indole ds-Gls is catalyzed by members of the SOT (EC 2.8.2.-) protein family. Glendening and Poulton (1990) partially purified a protein from *Lepidium sativum* L. that had PAPS-dependent ds-Gl SOT activity, however, at that time no molecular data were available. Recently, it was shown that three SOT proteins from *Arabidopsis* are involved in Gl biosynthesis catalyzing the sulfation of ds-Gls to the intact Gls (Varin and Spertini 2003, Piotrowski et al. 2004, Hirai et al. 2005). For most of the other enzymatic steps in Gl biosynthesis

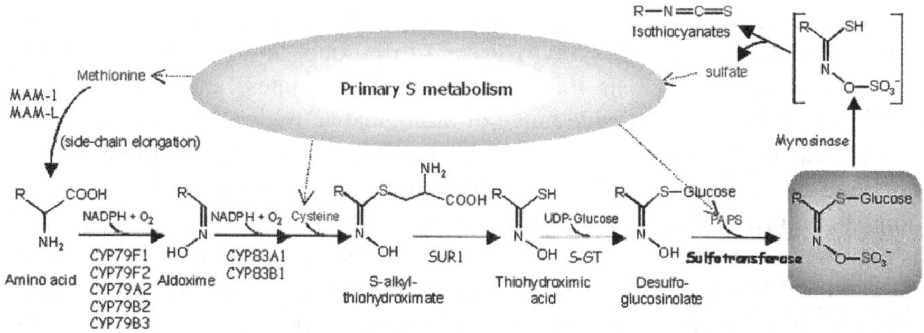

Fig. 7.2 Outline of glucosinolate biosynthesis and degradation. *MAM*, methylthioalkylmalate synthase; *CYP*, cytochrome P450; *SUR1*, a C-S lyase encoded by *SUPERROOT1* gene; *S-GT*, S-glucosyltransferase; *PAPS*, 3'-phosphoadenosine 5'-phosphosulfate. In *Arabidopsis*, methionine and tryptophan are the major precursors of Gls, but the side chain of methionine is first elongated by a cycle involving MAM-1 and MAM-L. Isoforms of CYP79s and CYP83s catalyze the initial reactions of the core biosynthetic pathway, followed by SUR1, S-GT, and sulfotransferase (Hirai et al. 2005, with permission of J Biol Chem)

candidates have been isolated and partially characterized (Wittstock and Halkier 2002, Douglas Grubb et al. 2004, Mikkelsen et al. 2004). The complete biosynthetic pathway in *Arabidopsis* is presented in Fig. 7.2. Interestingly, the enzymes involved are predicted to be localized in a variety of cell compartments according to different computer programs (Douglas Grubb et al. 2004, Mikkelsen et al. 2004, Klein et al. 2006). The proteins elongating the side chains of methionine might be localized in the plastids because the MAM proteins contain putative transit peptides. For the CYP79 proteins which catalyze the formation of aldoximes with different substrates, a localization in the ER or with a lower probability in mitochondria (e.g., CYP79B2, At4g39950) can be assumed, whereas CYP83 proteins (CYP83A1, At4g13770 and CYP83B1, At4g31500) catalyzing aldoxime oxidizing reactions might be localized in the ER or, with a lower probability, in the cytoplasm (Klein et al. 2006). The subsequent C-S-lyase protein(s) converting the *S*-alkylthiohydroximates into thiohydroximic acids are either localized in the cytoplasm (SUR1, At2g20610) (Mikkelsen et al. 2004) or, a second putative C-S lyase (At5g36160), in the ER (Hirai et al. 2005). The last step preceding the sulfotransferase reaction catalyzed by UDP-glucose:thiohydroximate glucosyltransferase (At1g24100) is predicted to be localized in the ER. In summary, one has to postulate a reaction sequence via the following compartments: plastids – ER – cytoplasm – ER – cytoplasm. Finally, the mature Gls are stored in the vacuole. Therefore several transport processes have to be assumed for the metabolic flow through the complete Gl biosynthetic pathway. To our knowledge so far no carrier proteins or transporters have been described on the molecular level which transport Gl intermediates.

Recently, the substrates for three AtSOT proteins (AtSOT16, AtSOT17, and AtSOT18) were predicted and then verified by different means (screening of many sulfated compounds, combining of knowledge, and integration of metabolomics and transcriptomics) (Varin and Spertini 2003, Piotrowski et al. 2004, Hirai et al. 2005). The novel strategy to identify functions of unknown genes was followed by Kazuki Saito and his group, Chiba, Japan. The metabolomes and the transcriptomes of sulfur-starved *Arabidopsis* plants were analyzed in nontargeted ways by Fourier-transform ion cyclotron mass spectrometry and DNA microarray. Approximately 2,000 metabolites and 21,000 genes were classified by batch-learning self-organizing map analysis according to their time-dependent changes. Clustering of the metabolomes and the transcriptomes indicated that the accumulation of Gls behaves coordinately with the expression of the genes responsible for their biosynthesis. Three previously uncharacterized putative *AtSOT* genes clustered together with known genes which were expected to be involved in Gl biosynthesis. In our laboratory in vitro enzymatic assays, using the respective recombinant gene products AtSOT16, AtSOT17, and AtSOT18, confirmed that these proteins catalyze the final step of Gl formation (Hirai et al. 2005, Klein et al. 2006).

The glucosylation and the sulfation reactions were assumed to be nonspecific with respect to the side chain (Halkier 1999) (Fig. 7.2). Therefore, it is an open question why three *ds-Gl SOT* genes have been conserved during evolution in *Arabidopsis* and maybe in other Brassicaceae species. It is also hypothesized that first the side chains are elongated to synthesize so-called parent Gls, then the glycone moity is developed and finally, several side chain modifications take place to produce the respective daughter Gl (Wittstock and Halkier 2002). However, it is still not clarified when the ds-Gls are sulfated by SOT proteins and whether there is a specificity for certain parent or daughter Gls.

2.3 Characteristics of *Arabidopsis* Desulfo-Glucosinolate Sulfotransferases

All three nuclear encoded ds-Gl AtSOT proteins are localized in the cytoplasm as demonstrated by transient expression of fusion constructs with the green fluorescent protein (GFP) in *Arabidopsis* protoplasts (Klein and Papenbrock 2004, Klein et al. 2006). It is remarkable that the in vitro pH optimum for the SOT reaction is 9.0 and therefore more alkaline than the average pH 7.5 in the cytoplasm. At pH 7.0 the enzyme reaction is expected to proceed at around 50% of the rate at pH 9.0 with variations for each AtSOT protein. One could assume that locally the pH in the cytoplasm might be higher.

To differentiate among the three ds-Gl AtSOT several authors investigated their expression under different conditions. Expression analysis by RT-PCR revealed that all three genes were expressed constitutively in leaves, flowers, and siliques (Varin and Spertini 2003). Later it was shown that the three *ds-Gl AtSOT* genes are indeed differentially expressed with the respective mRNAs accumulating to different extents under various conditions (Piotrowski et al. 2004, Klein et al. 2006). The mRNA levels of the three *AtSOT* genes differed in plant organs and tissues at different developmental stages and during a light/dark cycle. These expression patterns may be responsible for changes in Gl contents among organs, developmental stages, or environmental conditions as observed by Brown et al. (2003).

In experiments to test the effects of nutrient supply on Gl concentration, the sulfate concentrations applied were 1,500 µM (high) and 30 µM (low); in addition the nitrogen concentrations were varied (Hirai et al. 2004, 2005). It was shown by microarray analysis that all genes involved in Gl biosynthesis showed an apparently similar expression pattern under sulfur deficiency (Hirai et al. 2005), suggesting that the expression of these genes is controlled by the same regulatory mechanism. In contrast, the genes encoding the enzymes involved in primary sulfur metabolism, which are known to be regulated not only at mRNA accumulation level but also at enzymatic activity level, showed diverse expression levels (Hirai et al. 2005).

High (500 µM) and low (50 µM) sulfate concentrations in the medium did not influence the expression levels (Klein et al. 2006). In comparison to the conditions chosen by Hirai et al. (2005) the differences in the sulfur status were much smaller (10 times versus 50 times). The lower sulfate concentration was chosen because it represents the borderline for normal growth rates and should reflect the conditions on the field of sulfur-fertilized and nonfertilized *B. napus* plants (Ewald Schnug, Braunschweig, Germany, personal communication). Under biotic or abiotic stress, the differences in the *ds-Gl AtSOT* expression might be more pronounced.

3 Enzymology of Desulfo-Glucosinolate Sulfotransferases

3.1 Kinetic Parameters of Desulfo-Glucosinolate Sulfotransferases

The last step of the Gl biosynthesis is catalyzed by SOTs (Fig.7.2). It is thought that after formation of a core structure (the parent Gls), secondary modifications take place (Graser et al. 2001, Kliebenstein et al. 2001b). To date, little is known about these secondary modifications of parent Gls. As the Gl pattern differs among *Arabidopsis* ecotypes (Kliebenstein et al. 2001a), the investigation of the three ds-Gl SOTs from ecotype C24, which shows the broadest variety of Gls (Michael Reichelt, Max Planck Institute for Chemical Ecology, Jena, Germany, unpublished data), was most rational. In addition one exemplary SOT from the fully sequenced ecotype Col-0 was investigated. To determine if and how these three ds-Gl SOT proteins might influence the Gl pattern, different in vitro enzyme assays were performed.

K_m values for both AtSOT18 proteins from ecotypes C24 and Col-0 were determined. As exemplary substrates aromatic ds benzyl and two in vivo substrates, ds-3-methylthiopropyl Gl (ds3MTP) and ds-4-methylthiobutyl Gl (ds4MTB), were used, which are short aliphatic Gls deduced from methionine. The results showed that these enzymes differ in their affinity for the investigated substrates and the co-substrate PAPS (Klein et al. 2006, Klein and Papenbrock submitted). An explanation for different K_m values of both expressed proteins is the difference of two amino acids out of 350. An exchange in AtSOT18 of one of those two amino acids from the C24 ecotype to the respective amino acid from the Col-0 ecotype showed the same substrate specificities as the wild-type AtSOT18 protein from Col-0 (Klein et al. 2006). Interestingly, in humans interindividual variation in sulfation capacity may be important in determining an individual's response to xenobiotics, and recent studies have begun to suggest roles for SOT polymorphism in disease susceptibility (Gamage et al. 2006).

K_m values for AtSOT16 and 17 from ecotype C24 were determined as well with the same substrates mentioned above and revealed differences among these enzymes, but to less smaller extents. In conclusion, it shows that there are differences with respect to the kinetic behavior among ecotypes, but only small differences between the three AtSOTs within ecotype C24. Thus, one question remains: why are there three SOT proteins? In order to answer this question, the K_m values for more in vivo substrates need to be determined.

3.2 Substrate Specificities of Desulfo-Glucosinolate SOT Proteins

A leaf extract from *Arabidopsis* C24 contains mainly aliphatic Gls, derived from methionine. Aside from aliphatic ones, indolic Gls derived from tryptophan were found. These Gls originated from following parent Gls: 3MTP, 4MTB, 5-methylthiopentyl Gl (5MTP), 7-methylthioheptyl Gl (7MTH), 8-methylthiooctyl Gl (8MTO), and indol-3-yl-methyl Gl (I3M). Structures of all *Arabidopsis* Gls are shown in Reichelt et al. (2002). One-substrate measurements and measurements with a mixture of the mentioned parent Gls revealed a clear preference of AtSOT16 for I3M, the indolic Gl, followed by short methylthio Gls (3MTP and 4MTB). AtSOT17 and 18 favor long methylthio Gls, namely 7MTH and 8MTO. In summary, SOTs showed activity with (almost) all offered ds-Gls, but with noticable preferences. The results lead to the hypothesis that the investigated ds-Gl AtSOTs influence the Gl pattern in vivo. Comparison of ecotypes C24 and Col-0 shows that AtSOT18 from Col-0 seems to be less specific than AtSOT18 from C24. Obviously, the AtSOT18 proteins from the different ecotypes show a different behavior. Maybe this contributes to the different Gl patterns within the ecotypes (Klein and Papenbrock submitted).

The final aim of these kind of studies is to elucidate the function of ds-Gl SOTs *in planta*. Therefore, AtSOT proteins and a C24 leaf extract from *Arabidopsis* as substrate mixture were used to simulate the in vivo situation. Hence, this is a partially artificial and partially natural system. In conclusion, the substrate specificities found in pure in vitro assays could be confirmed. The other finding was that

Gls with secondary modifications were sulfated as well (Klein and Papenbrock submitted). Unfortunately, the knowledge on secondary modifications of parent Gls is limited (Graser et al. 2001). In future work it could be interesting to verify the general acceptance that parent ds-Gls and not ds-Gls with secondary modifications are sulfated (Kliebenstein et al. 2001b). However, assuming the general acceptance is right, no secondarily modified Gls would exist in a ds form to interact with the SOTs. Therefore, it is possible that artificially desulfated Gls with secondary modifications are sulfated, but with no in vivo relevance.

3.3 Speculations on Modulations of the Gl Pattern and Their Biological Effects in Case of Altered Expression of AtSOT16 to 18

The Gl-myrosinase system is an important antiherbivore defense in Gl-containing plants. Upon tissue damage, myrosinase and Gls come into contact and Gls are hydrolyzed. Many of these hydrolysis products are toxic to a wide range of organisms, such as bacteria, fungi, nematodes, and insects (Chew 1988, Louda and Mole 1991, Rask et al. 2000). In a previously published study 122 *Arabidopsis* ecotypes were analyzed to learn more about Gl hydrolysis products. It was found that each ecotype produced predominantly either isothiocyanates (ITC) or epithionitriles/nitriles, and to a minor extent oxazolidine-2-thione. This polymorphism is correlated with the expression of the epithiospecifier protein (ESP). ESP promotes the production of epithionitrile or nitrile, respectively, depending on the Gl side chain. In absence of ESP mainly ITCs are produced. *Arabidopsis* ecotype Col-0 revealed no detectable ESP expression, whereas ecotype C24 showed ESP expression (Lambrix et al. 2001). One can only speculate about changes in the Gl pattern due to modulation of AtSOT16, 17, and 18. In theory, by overexpression of AtSOT17 or 18, both preferred long chain ds-Gls from methionine (7MTH and 8MTO), nitrile, and epithionitrile production would be increased due to hydrolysis of the respective Gls. In contrast the production of oxazolidine-2-thione would be decreased, because the precursor Gl is a short Gl derived from methionine.

Numerous studies have shown that Gl hydrolysis products are toxic to a wide range of organisms. The biological effects of Gl hydrolysis products on insects were summarized (Wittstock et al. 2003). It was found that predominantly ITC is toxic to insects. Other hydrolysis products, nitriles, and thiocyanates were also found to be toxic to insects, but nothing is known about toxicity of epithionitriles and oxazolidine-2-thione. As oxazolidine-2-thione is known to cause goiter in mammals and has other harmful effects, this compound could act as a defense against mammalian herbivores (Wittstock et al. 2003).

In expression studies an increased mRNA amount of *AtSOT17* was found, especially in mature leaves in investigations on plant organ level and under sulfur deficiency. An increased mRNA accumulation could be observed in young plants for up to two weeks. Compared to *AtSOT17, AtSOT18* mRNA accumulated in five to six weeks old *Arabidopsis* plants, especially in the root and under sulfur deficiency (Hirai et al. 2005, Klein et al. 2006). In *Arabidopsis* leaves more than 80% of the total Gls were contributed due to aliphatic Gls. The proportion of aliphatic Gls decreased with age (Brown et al. 2003). In conclusion, modulation of the Gls pattern due to repression or overexpression of both AtSOT17 and 18 could help to find out whether the resulting hydrolysis products (nitriles and epithionitriles) are involved in plant defense. Thus, an overexpression of each protein may lead to an improved plant defense.

The hydrolysis of indolic Gls in the presence of ESP leads to elemental sulfur and indolyl-3-acetonitrile (IAN), which has auxin activity. Subsequently IAN could be converted to the auxin indol-3-acetic acid (IAA), a plant hormone (Mithen 2001, Halkier and Gershenzon 2006). Auxin is known to be involved in growth and development. IAN was found to act as a feeding deterrent against the locust *Schistocerca gregaria* (El Sayed et al. 1996), as a growth inhibitor of the plant-pathogenic fungus *Leptosphaeria maculans* (Mithen and Lewis 1986), and as a plant-pathogenic bacterium *Erwinia carotovora* (Brader et al. 2001).

An increased expression of *AtSOT16* gene can be observed at sulfur deficiency in young *Arabidopsis* plants up to two weeks and especially in roots (Hirai et al. 2005, Klein et al. 2006). A great proportion of indolic Gls, nearly half of the Gls content, was found in roots (Brown et al. 2003). Taken together, this leads to the hypothesis that overexpression of AtSOT16 may lead to increased production of IAN or IAA and elemental sulfur in roots. Under sulfur deficiency, increased elemental sulfur could be produced from indolic Gls and could be made available to the plant. In conclusion, hydrolysis products of indolic Gls as well as elemental sulfur can constitute to an improved defense-system against plant pathogens and herbivores.

4 Regulatory Aspects in Glucosinolate Biosynthesis

In general the availability of PAPS for sulfation in vivo depends on its synthesis, transport, degradation, and utilization as investigated in mammals (Klaassen and Boles 1997). PAPS is synthesized by phosphorylation of adenosine 5'-phosphate (APS) by APS kinase (AKN) In vitro tests have shown that excess APS inhibits APS kinase. In mammals PAPS is formed in the cytoplasm and can be degraded by two different pathways leading to the same end product, 5'-adenosine monophosphate (5'-AMP). There is no information about enzymatic PAPS degradation in plants (Klaassen and Boles 1997).

APS is also an intermediate in the biosynthetic pathway of cysteine starting from sulfate (Fig. 7.3). In *Arabidopsis* four ATP sulfurylase and four *AKN* genes have been identified. The subcellular localization of the gene products has not been investigated in detail. According to Rotte and Leustek (2000) plastid-localized ATP sulfurylase makes up 70% to 95% of the total enzyme activity in leaves. Preliminary results of subcellular localization studies of the four AKN proteins using fusion constructs with the green fluorescent proteins in transient transformation studies of protoplasts indicate

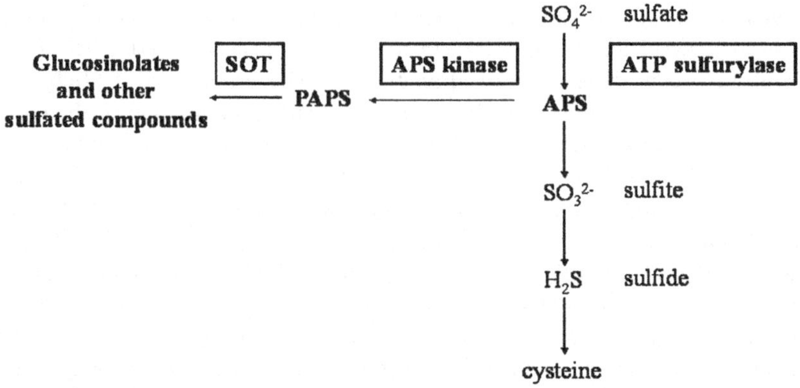

Fig. 7.3 3'-phosphoadenosine 5'-sulfate (APS) is an intermediate in sulfate assimilation and in 3'-phosphoadenosine 5'-phosphosulfate (PAPS) biosynthesis

that at least two AKN proteins are localized in the cytoplasm (Papenbrock, unpublished results), the same compartment where AtSOT proteins are localized. Both proximate enzymes using APS as substrate, AKN and APS reductase, have very low K_m values for APS in the low micromolar range (Suter et al. 2000, Lillig et al. 2001). To exclude competition on the same substrate, one could assume a spatial separation of APS pools in the cell. One might ask the question whether PAPS can be used up under certain conditions and therefore limit the process of sulfation reactions. It was shown that sulfation is a high-affinity, low-capacity enzymatic process in which the entire liver content of PAPS can be consumed in less than 2 min (Klaassen and Boles 1997). For plants there is no information so far on whether sulfate availability can influence the PAPS pool used for sulfation. The determination of the APS and PAPS pools in plants will be important for the understanding of the regulation of PAPS biosynthesis and therefore also the regulation of sulfation reactions.

Gl biosynthesis, or at least parts of the pathway, might be regulated via transcription factors. Recently, a transcription factor called OBP2 has been identified as a member of the DNA-binding-with-one-finger (DOF) transcription factors. OBP2 expression is induced in response to the generalist herbivore *Spodoptera littoralis* and by treatment with methyl jasmonate, both of which also trigger accumulation of indolic Gls, derivatives of tryptophan. In *Arabidopsis* indole-3-acetaldoxime produced from tryptophan by the activity of two cytochrome P450 enzymes, CYP79B2 and CYP79B3, serves as a precursor for indole Gl biosynthesis but is also an intermediate in the biosynthetic pathway of IAA. Another cytochrome P450 enzyme, CYP83B1, funnels indole-3-acetaldoxime into indolic Gls. Constitutive and inducible overexpression of OBP2 activates expression of CYP83B1. In addition, auxin concentration is increased in leaves and seedlings of OBP2 overexpression lines relative to wild-type, and plant size is diminished due to a reduction in cell size. Collectively, these data provide evidence that OBP2 is part of a regulatory network that regulates at least a part of Gl biosynthesis in *Arabidopsis* (Skirycz et al. 2006).

An example for regulation of Gl biosynthesis via a MYB transcription factor is the results of the investigation of an *Arabidopsis* activation-tagged line which was shown to be affected in the content of indolic and aliphatic Gls. The observed chemotype was caused by activation of the transcription factor gene *HIG1* (HIGH INDOLIC GLUCOSINOLATE 1, also referred to as MYB51). HIG1/MYB51 was shown to activate promoters of indolic Gl biosynthetic genes leading to increased accumulation of indolic Gls. HIG1/MYB51 seems to be a regulator of indolic Gl biosynthesis that also controls responses to biotic challenges (Gigolashvili et al. 2007). The challenge for the future will be to differentiate the regulatory function of different transcription factors on the whole Gl biosynthetic pathway or on single genes involved in Gl biosynthesis to be able to manipulate the Gl pattern in plants in a selective way.

5 Summary and Outlook

In the last years the knowledge about this exciting protein family of SOTs was constantly increasing. However, for many members the substrate specificities are still unknown. Therefore their role in biotic and abiotic stress response could not be clarified. Sophisticated analytical methods need to be applied to fully explore the pool of sulfated compounds in plants. In combination with specific expression profiling of the *SOT* genes in wild-type and mutant plants the functional roles might be elucidated. We begin to understand the role of the three AtSOT proteins in Gl biosynthesis. However, *in planta* substrate specificity of the single enzymes and their individual roles in Gl biosynthesis need to be clarified. The overall regulation of the three AtSOT proteins and of Gl biosynthesis, and breakdown of the different groups of Gl, are not well understood. Finally, the fundamental understanding of the role of sulfated compounds in the plant organism might help to improve agricultural sustainability and the nutritional value of crop plants for the human diet.

References

Bloem E, Haneklaus S, Peplow E, Sator C, Köhler T, Schnug E (2001) The effect of sulphur and nitrogen fertilisation on the glucotropaeolin content in *Tropaeolum majus* (L.). Proceedings der XXXVI. Vortragstagung Gewürz- und Heilpflanzen der DGQ, Jena Germany, pp. 185-90

Brader G, Tas E, Palva, ET (2001) Jasmonate-dependent induction of indole glucosinolates in *Arabidopsis* by culture filtrates of the nonspecific pathogen *Erwinia carotovora*. Plant Physiol 126:849-60

Brown PD, Tokuhisa JG, Reichelt M, Gershenzon J (2003) Variation of glucosinolate accumulation among different organs and developmental stages of *Arabidopsis thaliana*. Phytochemistry 62:471-81

Chew FS (1988) Biological effects of glucosinolates. In: Cutler HG (ed) Biologically active natural products, American Chemical Society, Washington DC, pp. 155-81

Coughtrie MW, Sharp S, Maxwell K, Innes NP (1998) Biology and function of the reversible sulfation pathway catalysed by human sulfotransferases and sulfatases. Chem Biol Interact 109:3-27

Douglas Grubb C, Zipp BJ, Ludwig-Muller J, Masuno MN, Molinski TF, Abel S (2004) *Arabidopsis* glucosyltransferase UGT74B1 functions in glucosinolate biosynthesis and auxin homeostasis. Plant J 40:893-908

El Sayed G, Louveaux A, Mavratzotis M, Rollin P, Quinsac A (1996) Effects of glucobrassicin, epiprogoitrin and related breadkown products on locusts feeding: *Schouwia purpurea* and desert locust relationships. Entomol Exp Appl 78:231-6

Gamage N, Barnett A, Hempel N, Duggleby RG, Windmill KF, Martin JL, McManus ME (2006) Human sulfotransferases and their role in chemical metabolism. Toxicol Sci 90:5-22

Gidda SK, Varin L (2006) Biochemical and molecular characterization of flavonoid 7-sulfotransferase from *Arabidopsis thaliana*. Plant Physiol Biochem 44:628-36

Gidda SK, Miersch O, Levitin A, Schmidt J, Wasternack C, Varin L (2003) Biochemical and molecular characterization of a hydroxyjasmonate sulfotransferase from *Arabidopsis thaliana*. J Biol Chem 278:17895-900

Gigolashvili T, Berger B, Mock HP, Muller C, Weisshaar B, Flugge UI (2007) The transcription factor HIG1/MYB51 regulates indolic glucosinolate biosynthesis in *Arabidopsis thaliana*. Plant J 2007 Apr 25; [Epub ahead of print]

Glendening TM, Poulton JE (1990) Partial purification and characterization of a 3'-phosphoadenosine 5'-phosphosulfate: desulfoglucosinolate sulfotransferase from cress (*Lepidium sativum*). Plant Physiol 94:811-18

Graser G, Oldham NJ, Brown PD, Temp U, Gershenzon J (2001) The biosynthesis of benzoic acid glucosinolate esters in *Arabidopsis thaliana*. Phytochemistry 57:23-32

Halkier BA (1999) Glucosinolates. In: Ikan R (ed): Naturally occurring glycosides: chemistry, distribution and biological properties, John Wiley & Sons Ltd, New York, pp. 193-223

Halkier BA, Gershenzon J (2006) Biology and biochemistry of glucosinolates. Annu Rev Plant Biol 57:303-33

Hirai MY, Yano M, Goodenowe DB, Kanaya S, Kimura T, Awazuhara M, Arita M, Fujiwara T, Saito K (2004) Integration of transcriptomics and metabolomics for understanding of global responses to nutritional stresses in *Arabidopsis thaliana*. Proc Natl Acad Sci USA 101:10205-10

Hirai MY, Klein M, Fujikawa Y, Yano M, Goodenowe DB, Yamazaki Y, Kanaya S, Nakamura Y, Kitayama M, Suzuki H, Sakurai N, Shibata D, Tokuhisa J, Reichelt M, Gershenzon J, Papenbrock J, Saito K (2005) Elucidation of gene-to-gene networks in *Arabidopsis* by integration of metabolomics and transcriptomics. J Biol Chem 280:25590-5

Klaassen CD, Boles JW (1997) The importance of 3'-phosphoadenosine 5'-phosphosulfate (PAPS) in the regulation of sulfation. FASEB J 11:404-18

Klarzynski O, Descamps V, Plesse B, Yvin JC, Kloareg B, Fritig B (2003) Sulfated fucan oligosaccharides elicit defense responses in tobacco and local and systemic resistance against tobacco mosaic virus. Mol Plant Microbe Interact 16:115-22

Klein M, Papenbrock J (2004) The multi-protein family of *Arabidopsis* sulphotransferases and their relatives in other plant species. J Exp Bot 55:1809-20

Klein M, Reichelt M, Gershenzon J, Papenbrock J (2006) The three desulfoglucosinolate sulfotransferase proteins in *Arabidopsis* have different substrate specificities and are differentially expressed. FEBS J 273:122-36

Kliebenstein DJ, Kroymann J, Brown P, Figuth A, Pedersen D, Gershenzon J, Mitchell-Olds T (2001a) Genetic control of natural variation in *Arabidopsis* glucosinolate accumulation. Plant Physiol 126:811-25

Kliebenstein DJ, Lambrix VM, Reichelt M, Gershenzon J (2001b) Gene duplication in the diversification of secondary metabolism: tandem 2-oxoglutarate-dependent dioxygenases control glucosinolate biosynthesis in *Arabidopsis*. Plant Cell 13:681-93

Kumar S, Tamura K, Nei M (2004) MEGA3: integrated software for Molecular Evolutionary Genetics Analysis and sequence alignment. Brief Bioinf 5:150-63

Lacomme C, Roby D (1996) Molecular cloning of a sulfotransferase in *Arabidopsis thaliana* and regulation during development and in response to infection with pathogenic bacteria. Plant Mol Biol 30:995-1008

Lambrix V, Reichelt M, Mitchell-Olds T, Kliebenstein DJ, Gershenzon J (2001) The *Arabidopsis* epithiospecifier protein promotes the hydrolysis of glucosinolates to nitriles and influences *Trichoplusia ni* herbivory. Plant Cell 13:2793-2807

Lillig CH, Schiffmann S, Berndt C, Berken A, Tischka R, Schwenn JD (2001) Molecular and catalytic properties of *Arabidopsis thaliana* adenylyl sulfate (APS)-kinase. Arch Biochem Biophys 392:303-10

Louda S, Mole S (1991) Glucosinolates: chemistry and ecology. In: Rosenthal GA, Berenbaum MR (eds): Herbivores: their interaction with secondary plant metabolites, vol. 1 The Chemical Participants. Academic Press, San Diego, pp. 123-64

Marsolais F, Varin L (1995) Identification of amino acid residues critical for catalysis and cosubstrate binding in the flavonol 3-sulfotransferase. J Biol Chem 270:30458-63

Marsolais F, Gidda SK, Boyd J, Varin L (2000) Plant soluble sulfotransferases: Structural and functional similarity with mammalian enzymes. Rec Adv Phytochem 34:433-56

Marsolais F, Boyd J, Paredes Y, Schinas AM, Garcia M, Elzein S, Varin L (2007) Molecular and biochemical characterization of two brassinosteroid sulfotransferases from *Arabidopsis*, AtST4a (At2g14920) and AtST1 (At2g03760). Planta 225:1233-44

Menard R, Alban S, de Ruffray P, Jamois F, Franz G, Fritig B, Yvin JC, Kauffmann S (2004) Beta-1,3 glucan sulfate, but not beta-1,3 glucan, induces the salicylic acid signaling pathway in tobacco and *Arabidopsis*. Plant Cell 16:3020-32

Mikkelsen MD, Naur P, Halkier BA (2004) *Arabidopsis* mutants in the C-S lyase of glucosinolate biosynthesis establish a critical role for indole-3-acetaldoxime in auxin homeostasis. Plant J 37:770-7

Mithen R (2001) Glucosinolates: biochemistry, genetics and biological activity. Plant Growth Regul 34:91-103

Mithen R, Lewis BG (1986) *In vitro* activity of glucosinolates and their products against *Leptosphaeria maculans*. Trans Br Mycol Soc 87:433-40

Morimitsu Y, Hayashi K, Nakagawa Y, Horio F, Uchida K, Osawa T (2000) Antiplatelet and anticancer isothiocyanates in Japanese domestic horseradish, wasabi. Biofactors 13:271-6

Niehrs C, Beisswanger R, Huttner WB (1994) Protein tyrosine sulfation, 1993: an update. Chem Biol Interact 92: 257-71

Piotrowski M, Schemenewitz A, Lopukhina A, Müller A, Janowitz T, Weiler EW, Oecking C (2004) desulfo-Glucosinolate sulfotransferases from *Arabidopsis thaliana* catalyzing the final step in biosynthesis of the glucosinolate core structure. J Biol Chem 49:50717-25

Rask L, Andreasson E, Ekbom B, Eriksson S, Pontoppidan B, Meijer J (2000) Myrosinase: gene family evolution and herbivore defense in Brassicaceae. Plant Mol Biol 42:93-113

Rausch T, Wachter A (2005) Sulfur metabolism: a versatile platform for launching defence operations. Trends Plant Sci 10:503-9

Reichelt M, Brown PD, Schneider B, Oldham NJ, Stauber E, Tokuhisa J, Kliebenstein DJ, Mitchell-Olds T, Gershenzon J (2002) Benzoic acid glucosinolate esters and other glucosinolates from *Arabidopsis thaliana*. Phytochemistry 59:663-71

Roche P, Debelle F, Maillet F, Lerouge P, Faucher C, Truchet G, Denarie J, Prome JC (1991) Molecular basis of symbiotic host specificity in *Rhizobium meliloti: nodH* and *nodPQ* genes encode the sulfation of lipo-oligosaccharide signals. Cell 67:1131-43

Rotte C, Leustek T (2000) Differential subcellular localization and expression of ATP sulfurylase and 5'-adenylylsulfate reductase during ontogenesis of *Arabidopsis* leaves indicates that cytosolic and plastid forms of ATP sulfurylase may have specialized functions. Plant Physiol 124:715-24

Rouleau M, Marsolais F, Richard M, Nicolle L, Voigt B, Adam G, Varin L (1999) Inactivation of brassinosteroid biological activity by a salicylate-inducible steroid sulfotransferase from *Brassica napus*. J Biol Chem 274:20925-30

Skirycz A, Reichelt M, Burow M, Birkemeyer C, Rolcik J, Kopka J, Zanor MI, Gershenzon J, Strnad M, Szopa J, Mueller-Roeber B, Witt I (2006) DOF transcription factor AtDof1.1 (OBP2) is part of a regulatory network controlling glucosinolate biosynthesis in *Arabidopsis*. Plant J 47:10-24

Smith DW, Johnson KA, Bingman CA, Aceti DJ, Blommel PG, Wrobel RL, Frederick RO, Zhao Q, Sreenath H, Fox BG, Volkman BF, Jeon WB, Newman CS, Ulrich EL, Hegeman AD, Kimball

T, Thao S, Sussman MR, Markley JL, Phillips GN Jr (2004) Crystal structure of At2g03760, a putative steroid sulfotransferase from *Arabidopsis thaliana*. Proteins 57:854-7

Suter M, von Ballmoos P, Kopriva S, den Camp RO, Schaller J, Kuhlemeier C, Schurmann P, Brunold C (2000) Adenosine 5'-phosphosulfate sulfotransferase and adenosine 5'-phosphosulfate reductase are identical enzymes. J Biol Chem 275:930-6

Varin L, Spertini D (2003) Desulfoglucosinolate sulfotransferases, sequences coding the same and uses thereof for modulating glucosinolate biosynthesis in plants. Patent WO 03/010318-A1, Concordia University, Canada (February, 6, 2003)

Varin L, DeLuca V, Ibrahim RK, Brisson N (1992) Molecular characterization of two plant flavonol sulfotransferases. Proc Natl Acad Sci USA 89:1286-90

Varin L, Marsolais F, Richard M, Rouleau M (1997) Biochemistry and molecular biology of plant sulfotransferases. FASEB J 11:517-25

Weinshilboum RM, Otterness DM (1994) Sulfotransferase enzymes. In: Kauffman FC (ed) Handbook of experimental pharmacology. Springer-Verlag, Berlin, pp. 45-78

Weinshilboum RM, Otterness DM, Aksoy IA, Wood TC, Her C, Raftogianis RB (1997) Sulfotransferase molecular biology: cDNAs and genes. FASEB J 11:3-14

Weisburger EK, Butrum R (2002) Chemoprevention of cancer in humans by dietary means. In: Berdanier CD et al. (eds) Handbook of nutrition and food, Boca Raton, CRC Press, pp. 1011-28

Wittstock U, Halkier BA (2002) Glucosinolate research in the *Arabidopsis* era. Trends Plant Sci 7:263-70

Wittstock U, Kliebenstein DJ, Lambrix V, Reichelt M, Gershenzon J (2003) Glucosinolate hydrolysis and its impact on generalist and specialist insect herbivores. Phytochemistry 37:101-25

Yamazoe Y, Nagata K, Ozawa S, Kato R (1994) Structural similarity and diversity of sulfotransferases. Chem Biol Interact 92:107-17

Zhang Y, Kensler TW, Cho CG, Posner GH, Talalay P (1994) Anticarcinogenic activities of sulforaphane and structurally related synthetic norbornyl isothiocyanates. Proc Natl Acad Sci USA 91:3147-50

Chapter 8
Response of Photosynthetic Organelles to Abiotic Stress: Modulation by Sulfur Metabolism

Basanti Biswal(✉), Mukesh K. Raval, Udaya C. Biswal and Padmanabha Joshi

Abstract Sulfur metabolism mediated modulation of plant response to various abiotic stress factors is the focus of this review. Since chloroplast is extremely sensitive to abiotic stress factors and at the same time a major location of sulfur assimilation, the organelle plays a major role in the modulation of stress response. The photosynthetic organelle coordinates carbon, nitrogen, and sulfur metabolic pathways and provides the essential precursors for synthesis of sulfur compounds. The abiotic stress factors like high light, low light, temperature extremes, drought, and UV radiations which the organelle experiences lead to creation of an oxidative environment and production of reactive oxygen species (ROS). Sulfur metabolites containing thiol residues with reversible oxidation-reduction potential effectively scavenge ROS in a series of biochemical reactions. Abiotic stress factors cause up- and downregulations of several stress-related genes. The stress signals, their transmission, and downstream signaling network regulating gene expression are complex. The stress-induced redox signals generated in chloroplast play a major role in different signal transduction systems and expression of stress-responsive genes in green plants.

1 Introduction

In recent years, the molecular physiology of sulfur assimilation has become one of the major studies in plant science. Sulfur assimilates not only play key roles in the primary metabolism of plants and provide structural components of essential cellular molecules, some of these assimilates act as signaling molecules for cellular communication with the environment. The synthesis of an initial metabolic sulfur compound like cysteine (Cys) involves sulfur uptake through specific transporters and its subsequent assimilation in different locations of the cell. The important secondary metabolic sulfur compounds like glutathione (GSH), phytochelatins

Basanti Biswal
School of Life Sciences, Sambalpur University, Jyotivihar-768019, Orissa, India
basanti_b@hotmail.com

N.A. Khan et al. (eds.), *Sulfur Assimilation and Abiotic Stress in Plants.*

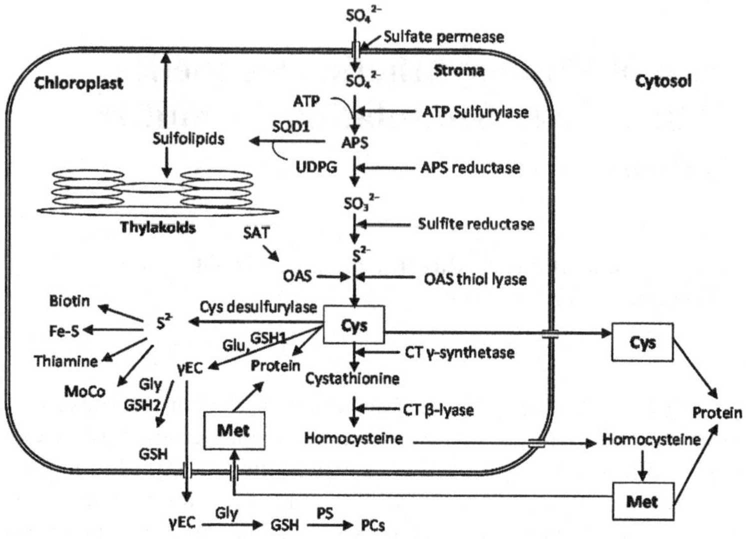

Fig. 8.1 A scheme for sulfur assimilation in chloroplast. ATP: adenosine triphosphate, APS: adenosine phosphosulfate, Cys: cysteine, CT: cystathionine, γEC: γ-glutamylcysteine, GSH1: γEC synthetase, GSH2: glutathione synthetase, Met: methionine, MoCo: molybdenum cofactor, OAS: O-acetyl serine, PC: phytochelatin, PS: phytochelatin synthase, SAT: serine acetyltransferase, SQD1: enzyme for catalysis of formation of sulfolipid- 6-sulfo-α-D-quinovosyl diacylglycerol, UDPG: uridine diphophoglucose. (Modified after Mullineaux and Rausch 2005, Pilon-Smits and Pilon 2006)

(PCs), sulfolipids, vitamins, and many other regulatory compounds are subsequently synthesized in different metabolic routes with Cys as the precursor. The assimilatory pathways of sulfur in chloroplast and sulfur metabolism as a part of cellular metabolic network for synthesis of other secondary sulfur metabolites are shown in Fig.8.1. The sulfur assimilatory pathway involves the assimilatory pathways of an carbon and nitrogen. It also needs involvement of an electron transport pathway that provides adenosine triphosphate (ATP) and reducing power. In this respect, cells of green plants are unique for developing a distinct metabolic combination of the pathways that effectively provide the requirements for the assimilation of this essential nutrient.

The level of crucial metabolites that regulates the sulfur assimilation process is dynamic and is sensitive to environmental changes. Therefore, the pool size of the assimilated sulfur in green plants depends not only on the availability of sulfur but also on the environmental conditions that affect the regulatory mechanism associated with the process of its assimilation. The plants, however, possess different sensory mechanisms that respond to environmental signals and tend to effectively modulate the pool. The pool size of some of the sulfur compounds, especially the compounds with thiol groups which are sensitive to oxidized environment, is important in sulfur metabolism. The thiol-containing sulfur metabolites are well known as potential modulators of the stress response. Therefore, green plants experiencing abiotic stress need these metabolites to develop effective adaptive mechanisms to counter the stress effect. The stress-induced increase in expression of sulfur transporters and enzymes involved in its assimilatory pathway have been recently investigated in many laboratories.

There are different channels of sulfur-mediated plant-adaptive mechanisms to counter the abiotic stress effects. The adaptation may involve repair of sulfur complexes, including incorporation of Fe-S cluster to apoproteins and stabilization of biomembranes with sulfolipids. The other channel of adaptation is chelation of heavy metals by sulfur compounds. The heavy metals are toxic to plants, and their exposure may induce O-acetyl serine(OAS) thiolyase gene (*atcys3A*) (Dominguez-Solis et al. 2001), ATP sulfurylase, adenosine phosphosulfate(APS) reductase, GSH, and PC synthetic pathways (Cobbett 2000, Pickering et al. 2000, Harada et al. 2002, Mendoza-Cozatl et al. 2005). PCs act as metal chelators. S^{2-}, being a soft base, can selectively ligate to soft acids like, Cu^{+}, Cd^{2+}, Hg^{2+}, As^{3+}. PC-metal chelates are excreted to vacuoles. The most important aspect of sulfur-mediated stress adaptation is modulation of stress effect by redox active thiol residues of its compounds. The redox active thiol residues can exist either in reduced or in oxidized forms, and the reversibility of the forms is an important property that provides a switch for initiating different downstream signaling cascades in the metabolic network for appropriate stress adaptation.

In addition to its participation as a modulator of oxidative stress induced by various abiotic stress factors, its deficiency in plants manifests in oxidative stress and enhances activity of enzymes like sulfate permease, ATP sulfurylase, and APS reductase participating in the sulfur assimilation and formation of various sulfur compounds (Gutierrez-Marcos et al. 1996, Leustek et al. 2000, Takahashi et al. 2000). ATP sulfurylase is limiting for sulfur uptake, as its over expression enhances sulfur uptake by plants (Pilon-Smits et al. 1999). It is encouraging to find that there is scope for genetic manipulation by upregulating the enzymes associated with sulfur assimilation and synthesis of redox state regulating sulfur compounds to combat stress (Link 2003).

For the last several years we have been working on photosynthesis of higher plants and the response of chloroplast to several abiotic stress factors. Since chloroplast is the major location of sulfur metabolism in green plants, the review focuses primarily on the possible participation of sulfur compounds in modulating the response of the photosynthetic organelle to the abiotic stress. The review very briefly describes the rationale for nature's selection of stress-sensitive chloroplasts for sulfur assimilation, structure and function of chloroplast, role of chloroplast in sulfur assimilation, oxidative stress induced by various abiotic stress factors and its modulation by chloroplast-mediated sulfur metabolism.

2 Stress-Sensitive Chloroplast as the Site for Major Sulfur Metabolism

Chloroplast contains the machine for synthesis of many sulfur compounds, including Cys and GSH. The synthesis of these compounds is directly linked to the availability of primary photochemical reaction products like ATP and reducing equivalents, and organic carbon skeletons from so-called dark reactions. At the same time, the organelle is known to be extremely sensitive to abiotic stress factors because of its unique structural and functional features. Evolution of O_2, presence of pigments, and photoelectron transport in different environmental conditions could create an oxidative environment and result in the production of various oxygen-free radicals. These conditions possibly have forced plants to develop internal redox control mechanisms which not only counter the stress effect but also exploit redox status as one of the important sensors in the cellular redox signaling network. In the chloroplast, the sulfur-rich compounds appear to play a major role in the background of potential of some of the redox-sensitive sulfur compounds to scavenge ROS and also reduce the stress-induced oxidative environment. At the same time, stress signals generated in the chloroplast are communicated to the cytoplasm through the redox sensing mechanism for expression of stress-responsive nuclear genes. It appears that retention of major sulfur metabolism in the oxygenic photosynthetic system is an evolutionary selection of nature.

3 Structure and Function of Chloroplasts

3.1 Structure and Primary Photochemistry

Chloroplasts are cellular organelles usually located in the green leaves of plants. A chloroplast consists of a continuous double membrane structure known as an envelope and the granular fluid matrix called stroma enclosed by the envelope. The stroma contains proteins, enzymes, and various metabolites. Chloroplast DNA strand is located in stroma. A membrane system known as lamellae is interspersed in the stroma. Lamellae stack at places to form piled sac like structures known as grana. Protein complexes, namely photosystem I (PS I), photosystem II (PS II), cytochrome b_6f (Cyt b_6f), and ATP synthase, are embedded in the membrane system. Photosystems absorb light and carry out electron transport. PS II oxidizes $2H_2O$ to O_2 and liberates $4H^+$, consequently reducing quinones Q_A^-/Q_B^-. Electron transport from the reduced quinone pool to PS I is mediated by Cyt b_6f complex. PS I generates reduced species NADPH at the end of the electron transport chain. The proton gradient created during photoelectron transport is utilized by ATPase to produce ATP from ADP and conserve energy in chemical form. Chloroplast, thus, is the solar powerhouse of green plants. It absorbs photon and converts it to electrical energy through a series of photoredox reactions. The energy conserved in ATP and redox potential gained by photoelectron transport is finally utilized for various metabolic processes in chloroplast, namely carbon assimilation, fatty acid synthesis, nitrogen assimilation, and sulfur assimilation.

3.2 *Carbon Assimilation*

Carbon is assimilated to produce sugar by a number of enzymes in the Calvin-Benson cycle. Carboxylation of ribulose-1,5-bisphosphate molecule by an enzyme ribulose-1,5-bisphosphate carboxylase-oxygenase (Rubisco) present in the stroma is the first step towards carbon assimilation (see Spreitzer and Salvucci 2002). Rubisco consists of 8 large subunits (LSU) and 8 small subunits (SSU) forming an oligomer L_8S_8 (Knight et al. 1990, Taiz and Zeiger 1998). Though the LSU may assemble as an octamer L_8 and contains the catalytic active site, it needs to incorporate SSUs into the complex for the normal carboxylase activity (Andrews 1988). Further, Mg^{2+}, CO_2, and an ATP-dependent activase play important roles in activation of Rubisco (Andrews and Lorimer 1987, Salvucci and Ogren 1996).

3.2.1 Regulation of Carbon Dioxide Fixation

3.2.1.1 Light

The pH of stroma increases from 7 to 8 with simultaneous increase in the level of Mg^{2+} upon exposure to light. Higher pH facilitates formation of HCO_3^- by dissolution of CO_2 in the alkaline aqueous medium. HCO_3^- carbamylates deprotonated ε-NH_2 group of Lys201 of Rubisco. It leads to Mg^{2+} binding to the active site along with HCO_3^- as a ligand to Mg^{2+}. The photosynthetic CO_2 assimilation is regulated by light through the change of pH of stroma (see Biswal et al. 2003).

3.2.1.2 Thioredoxin

It is a 12 kD protein, which contains a disulfide bridge. Reduced ferredoxin (Fd) in the presence of ferredoxin-thioredoxin reductase (FTR) reduces the disulfide group of thioredoxin to sulfyhydryl groups upon illumination. Reduced thioredoxin activates a number of enzymes in the Calvin-Benson cycle by reducing disulfide bridges in the enzymes (Raines et al. 1999). Activation of sedoheptulose-1,7-bisphosphatase (SBPase) is one of the examples (Raines et al. 1999).

3.2.1.3 Starch Synthesis

Starch is synthesized in chloroplast during exposure to light and broken down in dark. CO_2 fixation in the Calvin-Benson enzyme cycle yields fructose-1,6-bisphosphate. It is dephosphorylated to fructose-6-phosphate and then isomerizes to glucose-1-phosphate. ADP-glucose pyrophosphorylase adenylates glucose-1-phosphate. Starch synthase and branching enzymes act upon ADP-glucose to form starch (Emes and Tobin 1993).

3.3 Nitrogen Assimilation

Chloroplast is the site for assimilation of most of the NH_3 (Emes and Tobin 1993). Nitrate reductase reduces NO_3^- to NO_2^- in the cytoplasm. The NO_2^- is transported into the chloroplast by nitrite translocator (Brunswick and Cresswell 1988), where it is further reduced to NH_3 by ferredoxin-nitrite reductase (Crawford 1995). Glutamate dehydrogenase may catalyze the reaction between α-ketoglutarate and NH_3 to produce glutamic acid (Glu). Plastidic glutamine synthetase 2 (GS2) may act upon Glu and NH_3 to form glutamine (Gln) (Lam et al. 1995). Nitrogen could be incorporated into other amino acids by the action of transaminase once it is assimilated into Gln or Glu.

3.4 Sulfur Assimilation

Sulfur is an essential element for growth and physiological functioning of plants. Its content varies from 0.1% to 1% of the plant's dry weight (Pilon-Smits and Pilon 2006). Sulfur is taken up by plants in the form of sulfate through the roots and is then transported to shoots. Sulfate is transported across the envelope of chloroplast through sulfate permease. Its subsequent reduction and incorporation into organic compounds to produce Cys takes place in the chloroplasts as shown in Fig.8.1. Cys serves as the precursor of most other organic sulfur compounds in plants. About 70% of total organic sulfur in plant is present in the protein fraction as Cys and methionine (Met) residues. Cys and Met play a significant role in the structure and function of proteins. Thiol (glutathione), sulfolipids and secondary sulfur compounds, namely alliins, glucosinolates, and phytochelatins, are other forms of sulfur in plants. These compounds play an important role in physiology, food quality of crops, phytopharmaceutics, and protection against environmental stress and pests (see Schung, 1998, Grill et al. 2001, Abrol and Ahmad 2003).

3.5 Cross Talk between Carbon, Nitrogen, and Sulfur Metabolic Pathways

Sulfur and nitrogen assimilation are coordinated to maintain a ratio of reduced sulfur to reduced nitrogen of about 1:20. Reduced sulfur compounds stimulate nitrate reductase, and reduced nitrogen compounds stimulate ATP sulfurase and APS reductase, establishing cross talk between the two pathways (Reuveny et al. 1980, Koprivova et al. 2000). The reduced sulfur and nitrogen compounds also inhibit key enzymes of their own biosynthetic pathways by feedback mechanism. Sulfur uptake is not stimulated under nitrogen-limiting condition (Yamaguchi et al. 1999). Although cross talk between carbon and nitrogen pathways is the basis of most of the biochemical routes, the cross talk between carbon pathway and sulfur assimilation is also extensive and covers a complex network as depicted in Fig. 8.1.

The coordination of sulfur metabolism with the carbon and nitrogen assimilatory pathway is required for optimization of sulfur assimilation in chloroplasts, specifically when plants experience fluctuation in environmental conditions (Kopriva and Rennenberg 2004).

4 Sulfur Compounds in Chloroplasts

4.1 Sulfur Reduction in Chloroplast

Sulfate reduction predominantly takes place in the leaf chloroplasts (Pilon-Smits and Pilon 2006). Sulfate enters cytosol across sulfate permeases embedded in plasma membrane. The chloroplast envelope also contains sulfate permeases to import sulfate into stroma. Sulfate in stroma is activated to adenosine 5′-phosphosulfate (APS) prior to its reduction to sulfite. The activation is catalyzed by ATP sulfurylase, and APS is subsequently reduced to yield sulfite by APS reductase. GSH is the probable electron donor in this catalysis (Pilon-Smits and Pilon 2006). Sulfite is reduced to sulfide by sulfite reductase with Fd as a reductant.

4.2 Synthesis of Sulfur Compounds in Chloroplasts

Sulfolipids are synthesized from sulfite. A part of sulfite enters into the formation of sulfolipids. Sulfite is coupled to UDP-glucose catalyzed by sulfolipid-6-sulfo-α-D quinovosyl diacylglycerol (SQD1) enzyme to form UDP-sulfoquinovose, which is coupled to diacylglycerol to produce sulfolipid 6-sulfo-α-D-quinovosyl diacylglycerol (SQDG) (Pilon-Smits and Pilon 2006). The sulfolipids are unique to chloroplasts and are required for plastid functions including photosynthesis (Yu and Benning 2003).

Cysteine is the major form of organic sulfur compound in chloroplast. It is synthesized from OAS with incorporation of sulfide catalyzed by OAS lyase. Cys acts as source of formation of other sulfur compounds. Cys is transported from chloroplast to cytosol through an unknown transporter. Reaction of Cys with O-phosphohomoserine by cystathionine-γ-synthase yields cystathionine. Homocysteine is formed from cystathionine by the action of cystathionine-β-lyase. Homocysteine is transported across the envelope to cytosol, where it is converted to Met by the action of Met synthase. Met is transported into chloroplast. Cys and Met get incorporated into newly synthesized proteins in chloroplast as well as cytosol, as indicated in Fig. 8.1.

Cys-desulfurase acts on Cys to yield sulfide and Ala. Sulfide thus formed is utilized in synthesis of thiamine, biotin, molybdenum cofactor (MoCo), and Fe-S centers.

Fe-S clusters [2Fe-2S] or [4Fe-4S] are contained in proteins and enzymes associated with chloroplast function, namely PS I, Rieske protein of Cyt b_6f, Fd, import protein Tic55, and many enzymes in nitrogen and sulfur assimilation pathways. These centers are formed and incorporated into the apoproteins in chloroplast. Elements of *suf*-machinery, and some of *nif*- and *isc*- components, along with some special proteins like cystine lyase (C-DES) and P-loop ATPase (HCF101), are believed to exist in chloroplasts, which may be responsible for Fe-S cluster metabolism (Kessler and Papenbrock 2005). NifU proteins AtNFU1-3 are identified in stroma of *Arabidopsis* chloroplasts (Leon et al. 2003, Yabe et al. 2004). Suf proteins SufA-D and SufS are associated with Fe-S cluster biosynthesis. Complete Suf type proteins are believed to be present in chloroplast (Xu and Moller 2004). SufB, -C,-D and –S are identified in *Arabidopsis* chloroplasts (Xu and Moller 2004). AtNFU2 is supposed to carry the Fe-S cluster and transfer it to apoproteins. Mutant lacking in AtNFU2 shows drastic impairment of PS I activity and decreased level of Fd (Touraine et al. 2004, Yabe et al. 2004). HCF101 is associated with the formation of [4Fe-4S] cluster in PS I (Lezhneva et al. 2004, Stöckel and Oelmüller 2004).

Glutathione is a major sulfur compound maintaining the redox status of biosystems. GSH plays a role in chloroplastic gene expression via redox-mediated control of transcription (Link 2003). It is also synthesized in chloroplasts. Cys and Glu are acted upon by γ-glutamylcysteine synthetase (GSH1) to form γ-EC, which is bonded to glycine (Gly) by the action of glutathione synthetase (GSH2) to synthesize GSH (Mullineaux and Rausch 2005). However, the detailed mechanism and location of synthesis of GSH are under active investigation for further confirmation (Mullineaux and Rausch 2005).

Phytochelatins are polypeptides with a general structure $(\gamma\text{-GluCys})_n\text{Gly}$, where n varies from 2 to 5 (Cobbett 2000). Phytochelatins are synthesized from GSH by elongation of the γ-GluCys unit by phytochelatin synthase (PS). PCs chelate metal ions and detoxify excess metal ions, specifically heavy metal ions. PS is expressed constitutively, but it is activated by metal ions, especially Cd^{2+} or As^{3+} (Cobbett 2000, Pickering et al. 2000). PC-metal ion complex is transported to vacuoles by ABC-type transporters.

4.3 Significance of Sulfur Compounds in Structure and Function of Chloroplasts

Sulfur compounds like GSH or sulfur moieties in protein in its reduced state (oxidation state-2) are mostly responsible for maintaining the general redox status in chloroplasts. Electron transfer is essential for maintenance of redox status. This is performed by reversible –S-S- /SH couple or Fe-S clusters. A [2Fe-2S] is present in Rieske protein of Cytb_6f complex in the photoelectron transport chain of thylakoid. [4Fe-4S] cluster F_X occurs in dimeric subunits PsaA/B of PS I. Besides, [4Fe-4S] centers,

F_A and $F_{B,}$ occur in the PsaC subunit of PS I, and a [2Fe-2S] cluster occurs in Fd associated with PS I. These clusters act as photoelectron transfer units. One Cys and one Met along with two His residues participate in Cu binding in plastocyanin associated with PS I. Thioredoxin plays a regulatory role for change in conformations of FBPase and malate dehydrogenase (MDH) enzymes in chloroplasts through disulfide-thiol switch (Raines et al. 1999, Pilon-Smits and Pilon 2006).

5 Chloroplast under Abiotic Stress

A mature chloroplast exists in a nonequilibrium stationary state. It tends to resist perturbation by the external forces, including variations in the temperature, intensity of light, concentration of CO_2, salinity, and osmotic pressure. The appropriate internal changes in the system tend to restore the state as far as possible. The external changes are designated as stress, and the internal changes in the organelle in response to the stress collectively constitute the adaptation process. Processes like absorption of photon, electron transport, proton translocation, CO_2 fixation, and assimilation of other nutrients operate in a coordinated manner in a mature chloroplast. Therefore, response to stress also operates in an integrated pattern (see Biswal et al. 2003).

5.1 Chloroplast as the Sensor of Stress

Light may act as a major external environmental factor for green plants. The light is sensed mainly by PS I and PS II, and subsequently the photosignals are transduced through various components of the organelle in chloroplasts. Further, the changes in other environmental factors in the presence of light may also be sensed by the photosystems, and the consequence is observed primarily through photoinhibition of PS II of thylakoid (Biswal 1997). The sensitivity of photosystems, especially PS II, makes chloroplast the sensor of stress in plants (see Biswal et al. 2003). Operation of a feedback mechanism via the redox state of the components associated with the electron transport system of PS I and PS II is reported during stress acclimation of the photosynthetic apparatus (Anderson et al. 1995, Biswal and Biswal 1999).

A shift in the redox steady state of the chloroplast system due to physiological variation induced by the stress may lead to the reduction of O_2 to toxic oxygen free radicals. A strong oxidant generated on the donor side of PS II in the stress-impaired system may oxidize pigments, proteins, or lipids of thylakoid. These special characteristics associated with the thylakoid make the photosynthetic organelle a major stress sensor in green plants (see Biswal et al. 2003).

5.2 All Abiotic Stress Factors Lead to Oxidative Stress

The key feature of chloroplast under abiotic stress is nonutilization of energy by the disturbed sink, which may create redox or excitation pressure on the photosystems. The building up of the excitation pressure initiates the events associated with the stress adaptation or stress-induced damage of the system. Chloroplasts under abiotic stress exhibit the high light stress syndrome (see Biswal et al. 2003). Ultimately, almost all abiotic stress may lead to photoinhibitory damage in chloroplasts resulting in oxidative stress as per the scheme provided in Fig. 8.2.

Models for signal transduction pathways for adaptation to stress involving various components associated with the electron transport chain, redox-responsive protein kinase, and thiol-regulated enzymes are proposed, and chlorophyll (Chl) precursors may mediate in transducing the signals in these pathways (Mullineaux and Karpinski 2002). In response to stress, the chloroplast system may employ multiple antioxidants, enzymes, or chemical species as an adaptation strategy (Lee et al. 2007, Lee YP et al. 2007).

The details of responses of chloroplasts to abiotic stress factors like high light, water stress, temperature extremes, and UV radiations that ultimately generate oxidative stress, and their modulation by sulfur metabolism are provided below.

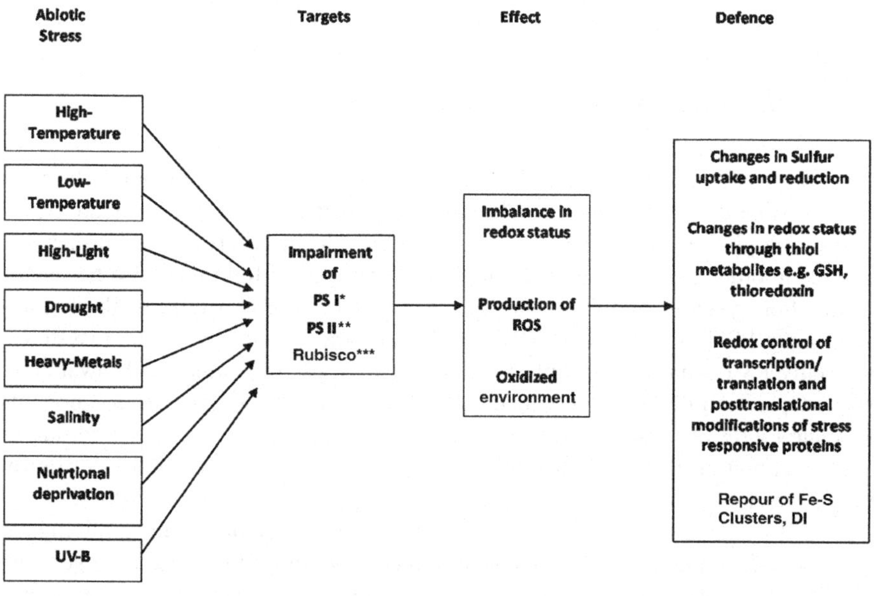

Fig. 8.2 A scheme showing that the abiotic stress leads to imbalance in the redox status of chloroplast. Consequently an oxidative stress results. Sulfur metabolism plays an important role in stress management in plant by providing a long-term and second line of defense to abiotic stress. * mark in the second box indicates the susceptibility of PSI, PS II, and Rubisco to abiotic stress factors. The higher the number, the greater the susceptibility of the system to stress

5.3 High Light- Induced Stress

5.3.1 PS II is the Target of High Light Stress

High light stress results when light is absorbed by photosystems of thylakoid, and its utilization in CO_2 fixation are not properly coordinated. Normally, exposure to light intensity between 1000-2000 μM $m^{-2}s^{-1}$ for a plant is considered as high light in experimental conditions. PS II is the main target of high light stress. The light absorbed by the special pair pigment P680 of the PS II reaction center (RC II) leads to the formation of a charge pair $Pheo^-$ (pheophytin) and $P680^+$. The latter is reduced by Y_Z (Tyr161, D1), yielding Y_Z^+. Both $P680^+$ and Y_Z^+ are strong oxidants. Y_Z^+ oxidizes O_2 evolving complex (OEC). The OEC has a very delicate structure and is a soft target to stress. The instability of metal cluster Mn_4Ca in OEC under stress makes the complex prone to damage. Damaged OEC does not effectively donate electrons to Y_Z^+ and $P680^+$. The life-times of these strong oxidants consequently become longer, and they may oxidize pigments, lipids, and amino acids of proteins in the vicinity of PS II (Andersson and Barber 1996, Biswal et al. 1997). D1 protein is oxidized by the strong oxidant $P680^+$ and is subsequently degraded (Jegerschold and Styring 1996). The charge recombination process between Q_A^-/Q_B^- and long-lived $P680^+$ brings about the formation of $^3P680$ and the subsequent production of highly toxic $^1O_2$. Since the half-life of $^1O_2$ is short, its primary target is PS II, particularly its reaction center proteins, especially D1 (Keren et al. 1997). D1 protein is prone to oxidation at the stromal side of the second helix and lumenal side of the fourth helix, probably due to $^1O_2$ produced at RC II as revealed by the mass spectrometric studies (Barber and Sharma 2000). Oxidation of D1 protein may lead to proteolytic degradation of the protein (Barber and Sharma 2000).

5.3.2 Turnover of Reaction Center II Proteins

The turnover of D1 protein of RC II during high light stress has been considered as one of the major adaptive responses of PS II to stress. In addition to high light, the light-dependent turnover of the protein has been reported during water stress (Giardi et al. 1996, Deo and Biswal 2001) and nutritional stress (Godde and Hefer 1994). The association of sulfur metabolism with amelioration of stress is indicated by the fact that D1 biosynthesis and reassembly require sulfate uptake (Vasilikiotis and Melis 1994, Melis and Chen 2005).

5.3.3 Gene Expression in Response to High Light

The redox signaling system is believed to regulate *lhcb* gene expression at varying light intensities (Durnford and Falkowski 1997). The level of D1 protein is enhanced under high light (Shapira et al. 1997) and decreased under low-light intensity

(Sailaja and RamaDas 1995). A distant relative of *lhca* is also reported to express under low- or high light stress. A novel one-transmembrane helix containing protein Ohp2 is identified, which is a distant relative of light-harvesting proteins associated with PS I (Andersson et al. 2003). Light stress triggers *ohp2* gene expression. Other stress conditions did not upregulate the expression of *ohp2*. The accumulation of Ohp2 might be a novel photoprotective strategy induced within PS I in response to light stress.

Proteomic analysis of *Arabidopsis* exposed to normal (100 μM $m^{-2}s^{-1}$) high light (1000 μM $m^{-2}s^{-1}$) conditions implicates 64 proteins associated with high light stress (Phee et al. 2004). Out of the 52 proteins selected for analysis, 35 photosynthetic proteins are downregulated, 14 proteins including heat shock proteins (HSPs) and dehydroascorbate reductase (DHAR) are upregulated, and 3 novel proteins are synthesized with unknown functions.

5.4 Water Stress

The term water stress is used for a water deficit condition and is very often referred to as drought stress. Under drought stress leaf stomata are closed to restrict CO_2 assimilation and H_2O loss (Cornic 1994), resulting in the inhibition of photosynthesis (Bradford and Hsiao 1982). Although the stress causes alterations in different biochemical pathways (Graan and Boyer 1990, Lauer and Boyer 1992, Lu and Zhang 1998, Ramachandra Reddy et al. 2004), PS II is known to be the primary target of drought stress. OEC of PS II (Canaani et al. 1986, Toivonen and Vidaver 1988) and the PS II reaction center proteins and their turnover pathways (He et al. 1995, Giardi et al. 1996) are reported to be affected. The stress also induces reduction in the levels and activities of the enzymes of the photosynthetic carbon reduction cycle, including the key enzyme, Rubisco (Ramachandra Reddy et al. 2004). Under stress conditions the differential loss in the efficiency of photoelectron transport and carbon reduction cycle may result in imbalance in the redox state of photosynthetic organelles. This situation still worsens if the light- driven electron transport proceeds unabated.

The effect of water stress is known to increase the level of H_2O_2 and $O_2^{\bullet-}$ radicals and to enhance the activities of SOD, ascorbate peroxidase, and glutathione reductase (GR) (Jiang and Zhang 2002). However, Mahan and Wanjura (2005) have observed an increase in the amount of ascorbate and ascorbate peroxidase activity without any change in GSH concentration in field-grown cotton. They have argued that the GSH metabolism in field-grown cotton is enough to meet the challenge of water-stress-mediated oxidative damage. On the other hand, species-dependent contradictory reports on the level of GSH under water stress are available. While the level of GSH is found to increase by 3.2 times in pea (Iturbe-Ormataexe et al. 1998), it declines two-fold in rice (Srivalli et al. 2003). The content in ascorbate and reduced GSH is also known to increase in response to the stress. Water stress stimulates the *de novo* synthesis of GR (Pastori and Trippi 1992). In general there is a higher level of GSH and ascorbate in water deficit plants. An elevated GSH content is usually correlated with the increased adaptive response of plants to abiotic stress (Kocsy et al. 1996, Okane et al. 1996)

Therefore, most of what we know about plant defense associated with sulfur metabolism is determined by the synthesis and reduction level of GSH. In association with Cys, GSH induces the accumulation of transcripts of Cu/Zn superoxide dismutase (SOD) enzyme and elevates the activities of SOD and ascorbate-GSH cycle (McKersie et al. 1996, Hawkesford and De Kok 2006), which are known to scavenge the ROS. However, under acute stress conditions there is a drop in GSH concentration, resulting in an oxidized redox state that initiates the degradation of the organelle (Tausz et al. 2004).

In addition to the modulation of drought-induced oxidative stress, sulfur metabolism may be involved in maintaining structural integrity of biomembranes. In response to water stress, there is an increase in accumulation of SQDG in drought-resistant plants (Okanenko and Taran 1998). These lipids, besides stabilizing light harvesting complex II under stress conditions, are localized as the prosthetic group at the surface of the D_1/D_2 reaction center and help in holding them together (Sigrist et al.1988). Therefore, an enhanced level of SQDG accumulation in water stress could help in maintaining membrane fluidity and stabilizing PS II organization.

5.5 *Temperature Stress*

5.5.1 Low Temperature Targets Photosystem I

PS I is the target of stress under low temperature (4 °C), at low light (100-200 µmol $m^{-2}s^{-1}$), and in the presence of oxygen (Terashima et al. 1994, Sonoike 1996). Chilling stress inhibits PS I activity but does not significantly inhibit PS II activity (Terashima et al. 1994). Chilling impairs iron- sulfur centers: F_A/F_B and F_X of PS I (Sonoike 1996). The impairment may be caused by $O_2^{\bullet -}$ generated on the reducing side of PS I (Miyake and Asada, 1992, Ogawa et al. 1995). Finally, a serine type protease may be involved in proteolysis of the subunit B of PS I and may cleave the protein to yield 45, 51 and 18 kD fragments (Sonoike et al. 1997). Though the event is similar to the D1 degradation of PS II, PsaB turnover rate is extremely slow compared to that of D1 and requires several days (Sonoike 1996).

5.5.2 High-Temperature Stress

The high temperature stress impairs PS II (Vani et al. 2001). Temperature in the range of 45-60 °C inactivates Rubisco (Li et al. 2002). The inactivation may be related to the thermal denaturation of activase (Salvucci et al. 2001). High temperature may lead to sequestration of Rubisco activase to thylakoid membrane. SBPase activity prevents the sequestration of the activase and maintains the activity of Rubisco. Over expression of SBPase enhances photosynthesis against high-temperature stress (Feng et al. 2007). There is also report that heat may induce expression of a novel gene for activase, and this novel form of activase may play a role in the acclimation process (Law et al. 2001). Inactivation of PS II and Rubisco due to heat stress results in similar stress-induced manifestations, as in case of high light stress. The link is suggested by the fact that the plastid-specific heat shock proteins are linked to the protection of the photosynthetic organelle during light and heat stress (Biswal 1997, Schroda et al. 1999, Lee BH et al. 2000). Heat stress (42 °C) in rice induces expression of the *oshsp26* gene, which encodes chloroplast-localized small heat shock protein (smHSP). This gene is also expressed when rice plant is treated with oxidative agents like methyl viologen in light or H_2O_2 in light (or dark). The results indicate the protective role of smHSP against both oxidative and heat stresses (Lee BH et al. 2000).

The nature of the function of heat shock proteins to protect chloroplasts is not clear. However, there is a report that conserved Met in chloroplast Hsp21 may be involved in the formation of groove for sequence-independent recognition of hydrophobic domains in peptides (Sundby et al. 2005). Hsp21 may bind to hydrophobic domains of proteins, preventing its aggregation during stress. Debel et al. (1997) have observed the accumulation of a 23 kD nuclear encoded heat shock protein in the mitochondria under light stress. These authors have proposed a possible coupling of the chloroplast and mitochondrial functions in protecting the photosynthetic organelle against the stress.

5.6 UV Radiation

A lot of work has been conducted recently to understand the mechanisms of damage and repair of the photosynthetic apparatus of green leaf exposed to UV radiation (McKenzie et al. 2003). The radiation is known to affect the growth, photosynthetic function, and alteration in gene expression (Bornman 1989, Jordan 1996, Teramura and Ziska 1996). The radiation reduces the activities of chloroplast ATPase (Brosche and Strid 2003), Rubisco (Jordan et al. 1992) and violaxanthin de-epoxidase (Pfundel et al. 1992, Bischof et al. 2002). The PS II of photosynthetic apparatus is affected the most among all thylakoid complexes. The radiation damages the D_1/D_2 reaction center proteins (Friso et al. 1994 a, b), OEC and impairs oxidizing as well as reducing sides of PS II at multiple sites (Jordan 1996, Teramura and Ziska 1996). In the process, the radiation, through dismantling of thylakoid complex and loss of Rubisco, generates ROS, which are considered to be the sensor of UV-B responses (Brosche and Strid 2003)

The ascorbate-GSH cycle, which regenerates a large pool of ascorbate, contributes enormously to the management of oxidative stress (Salin 1987, Kunert and Foyer

1994). The enhancement in the level of ascorbate, GSH (Takeuchi et al. 1996, Jansen et al. 1998), and in the activities of SOD (Rao et al. 1996, Jansen et al. 1996), GR (Jansen et al. 1996, Rao et al. 1996), and ascorbate peroxidase (Landry et al. 1995, Takeuchi et al. 1996) in response to UV-B stress could, therefore, be considered as an adaptive process involving metabolic adjustment. In the process of adjustment it is not necessary that all these parameters increase. Even a significant rise in the ratio of reduced GSH to oxidized GSH without any alteration in ascorbate content in response to UV-B stress is sufficient to maintain the redox status of the cell to meet the challenge posed by oxyfree radicals (Costa et al. 2002). Rao and Ormrod (1995), on the other hand, have observed that *Arabidopsis thaliana*, which possesses an effective free radical scavenging system, is susceptible to UV-B stress when its flavonoid biosynthesis is blocked. But the observation made by A-H-Mackerness et al. (1998) that the antioxidant enzyme activities depend on the rate of free radical generation make us believe that the enzymatic activities in plants depend on efficient organization of different metabolic processes, including sulfur metabolism.

6 Sulfur Metabolism Mediated Modulation of Oxidative Stress

Chloroplasts under various kinds of stresses, as discussed above, ultimately experience oxidative stress. The plants, therefore, develop a second line of defense mechanism against ROS, like $O_2^{\bullet-}$, H_2O_2, $^{\bullet}OH$, and $^1O_2$. Disulfide/thiol exchange reactions involving the GSH pool and the production of H_2O_2 are believed to be crucial for regulation of stress adaptation at the molecular level (Foyer et al. 1997). The oxidative stress induces a remarkable increase in the transcripts of two ascorbate peroxidase genes. These enzymes are cytoplasmic and are involved in regulating the redox level of the GSH pool. SODs, ascorbate peroxidase, DHAR, and GSH are associated with defense mechanisms against stress. It is important to note that sulfur deprivation leads to oxidative stress in plants (Pilon-Smits and Pilon 2006).

The origins and factors responsible for production of ROS in chloroplast in light are complex. The studies on photosynthetic response to various abiotic stress factors however, indicate that Rubisco is the most susceptible component of the organelle to the stress. This consequently results a loss in the efficiency of Calvin-Benson cycle compared to the loss in primary photochemical reaction associated with the thylakoid membrane. A weak sink (Calvin-Benson cycle) for photoelectron transport leads to accumulation of excess unused quanta and excess electrons in the organelle. Under this condition, O_2 through the components of PS I receives the electrons and is reduced to $O_2^{\bullet-}$, which subsequently, through a series of redox reactions involving ascorbate-GSH cycle, are converted to H_2O as summarized in Fig.8.3.

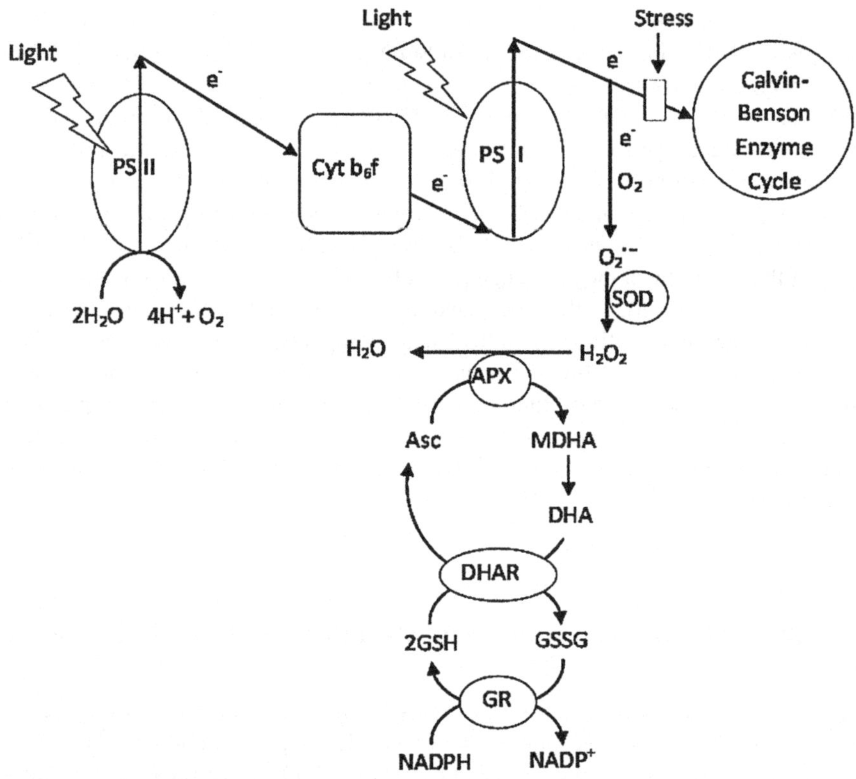

Fig. 8.3 A scheme depicting glutathione-mediated scavenging of reactive oxygen species produced under stress. APX: ascorbate peroxidase, DHA: dehydroascorbate, DHAR: dehydroascorbate reductase, GSH: Glutathione, GSSG: oxidized glutathione, GR: glutathione reductase; MDHA: monodehydroascorbate, $NADP^+$: nicotinamide adenine dinucleotide phosphate (oxidized), NADPH: reduced $NADP^+$, SOD: superoxide dismutase. (Modified after Asada 2000.)

7 Stress-Induced Redox Signals Generated in Chloroplasts, Their Transmission and Gene Expression

Light-induced redox signals generated in chloroplast are known to participate in the expression of photosynthetic genes (see Biswal et al. 2003). The signals not only modulate gene expression in chloroplast but are also transmitted to cytoplasm for expression of stress-related nuclear genes.

7.1 Signal Transduction and Modulation of Plastid Gene Expression

The best example of stress-induced plastid gene expression is the high light-induced turnover of D1 protein of PS II of thylakoid. Although the precise mechanism of

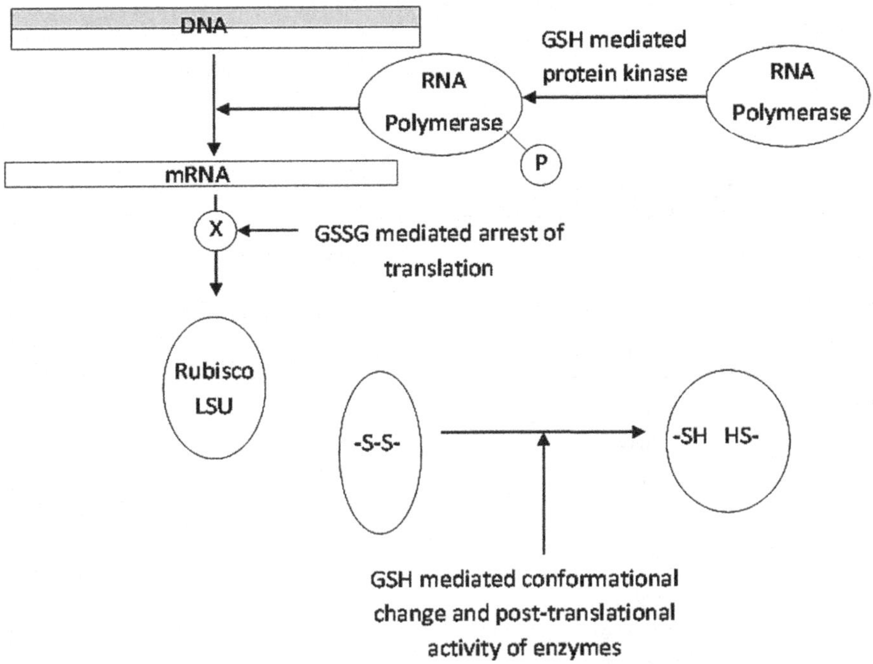

Fig. 8.4 A schematic representation of glutathione-mediated redox regulation of expression of chloroplastic gene. GSH: glutathione, GSSG: oxidized glutathione, P: phosphate, Rubisco LSU: large subunit of Rubisco. X represents blockage of pathway

redox signaling for synthesis of D1 protein is not known, high light is known to activate the *psbA* gene that codes for D1 protein through a signal induced by a high level of reduced quinone, which on oxidation leads to the inactivation of gene transcription. This suggests plastid gene expression to be under the control of a kind of redox signaling system. The involvement of redox-regulated thylakoid protein kinase and subsequent downstream signaling to the level of gene expression could be a possibility. Similarly, redox control of translation of *psbA* is known to remain under the control of Fd-thioredoxin system (see Mullineaux and Rausch 2005).

RNA polymerase phosphorylation through GSH-mediated signaling, as depicted in Fig.8.4, may be considered as one of the thiol redox controls of plastid gene expression. The ROS-induced enhancement in the oxidized pool of GSH is likely to arrest translation of the Rubisco large subunit (Irihimovitch and Shapira 2000).

7.2 *Chloroplast to Nucleus Signal Transduction is Primarily Redox Controlled*

Nuclear gene expression through a plastid signal generated by the photooxidative environment of the organelle induced by high light stress has long been known

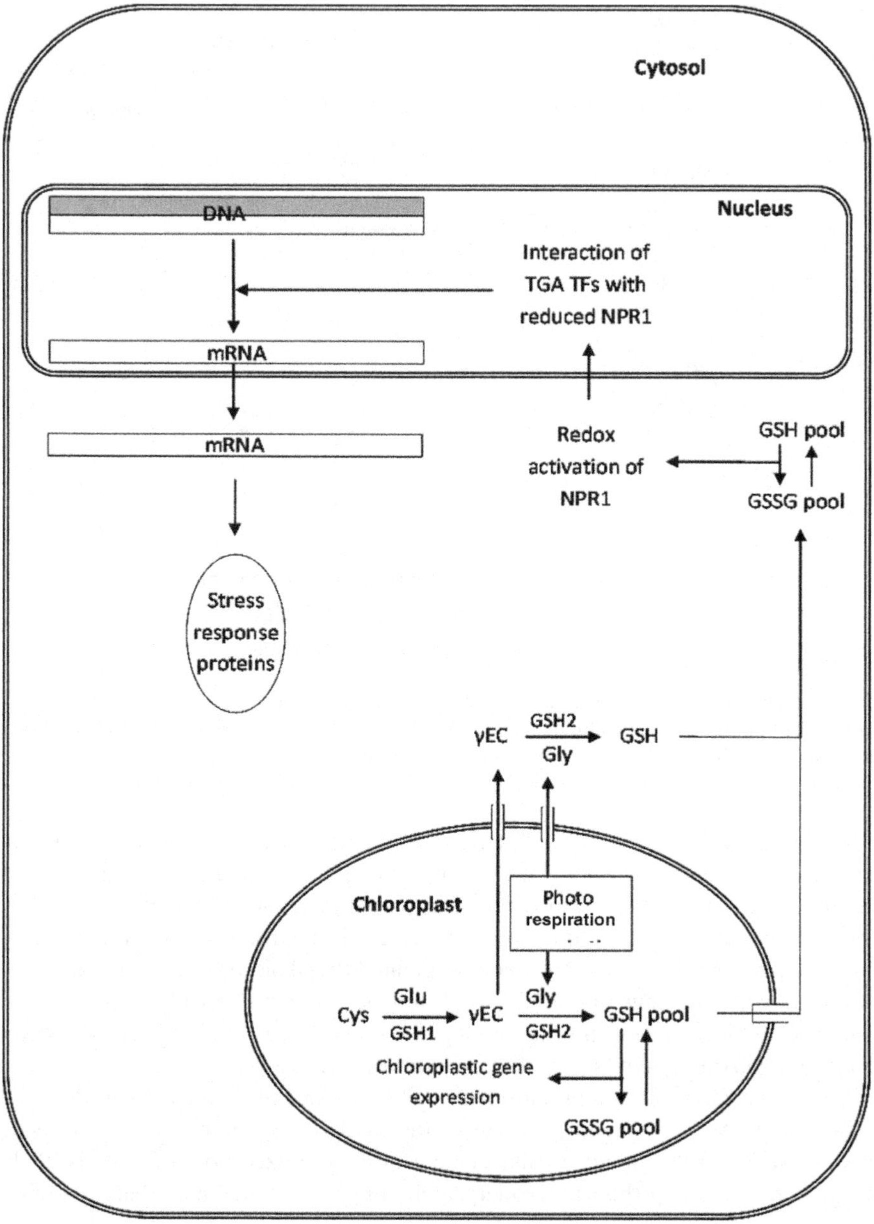

Fig. 8.5 A scheme representing the stress-responsive nuclear gene expression with a redox signal from chloroplast. Cys: cysteine, γEC: γ-glutamylcysteine, Gly: glycine, GSH1: γEC synthetase, GSH2: glutathione synthetase, GSH: glutathione, GSSG: oxidized glutathione, NPR1: nonexpresser of pathogenesis-related genes- a stress related protein, TGA TFs: transcription factors. Gly produced by photorespiration is provided for GSH synthesis, which results in photosynthetic regulation of GSH pool

(see Biswal et al. 2003). As discussed earlier, chloroplast is the most sensitive organelle to abiotic stress. The abiotic stress factors affecting the organelle produce a cellular oxidative environment, which may alter the redox status of cytoplasm. GSH possibly plays a major role in the chloroplast-to-nucleus communication system during oxidative stress. The redox state of cytoplasmic GSH system significantly affects gene expression through various redox-sensitive regulatory proteins, including NPR1, which are transported to the nucleus, interact with the transacting factors, and regulate the expression of stress-responsive genes. In this background, the pool of GSH in cytoplasm, which plays important role in nuclear gene expression, is modulated by the activity of chloroplast. Chloroplast provides the organic carbon skeleton for synthesis of GSH in cytoplasm, which also imports precursors of GSH namely, γEC from chloroplast. It is therefore likely that stress-induced perturbation in photosynthetic activity of chloroplast brings a change in the GSH pool and its redox status in cytoplasm and, consequently, alteration in nuclear gene expression. Fig.8.5 shows the possible transmission of plastid-specific redox signal and regulation of expression of nuclear genes.

8 Conclusion and the Future

1. Some of the fundamental features of sulfur uptake and its subsequent assimilation are known. The expression and activities of the enzymes that participate in its metabolism are now better understood in the light of information available with *Arabidopsis thaliana* as a model system. At the same time, extensive use of advanced techniques like DNA microarray, proteomics, metabolomics, and techniques relating to production of transgenics have helped to expand our understanding of the importance of sulfur assimilates in plant metabolism in general. However, the regulatory mechanisms of its metabolism specifically in photosynthetic tissue still remain complex and need attention.
2. It is almost clear that the redox changes induced by abiotic stress factors lead to changes in expression of stress-responsive genes, but the precise nature and identification of signal transduction pathways are poorly understood. Although some of the redox signaling systems mediated by redox-active thiols are known, their precise origin in chloroplast, subsequent transmission, and their cross talk with other cellular transduction systems for plastid and nuclear gene expression have to be worked out.
3. More detailed studies are required to understand the stress response modulated by sulfur metabolism at molecular and mechanistic levels in order to develop an effective strategy to raise transgenic species for stress resistance. It is necessary to identify the stress-relevant genes and the redox active proteins and define their functions. Once the mechanism at the level of genes is understood, its manipulation in the direction of biotechnological application will be possible. It is important to note that redox homeostasis is a very complex and interwoven metabolic network. Careful and judicious manipulation can only lead to biotechnological manipulation for producing stress resistance traits without a gross misbalance in the network.

References

Abrol YP, Ahmad A (2003) Sulphur in Plants. Kluwer Academic Publishers, Dordrecht, The Netherlands

A-H-Mackerness S, Surplus SL, Jordan BR, Thomas B (1998) Effects of supplementary UV-B on photosynthetic transcript at different stages of leaf development and light levels in pea (*Pisum sativum* L.): role of ROS and antioxidant enzymes. Photochem Photobiol 68:88-96

Anderson JM, Chow WS, Park Y (1995) The grand design of photosynthesis: acclimation of the photosynthetic apparatus to environmental cues. Photosynth Res 46:129-39

Andersson B, Barber J (1996) Mechanisms of photodamage and protein degradation during photoinhibition of photosystem II. In: Baker NR (ed) Photosynthesis and the environment. Kluwer Academic Publishers, Dordrecht, The Netherlands, pp. 101-21

Andersson U, Heddad M, Adamska I (2003) Light stress-induced one-helix protein of the chlorophyll a/b-binding family associated with photosystem I. Plant Physiol 132:811-20

Andrews TJ (1988) Catalysis by cyanobacterial ribulose bisphosphate carboxylase large subunits in the complete absence of small subunits. J Biol Chem 263:12213-19

Andrews TJ, Lorimer GH (1987) Rubisco: structure, mechanisms, and prospects for improvement. In: Hatch MD, Boardman NK (eds) The biochemistry of plants, vol. 10. Academic Press, San Diego, pp. 131-218

Asada K (2000) The water-water cycle as alternative photon and electron sinks. Phil Trans R Soc Lond B 355:1419-31

Barber J, Sharma J (2000) Application of mass spectrometry to the study of photosystem II. In: Yunus M, Pathre U, Mohanty P (eds), Probing photosynthesis: mechanisms, regulation and adaptation, Taylor & Francis, London, pp. 413-25

Bischof K, Krabs G, Hanelt D, Wiencke C (2002) Solar ultraviolet radiation affects the activity of ribuluse-1,5 bisphosphate carboxylase oxygenase and composition of photosynthetic and xanthophyll cycle pigments in intertidal green alga *Ulva lactuca* L. Planta 5:502-9

Biswal B (1997) Chloroplasts, pigments and molecular responses of photosynthesis under stress. In: Pessarakli M (ed), Handbook of photosynthesis, Marcel Dekker Inc, New York, pp. 877-85

Biswal B, Biswal UC (1999) Photosynthesis under stress: stress signals and adaptive response of chloroplasts. In: Pessarakli M (ed), Handbook of plant and crop stress, Marcel Dekker Inc, New York, pp. 315-36

Biswal UC, Biswal B, Raval MK (2003) Chloroplast biogenesis: from proplastid to gerontoplast, Kluwer Academic Publishers, Dordrecht, The Netherlands

Biswal UC, Raval MK, Biswal B, Joshi PN (1997) Photosystem II and its response to stress. In: Purohit SS (ed), Agro's annual review of plant physiology, Agro Botanical Publishers, Bikaner, India, pp. 36-71

Bornman JF(1989) Target sites of UV-B radiation in photosynthesis of higher plants. J Photochem Photobiol (B: Biology) 4:145-58

Bradford KJ, Hsiao TC (1982) Stomatal behaviour and water relations of waterlogged tomato plants. Plant Physiol 70:1508-13

Brosche M, Strid A (2003) Molecular events following perception of UV-B radiation by plants. Physiol Plant 117:1-10

Brunswick P, Cresswell CF (1988) Nitrate uptake into intact pea chloroplast II. Influence of electron transport regulates uncoupler ATPase and anion uptake inhibitors and protein binding reagents. Plant Physiol 86:384-9

Canaani C, Havaux M, Malkin S (1986). Hydroxylamine, hydrazine and methylamine donate electrons to the photo-oxidizing side of PSII in leaves inhibited in oxygen evolution due to water stress. Biochim Biophys Acta 851:151-5

Cobbett SC (2000) Phytochelatins and their roles in heavy metal detoxification. Plant Physiol 123:825-32

Cornic G (1994) Drought stress and high light effect on photosynthesis. In: Baker NR, Boyer JR (eds) Photoinhibition of photosynthesis, Oxford Bios Scientific Publishers pp. 297-313

Costa H, Gallego SM, Tomaro ML (2002) Effect of UV-B radiation on antioxidant defense system in sunflower cotyledons Plant Sci 162:939-45

Crawford NM (1995) Nitrate: nutrient and signal for plant growth. Plant Cell 7:859-68

Debel K, Sierralta WD, Braun HP, Schmitz UK, Kloppstech K (1997) The 23 kDa light stress regulated heat shock protein of *Chenopodium rubrum L.* is located in the mitochondria. Planta 201:326-33

Deo PM, Biswal B (2001) Response of senescing cotyledons of clusterbean to water stress in moderate and low light: possible photoprotective role of ß-carotene. Physiol Plant 112:47-54

Domínguez-Solís, JR, Gutiérrez-Alcalá G, Vega JM, Romero LC, Gotor C (2001) The cytosolic *O*- acetylserine (thiol) lyase gene is regulated by heavy metals and can function in cadmium tolerance. J Biol Chem 276:9297-9302

Durnford DG, Falkowski PG (1997) Chloroplast redox regulation of nuclear gene transcription during photoacclimation. Photosynth Res 53:229-41

Emes MJ, Tobin AK (1993) Control of metabolism and development in higher plant plastids. Int Rev Cytol. 145:149-216

Feng L, Wang K, Li Y, Tan Y, Kong J, Li H, Li Y, Zhu Y (2007) Overexpression of SBPase enhances photosynthesis against high temperature stress in transgenic rice plants. Plant Cell Rep 26:1635-1646

Foyer CH, Lopez-Delgado H, Dat JF, Scott IM (1997) Hydrogen peroxide and glutathione-associated mechanisms of acclimatory stress tolerance and signaling. Physiol Plant 100:241-54

Friso G, Barbato R, Giacometti GM, Barbar J (1994a) Degradation of D_2 protein due to UV-B irradiation of the reaction centre of PS II. FEBS Lett 339:217-21

Friso G, Spetea C, Giacometti GM, Vass I, Barbato R (1994b) Degradation of photosystem II reaction centre D_1 protein induced by UV-B radiation in isolated thylakoids. Identification and characterization C and N terminal breakdown product. Biochim Biophys Acta 1184:78-84

Giardi MT, Cona A, Geiken B, Kucera T, Masojidek J, Mattoo AK (1996) Long term drought stress induces structural and functional reorganisation of photosystem II. Planta 199:118-25

Godde D, Hefer M (1994) Photoinhibition and light dependent turnover of D1 reaction centre polypeptide of PSII are enhanced by mineral stress conditions. Planta 193:290-9

Graan T, Boyer JS (1990) Very high CO_2 partially restores photosynthesis in sunflower at low water potentials. Planta 181:378-84

Grill D, Tausz M, De Kok LJ (2001) Significance of glutathione to plant adaptation to the environment. Kluwer Academic Publishers, Dordrecht

Gutierrez-Marcos JF, Roberts MA, Campbell EI, Wray JL (1996) Three members of a novel small gene-family from *Arabidopsis thaliana* able to complement functionally an *Escherichiacoli* mutant defective in PAPS reductase activity encode proteins with a thioredoxin-like domain and 'APS reductase' activity. Proc Natl Acad Sci USA 93:13377-82

Harada E, Yamaguchi Y, Koizumi N, Sano H (2002) Cadmium stress induces production of thiol compounds and transcripts for enzymes involved in sulfur assimilation pathways in *Arabidopsis*. J Plant Physiol 159:445-8

Hawkesford MJ, De Kok LJ (2006) Managing sulfur metabolism in plants. Plant Cell Environ 29:382-95

He JX, Wang L, Liang H (1995) Effects of water stress on photochemical function and protein metabolism of photosystem II in wheat leaves. Physiol Plant 93:771-7

Irihimovitch V, Shapira M (2000) Glutathione redox potential modulated by reactive oxygen species regulates translation of Rubisco large subunit in the chloroplast. J Biol Chem 275:16289-95

Iturbe-Ormataexe I, Escuredo PR, Arresse-Igor C, Becana M (1998) Oxidative damage in pea plants exposed to water deficit or paraquat. Plant Physiol. 111:1145-52

Jansen MAK, Gaba V, Greenberg B (1998) Higher plants and UV-B radiation: balancing damage, repair and acclimation, Trends Plant Sci 3:131-5

Jansen MAK, Babu TS, Heller D, Gaba V, Mattoo AK, Edelman M (1996) Ultraviolet-B effect on *Spirodela oligorrhiza*: induction of different protection mechanisms. Plant Sci 115:217-23

Jegerschold C, Styring S (1996) Spectroscopic characterisation of intermediate steps involved in donor side induced photoinhibition of photosystem II. Biochemistry 35:7794-7801

Jiang M, Zhang J (2002) Water stress-induced abscisic acid accumulation triggers the increased generation of reactive oxygen species and up-regulates the activities of antioxidant enzymes in maize leaves. J Exp Bot 53:2401-10

Jordan BR (1996) The effects of ultraviolet-B radiation on plants: a molecular perspective. Adv Bot Res 22:97-162

Jordan BR, He J, Chow WS, Anderson JM (1992) Changes in the mRNA levels and polypeptide subunits of rebulose-1,5-biphosphate carboxylase in response to UV-B radiation. Plant Cell Environ 15:91-8

Keren N, Berg A, van Kan PJM, Levanon H, Ohad I (1997) Mechanism of photosystem II photoinactivation and D1 protein degradation at low light: the role of back electron flow. Proc Natl Acad Sci USA 94:1579-84

Kessler D, Papenbrock J (2005) Iron–sulfur cluster biosynthesis in photosynthetic organisms. Photosynth Res 86:391-407

Knight S, Andersson I, Branden CI (1990) Crystallographic analysis of ribulose 1,5-bisphosphate carboxylase from spinach at 2.4 Å resolution. Subunit interactions and active site. J Mol Biol 215:113-60

Kocsy G, Brunner M, Ruegsegger A, Stamp P, Brunold C (1996) Gluotathione synthesis in maize genotypes with different sensitivities to chilling. Planta 198:365-70

Kopriva S, Rennenberg H (2004) Control of sulphate assimilation and glutathione synthesis: interaction with C and N metabolism. J Exp Bot 55:1831-42

Koprivova A, Suter M, op den Camp R, Brunold C, Kopriva S (2000) Regulation and sulfate assimilation by nitrogen in *Arabidopsis*. Plant Physiol 122:737-46

Kunert KJ and Foyer CH (1994) Thiol/disulphide exchange in plants. In: De Kok LJ, Stulen I, Rennenberg H, Brunhold C, Rausen W (eds), Sulfur nutrition and assimilation in higher plants: regulatory, agricultural and environmental aspects, SPB Academic Publishers, The Hague, pp. 132-64

Lam HM, Coschigano K, Schultz C, Melo-Oliveira R, Tjaden G, Oliveira I, Ngai N, Hsieh MH, Coruzzi G (1995) Use of *Arabidopsis* mutants and genes to study amide amino acid biosynthesis. Plant Cell 7:887-98

Landry LG, Chapple CCS, Last RL (1995) *Arabidopsis* mutants lacking phenolic sunscreens exhibit enhanced ultraviolet-B injury and oxidative damage. Plant Physiol 109:1159-66

Lauer MJ, Boyer JS (1992) Internal CO_2 measure directly in leaves: abscisic acid and low leaf water potential cause opposing effects. Plant Physiol 98:1010-16

Law RD, Crafts-Brandner SJ, Salvucci ME (2001) Heat stress induces the synthesis of a new form of ribulose-1,5-bisphosphate carboxylase/oxygenase activase in cotton leaves. Planta 214:117-25

Lee BH, Won SH, Lee HS, Miyao M, Clung WI, Kim IJ, Jo J (2000) Expression of the chloroplast-localized small heat shock protein by oxidative stress in rice. Gene 245:283-90

Lee SH, Ahsan N, Lee KW, Kim DH, Lee DG, Kwak SS, Kwon SY, Kim TH, Lee BH (2007) Simultaneous over expression of both CuZn SOD and ascorbate peroxidase in transgenic tall fescue plants confers increased tolerance to a wide range of abiotic stresses. J Plant Physiol 164:1626–1638

Lee YP, Kim SH, Bang JW, Lee HS, Kwak SS, Kwon SY (2007) Enhanced tolerance to oxidative stress in transgenic tobacco plants expressing three antioxidant enzymes in chloroplasts. Plant Cell Rep 26:591-8

Leon S, Touraine B, Ribot C, Briat JF, Lobreaux S (2003) Iron–sulphur cluster assembly in plants: distinct NFU proteins in mitochondria and plastids from *Arabidopsis thaliana*. Biochem J 371:823-30

Leustek T, Martin MN, Bick JA, Davies JP (2000) Pathways and regulation of sulfur metabolism revealed through molecular and genetic studies. Annu Rev Plant Physiol Plant Mol Biol 51:141-65

Lezhneva L, Amann K, Meurer J (2004) The universally conserved HCF101 protein is involved in assembly of [4Fe–4S]-cluster-containing complexes in *Arabidopsis thaliana* chloroplasts. Plant J 37:174-85

Li, G, Mao H, Ruan X, Xu Q, Gong Y, Zhang X, Zhao N (2002) Association of heat-induced conformational change with activity loss of Rubisco. Biochem Biophys Res Commun 290:1128-32

Link G (2003) Redox regulation of chloroplast transcription. Anti-oxidant Redox Signal 5:79-87
Lu C, Zhang J (1998) Effects of water stress on photosynthesis, chlorophyll fluorescence and photoinhibition in wheat plants. Austr J Plant Physiol 25:883-92
Mahan JR, Wanjura DF (2005) Seasonal patterns of glutathione and ascorbate metabolism in field-grown cotton under water Stress. Crop Sci 45:193-201
McKenzie RL, Bjorn LO, Bais A, Ilyasd M (2003) Changes in biologically active ultraviolet radiation reaching the Earth's surface. Photochem. Photobiol Sci 2:5-15
McKersie BD, Bowley S, Harjanto E, Leprince O (1996) Water-deficit tolerance and field performance of transgenic alfalfa overexpressing superoxide dismutase. Plant Physiol 111:1177-81
Melis A, Chen HC (2005) Chloroplast sulfate transport in green algae: genes, proteins and effects. Photosynth Res 86:299-307
Mendoza-Cozatl D, Loza-Tavera H, Hernandez-Navarro A, Moreno-Sanchez R (2005) Sulfur assimilation and glutathione metabolism under cadmium stress in yeast, protests and plants. FEMS Microbiol Rev 29:653-71
Miyake C, Asada K (1992) Thylakoid-bound ascorbate peroxidase in spinach chloroplasts and photoreduction of its primary oxidation product monodehydroascorbate radicals in thylakoids. Plant Cell Physiol 33:541-53
Mullineaux PM, Karpinski S (2002) Signal transduction in response to excess light: getting out of the chloroplast. Curr Opin Plant Biol 5:43-8
Mullineaux PM, Rausch T (2005) Glutathione, photosynthesis and the redox regulation of stress-responsive gene expression. Photosynth Res 86:459-74
Ogawa K, Kanematsu S, Takabe K, Asada K (1995) Attachment of CuZn-superoxide dismutase to thylakoid membranes at the site of superoxide generation (PS I) in spinach chloroplast: detection by immuno-gold labeling after rapid freezing and substitution method. Plant Cell Physiol 36:565-73
Okane D, Gill V, Byod P, Burdon R (1996)Chilling oxidative stress and antioxidant responses in *Arabidopsis thaliana* Callus. Planta 198:371-7
Okanenko A, Taran N (1998). Impact of heat stress on cereal lipid composition. In: De Kok LJ, Stulen I (eds), Responses of plant metabolism in air pollution and global change, Backhuys Publishers, Leiden, pp. 391-4
Pastori GM, Trippi VS (1992) Oxidative stress induces high rate of glutathione reductase synthesis in drought resistant maize strain. Plant Cell Physiol 33:957-61
Pfundel EE, Pan RS, Dilley A (1992) Inhibition of violaxanthin deepoxidation by UV- B radiation in isolated chloroplasts and intact leaves. Plant Physiol 98:1372-80
Phee BK, Cho JH, Park S, Jung JH, Lee YH, Jeon JS, Bhoo SH, Hahn TR (2004) Proteomic analysis of the response of *Arabidopsis* chloroplast proteins to high light stress. Proteomics 4:3560-8
Pickering IJ, Prince RC, George MJ, Smith RD, George GN, Salt DE (2000) Reduction and coordination of arsenic in Indian mustard. Plant Physiol 122:1171-7
Pilon-Smits EAH, Hwang S, Lytle CM, Zhu Y, Tai JC, Bravo RC, Chen Y, Leustek T, Terry N (1999) Overexpression of ATP sulfurase in Indian mustard leads to increased selenate uptake, reduction and tolerance. Plant Physiol 119:123-32
Pilon-Smits EAH, Pilon M (2006) Sulfur metabolism in plastids. In: Wise RR, Hoober JK (eds), The Structure and function of plastids, Springer, The Netherlands, pp 387-402
Raines CA, Lloyd JC, Dyer TA (1999) New insights into the structure and function of sedoheptulose-1,7-bisphosphatase; an important but neglected Calvin cycle enzyme. J Exp Bot 50:1-8
Ramachandra Reddy A, Chaitanva KV, Vivekanandan M (2004) Drought-induced responses of photosynthesis and antioxidant metabolism in higher palnts. J Plant Physiol 161:1189-1202
Rao MV, Ormrod DP (1995) Impact of UVB and O_3 on the oxygen free radical scavenging system in *Arabidopsis thaliana* genotypes differing in flavonoid biosynthesis. Photochem Photobiol 62:719-26
Rao MV, Paliyath G, Ormrod DP (1996) Ultraviolet-B and ozone induced biochemical changes in antioxidant enzymes in *Arabidopsis thaliana*. Plant Physiol 110:125-36
Reuveny Z, Dougall DK, Trinity PM (1980) Regulatory coupling of nitrate and sulfate assimilation pathways in cultured tobacco cells. Proc Natl Acad Sci USA 77:6670-2

Sailaja MV, Ramadas VS (1995) Photosystem II acclimation to limiting growth light in fully developed leaves of *Amaranthus hypochondriacus* L. and NAD-ME C_4 plant. Photosynth Res 46:227-33

Salin ML 1987. Toxic oxygen species and protective systems of the chloroplasts. Plant Physiol 72:681-9

Salvucci ME, Ogren WL (1996) The mechanism of Rubisco activase: insight from studies of the properties and structure of the enzyme. Photosynth Res 47:1-11

Salvucci ME, Osteryoung KW, Crafts-Brandner SJ, Vierling E (2001) Exceptional sensitivity of Rubisco activase to thermal denaturation *in vitro* and *in vivo*. Plant Physiol 127:1053-64

Schung, E. (1998) Sulfur in agroecosystems. Kluwer Academic Publishers, Dordrecht, The Netherlands

Schroda M, Vallon O, Wollman FA, Beck CF (1999) A chloroplast-targeted heat shock protein 70 (HSP70) contributes to the photoprotection and repair of photosystem II during and after photoinhibition. Plant Cell 11:1165-78

Shapira M, Lers A, Heifetz PB, Irihimovitz V, Osmond CB, Gillham NW, Boynton JE (1997) Differential regulation of chloroplast gene expression in *Chlamydomonas reinhardtii* during photoacclimation: light stress transiently suppresses synthesis of Rubisco LSU protein while enhancing synthesis of the PSII D1 protein. Plant Mol Biol 33:1001-11

Sigrist M, Zwillenberg C, Giroud Ch, Eichenberger W, Boschetti A (1988) Sulfolipid associated with the light-harvesting complex associated with photosystem II apoproteins of *Chlamydomonas reinhardii*. Plant Sci 58:15-23

Sonoike K (1996) Photoinhibition of photosystem I: its physiological significance in the chilling sensitivity of plants. Plant Cell Physiol 37:239-47

Sonoike K, Kamo M, Hihara Y, Hiyama T, Enami I (1997) The mechanism of the degradation of *psaB* gene product, one of the photosynthetic reaction center subunits of photosystem I, upon photoinhibition. Photosynth Res 53:55-63

Spreitzer RJ, Salvucci ME (2002) RUBISCO: structure, regulatory interactions, and possibilities for a better enzyme. Annu Rev Plant Physiol Plant Mol Biol 53:449-75

Srivalli B, Sharma G, Khanna-Chopra R (2003) Antioxidative defense system in an upland rice cultivar subjected to increasing intensity of water stress followed by recovery. Physiol Plant 119:503-12

Stöckel J, Oelmüller R (2004) A novel protein for Photosystem I biogenesis. J Biol Chem 279:10243-251

Sundby C, Harndahl U, Gustavsson N, Ahrman E, Murphy D (2005) Conserved methionines in chloroplasts. Biochim Biophys Acta 1703:191-202

Taiz L, Zeiger E (1998) Plant Physiology, 2nd ed. The Benjamin/Cummings Publication Co Inc, California

Takahashi H, Watanabe-Takahashi A, Smith FW, Blake-Kalff M, Hawkesford MJ, Saito K (2000) The roles of three functional sulphate transporters involved in uptake and translocation of sulphate in *Arabidopsis thaliana*. Plant J 23:171-82

Takeuchi Y, Murakami M, Nakajima N, Kondo N, Nikaido O (1996) Induction and repair of damage to DNA in cucumber cotyledons irradiated with UV-B. Plant Cell Physiol 37:181-7

Tausz M, Helena S, Grill D (2004) The glutathione system as a stress marker in plant ecophysiology: is a stress response concept valid. J Exp Bot 55:1955-62

Teramura AH, Ziska LH (1996) In: Baker NR(ed) Photosynthesis and environment, Kluwer Academic publishers, The Netherlands, pp. 435-50

Terashima I, Funayama S, Sonoike K (1994) The site of photoinhibition in leaves of *Cucumis sativus* L. at low temperatures is photosystem I, not photosytem II. Planta 193:300-6

Toivonen P, Vidaver W (1988) Variable chlorophyll a fluorescence and CO_2 uptake in water stressed white spruce seedlings. Plant Physiol 86:744-8

Touraine B, Boutin JP, Marion-Poll A, Briat JF, Peltier G, Lobreaux S (2004) Nfu2: a scaffold protein required for [4Fe–4S] and ferredoxin iron–sulphur cluster assembly in *Arabidopsis* chloroplasts. Plant J 40:101-11

Vani B, Saradhi, PP, Mohanty P (2001) Characterization of high temperature induced stress impairments in thylakoids of rice seedlings. Indian J Biochem Biophys 38:220-9

Vasilikiotis C, Melis A (1994) Photosystem II reaction center damage and repair cycle: chloroplast acclimation strategy to irradiance stress. Proc Natl Acad Sci USA 91: 7222-6

Xu XM, Moller SG (2004) AtNAP7 is a plastidic SufC-like ATP-binding cassette/ATPase essential for *Arabidopsis* embryogenesis. Proc Natl Acad Sci USA 101: 9143-8

Yabe T, Morimoto K, Kikuchi S, Nishio K, Terashima I, Nakai M (2004) The *Arabidopsis* chloroplastic NifU-like protein CnfU, which can act as an iron-sulfur cluster scaffold protein, is required for biogenesis of ferredoxin and Photosystem I. Plant Cell 16:993-1007

Yamaguchi Y, Nakamura T, Harada E, Koizumi N, Sano H (1999) Differential accumulation of transcripts encoding sulfur assimilation enzymes upon sulfur and/or nitrogen deprivation in *Arabidopsis thaliana*. Biosci Biotechnol Biochem 63:762-6

Yu B, Benning C (2003) Anionic lipids are required for chloroplast structure and function in *Arabidopsis*. Plant J 36:762-70

Chapter 9
Modified Levels of Cysteine Affect Glutathione Metabolism in Plant Cells

B. Zechmann(✉), M. Müller and G. Zellnig

Abstract Cysteine, the initial product of sulfate assimilation, is supposed to be the rate-limiting factor for glutathione (γ-glutamyl-cysteinyl-glycine) synthesis in nonstressed plants. In plants the assimilation of sulfate to sulfide exclusively takes place in plastids, whereas the synthesis of cysteine is carried out in plastids as well as in mitochondria and the cytosol. Glutathione synthesis, on the other hand, is thought to take place in plastids and the cytosol. Considering the above described pathways, the availability of cysteine, especially in plastids and/or cytosol, is essential for glutathione synthesis. Glutathione degradation into its constituents is thought to take place at the plasmalemma, the tonoplast, and within vacuoles and the apoplast. In this chapter we describe how the artificial elevation of cysteine in *Cucurbita pepo* (L.) plants by L-2-oxothiazolidine-4-carboxylic acid (OTC) changes glutathione and its precursor contents in single cells and organelles, thus giving a deeper insight into glutathione synthesis and degradation on the cellular level.

1 Introduction

Cysteine, the initial product of sulfate assimilation in plants, is a component of glutathione, which is the most important sulfur-containing antioxidant and redox buffer in plants. Glutathione is a tripeptide thiol (γ-glutamyl-cysteinyl-glycine) and has numerous functions in plants. It is involved in the uptake, transport, and storage of reduced sulfur and in the detoxification of reactive oxygen species, xenobiotics, heavy metals, etc., substances which are able to damage cellular components (Noctor et al. 1998, Tausz 2001, Blokhina et al. 2003, Kopriva and Rennenberg 2004, Tausz et al. 2004). On the cellular level glutathione is a key regulator of redox signaling, which can control gene expression, transcription, and translation. Due to its roles in redox signaling glutathione might also be involved in plant metabolism and development and might even be involved in the activation of cell death (Foyer

B. Zechmann
University of Graz, Institute of Plant Sciences, Schubertstrasse 51, 8010 Graz, Austria
bernd.zechmann@uni-graz.at

N.A. Khan et al. (eds.), *Sulfur Assimilation and Abiotic Stress in Plants.*

et al. 2001). Glutathione occurs within plants as reduced glutathione (GSH) and as glutathione disulfide (GSSG) – the oxidative form of glutathione. GSH contains one free sulfhydryl group (also known as thiol or SH-group), which is provided by cysteine, whereas in GSSG this group is connected by a disulfide bridge (Wonisch and Schaur 2001). Various free radicals like oxygen-, carbon-, or nitrogen-centered radicals and oxidants, also known as reactive oxygen species (ROS), are able to oxidize GSH to GSSG. During this reaction GSH donates hydrogen atoms, which leads to the formation of thiyl (GS) radicals. Through dimerization of these radicals the disulfide GSSG is formed; (GSH + X => GS + XH; GS + GS => GSSG). To maintain the protection of the cell against ROS and other radicals, GSSG is permanently reduced by glutathione reductase to GSH by using NADPH (GSSG + NADPH +H^+ => 2 GSH + $NADP^+$; Wonisch and Schaur 2001). In nonstress situations glutathione occurs in 90% in its reduced form. Deficiency of GSH within plant cells during oxidative stress situations demonstrates the need of the plant for cellular protection against ROS or other radicals, whereas high levels of GSSG indicate oxidative stress within the plant. However, glutathione levels within cells are not just influenced by oxidation and reduction processes, but are even more the product of an equilibrium between synthesis, degradation, use, and short- and long-distance transport of glutathione and its precursors (Foyer et al. 2001). Especially, the availability of glutathione precursors in plastids and the cytosol are interfering with glutathione synthesis. Cysteine, for example, is supposed to be the rate-limiting factor during glutathione synthesis in nonstressed plants (Kopriva and Rennenberg 2004, Kopriva and Koprivova 2005), which can lead to a decreased ability of the plant to fight oxidative stress.

In this chapter we present data on how the artificial elevation of cysteine affects the localization of glutathione and its precursors within single cells and organelles and discuss these changes with respect to glutathione synthesis and degradation.

2 Compartmentation of Cysteine and Glutathione Synthesis

Cysteine is synthesized in plastids, mitochondria, and the cytosol after the reduction of sulfate to sulfide, which exclusively takes place in plastids (e.g., Kopriva 2006, Wirtz and Hell 2007, Fig. 9.1). After the uptake of sulfate by cells, it is activated in plastids and the cytosol by ATP sulfurylase to adenosine 5-phosphosulfate, which is then exclusively reduced in plastids to sulfite by adenosine 5-phosphosulfate reductase. Sulfite is then further reduced in plastids to sulfide (catalyzed by sulfite reductase) and is then incorporated in plastids, mitochondria, and the cytosol into the amino acid skeleton of O-acetylserine (OAS) to form cysteine by OAS thiolyase. OAS, on the other hand, is formed from serine by serine acetyltransferase in plastids, mitochondria, and the cytosol (e.g., Kopriva 2006, Wirtz and Hell 2007, Fig. 9.1). Within plant cells cysteine can then be found in many proteins and peptides, such as glutathione.

Glutathione is synthesized out of its constituents cysteine, glutamate, and glycine in two ATP-dependent steps. In the first step cysteine and glutamate are linked

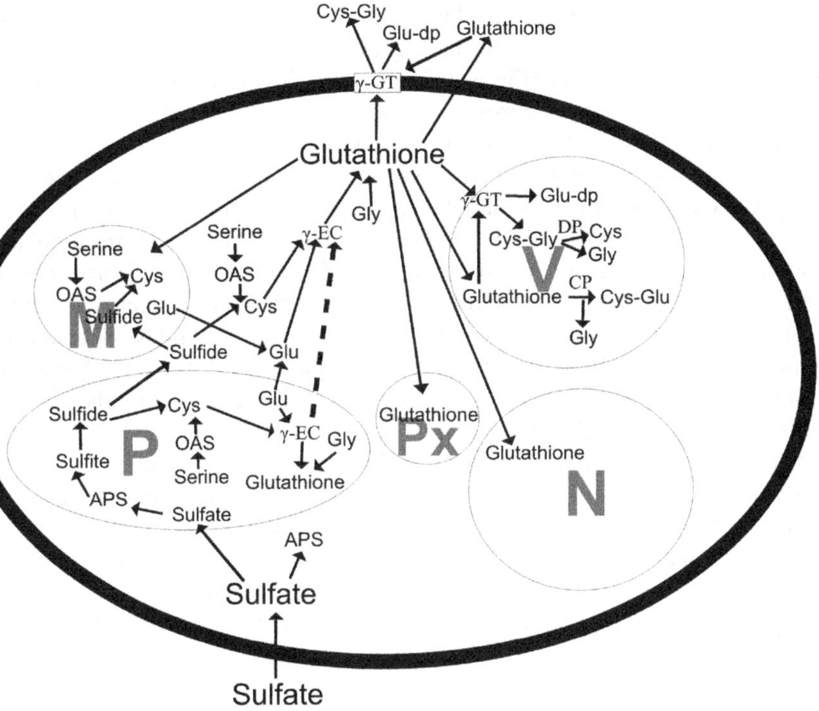

Fig. 9.1 Schema of a plant cell showing the current opinion about the compartmentation of sulfur assimilation and the synthesis of cysteine and glutathione and the degradation of the latter. After sulfate uptake into plastids (P) it is activated to adenosine 5-phosphosulfate (APS) and then reduced to sulfite and subsequently to sulfide. Sulfide is then incorporated in plastids, mitochondria (M), and the cytosol into the amino acid skeleton of O-acetylserine (OAS), which is formed out of serine in those cell compartments. This process leads to the formation of cysteine (cys), which is linked with glutamate (glu) either exclusively in plastids as shown for *Arabidopsis* plants or in both plastids and the cytosol in other plants species to form γ-glutamyl cysteine (γ -EC). The latter is then linked with glycine (gly) in plastids and the cytosol to form the tripeptide glutathione. This would imply a transport of γ –EC from plastids into the cytosol (dashed arrow) in plants where the first step of glutathione synthesis takes place exclusively in plastids. Whereas glycine synthesis is thought to take place in plastids, peroxisomes and the cytosol (Notcor et al. 1997b, Bauwe and Kolukisaoglu 2003), the synthesis of glutamate is thought to occur only in plastids and mitochondria (Suzuki and Knaff 2005). Therefore glutamate has to be exported by these cell compartments into the cytosol in order to allow glutathione synthesis in this cell compartment. Glutathione was also detected in nuclei (N) and peroxisomes (Px). Nevertheless, clear evidence for the import of glutathione into these organelles is still lacking. Glutathione degradation is thought to take place at the tonoplast, plasmalemma, or in vacuoles and the apoplast. Glutathione is degraded by γ-glutamyl transpeptidase (γ-GT) which transfers glutamate to other dipeptides (Glu-dp) or by carboxypeptidase (CP) that is located within vacuoles and removes glycine from glutathione. The remaining dipeptide (cys-glu) might then be metabolized by dipeptidases (DP) to the component amino acids

together to form γ-glutamyl cysteine. This reaction is catalyzed by γ-glutamyl cysteine synthetase, and seems to take place either exclusively in plastids, as demonstrated in leaves of *Arabidopsis* plants (Wachter et al. 2005), or in both plastids and the cytosol in other plant species (Gomez et al. 2004, Kopriva 2006, Fig. 9.1).

In the second step, glycine is added to γ-glutamyl cysteine catalyzed by glutathione synthetase to form the final product, glutathione. This reaction takes place in plastids and in the cytosol (Fig. 9.1), which are therefore considered as the main compartments of glutathione synthesis (Gomez et al. 2004, Sugiyama et al. 2004, Wachter et al. 2005, Zechmann et al. 2006a).

Degradation of glutathione is thought to take place at the plasmalemma, at the tonoplast, and within vacuoles and the apoplast. Glutathione is, therefore, degraded by either γ-glutamyl transpeptidase, which can be found at the plasmalemma, at the tonoplast, or within vacuoles, and the apoplast (Foyer et al. 2001, Storozhenko et al. 2002, Shaw et al. 2005, Ohkama-Ohtsu et al. 2007a, b) and transfers glutamate from glutathione to other dipeptides, or by carboxypeptidase located within vacuoles (Wolf et al. 1996) which removes glycine from glutathione (Fig. 9.1). The remaining dipeptides are then metabolized by a dipeptidase to the component amino acids (Foyer et al. 2001).

Considering the above described compartmentation of glutathione synthesis within single plant cells, glutathione production strongly depends on the availability of (1) cysteine and glutamate in plastids and/or the cytosol and (2) γ-glutamyl cysteine and glycine in plastids and the cytosol. Therefore, in order to gain a deeper insight into glutathione synthesis and its degradation, it is essential to determine contents of glutathione and its precursors on the cellular level rather than in whole organs.

3 Detection of Subcellular Glutathione and Its Precursors' Contents

Methods for detecting glutathione and its precursors simultaneously in single cells and organelles in one experiment are rare. Using biochemical methods, glutathione has so far been detected in different plant species within mitochondria, peroxisomes, and plastids (Jiménez et al. 1997, 1998, Kuzinak and Sklodowska 2001, 2004, 2005) and in the apoplast (Vanacker et al. 1998 a, b, c, Ohkama-Ohtsu et al. 2007a). Nevertheless, biochemical methods include the isolation of organelles from a rather large amount of plant material. This can lead to a loss or accumulation of glutathione, from or within organelles, and possible contamination with non-organelle-specific glutathione during the extraction and isolation of organelles (Noctor et al. 2002, Chew et al. 2003). Light microscopical studies have revealed that glutathione occurs in the cytosol and nuclei of different plant species (e.g., Müller et al. 2005). Nevertheless, this method is limited by the optical resolution of the light microscope (about 200 nm) and by the ability of the stain to infiltrate different cell compartments, e.g., chloroplasts (Hartmann et al. 2003, Müller et al. 2005). So far, it has not been possible to obtain more information about the presence of glutathione and its precursors in mitochondria, peroxisomes, and chloroplasts within plant cells by using this method. A powerful alternative to the above described methods is the localization of glutathione and its precursors with electron microscopical techniques, allowing for their detection in all cell compartments simultaneously (Müller et al. 2004, Zechmann et al. 2006a, b).

With these methods, it was possible to detect differences in the compartmentation of glutathione and its precursors within cells of leaves and roots of *Cucurbita pepo* plants. Glutathione contents within cells were always found to be highest in mitochondria and lowest in plastids (Müller et al. 2004, Zechmann et al. 2006a). Cysteine, on the other side, was found in similar contents in all cell compartments except mitochondria, which contained the lowest cysteine levels in all organs. Glutamate levels were found in similar concentrations in all cell compartments except the cytosol, which showed lowest glutamate contents. Glycine labeling was similar in all cell compartments except the vacuole, which contained very low amounts of gold particles bound to glycine. Whereas glutathione and glutamate were not detected in vacuoles and the apoplast, cysteine and glycine were found in vacuoles but appeared to be absent in the apoplast in cells of *Cucurbita pepo* plants (Müller et al. 2004, Zechmann et al. 2006a, b).

As it has not been possible so far to detect glutathione in vacuoles with the immunogold labeling approach or with the above described biochemical and light microscopical methods, it seems that the distribution of glutathione within plant cells is limited to the cytosol and to organelles (Foyer and Rennenberg 2000, Rennenberg 2001, Müller et al. 2004). Nevertheless, carboxypeptidase, an enzyme that removes glycine from glutathione, and an isoform of γ-glutamyl transpeptidase, which transfers glutamate from glutathione to other dipeptides, have also been detected within vacuoles (Wolf et al. 1996, Ohkama-Ohtsu et al. 2007b). These results indicate that glutathione could also occur in vacuoles, where it is then degraded by the above mentioned enzymes. Therefore it seems that (1) glutathione degradation occurs at the tonoplast as described above, or (2) glutathione rapidly degrades within vacuoles to its constituents, or (3) its concentration was too low to detect.

Neither glutathione nor its precursors were detected in the apoplast with the above described microscopical methods in seedlings and plants of *Cucurbita pepo* (Müller et al. 2004, Zechmann et al. 2006b), even during pathogen attack (Zechmann et al. 2005, 2007b). These results are similar to what was found with biochemical methods in leaves of oat, barley (Vanacker et al. 1998a, b), and *Arabidopsis* (Ohkama-Ohtsu et al. 2007a), which showed apoplastic glutathione contents to be extremely low in comparison to the total glutathione pool found in the investigated organ. Therefore, it seems that glutathione in the apoplast does not play an important role in the protection of plants, even during stress situations. Further, glutathione was not detected with the immunogold labeling approach in the apoplast, suggesting that if glutathione degradation occurs at the plasmalemma (Storozhenko et al. 2002) or in the apoplast (Ohkama-Ohtsu et al. 2007a), then either (1) the evolving degradation products, such as glutamate, cysteine, and glycine, would not be deposited into the apoplast, but would be rapidly relocated inside the cells, or (2) the constitutents of glutathione would be rapidly linked with other amino acids in the apoplast after their degradation, or (3) contents of glutathione and its precursors would be too low to be detected.

4 Artificial Elevation of Cysteine and Glutathione

There are numerous ways to artificially elevate cysteine and, subsequently, glutathione contents in plants, which have led to a deeper understanding of how compartment-specific alterations in glutathione synthesis affect metabolism and defense (Maughan and Foyer 2006). The development of mutants that overexpress γ-glutamyl cysteine synthetase or serine acetyltransferase led to significantly increased levels of glutathione in *Arabidopsis, Nicotiana*, and *Populus* plants (Noctor et al. 1998, Noctor et al. 2002) and to both elevated glutathione and cysteine levels in *Nicotiana* plants (Wirtz and Hell 2007). Nevertheless, genetic engineering can also lead to negative effects in both the ultrastructure and phenotype. In *Nicotiana* plants the overexpression of γ-glutamyl cysteine synthetase induced swollen thylakoids and spherical vesicles within stromal thylakoids of chloroplasts. Additionally, the phenotype developed signs of chlorosis and necrosis depending on different light intensities (Creissen et al. 1999). To avoid such negative effects, the treatment of plants with L-2-oxothiazolidine-4-carboxylic acid (OTC) has become a powerful alternative for increasing cysteine and subsequently glutathione levels in plants. OTC is a cysteine precursor and proline analogue that has been proven to artificially increase glutathione levels in spinach plants (Hausladen and Kunert 1990), tobacco plants (Gullner et al. 1999), pea leaves (Gullner and Dodge 2000), leaves of poplar (Kömives et al. 2003), and pumpkin plants without changing its phenotype or the ultrastructure of roots and leaves (Zechmann et al. 2007a). Treatment of OTC is known to enhance glutathione levels by either reducing its degradation due to inhibition of the metabolism of 5-oxo-L-proline by OTC, which instead of 5-oxo-L-proline is used as a substrate by 5-oxo-L-prolinase, or by serving as an intracellular delivery system for cysteine, the precursor of reduced glutathione, via S-carbocysteine, the product of hydrolysis of OTC by 5-oxo-L-prolinase (Williamson and Meister 1981, Hausladen and Kunert 1990, Farago and Brunold 1994). For the experiments described here we cultivated and treated seedlings of *Cucurbita pepo* (L.), as described recently (Zechmann et al. 2007a). We also used a well-established immunogold labeling method combined with computer-supported image analysis, which allows the detection of changes in glutathione and precursor contents in all cellular compartments simultaneously during one experiment with the transmission electron microscope (Zechmann et al. 2006a, b). Since cysteine and subsequently γ-glutamyl cysteine are usually the rate-limiting factors during glutathione synthesis (Noctor et al. 2002, Kopriva and Rennenberg 2004), the present experiments were aimed at finding out which precursors limit glutathione synthesis when cysteine occurs in excess and in which cell compartments cysteine accumulates or relocates when high concentrations of it occur within plant cells.

5 Effects of Artificial Elevation of Cysteine on Levels of Glutathione and Its Precursors

The treatment with 1 mM OTC for 48 hours significantly increased cysteine levels only in plastids (113% in roots, 55% in cotyledons), mitochondria (98.5% in roots, 52% in cotyledons), and the cytosol (21.2% in roots, 29% in cotyledons; Fig. 9.2a, e, and Fig. 9.3a; Table 9.1A), demonstrating that these cell compartments are indeed the main centers of cysteine synthesis in leaves and roots. Subsequently, glutathione levels in these organs were markedly elevated in all investigated cell compartments (up to 145.5% in peroxisomes of cotyledons; Fig. 9.2b, f, and Fig. 9.3b; Table 9.1B), indicating that cysteine is the limiting factor of glutathione synthesis in these plants. The observed increase of glutathione levels after OTC-treatment was slightly lower than what was measured with biochemical methods in pea leaf discs and poplar leaves, where the treatment with 1 mM OTC for 48 hours induced an increase in glutathione levels of 183% and 2.5-fold (150%), respectively (Gullner and Dodge, 2000, Kömives et al. 2003). In these experiments leaf discs of pea and petioles of poplar leaves were immersed directly into the OTC-solution, whereas in the present study OTC was applied through the roots. Since OTC was taken up by the roots and transported to different organs, a delayed activation of glutathione synthesis, especially as observed in the first true leaves, appears as a logical consequence.

Nevertheless, roots which were directly immersed into the OTC-solution showed a higher increase of cysteine levels after OTC-treatment in mitochondria and plastids than what was found in cotyledons (Fig. 9.3a). Immunogold labeling of glutamate revealed that it was increased (up to 120% in mitochondria) in roots (Fig. 9.4a, Table 9.1C), whereas glycine was decreased (up to 38% in plastids) in all cell compartments except nuclei (Fig. 9.4b, Table 9.1D). In contrast to roots, cotyledons showed a decrease of glutamate (up to 26% in nuclei) and glycine (up to 59% in mitochondria) levels after OTC-treatment (Fig. 9.2c, g, d, h, and Fig. 9.4a, b). The above described OTC-induced changes in glutathione and its precursor contents were also found in plastids and the cytosol. These results support the current opinion that these cell compartments are the main centers of glutathione synthesis (Sugiyama et al. 2004, Wachter et al. 2005, Zechmann et al. 2006a). Therefore, it can be concluded that if cysteine occurs in excess within plants, glycine becomes the limiting factor of glutathione synthesis in roots, and both glycine and glutamate limit glutathione production in leaves as presented here for cotyledons.

Similar results have also been found in leaf discs of poplar overexpressing γ-glutamyl cysteine synthetase. When cysteine was excessively applied to this plant material, the availability of glutamate became the limiting factor for the production of γ-glutamyl cysteine and subsequently glutathione (Noctor et al. 1996, Noctor et al. 1997a). That glycine supply can be insufficient to support maximum rates of glutathione synthesis has also been suggested for leaves in the dark or in the absence of photorespiration (Noctor et al. 1997a, b) and during stress situations such as pathogen attack (Zechmann et al. 2007b).

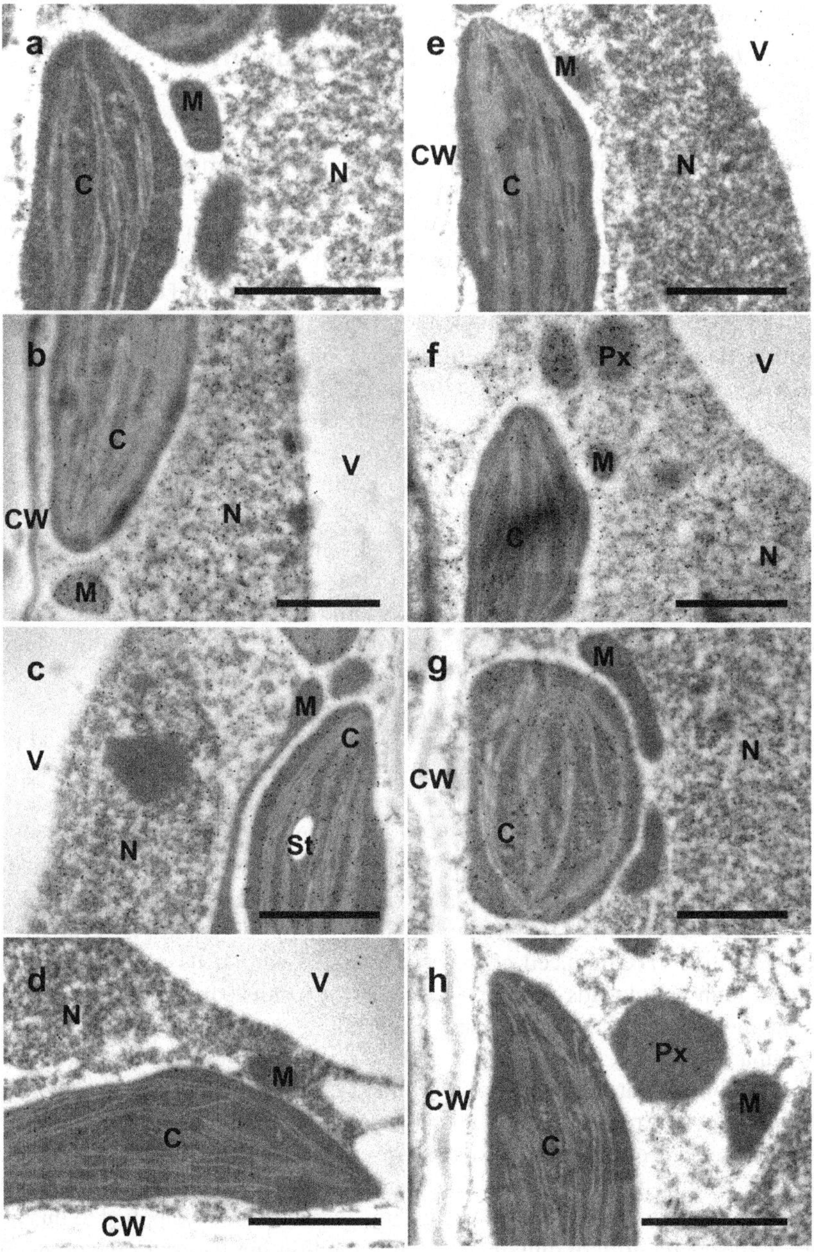

Fig. 9.2 Transmission electron micrographs of mesophyll cells from cotyledons from control plants (**a-d**) and plants treated with 1 mM OTC for 48 hours (**e-h**) with immunogold labeling of cysteine (**a, e**), glutathione (**b, f**), glutamate (**c, g**), and glycine (**d, h**) according to Zechmann et al. 2006a, b. Sections were post-stained with uranyl-acetate for 30 s. Bars: 1 µm. C = chloroplasts with or without starch (St), CW = cell walls, M = mitochondria, N = nuclei, Px = peroxisomes, V = vacuoles

Table 9.1 Subcellular glutathione and precursor contents*

A	Gold particles bound to cysteine per µm^2		
Cell structures	First true leaves	Cotyledons	Roots
Mitochondria	1.9 ± 0.8 a	3 ± 0.5 ab	3.2 ± 0.5 ac
Plastids	4.9 ± 0.3 de	3.7 ± 0.3 bfgh	2.6 ± 0.4 ac
Nuclei	4.2 ± 0.4 eb	5 ± 0.5 degi	3.3 ± 0.2 bcg
Peroxisomes	2.8 ± 0.9 ach	2.4 ± 0.6 ac	n.d.
Cytosol	3.4 ± 0.3 bci	4 ± 0.3 bd	4.4 ± 0.2 defi
Vacuoles	3.9 ± 0.2 bgh	2.1 ± 0.2 ac	n.d.
B	Gold particles bound to glutathione per µm^2		
Cell structures	First true leaves	Cotyledons	Roots
Mitochondria	350.1 ± 11.6 a	276.7 ± 7 b	169.3 ± 4.2 c
Plastids	57.1 ± 2.2 de	27.7 ± 0.8 f	30.9 ± 1.2 f
Nuclei	122.5 ± 3.8 g	73.9 ± 2.5 h	50.4 ± 1.8 i
Peroxisomes	97.4 ± 6 g	65.6 ± 5.9 dh	n.d.
Cytosol	53.0 ± 2.5 ei	56.3 ± 3.1 di	41 ± 1.4 j
C	Gold particles bound to glutamate per µm^2		
Cell structures	First true leaves	Cotyledons	Roots
Mitochondria	86.8 ± 8.8 ab	47.4 ± 2.8 c	22.3 ± 1.6 d
Plastids	138 ± 6.7 e	83.5 ± 2.3 a	20.9 ± 2.1 d
Nuclei	64.1 ± 2.7 b	36.8 ± 2 f	15.9 ± 1 g
Peroxisomes	120 ± 11.2 ae	46.8 ± 2.5 h	n.d.
Cytosol	33.4 ± 2 fi	28.6 ± 1.9 i	11.9 ± 0.7 g
D	Gold particles bound to glycine per µm^2		
Cell structures	First true leaves	Cotyledons	Roots
Mitochondria	7.4 ± 1.3 abc	8.2 ± 1.1 a	3.7 ± 0.5 d
Plastids	4.2 ± 0.4 befhi	4.3 ± 0.4 befgh	3.3 ± 0.5 dfg
Nuclei	7 ± 0.5 c	5.9 ± 0.4 ach	4.2 ± 0.3 acefi
Peroxisomes	6.6 ± 1.4 acf	3.1 ± 0.8 dgi	n.d.
Cytosol	4.6 ± 0.3 ae	4.3 ± 0.2 agh	4.8 ± 0.2 aef
Vacuoles	1.2 ± 0.1 i	1 ± 0.1 i	n.d.

*Values are means with standard errors and document the amounts of gold particles bound to (A) cysteine, (B) glutathione, (C) glutamate, and (D) glycine per square micrometer in different cell compartments of cells from first true leaves, cotyledons, and roots. Significant differences between the samples are indicated by different lowercase letters; samples which are significantly different from each other have no letter in common. $P<0.05$ was regarded significant analyzed by Kruskal-Wallis test followed by post hoc comparison according to Conover. $n>20$ for peroxisomes and vacuoles and $n>60$ for other cell structures. n.d. = not determined

The first true leaves showed significantly elevated glutathione levels after OTC-treatment only in plastids (24.5%) and the cytosol (93%), whereas glutathione levels in other cell compartments remained unchanged (Fig. 9.3b). Since glutathione synthetase is located in the cytosol and in plastids, and the latter do not depend on the import of cytosolic glutathione (Wachter et al. 2005), these results also indicate that plastids, together with the cytosol, are the main centers of glutathione synthesis (Noctor and Foyer 1998, Foyer et al. 2001, Sugiyama et al. 2004). After its production, glutathione must then be relocated into other organelles like mitochondria, nuclei, or peroxisomes, which depend on the import of cytosolic glutathione (Wachter et al. 2005).

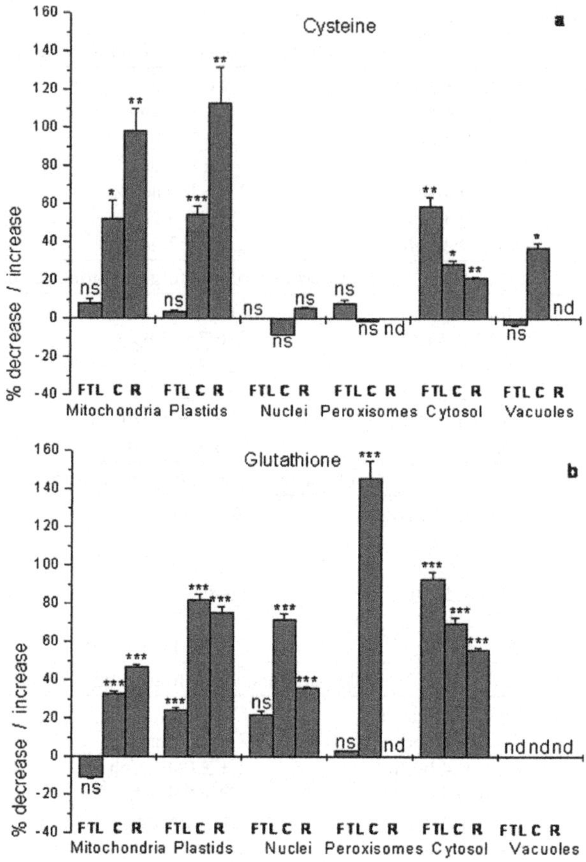

Fig. 9.3 Values are means ± standard errors and document the percentage of increase or decrease of the amount of gold particles per µm^2 bound to cysteine (**a**) and glutathione (**b**) in 1 mM OTC-treated cells (48 hours) of first true leaves (FTL), cotyledons (C), and roots (R) in comparison to the control (values given in Table 9.1A, B). nd=not determined; n > 20 for peroxisomes and vacuoles and n > 60 for other cell structures. Significant differences between organelles of control and OTC-treated cells were calculated with the Mann Whitney U-test. ns if p>0.05, (*) if p<0.05, (**) if p<0.01, and (***) if p<0.001

Cysteine levels were only increased in the cytosol (59%) of first true leaves (Fig. 9.3a). Therefore, it seems that the weaker increase of glutathione levels during OTC-treatment in this organ, in comparison to roots and cotyledons, is caused by a weaker induction of cysteine synthesis, which is most probably due to the longer transport of OTC from roots into the first true leaves than, e.g., cotyledons.

Cotyledons, which showed the strongest increase in glutathione and a strong increase in cysteine levels after OTC-treatment in most cell compartments, also contained a significantly higher amount of cysteine in vacuoles (Fig. 9.3a). Therefore, these results demonstrate that vacuoles are used as a sink for cysteine when concentrations within cells become too high, like during the present OTC-treatment.

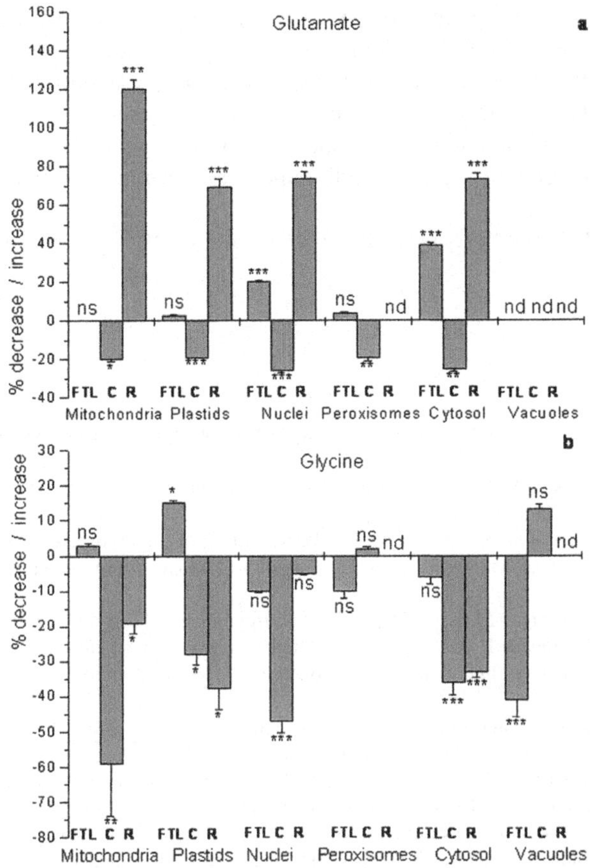

Fig. 9.4 Values are means ± standard errors and document the percentage of increase or decrease of the amount of gold particles per μm² bound to glutamate (**a**) and glycine (**b**) in 1 mM OTC-treated cells (48 hours) of first true leaves (FTL), cotyledons (C), and roots (R) in comparison to the control (values given in Table 9.1C, D). nd = not determined; n > 20 for peroxisomes and vacuoles and n > 60 for other cell structures. Significant differences between organelles of control and OTC-treated cells were calculated with the Mann Whitney U-test. ns if $p>0.05$, (*) if $p<0.05$, (**) if $p<0.01$, and (***) if $p<0.001$

6 Conclusion

From the studies presented in this chapter, we can conclude that glutathione synthesis takes place exclusively in plastids and the cytosol, and that cysteine is the limiting factor during glutathione synthesis in nonstressed plants. However, if cysteine occurs in excess within plants, then glycine in roots, and both glycine and glutamate in leaves, become the limiting factors for glutathione synthesis, as demonstrated here for roots and cotyledons. Additionally, vacuoles act as a sink for cysteine when it occurs in excess within plant cells. As both cysteine and glycine are found to be present in vacuoles, it appears that glutathione degradation takes place either at the tonoplast or within vacuoles by γ-glutamyl transpeptidase and/or carboxypeptidase releasing these two amino acids into the vacuoles. The present investigations support and extend the current knowledge about glutathione metabolism on the cellular level, which is essential for understanding the stress responses of plant cells.

Acknowledgements This work was supported by the Austrian Science Fund (FWF; P16273 and P18976).

References

Bauwe H, Kolukisaoglu Ü (2003) Genetic manipulation of glycine decarboxylation. J Exp Bot 54(387):1523-35

Blokhina O, Virolainen E, Fagerstedt KV (2003) Antioxidants, oxidative damage and oxygen deprivation stress: a review. Ann Bot 91:179-94

Chew O, Whelan J, Millar AH (2003) Molecular definition of the ascorbate-glutathione cycle in *Arabidopsis* mitochondria reveals dual targeting of antioxidant defenses in plants. J Biol Chem 278:46869-77

Creissen G, Firmin J, Fryer M, Kular B, Leyland N, Reynolds H, Pastori G, Wellburn F, Baker N, Wellburn A, Mullineaux P (1999) Elevated glutathione biosynthetic capacity in the chloroplasts of transgenic tobacco plants paradoxically causes increased oxidative stress. Plant Cell 11:1277-91

Farago S, Brunold C (1994) Regulation of thiol contents in maize roots by intermediates and effectors of glutathione synthesis. J Plant Physiol 144:433-7

Foyer CH, Rennenberg H (2000) Regulation of glutathione synthesis and its role in abiotic and biotic stress defence. In: Brunold C, Rennenberg H, De Kok LJ, Stulen I, Davidian JC (eds) Sulfur nutrition and sulfur assimilation in higher plants, Paul Haupt Verlag, Bern, pp 127-53

Foyer CH, Theodoulou FL, Delrot S (2001) The functions of inter- and intracellular glutathione transport systems in plants. Trends Plant Sci 6/10:486-92

Gomez LD, Vanacker H, Buchner P, Noctor G, Foyer CH (2004) Intercellular distribution of glutathione synthesis in maize leaves and its response to short-term chilling. Plant Physiol 134:1662-71

Gullner G, Tóbiás I, Fodor J, Kömives T (1999) Elevation of glutathione level and activation of glutathione-related enzymes affect virus infection in tobacco. Free Radical Res 31:155-61

Gullner G, Dodge AD (2000) Accumulation of glutathione in pea leaf discs exposed to the photooxidative herbicides acifluorfen and 5-aminolevulinic acid. J Plant Physiol 156:111-17

Hartmann TN, Fricker MD, Rennenberg H, Meyer AJ (2003) Cell-specific measurement of cytosolic glutathione in poplar leaves. Plant Cell Environ 26:965-75

Hausladen A, Kunert KJ (1990) Effects of artificially enhanced levels of ascorbate and glutathione on the enzymes monodehydroascorbate reductase, dehydroascorbate reductase, and glutathione reductase in spinach (*Spinacia oleracea*). Physiol Plant 79:384-8

Jiménez A, Hernández JA, del Río LA, Sevilla F (1997) Evidence for the presence of the ascorbate-glutathione cycle in mitochondria and peroxisomes of pea leaves. Plant Physiol 114:275-84

Jiménez A, Hernández JA, Pastori G, del Río LA, Sevilla F (1998) Role of the ascorbate-glutathione cycle of mitochondria and peroxisomes in the senescence of pea leaves. Plant Physiol 118:1327-35

Kömivés T, Gullner G, Rennenberg H, Casida JE (2003) Ability of poplar (*Populus spp.*) to detoxify chloroacetanilide herbicides. Water Air Soil Poll Foc 3:277-83

Kopriva S, Rennenberg H (2004) Control of sulphate assimilation and glutathione synthesis: interaction with N and C metabolism. J Exp Bot 55:1831-42

Kopriva S, Koprivova A (2005) Sulfate assimilation and glutathione synthesis in C4 plants. Photosynth Res 86:363-73

Kopriva S (2006) Regulation of Sulfate Assimilation in *Arabidopsis* and Beyond. Ann Bot 97:479-95

Kuzniak E, Sklodowska A (2001) Ascorbate, glutathione and related enzymes in chloroplasts of tomato leaves infected by *Botrytis cinerea*. Plant Sci 160:723-31

Kuzniak E, Sklodowska A (2004) Comparison of two methods for preparing mitochondria from tomato leaves to study the ascorbate-glutathione cycle activity. Biol Plant 48:537-42

Kuzniak E, Sklodowska A (2005) Compartment-specific role of the ascorbate-glutathione cycle in the response of tomato leaf cells to *Botrytis cinerea* infection. J Exp Bot 56:921-33

Maughan S, Foyer CH (2006) Engineering and genetic approaches to modulating the glutathione network in plants. Physiol Plant 126:382-97

Müller M, Zechmann B, Zellnig G (2004) Ultrastructural localization of glutathione in *Cucurbita pepo* plants. Protoplasma 223:213-19

Müller M, Zellnig G, Urbanek A., Zechmann B (2005) Recent developments in methods intracellulary localizing glutathione within plant tissues and cells (a Minivreview). Phyton (Horn) Austria 45:45-55

Noctor G, Strohm M, Jouanin L, Kunert KJ, Foyer CH, Rennenberg H (1996) Synthesis of glutathione in leaves of transgenic poplar overexpressing y-glutamylcysteine synthetase. Plant Physiol 112:1071-8

Noctor G, Arisi ACM, Jouanin L, Valadier MH, Roux Y, Foyer CH (1997a) Light-dependent modulation of foliar glutathione synthesis and associated amino acid metabolism in poplar overexpressing c-glutamylcysteine synthetase. Planta 202:357-69

Noctor G, Arisi ACM, Jouanin L, Valadier MH, Roux Y, Foyer CH (1997b) The role of glycine in determining the rate of glutathione synthesis in poplar. Possible implications for glutathione production during stress. Physiol Plant 100:225-63

Noctor G, Arisi A-CM, Jouanin L, Kunert KJ, Rennenberg H, Foyer CH (1998) Glutathione: biosynthesis and metabolism explored in transformed poplar. J Exp Bot 49:623-47

Noctor G, Foyer CH (1998) Ascorbate and glutathione: keeping active oxygen under control. Annu Rev Plant Physiol Plant Mol Biol 49:229-79

Noctor G, Gomez L, Vanacker H, Foyer CH (2002) Interactions between biosynthesis, compartmentation and transport in the control of glutathione homeostasis and signalling. J Exp Bot 53/372:1283-1304

Ohkama-Ohtsu N, Radwan S, Peterson A, Zhao P, Badr AF, Xiang C, Oliver DJ (2007a) Characterization of the extracellular c-glutamyl transpeptidases, GGT1 and GGT2, in *Arabidopsis*. Plant J 49:865-77

Ohkama-Ohtsu N, Zhao P, Xiang C, Oliver DJ (2007b) Glutathione conjugates in the vacuole are degraded by γ-glutamyl transpeptidase GGT3 in *Arabidopsis*. Plant J 49:878-88

Rennenberg H (2001) Glutathione - an ancient metabolite with modern tasks. In: Grill D, Tausz M, De Kok LJ (eds) Significance of glutathione to plant adaptation to the environment, Kluwer Academic Publishers, Dordrecht, Boston, London, pp. 1-11

Shaw ML, Pither-Joyce MD, McCallum JA (2005) Purification and cloning of a γ-glutamyl transpeptidase from onion (*Allium cepa*). Phytochem 66:515-22

Storozhenko S, Belles-Boix E, Babiychuk E, Hérouart D, Davey MW, Slooten L, van Montagu M, Inzé D, Kushnir S (2002) γ-glutamyl transpeptidase in transgenic tobacco plants. Cellular localization, processing, and biochemical properties. Plant Physiol 128:1109-19

Sugiyama A, Nishimura J, Mochizuki Y, Inagaki K, Sekiya J (2004) Homoglutathione synthesis in transgenic tobacco plants expressing soybean homoglutathione synthetase. Plant Biotech 21:79-83

Suzuki A, Knaff DB (2005) Glutamate synthase: structural, mechanistic and regulatory properties, and role in the amino acid metabolism. Photosynth Res 83:191-217

Tausz M (2001) The role of glutathione in plant response and adaptation to natural stress. In: Grill D, Tausz M, De Kok LJ (eds) Significance of glutathione to plant adaptation to the environment, Kluwer Academic Publishers, Dordrecht, Boston, London, pp. 101-22

Tausz M, Šircelj H, Grill D (2004) The glutathione system as a stress marker in plant ecophysiology: is a stress-response concept valid? J Exp Bot 55:1955-62

Vanacker H, Carver TLW, Foyer CH (1998a) Pathogen-induced changes in the antioxidant status of the apoplast in barley leaves. Plant Physiol 117:1103-14

Vanacker H, Foyer CH, Carver TLW (1998b) Changes in apoplastic antioxidants induced by powdery mildew attack in oat genotypes with race non-specific resistance. Planta 208:444-52

Vanacker H, Harbinson J, Ruisch J, Carver TLW, Foyer CH (1998c) Antioxidant defences of the apoplast. Protoplasma 205:129-40

Wachter A, Wolf S, Steininger H, Bogs J, Rausch T (2005) Differential targeting of GSH1 and GSH2 is achieved by multiple transcription initiation: implications for the compartmentation of glutathione biosynthesis in the Brassicaceae. Plant J 41:15-30

Williamson JM, Meister A (1981) Stimulation of hepatic glutathione formation by administration of l-2-oxothiazolidine-4-carboxylate, a 5-oxo-l-prolinase substrate. Proc Natl Acad Sci USA 78:936-9

Wirtz M, Hell R. (2007) Dominant-negative modification reveals the regulatory function of the multimeric cysteine synthase protein complex in transgenic tobacco. Plant Cell 19:625-39

Wolf AE, Dietz KJ, Schröder P (1996) Degradation of glutathione S-conjugates by a carboxypeptidase in the plant vacuole. FEBS Letters 384:31-4

Wonisch W, Schaur RJ (2001) Chemistry of glutathione. In: Grill D, Tausz M, De Kok LJ (eds) Significance of glutathione to plant adaptation to the environment, Kluwer Academic Publishers, Dordrecht, Boston, London, pp. 13-26

Zechmann B, Zellnig G, Müller M (2005) Changes in the subcellular distribution of glutathione during virus infection in *Cucurbita pepo* (L.). Plant Biol 7:49-57

Zechmann B, Müller M, Zellnig G (2006a) Intracellular adaptations of glutathione content in *Cucurbita pepo* (L.) induced by reduced glutathione and buthionine sulfoximine treatment. Protoplasma 227:197-209

Zechmann B, Zellnig G, Müller M (2006b) Immunocytochemical localization of glutathione precursors in plant cells. J Electron Microsc 55:173-81

Zechmann B, Zellnig G, Urbanek-Krajnc A, Müller M (2007a) Artificial elevation of glutathione affects symptom development in ZYMV-infected *Cucurbita pepo* L. plants. Arch Virol 152:747-62

Zechmann B, Zellnig G, Müller M (2007b) Virus-induced changes in the subcellular distribution of glutathione precursors in *Cucurbita pepo* (L.). Plant Biol DOI: 10.1055/s-2006-924670

Chapter 10
Role of Glutathione in Abiotic Stress Tolerance

S. Srivalli and Renu Khanna-Chopra(✉)

Abstract Reduced glutathione (GSH) is the most abundant source of nonprotein thiols in plant cells. It has multiple functions in plants, including cell differentiation, cell death and senescence, pathogen resistance, enzymatic regulation, as an antioxidant, in formation of phytochelatins, detoxification of xenobiotics, and also acts as a storage and transport form of reduced sulfur. Along with its oxidized form (GSSG), it acts as a redox couple important for maintaining cellular homeostasis, playing a key role in diverse signaling systems in plants. Biosynthesis of glutathione is stimulated under oxidative stress conditions, as GSH gets converted to GSSG. It also acts as redox sensor of environmental cues and coordinates cellular responses in several ways. Altered levels of glutathione can affect plants as, for instance, by increasing glutathione biosynthesis, which has been shown to enhance resistance to oxidative stress. The aim of this review is to give an overview of the role of glutathione in abiotic stresses such as drought, salinity, and extremes of temperatures and also of some recent advances in this area.

1 Introduction

Reduced glutathione (GSH), a low molecular weight tripeptide (γ-glutamylcysteinyl-Gly), is the most abundant source of nonprotein thiols in plant cells. In fact, GSH is widespread, being present in almost all eukaryotes with the exception of those that do not have mitochondria or chloroplasts, but its production among prokaryotes is restricted to cyanobacteria and proteobacteria and a few strains of gram-positive bacteria (Masip et al. 2006). It has multiple functions in plants and plays a crucial role as an antioxidant in cellular defense and protection. In fact, GSH has been associated with several growth- and development-related events in plants, including cell differentiation, cell death and senescence, pathogen resistance and enzymatic regulation

Renu Khanna-Chopra
Stress Physiology & Biochemistry Laboratory, Water Technology Centre,
Indian Agricultural Research Institute (IARI), New Delhi - 110012, India
renn_wtc@rediffmail.com

N.A. Khan et al. (eds.), *Sulfur Assimilation and Abiotic Stress in Plants.*

(Ogawa 2005). This ability is primarily due to the chemical reactivity of the thiol group and an oxidation reduction potential of –0.23 V that allows it to act as an electron acceptor and donor for various biological reactions (Xiang et al. 2001).

GSH can function as an antioxidant in many ways. It is part of the ascorbate-glutathione cycle, which is essential in scavenging hydrogen peroxide (H_2O_2), where it acts as the reducing agent that recycles ascorbate from its oxidized (dehydroascorbate; DHA) to its reduced form by the enzyme dehydroascorbate reductase. It can also do so in a nonenzymatic manner at alkaline pH as in stroma of chloroplasts. The regeneration from oxidized glutathione (GSSG) to GSH is catalyzed by glutathione reductase, and NADPH is used as the reducing power (Fig. 10.1). It also plays an indirect role in protecting membranes by maintaining α-tocopherol and zeaxanthin in the reduced state. It can also function directly as a free radical scavenger by reacting with superoxide, singlet oxygen, and hydroxyl radicals. GSH protects proteins against denaturation caused by the oxidation of protein thiol groups under stress. In addition, GSH is a substrate for glutathione peroxidase (GPX) and glutathione-*S*-transferases (GST), which are also involved in the removal of reactive oxygen species (ROS) (Noctor et al. 2002).

Other functions for GSH include the formation of phytochelatins, which have an affinity for heavy metals and are transported as complexes into the vacuole, thus allowing the plants to have some level of resistance to heavy metals (Sharma and Dietz 2006). GSH is also involved in the detoxification of xenobiotics and acts as a storage and transport form of reduced sulfur. In developing endosperm of wheat, however, it has been reported that GSH is the main transport form, as sulfur is normally limiting to the plants. In S-rich conditions sulfate is the major form of S to the endosperm (Fitzgerald et al. 2001). Along with its oxidized form (GSSG), it acts as a redox couple important for maintaining the cellular homeostasis, playing a key role in diverse signaling systems in plants (Noctor 2006).

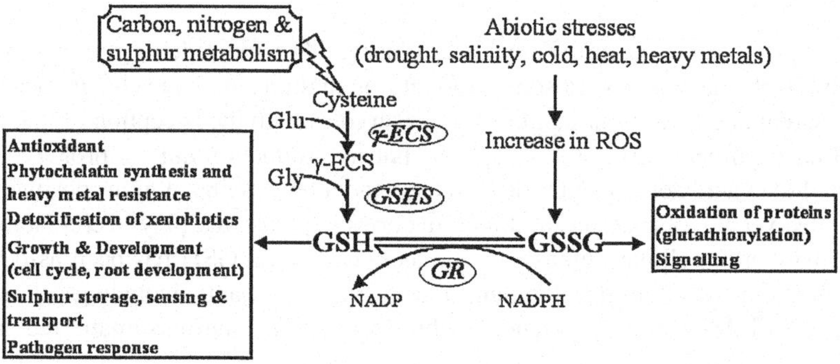

Fig. 10.1 An overview of glutathione biosynthesis and its multiple roles in plant cells. γ-ECS - γ-Glutamyl-Cysteine synthetase; Glu - glutamic acid; GR – glutathione reductase; GSHS - glutathione synthetase; Gly – glycine; GSSG – oxidized glutathione; NADP – nicotinamide adenine dinucleotide phosphate; GSH-reduced glutathione, NADPH – reduced nicotinamide adenine dinucleotide phosphate

GSH is synthesized in cytosol as well as chloroplasts in an ATP-dependent two-step reaction catalyzed by the enzymes γ-glutamylcysteine synthetase (γ-ECS) and glutathione synthetase (GSHS) (Fig. 10.1). Both the enzymes have been extracted from plant tissues and cloned. γ-ECS is subject to regulation through feedback inhibition and is the rate-limiting step for glutathione biosynthesis. This is controlled primarily through the level of available L-cysteine (May et al. 1998). Biosynthesis of glutathione is stimulated under oxidative stress conditions as GSH gets converted to GSSG (Grene 2002, Fig. 10.1).

Environmental stresses cause increased ROS levels in plants, and the response of glutathione can be crucial for adaptive responses. Glutathione has been traditionally used as a marker for various stresses (Mittova et al. 2003, Navari-Izzo et al. 1997, Srivalli et al. 2003). It also acts as redox sensor of environmental cues and coordinates cellular responses in several ways, including induction of antioxidant defenses (Hérouart et al. 1993), mediating signaling pathways and regulating gene expression. Altered levels of glutathione can affect plants as, for instance, by increasing glutathione biosynthesis, which has been shown to enhance resistance to oxidative stress (Arisi et al. 1998, Zhu et al. 1999). This review gives an overview on the role of glutathione in abiotic stresses and some recent advances made in this area.

2 Biosynthesis, Compartmentation, and Transport

Glutathione is synthesized from standard amino acids in two steps. From glutamic acid and cysteine, γ-glutamyl-cysteine (γ-EC) synthetase (γ-ECS) produces γ-EC in an ATP-dependent reaction. Glutathione synthetase (GSHS) catalyzes the ATP-dependent reaction between γ-EC and glycine to form GSH. In *Arabidopsis* γ-ECS and GSHS enzymes are encoded by two different genes, *GSH1* and *GSH2*, respectively. However, in *Streptococcus agalactiae*, a novel enzyme was isolated which catalyzes both the activities (Janowiak and Griffith 2005). This bifunctional enzyme encodes an 85 kDa protein with similar specific activities of both γ-ECS and GSHS. In *Arabidopsis*, γ-ECS seems to be exclusively localized in plastids as compared to other plants, where it is also cytosolic. GSHS is dually targeted to plastids and cytosol from a single gene (Wachter et al. 2005), whereas multiple copies of both genes exist for other plants. Alternative splicing of *GSHS* pre-mRNA can produce GSHS isozymes targeted to the cytosol and plastid (Skipsey et al. 1999). GSH synthesis is also differentially localized in C4 plants such as maize with mesophyll cells having a higher GSHS activity as compared to bundle sheath cells (Burgener et al. 1998). Glutathione reductase was found only in the mesophyll cells of maize (Pastori et al. 2000). Thus, species-specific differences exist for the intercellular localization of the enzymes involved in GSH biosynthesis.

Glutathione has been detected in virtually all cell compartments such as cytosol, chloroplasts, endoplasmic reticulum, vacuoles, and mitochondria (Noctor and Foyer 1998) and occurs predominantly in the reduced form. The concentration of GSH is the highest in the chloroplasts (1-4 mM). The biosynthetic pathway of glutathione is well established and, unlike ascorbate biosynthesis, is similar in plants, animals, and microorganisms. In two ATP-dependent steps, catalyzed by γ-glutamylcysteine synthetase (γ-ECS) and glutathione synthetase (GSHS), the constituent amino acids are linked to form the complete tripeptide. These steps occur in both chloroplastic and nonchloroplastic compartments, and the variations in glutathione concentration and redox state may be crucial in signaling (Noctor et al. 2002).

Compartmentation and transport are important for the various biological functions of glutathione. GSH is the major transport form of reduced sulfur, both in the xylem and the phloem (Herschbach et al. 2000). Evidence also exists to show that photosynthetic cells are able to export glutathione synthesized in the chloroplasts from the cell (Noctor et al. 2002). Chloroplasts are also able to take up GSH from the cytosol, and the rate of uptake is similar to its synthesis (Kumar et al. 2003). A GS–conjugate transporter is known to operate at the tonoplast (Foyer et al. 2001, Rea et al. 1998). Studies on tobacco and bean protoplasts have indicated that glutathione transport in leaf tissues was mediated with proton cotransport (Jamai et al. 1996, Schneider et al. 1992). However, it was observed in studies on a broad range of plant species, including Arabidopsis, *Beta, Vigna*, and *Zea*, that transport of GS conjugates across the vacuolar membrane was MgATP-dependent and not H+-dependent (Li et al. 1995) and the broad distribution of this vacuolar transporter showed that it plays a general role in the vacuolar sequestration of glutathione-conjugable cytotoxic agents. Little however, is known about the uptake of glutathione across the plasma membrane, the plastid envelope, the mitochondrial envelope, or the endoplasmic reticulum membrane and the localization of the transporters (Maughan and Foyer 2006).

The first successful identification of a glutathione transporter (HGT1) was done in yeast, which had high affinity to GSH, GSSG, GS-NEM (glutathione-N-ethylmaleimide conjugate (Bourbouloux et al. 2000). Preliminary characterization of a glutathione transporter BjGT1, in *Brassica juncea*, showed that it was able to mediate uptake of labeled glutathione in yeast, but the specificity of this transporter was not studied (Bogs et al. 2003). A rice glutathione transporter, OsGT1, which is able to transport GSH, GSSG and GS conjugates, and whose structural features significantly differ from BjGT1 and HGT1, has also been characterized (Zhang et al. 2004). In this respect, the transport properties of OsGT1 also differ from that of the AtMRP1 and AtMRP2 transporters that mediate transfer of GS conjugates into the vacuoles but have very low affinity for GSH (Tommasini et al. 1993, Rea et al. 1998). OsGT1 corresponded to a new family of proteins, which are able to mediate the transport of GS conjugates in plants, in addition to the tonoplast ABC transporters of the MRP subclass (Rea et al. 1998, Tommasini et al. 1998). The fact that expression of OsGT1 in yeast confers the ability to take up exogenous glutathione and the lack of targeting sequence for organelles suggests that the protein is targeted to the plasma membrane. OsGT1 is only weakly expressed in various parts of rice plants, which suggested that it does not play a key role in sulfur transport under normal growth conditions and its role may be to retrieve GSH, GSSG, GS conjugates under various stresses (Zhang et al. 2004).

3 Role in Abiotic Stresses

Abiotic stresses, in general, elicit an oxidative response via increases in ROS in plant cells. These include stresses such as drought stress, salinity, UV-B, temperature extremes, and pollutants (lnzé and Van Montagu 1995). These stresses are serious threats to agriculture, as they cause morphological, molecular, and biochemical changes which affect plant growth and productivity. In fact, different abiotic stresses are often interconnected and cause similar cellular damage. The plant responses to these stresses are complex and multifold. To prevent oxidative damage, the antioxidant response can be crucial in plants. Although glutathione has multiple roles in plant cells, its function as an antioxidant is an important one, in that the role of glutathione in abiotic stresses is critical in both prokaryotes and eukaryotes. In fact, a glutathione-deficient mutant of *Rhizobium* has been shown to be sensitive to osmotic and oxidative stresses and even weak organic acids (Riccillo et al. 2000). Glutathione restores responses to these stresses almost to wild-type levels.

3.1 Marker for Abiotic Stresses: Response to Various Stresses

The response of glutathione has been studied in various abiotic stresses due to its role as an antioxidant. Rather, GSH levels and GSH/GSSG ratios have been often indicative of the stress faced by the plant. In fact, its role as a marker of abiotic stresses and the rationale for its use has been highlighted earlier (Tausz et al. 2004). The response of glutathione levels and the redox state have often varied in different studies. Here, we focus on these responses to certain important environmental stresses to see if any general consensus emerges from these studies and how valid it is to interpret stress levels through responses of glutathione.

3.1.1 Drought Stress

Plant resistance against drought is often through the accumulation of glutathione, which acts as a reductant. It has been observed in *Boea hygroscopica*, a dessication-tolerant plant species that constitutive GSH increased by almost 50 times (Sgherri et al. 1994). This may then protect the plant through protection of sulfydryl groups of proteins (Navari-Izzo et al. 1997). High levels of GSH during drought stress are achieved through high activity of glutathione reductase (GR) (Gamble and Burke 1984). However, this may not always be the case. The GSH/GSSG ratio is likely to be influenced by the duration and severity of stress and the plant species (Smirnoff 1993). This is highlighted by a series of studies in wheat. When wheat plants (*T. durum* cv. Ofanto, drought tolerant) were subjected to two phases of 12 days water stress with rewatering after every phase, GSH/GSSG showed an increase as compared to the controls (Menconi et al. 1995). However, ascorbate to dehydroascorbate

(AsA/DHA) ratio remained unchanged. GR declined only in the second phase of stress by 27% of the controls. In another study, this cultivar was compared with a drought-sensitive cultivar, Adamello, and subjected to 35 days of water stress and then rewatered for 4 days (Loggini et al. 1999). Drought stress caused a 30% decline in total glutathione and GSH in both the cultivars, and full recovery was not possible after rewatering. In Adamello, GR increased to 136% of the controls during drought stress, whereas no change was observed in the cv. Ofanto. These differences were attributed to the ability of the tolerant cultivar to acclimate to drought stress, whereas Adamello, the sensitive cultivar, had to increase its antioxidative defense mechanism to prevent irreversible damage due to drought stress. In contrast, Lascano et al. (2001) found no significant differences in the four drought-tolerant wheat varieties after one month of drought stress in the field but observed an increase in glutathione in the leaves within 48 h of a short-term osmotic stress. Another study on a short-term drought stress treatment of 24 hr showed an increase in GSH and GR activity in wheat leaves (Bartoli et al. 1999). Thus, no consistent picture emerges for GSH levels in drought-stressed wheat, as the water stress intensities have varied across these studies.

In nature, crops experience multiple cycles of stress and often more than one kind. Our interest was not just to study the response to drought stress but also to observe the performance of the plants to multiple cycles of drought stress. Thus, wheat was subjected to drought acclimation studies in our lab. Drought acclimation is a phenomenon wherein plants are able to respond to water stress better after having already experienced a milder form of the stress. There was a seven- and fivefold increase in superoxide radicals in leaves and roots, respectively, of wheat seedlings experiencing severe stress for the first time as compared to the three- and twofold increase in leaves and roots, respectively, of acclimated plants (Selote et al. 2004). This oxidative stress tolerance of acclimated plants was due to coordinated antioxidant defense at both the cellular and subcellular levels (Selote and Khanna-Chopra 2006). Acclimated plants had higher GSH levels and GR activity in crude extracts and in the chloroplasts. On comparison of a drought-tolerant wheat cultivar C306 with a drought-susceptible one, Moti, it was observed that C306 had higher tolerance to oxidative stress with higher GR activity under stress (Khanna-Chopra and Selote 2007). Moti showed a significant decline in the GSH/GSSG ratio as compared to C306 under water stress. It was concluded that genotypic differences to water stress in wheat could be partly attributed to the ability of plants to acclimate and induce antioxidant defense. A similar drought acclimation study was conducted in an upland rice cv. Tulsi, which was subjected to three cycles of water stress of increasing stress intensity followed by rewatering (Srivalli et al. 2003). It was observed that there was an increase in the GSH levels and GR activity in the second cycle of water stress as compared to the first cycle. This was not so in the third cycle of stress, which was the most severe form.

In general, it has been observed that there is a change in the redox state of glutathione under stress followed by increase in the synthesis and a highly reduced redox state of glutathione under acclimation. During severe stress, this ratio becomes more in the oxidized state with loss and degradation of glutathione.

However, the response should be seen in totality along with the other components of the antioxidant system. The second point to consider would be that GSH levels are affected by the sulfur nutritional status of the plant as well.

3.1.2 Salinity Stress

Soil salinity is another major abiotic stress affecting agricultural productivity. Increased salinization of arable land is expected to have devastating global effects, resulting in 30% land loss within the next 25 years, and up to 50% by the year 2050 (Wang et al. 2003). Hence, increasing the yield of crop plants in soils, including salinized regions, is essential for feeding the world and strategies have to be evolved to improve salinity stress tolerance (Blumwald et al. 2004). Adaptation of the plants to high salinity involves osmotic adjustment and the compartmentation of toxic ions and tolerating oxidative stress. Hence, the response of plants to salinity stress is multigenic in nature.

Enhanced activities of antioxidant enzymes have been reported in tolerant cultivars of pea (Hernández et al. 2000) and foxtail millet (Sreenivasulu et al. 2000) under salt stress, whereas this increase was not observed in the sensitive cultivars. Increase in GR activity has been reported in other plants such as lentils (Bandeoglu et al. 2004) and a tolerant cultivar of wheat (Sairam et al. 2004). A NaCl-tolerant cell line of cotton exhibited significantly higher activities of GR (287%), γ-ECS (224%) and GST (500%) than the control plants (Gossett et al. 1996). Buthionine sulfoximine (BSO), an inhibitor of GSH, reduced growth of the control lines by almost 94%, which was less in the NaCl-tolerant cell line; exogenous GSH restored growth in both the cell lines. A threefold increase in cysteine and glutathione was observed in wild-type plants of *Brassica napus* L. canola, but not in the salt-tolerant transgenic overexpressing a vacuolar Na+/H+ antiporter (Ruiz and Blumwald 2002). This could be a possible protective mechanism against salt-induced oxidative damage. Salt-induced oxidative damage led to necrotic lesions in the minor veins of pea leaves, as oxidative stress was higher in the apoplasts (Hernández et al. 2001). The GSH/GSSH ratio declined under salt stress, and there was no GR activity in the apoplasts; the sensitive cultivar Lincoln was more affected than the tolerant cv Puget with reduction in growth response. Continuous exposure to salt stress in rice seedlings made them more tolerant when the GSG/GSSG levels returned to normal values after an initial decline (Fadzilla et al. 1997).

Rice, a salt-sensitive cereal crop, is more affected when the salt stress is at the young seedling stages and less so at reproduction. However, it has genetic variation for salt stress, and it was observed that salinity stress induced a number of genes in the more sensitive genotype of *indica* rice IR29, including GSTs, which was not so in the tolerant FL478 (Walia et al. 2005). Among the 126 salinity-tolerant cDNAs identified and isolated from the roots of a mangrove plant *Bruguiera cylindrica*, using subtractive hybridization, one was a putative GST (Wong et al. 2007). Glutathione peroxidase (GPX) is also a major enzyme, which can reduce alkyl and lipid hydroperoxides to protect the cells from lipid peroxidation. Salt stress increased levels of *gpx-1* and *gpx-2* transcripts in *Synechocystis* PCC 6803, showing their importance in protecting the cellular machinery against oxidative damage (Gaber et al. 2004). Transgenic tobacco plants overexpressing both GST and GPX displayed improved seed germination and seedling growth under stress (Roxas et al. 1997; Table 10.1). In addition to higher activities of GST/GPX, the transgenic seedlings also showed higher levels of glutathione and ascorbate than wild-type plants (Roxas et al. 2000; Table 10.1).

Genetic manipulation of crop plants for enhancing abiotic stress tolerance is essential for sustainable agriculture. Methylglyoxal (MG), a cytotoxic byproduct of carbohydrate and lipid metabolism and a substrate of glyoxalase I, exists in micromolar concentrations in plants and increases several fold in response to salinity, drought, and cold stress conditions (Yadav et al. 2005a). Transgenic tobacco plants underexpressing glyoxalase I showed accumulation of MG, resulting in inhibition

Table 10.1 Transgenic plants overexpressing genes related to glutathione metabolism

Gene	Protein targeted to	Transgenic plant	Stress tolerance	Reference
E. coli gor, a gene encoding GR	Chloroplast	Tobacco	Paraquat and SO_2 tolerance	Aono et al. (1993)
	Cytosol	Tobacco	Paraquat tolerance and enhanced reduction of ascorbate	Foyer et al. (1991)
	Chloroplast	Poplar	Increased resistance to photoinhibition	Foyer et al. (1995)
γ-ECS gene (encodes γ-glutamylcysteine synthetase)	Not identified	*Arabidopsis*	No significant difference from wild type with respect to heavy metal resistance	Xiang et al. (2001)
	Cytosol	Poplar	Increased GSH synthesis	Noctor et al. (1996)
	Chloroplast	Poplar	Increased GSH levels, upregulation of specific amino acids	Noctor et al. (1998)
	Chloroplast	Tobacco	Increased GSH levels but, exhibited light-dependent chlorosis and necrosis	Creissen et al. (1999)
GSHS gene (encodes GSHS)	Cytosol	Poplar	No increase in foliar GSH levels	Strohm et al. (1995)
E. coli gshII gene (encoding GSHS)	Cytosol	Indian mustard	Enhances cadmium tolerance	Zhu et al. (1999)
Nt107 gene (encodes an enzyme with both GST/GPX activity)	Cytosol	Tobacco	Germination & seedling growth under chilling and salt stress	Roxas et al. (1997)
Nt107 gene	Chloroplast	Tobacco	Cold and salt tolerance	Roxas et al. (2000)

of seed germination; exogenous application of GSH reduced MG levels, demonstrating the critical role of glyoxalase I and GSH. Double transgenic plants overexpressing both glyoxalase I and II conferred improved salinity tolerance and were able to grow, flower, and set normal viable seeds under continuous salinity stress conditions (200 mM NaCl) with only a 5% loss in total productivity (Singla-Pareek et al. 2003). Glyoxalase enzymes are important for the GSH-based detoxification of MG, and their overexpression leads to further enhanced levels of GSH. Under salinity stress, GSH levels and the GSH/GSSH ratio were higher in the transgenic plants overexpressing glyoxalase enzymes than wild-type plants (Yadav et al. 2005b). Exogenous application of GSSG retarded the growth of wild-type plants and not the transgenic ones. The multifunctional role of GSH was clear when these transgenics overexpressing glyoxalase enzymes were grown in the presence of 5 mM $ZnCl_2$ (Singla-Pareek et al. 2006). They were able to grow normally and set viable seeds, and it was observed that the roots were the major sink for excess zinc accumulation. Increase in phytochelatins (GSH polymerization), maintenance of GSH homeostasis. and multiple stress tolerance towards other heavy metals such as cadmium and lead were also observed.

3.1.3 Temperature Stress

Temperature extremes are among the primary causes of reduced crop yields. Improved understanding of plant responses to temperature can lead to more reliable predictions of crop yield and may help in designing crops that are better matched to the environment (Atkinson and Porter 1996). Heat stress damages cellular structure and metabolic pathways and contributes to secondary water stress. Grain yield reductions of between 4%-10% have been estimated for every rise of 1 °C in mean temperature, but these reductions could be greater, particularly in the grain-filling stages due to increased rate of leaf senescence. Although there have been several reports on the oxidative stress and the response of antioxidant defense mechanisms in heat-stressed plants (Anderson and Padhye 2004, Dat et al. 1998), there have been fewer reports focusing on glutathione. Treatment of maize roots to heat shock temperatures of 40 °C resulted in decrease of cysteine levels and increase in GSH levels (Nieto-Sotelo and Ho 1986). There was an increase in the activity of the GSH synthesizing capacity in maize root cells, which was related to the cells' capacity to cope with heat stress conditions. Accumulation of GSH has also been observed in heat-stressed tomato seedlings (Rivero et al. 2004). In wheat, it was established that heat stress induced accumulation of GSH levels and increased the activity of the enzymes involved in GSH synthesis and the GSH/GSSG ratio (Kocsy et al. 2002). This was co-related to the frost sensitivity of the wheat genotypes as well as with the accumulation being higher in the sensitive genotypes. In fact, heat stress increased GSH levels in the flag leaf of two wheat genotypes with contrasting behavior in heat tolerance at all the stages during grain development (Chauhan 2005). Overall, GSH levels and GSH/GSSG ratio declined during grain development in flag leaves of wheat in both control and heat-stressed plants (Table 10.2). The

Table 10.2 GSH (reduced glutathione) and GSH/GSSG (reduced/oxidized glutathione) ratio in the flag leaf of wheat genotypes during grain development under normal and heat stress environments (Chauhan 2005)

	GSH (μmolg^{-1}FW)		GSH/GSSG	
Days after anthesis	Hindi62	PBW343	Hindi62	PBW343
Normal stress environment				
A	3.40 ± 0.16	5.39 ± 0.29	2.14 ± 0.17	1.7 ± 0.12
14	3.58 ± 0.07	4.78 ± 0.18	2.17 ± 0.33	1.30 ± 0.17
28	2.20 ± 0.13	1.62 ± 0.23	1.51 ± 0.23	0.49 ± 0.08
35	1.48 ± 0.09	0.34 ± 0.11	1.40 ± 0.18	0.20 ± 0.06
Heat stress environment				
A	9.24 ± 0.81	7.10 ± 0.15	2.59 ± 0.29	0.77 ± 0.05
10	7.22 ± 0.15	7.94 ± 0.39	2.44 ± 0.25	0.99 ± 0.13
20	3.83 ± 0.06	4.52 ± 0.30	1.20 ± 0.03	0.47 ± 0.04
25	1.00 ± 0.14	0.58 ± 0.07	0.29 ± 0.04	0.14 ± 0.02

rate of decline was, however, higher under a heat stress environment. Field studies have also been conducted in cotton seedlings to observe the temperature variations on oxidative stress (Mahan and Mauget 2005). There was no alteration in the GSH levels and GR activity to low or high temperatures which could have been an acclimatory response of the plants.

Low temperature or chilling stress is also a major abiotic stress that limits crop productivity. Many crops are susceptible to injury when air temperatures fall below a nonfreezing critical threshold (Kocsy et al. 2000a). The duration of chilling stress can have an important effect on chilling-sensitive plants. Increase in GSH levels and/or GR activity during chilling stress has been observed in many plant species such as tomato (Walker and McKersie 1993), *Sorghum* (Badiani et al. 1997), wheat (Kocsy et al. 2000b), jack pine (Zhao and Blumwald 1998), and poplars (Foyer et al. 1995). In fact, in wheat, chromosome 5A is involved in the regulation of GSH accumulation during cold hardening (Kocsy et al. 2000b).

Important food crops such as rice and maize are very sensitive to chilling stress. In maize, it has been shown that chilling stress increased the cysteine and GSH levels in the chilling-tolerant genotype as compared to the chilling-sensitive type (Kocsy et al. 1996). The importance of GSH in protection against chilling injury in maize was shown by the use of herbicide safeners on chilling-sensitive lines (Kocsy et al. 2001). This increased the pools of GSH and its precursors, cysteine and γ-glutamyl-cysteine, and there was an increase in GR activity as well, which increased the relative protection considerably. A long duration chilling stress experiment in maize showed increased activities of all antioxidant enzymes, including GR, under stress (Hodges et al. 1997). This could be due to acclimation as a 34%-47% increase in GR activity and induction in new isoforms of GR have been shown in maize during acclimation to chilling stress (Anderson et al. 1995, Prasad 1996). In fact, when transgenic cotton overproducing GR was field-grown, there was no significant difference to wild plants as GR activity doubled in wild-type cotton during slow chilling exposure in the field (Logan et al. 2003). Genetic transformation studies in maize for freezing tolerance showed the upregulation of three genes, including GSTs, under both normal and cold-acclimated conditions (Wang 2005). GSTs are involved in the oxidative signaling pathway and contribute to the genetic acclimation towards freezing tolerance in maize. Chilling stress and cold acclimation studies in rice showed increase in several antioxidant enzymes, including GR (Kuk et al. 2003, Oidaira et al. 2000). Proteome analysis of chilling stress in rice showed the upregulation of cysteine synthase (Yan et al. 2006). This enzyme is responsible for the final step in cysteine biosynthesis, a key limiting step in GSH production. Thus, GSH and GR are important for resistance to chilling stress.

3.2 Role of Glutathione in Redox Homeostasis and Signaling

The multifunctional role of glutathione in leaf cells depends on its concentration and/or the redox state of the leaf glutathione pools (May et al. 1998). The number of important and critical functions include defense against abiotic, biotic stresses and systemic acquired resistance, signaling, control of plant growth and development, cell death, sulfur storage and sensing (Maughan and Foyer 2006). Modulation of glutathione content transmits information through diverse signaling mechanisms, which include establishment of an appropriate thiol/disulphide exchange. Moreover, glutathione also participates in calcium signaling in plants by stimulating calcium release into cytosol (Gomez et al. 2004).

Glutathione accumulates to high concentrations, especially in stress situations. Increase in glutathione concentrations during stress offsets stress-initiated oxidation of glutathione and causes changes in gene expression directly or through interaction with regulatory proteins and/or transcription factors. This increase is equally important in signal transduction and defense against ROS and is through a multi-level control mechanism, which includes coordinate activation of genes encoding glutathione biosynthetic enzymes and GR (Xiang and Oliver 1998). Thus, glutathione acts as a redox sensor of environmental cues, and increase in glutathione helps plant resistance to oxidative stress.

The glutathione redox couple is an information-rich redox buffer that interacts with numerous cellular components (Foyer and Noctor 2005). Redox homeostasis occurs through ROS-antioxidant interaction and is a metabolic interface between stress perception and physiological and molecular adaptive responses of plant cells. This could be either induction of acclimatory processes or programmed cell death. Since increase in intracellular ROS to most stress responses is variable, the changes in glutathione pools are important in redox signaling (Ball et al. 2004, Gomez et al. 2004, Mou et al. 2003). Glutathione status may be perceived by the cell through several mechanisms, including protein glutathionylation, i.e., formation of disulphide bonds between glutathione and cysteine residues. This may be driven by an increase in GSSG or ROS or may be enzyme catalyzed, i.e., glutaredoxins (Lemaire 2004). This protects proteins from an irreversible damage to cysteinyl residues.

Many stresses cause an initial increase in GSSG, which in turn stimulates GSH synthesis to mainatain cellular homeostasis, often leading to an increase in total glutathione. Deviations from a high GSH/GSSG ratio during stress can trigger programmed cell death. The activity of GR is also essential in maintaining a high GSH/GSSG ratio (Foyer et al. 1991). The important factors limiting GSH synthesis are cysteine availability and γ-ECS activity. GSH is 20-fold more abundant than free cysteine, and there is a strong interaction between sulfur assimilation and glutathione synthesis (Noctor 2006).

The biosynthesis of GSH is not only regulated by the availability of sulfur, nitrogen, and carbon but also the availability of glutamic acid (the primary product of the GS-GOGAT pathway of nitrogen assimilation) and glycine (an intermediary in photorespiration). Rather, it has been shown in transformed poplar plants overexpressing γ-ECS gene that feeding glycine increased GSH synthesis in both chloroplasts and cytosol (Noctor et al. 1999). Hence, a complex regulatory cross talk between nitrogen, sulfur, and carbon metabolism is required for the control of GSH biosynthesis (Kopriva 2006, Kopriva and Rennenberg 2004). Besides availability of its constituent amino acids, GSH biosynthesis is also regulated by the transcriptional control of the enzymes involved in its synthesis and their activity and hormonal control.

3.3 Altered Levels of Glutathione Elicit Different Responses in Different Plants

A γ-glutamylcysteine synthetase (γ-ECS) mutant of Arabidopsis, *rml1*, produced no detectable amount of glutathione. This mutant could survive only on culture medium supplemented with GSH. It was defective in root development and produced only a small shoot (May et al. 1998). Another γ-ECS mutant of Arabidopsis, *gsh1*, conferred a recessive embryo-lethal phenotype and could develop and mature only when exogenous GSH was added (Cairns et al. 2006). Arabidopsis *cad2* mutant produces only a 30% wild-type GSH level and showed growth similar to that of wild-type plants under normal environment, but is hypersensitive to heavy metal Cd stress (Howden et al. 1995). This mutant was deficient in γ-glutamylcysteine synthetase enzyme (Cobbett et al. 1998). A point mutation in the *GSH1* gene of *Arabidopsis* (*ROOT MERISTEMLESS 1* mutant, homologous to *cad2*) accumulates approximately 3% of wild-type level of GSH. This mutant is impaired in root meristem formation, growth, and fertility (Vernoux et al. 2000). Another γ-glutamylcysteine synthetase 1 (*GSH1*), which encodes choloroplastic γ-ECS, mutant *rax 1-1* of *Arabidopsis* has 50% of wild-type GSH (Ball et al. 2004). A direct link was established in this study between glutathione biosynthesis and expression of ascorbate peroxidase gene under nonstressed conditions using this mutant.

Increasing glutathione biosynthetic capacity has been shown to enhance resistance to oxidative stress in *Brassica juncea* (Zhu et al. 1999). Transgenic tobacco overexpressing *gor*, a gene encoding GR, was more resistant to paraquat and sulfur dioxide as compared to nontransformed plants (Aono et al. 1993). Interestingly, the enhanced resistance to oxidative stress was due to increased reduction of the ascorbate pool rather than increase in the levels of GSH (Foyer et al. 1991). In contrast, it was shown in transgenic poplar overexpressing *gor* that the enhanced protection against oxidative stress was due to increase in levels of total pool of glutathione as well as GSH/GSSG ratio (Foyer et al. 1995). However, it was concluded that although overexpression of GR enzyme was beneficial to the plants, there was no direct correlation between enzyme levels and stress tolerance (Creissen et al. 1996). GSSG was shown to stimulate tolerance towards salinity and chilling stress in transgenic tobacco plants (Roxas et al. 1997).

Transgenic manipulation in *Arabidopsis* of γ-ECS gene in sense and antisense orientation produced GSH levels of 200% and 3% to that of the wild-type level. Although antisense plants showed hypersensitivity to heavy metals, Cu and Cd, and photooxidative and ozone stress, sense plants did not show any significant differences from wild-type (Xiang et al. 2001). Transgenic poplar plants overexpressing a cytosol-targeted γ-ECS gene showed three- to fourfold increase in GSH levels (Noctor et al. 1996), whereas transgenic poplar plants overexpressing a cytosol-targeted GSHS gene did not show any increase in foliar GSH levels (Strohm et al. 1995). However, transgenic tobacco overexpressing chloroplast targeted γ-ECS contained elevated glutathione levels but exhibited light-dependent chlorosis and necrosis (Creissen et al. 1999). This was attributed to the increased levels of GSSG and a failure of the redox sensing mechanism in chloroplasts of the transgenic plants. This was not the case in transgenic poplar plants overexpressing a chloroplast targeted γ-ECS, which showed both increased GSH synthesis and upregulation of specific amino acids in the chloroplasts (Noctor et al. 1998).

Thus, alteration in reduced glutathione levels affects growth and development of plants in addition to their role in stress tolerance. This has been summarized in Table 10.1.

3.3.1 Hormonal Control

Arabidopsis plants treated with cadmium or copper responded by increasing transcription of the genes involved in GSH synthesis (Xiang and Oliver 1998). Jasmonic acid also activated the same set of genes in a coordinated manner with heavy metals but did not alter the glutathione content in unstressed plants, which supports the idea that the glutathione concentration is controlled at multiple levels. It has been observed that GSH levels increased on treatment with abscisic acid (ABA) in maize (Jiang and Zhang 2001). However, it is not clear that the increased GSH levels are in direct response to ABA or the result of enhanced oxidative stress (Guan et al. 2000). The effect of salicylic acid on GSH biosynthesis is also not clear (Kopriva 2006). Thus, the role of phytohormones in the regulation of GSH synthesis needs to be investigated in detail (Kopriva and Rennenberg 2004).

4 Conclusions and Future Work

Glutathione has a central role in plant responses to abiotic stresses. The pathway of glutathione biosynthesis from glutamate, cysteine, and glycine is well established, and work on several mutants has shown the role of GSH in various aspects of plant growth and development. Besides its multiple functions, especially as an antioxidant, it functions as a key signaling compound for stress perception. Hence, the redox status of glutathione is important in maintaining cellular homeostasis. Environmental stresses perturb this ratio. Departures from a typical high GSH/GSSG ratio can even trigger plant cell death. The ability of the plant cells to restore this ratio may be accompanied by increase in the total pool of glutathione and would be subject to regulation by multiple controls such as a high GR activity. Identification of other factors could provide us tools for manipulation of plant stress responses.

The value of glutathione as a universal stress marker remains unclear, as the observed response could be at a dynamic stage (Noctor 2006). Differences also exist for different plant species, duration, and stress intensity. It has been observed frequently that when one component of the antioxidant defense system becomes limiting, plants respond by upregulating another antioxidant. The importance of the glutathione system in relation to other components of the antioxidant defense system has to be established. However, they are an integral part in assessing the stress response of plants, especially when seen in totality with other physiological responses. Similarly, plant responses to glutathione levels under acclimation conditions of various environmental stresses would give us more information regarding tolerance to abiotic stresses.

Glutathione pools within the cellular compartments are connected by transport across membranes. There is little information about the uptake of glutathione across the plasma membrane, the plastid envelope, the mitochondrial envelope, or the endoplasmic reticulum membrane and the localization of the transporters. It needs to be understood that most of the work regarding the functional aspects of glutathione have been done in model plant systems such as *Arabidopsis*, and these cannot be directly extrapolated to crops. In fact, more studies are needed on various crops under field conditions.

References

Anderson JA, Padhye SR (2004) Protein aggregation, radical scavenging capacity, and stability of hydrogen peroxide defense systems in heat-stressed *Vinca* and sweet pea leaves. J Am Soc Hort Sci 129:54-9

Anderson MD, Prasad TK, Stewart CR (1995) Changes in isozyme profiles of catalase, peroxidase, and glutathione reductase during acclimation to chilling in mesocotyls of maize seedlings. Plant Physiol 109:1247-57

Aono M, Kubo A, Saji H, Tanaka K, Kondo N (1993) Enhanced tolerance to photooxidative stress of transgenic *Nicotiana tabacum* with high chloroplastic glutathione reductase activity. Plant Cell Physiol 34:129-35

Arisi ACM, Cornic G, Jouanin L, Foyer CH (1998) Overexpression of iron superoxide dismutase in transformed poplar modifies the regulation of photosynthesis at low CO_2 partial pressures or following exposure to the prooxidant herbicide methyl viologen. Plant Physiol 117:565-74

Atkinson D, Porter JR (1996) Temperature, plant development and crop yields. Trends Plant Sci 1:199-24

Badiani M, Paolacci AR, Fusari A, D'ovidio R, Scandalios JG, Porceddu E, Sermanni GG (1997) Non-optimal growth temperatures and antioxidants in the leaves of Sorghum bicolor (L.) Moench. II. Short-term acclimation. J Plant Physiol 151:409-421

Ball L, Accotto GP, Bechtold U, Creissen G, Funck D, Jimenez A, Kular B, Leyland N, Mejia-Carranza J, Reynolds H, Karpinski S, Mullineaux PM (2004) Evidence for a direct link between glutathione biosynthesis and stress defense gene expression in Arabidopsis. Plant Cell 16:2448-62

Bandeoglu E, Eyidogan F, Yücel M, Öktem HA (2004) Antioxidant responses of shoots and roots of lentil to NaCl-salinity stress. Plant Growth Regul 42:69-77

Bartoli CG, Simontacchi M, Tambussi E, Beltrano J, Montaldi E, Puntarulo S (1999) Drought and watering-dependent oxidative stress: effect on antioxidant content in *Triticum aestivum* L. leaves. J Exp Bot 50:375-83

Blumwald E, Grover A, Good AG (2004) Breeding for abiotic stress resistance: Challenges and Opportunities. In: New directions for a diverse planet. Proceedings of the 4th International Crop Science Congress, 26 Sep – 1 Oct 2004, Brisbane, Australia

Bogs J, Bourbouloux A, Cagnac O, Wachter A, Rausch T, Delrot S (2003) Functional characterization and expression analysis of a glutathione transporter, BjGT1, from *Brassica juncea*: evidence for regulation by heavy metal exposure. Plant Cell Environ 26:1703-11

Bourbouloux A, Shahi P, Chakladar A, Delrot S, Bachhawat AK (2000) Hgt1p, a high affinity glutathione transporter from the yeast *Saccharomyces cerevisiae*. J Biol Chem 275:13259-65

Burgener M, Suter M, Jones S, Brunold C (1998) Cyst(e)ine is the transport metabolite of assimilated sulfur from bundle-sheath to mesophyll cells in maize leaves. Plant Physiol 116:1315-22

Cairns NG, Pasternak M, Wachter A, Cobbett CS, Meyer AJ (2006) Maturation of Arabidopsis seeds is dependent on glutathione biosynthesis within the embryo. Plant Physiol 141:446-55

Chauhan S (2005) Physiological and molecular basis of heat tolerance with emphasis on oxidative stress metabolism in wheat. PhD Thesis. HNB Garhwal University, Srinagar, Uttaranchal, India

Cobbett CS, May MJ, Howden R, Rolls B (1998) The glutathione-deficient, cadmium-sensitive mutant, cad2-1, of *Arabidopsis thaliana* is deficient in γ-glutamylcysteine synthetase. Plant J 16:73-8

Creissen G, Broadbent P, Stevens R, Wellburn AR, Mullineaux P (1996) Manipulation of glutathione mentaboilsm in transgenic plants. Biochem. Soc. Trans 24:465-9

Creissen G, Firmin J, Fryer M, Kular B, Leyland N, Reynolds H, Pastori G, Wellburn F Baker N, Wellburn A, Mullineaux P (1999) Elevated glutathione biosynthetic capacity in the chloroplasts of transgenic tobacco plants paradoxically causes increased oxidative stress. Plant Cell 11:1277-91

Dat JF, Lopez-Delgado H, Foyer CH, Scott IM (1998) Parallel changes in H_2O_2 and catalse during thermotolerance induced by salicylic acid or heat acclimation in mustard seedlings. Plant Physiol 116:1351-7

Fadzilla NM, Finch RP, Burdon RH (1997) Salinity, oxidative stress and antioxidant responses in shoot cultures of rice. J Exp Bot 48:325-31

Fitzgerald MA, Ugalde TD, Anderson JW (2001) Sulphur nutrition affects delivery and metabolism of S in developing endosperms of wheat. J Exp Bot 52:1519-26

Foyer CH, Lelandais M, Galap C, Kunert KJ (1991) Effects of elevated cytosolic glutathione reductase activity on the cellular glutathione pool and photosynthesis in leaves under normal and stress conditions. Plant Physiol. 97:863-72

Foyer CH, Noctor G (2005) Redox homeostasis and antioxidant signaling: A metabolic interface between stress perception and physiological responses. Plant Cell 17:1866-75

Foyer CH, Souruau N, Perret S, Lelandais M and Kunert K-J, Provust C, Jouanin L (1995) Overexpression of glutathione reductase but not glutathione synthetase leads to increases in antioxidant capacity and resistance to photoinhibition in poplar trees. Plant Physiol 109:1047-57

Foyer CH, Theodoulou FL, Delrot S (2001) The functions of inter- and intracellular glutathione transport systems in plants. Trends Plant Sci. 6:486-92

Gaber A, Yoshimura K, Tamoi M, Takeda T, Nakano Y, Shigeoka S (2004) Induction and functional analysis of two reduced nicotinamide adenine dinucleotide phosphate-dependent glutathione peroxidase-like proteins in *Synechocystis* PCC 6803 during the progression of oxidative stress. Plant Physiol 136:2855-61

Gamble PE, Burke JJ (1984) Effect of water stress on the chloroplast antioxidant system I. Alterations in glutathione reductase activity. Plant Physiol 76:615-21

Gomez LD, Noctor G, Knight MR, Foyer CH (2004) Regulation of calcium signalling and gene expression by glutathione. J Exp Bot 55:1851-9

Gossett DR, Banks, SW, Millhollon EP, Lucas MC (1996) Antioxidant response to NaCl stress in a control and an NaCl-tolerant cell line grown in the presence of paraquat, buthionone sulfoximine, and exogenous glutathione. Plant Physiol 112:803-9

Grene R (2002) Oxidative stress and acclimation mechanisms in plants. In: Somerville CR, Meyerowitz EM (eds) The Arabidopsis book, American Society of Plant Biologists, Rockville, MD, pp. 1-19

Guan L M, Zhao J, Scandalios JG (2000) *Cis*-elements and *trans* factors that regulate expression of the maize *Cat1* antioxidant gene in response to ABA and osmotic stress: H_2O_2 is the likely intermediary signaling molecule for the response. Plant J 22:87-95

Hernández JA, Ferrer MA, Jime'nez A, Barcelo' AR, Sevilla F (2001) Antioxidant systems and O_2^-/H_2O_2 production in the apoplast of pea leaves. Its relation with salt-induced necrotic lesions in minor veins. Plant Physiol 127:817-31

Hernández JA, Jime'nez A, Mullineaux P, Sevilla F (2000) Tolerance of pea leaves (*Pisum sativum*) to long term salt stress is associated with induction antioxidant defences. Plant Cell Biol 23:853-62

Hérouart D, Van Montagu M, Inze D (1993) Redox-activated expression of the cytosolic copper/zinc superoxide dismutase gene in *Nicotiana*. Proc Natl Acad Sci USA 90:3108-12

Herschbach C, van der Zalm E, Schneider A, Jouanin L, Kok LJD, Rennenberg H (2000) Regulation of sulfur nutrition in wild-type and transgenic poplar over-expressing γ-glutamylcysteine synthetase in the cytosol as affected by atmospheric H_2S. Plant Physiol 124:461-74

Hodges DM, Andrews CJ, Johnson DA, Hamilton RI (1997) Antioxidant enzyme responses to chilling stress in differentially sensitive inbred maize lines. J Exp Bot 48:1105-13

Howden R, Andersen CR, Goldsbrough PB, Cobbett CS (1995) A cadmium-sensitive, glutathione-deficient mutant of *Arabidopsis thaliana*. Plant Physiol 107:1067-73

Inzé D, Van Montagu, M (1995) Oxidative stress in plants. Curr Opin Biotech 6:153-158

Jamai A, Tommasini R, Martinoia E, Delrot S (1996) Characterization of glutathione uptake in broad bean leaf protoplasts. Plant Physiol 111:1145-52

Janowiak BE, Griffith OW (2005) Glutathione synthesis in *Streptococcus agalactiae*: One protein accounts for {gamma}-glutamylcysteine synthetase and glutathione synthetase activities. J Biol Chem 280:11829-39

Jiang M, Zhang J (2001) Effect of abscisic acid on active oxygen species, antioxidative defence system and oxidative damage in leaves of maize seedlings. Plant Cell Physiol 42:1265-73

Khanna-Chopra R, Selote DS (2007). Acclimation to drought stress generates oxidative stress tolerance in drought-resistant than – susceptible wheat cultivar under field conditions. Environ Exp Bot 60:276-83

Kocsy G, Ballmoos P von, *Rüegsegger* A, Szalai G, Galiba G, Brunold C (2001) Increasing the glutathione content in a chilling-sensitive maize genotype using safeners increased protection against chilling-induced injury. Plant Physiol 127:1147-56

Kocsy G, Brunner M, Rü*egsegger* A, Stamp P, Brunold C (1996) Glutathione synthesis in maize genotypes with different sensitivity to chilling. Planta 198:365-70

Kocsy G, Galiba G, Brunold C (2000a) Role of glutathione in adaptation and signaling during chilling and cold acclimation in plants. Physiol Plant 113:158-64

Kocsy G, Szalai G, Galiba G (2002) Effect of heat stress on glutathione biosynthesis in wheat. Acta Biol Szeged 46:71-2

Kocsy G, Szalai G, Va'gu'jfalvi A, Ste'hli L, Orosz G, Galiba G (2000b) Genetic study of glutathione accumulation during cold hardening in wheat. Planta 210:295-301

Kopriva S (2006) Regulation of sulfate assimilation in Arabidopsis and beyond. Ann Bot 97:479-95

Kopriva S, Rennenberg H (2004) Control of sulphate assimilation and glutathione synthesis: interaction with N and C metabolism. J Exp Bot 55:1831-42

Kuk YI, Shin JS, Burgos NR, Hwang TE, Han O, Cho BH, Jung S, Guh JO (2003) Antioxidative enzymes offer protection from chilling damage in rice plants. Crop Sci 43:2109-17

Kumar S, Singla-Pareek SL, Reddy MK, Soproy SK (2003) Glutathione: Biosynthesis, homeostasis & its role in abiotic stresses. J Plant Biol 30:179-87

Lascano HR, Antonicelli GE, Luna CM, Melchiorre MN, Gomez LD, Racca RW, Trippi VS, Casano LM (2001) Antioxidant system response of different wheat cultivars under drought: field and *in vitro* studies. Aus J Plant Physiol 28:1095-1102

Lemaire SD (2004) The glutaredoxin family of oxygenic photosynthetic organisms. Photosynth Res 79:305-18

Li Z-S, Zhao Y, Rea PA (1995) Magnesium adenosine 5′-triphosphate-energized transport of glutathione-*S*-conjugates by plant vacuolar membrane vesicles. Plant Physiol 107:1257-68

Logan BA, Monteiro G, Kornyeyev D, Payton P, Allen RD, Holaday AS (2003) Transgenic overproduction of gluatathione reductase does not protect cotton, *Gossypium hirsutum* (Malvaceae), from photoinhibition during growth under chilling conditions. Amer J Bot 90:1400-3

Loggini B, Scartazza A, Brugnoli E, Navari-Izzo F (1999) Antioxidative defense system, pigment composition, and photosynthetic efficiency in two wheat cultivars subjected to drought. Plant Physiol. 119:1091-1100

Mahan JR, Mauget SA (2005) Antioxidant metabolism in cotton seedlings exposed to temperature stress in the field. Crop Sci. 45:2337-45

Masip L, Veeravalli K, Georgiou G (2006) The many faces of glutathione in bacteria. Antiox Red Signal 8:753-62

Maughan S, Foyer CH (2006) Engineering and genetic approaches to modulating the glutathione network in plants. Physiol Plant 126:382-97

May MJ, Vernoux T, Leaver C, Van Montagu M, Inzé D (1998) Glutathione homeostasis in plants: implications for environmental sensing and plant development. J Exp Bot 49:649-67

Menconi M, Sgherri CLM, Pinzino C, Navari-Izzo F (1995) Activated oxygen production and detoxification in wheat plants subjected to a water deficit programme. J Exp Bot 46:1123-30

Mittova V, Theodoulou FL, Kiddle G, Gomez L, Volokita M, Tal M, Foyer CH, Guy M. (2003) Coordinate induction of glutathione biosynthesis and glutathione-metabolizing enzymes is correlated with salt tolerance in tomato. FEBS Lett 554:417-21

Mou Z, Fan W, Dong X (2004) Inducers of plant systemic acquired resistance regulate NPR1 function through redox changes. Cell 113:935-44

Navari-Izzo F, Meneguzzo S, Loggini B, Vazzana C, Sgherri CLM (1997) The role of the glutathione system during dehydration of *Boea hygroscopica*. Physiol Plant 99:23-30

Nieto-Sotelo J, Ho T-H D (1986) Effect of heat shock on the metabolism of glutathione in maize roots. Plant Physiol 82:1030-5

Noctor G (2006) Metabolic signalling in defence and stress: the central roles of soluble redox couples. Plant Cell Environ 29:409-25

Noctor G, Foyer CH (1998) Ascorbate and glutathione: keeping active oxygen under control. Annu Rev Plant Physiol Plant Mol Biol 49:249-79

Noctor G, Strohm M, Jouanin L, Kunert K-J, Foyer CH, Rennenberg H (1996) Synthesis of glutathione in leaves of transgenic poplar overexpressing gamma-glutamylcysteine synthetase. Plant Physiol. 112:1071-8

Noctor G, Arisi ACM, Jouanin L, Kunert KJ, Rennenberg H, Foyer CH (1998) Glutathione: biosynthesis, metabolism and relationship to stress tolerance explored in transformed plants. J Exp Bot 49:623-47

Noctor G, Arisi ACM, Jouanin L, Foyer CH (1999) Photorespiratory glycine enhances glutathione accumulation in both the chloroplastic and cytosolic compartments. J Exp Bot 50:1157-67

Noctor G, Gomez L, Vanacker H and Foyer CH (2002) Interactions between biosynthesis, compartmentation and transport in the control of glutathione homeostasis and signaling. J Exp Bot 53:1283-1304

Ogawa K (2005) Glutahione-associated regulation of plant growth and stress responses. Antiox Red Signal 7:973-81

Oidaira H, Sano S, Koshiba T, Ushimaru T (2000) Enhancement of antioxidative enzyme activities in chilled rice seedlings. J Plant Physiol 156:811-13

Pastori G, Foyer CH, Mullineaux P (2000) Low temperature-induced changes in the distribution of H_2O_2 and antioxidants between the bundle sheath and mesophyll cells of maize leaves. J Exp Bot 51:107-13

Prasad TK (1996) Mechanisms of chilling-induced oxidative stress injury and tolerance in developing maize seedlings: changes in antioxidant system, oxidation of proteins and lipids, and protease activities. Plant J 10:1017-26

Rea PA, Li ZS, Lu YP, Drozdowicz YM, Martinoia E (1998) From vacuolar GS-X pumps to multispecific ABC transporters. Annu Rev Plant Physiol Plant Mol Biol 49:727-60

Rivero RM, Ruiz JM, Romero L (2004) Oxidative metabolism in tomato plants subjected to heat stress. J Hort Sci Biotech 79:560-4

Roxas V P, Lodhi SA, Garrett DK, Mahan JR, Allen RD (2000) Stress tolerance in transgenic tobacco seedlings that overexpress glutathione S-transferase/glutathione peroxidase. Plant Cell Rep 41:1229-34

Roxas V P, Smith R K, Allen ER, Allen RD (1997) Overexpression of glutathione S-transferase/ glutathione peroxidase enhances the growth of transgenic tobacco seedlings during stress. Nature Biotech 15:988991

Ruiz JM, Blumwald E (2002) Salinity-induced glutathione synthesis in *Brassica napus*. Planta 214:965-9

Sairam RK, Srivastava GC, Agarwal S, Meena RC (2004) Differences in antioxidant activity in response to salinity stress in tolerant and susceptible wheat genotypes. Biol Plant 49:85-91

Schneider A, Martini N, Rennenberg H (1992) Reduced glutathione (GSH) transport in cultured tobacco cells. Plant Physiol Biochem 30:29-38

Selote D, Khanna-Chopra R (2006) Drought-acclimation confers oxidative stress tolerance by inducing coordinated antioxidant defense at cellular and sub-cellular level in leaves of wheat seedlings. Physiol Plant 127:494-506

Selote D, Bharti S, Khanna-Chopra R (2004). Drought acclimation reduces O_2^- accumulation and lipid peroxidation in wheat seedlings. Biochem Biophys Res Commun 314:724-9

Sgherri CLM, Loggini B, Bochicchio A, Navari-Izzo F (1994) Antioxidant system in *Boea hygroscopica*: changes in response to desiccation and rehydration. Phytochem 37:377-81

Sharma SS, Dietz K-J (2006) The significance of amino acids and amino acid-derived molecules in plant responses and adaptation to heavy metal stress. J Exp Bot 57:711-26

Singla-Pareek SL, Reddy MK, Sopory SK (2003) Genetic engineering of the glyoxalase pathway in tobacco leads to enhanced salinity tolerance. Proc Natl Acad Sci USA 100:14672-7

Singla-Pareek SL, Yadav SK, Pareek A, Reddy MK, Sopory SK (2006) Transgenic tobacco overexpressing glyoxalase pathway enzymes grow and set viable seeds in zinc-spiked soils. Plant Physiol 140:613-23

Skipsey M, Andrews CJ, Townson J, Jepson I, Edwards R (1999) Isolation of cDNA (AJ243813) and genomic clones (AJ243812) of glutathione synthetase containing plastidic targeting sequences from *Arabidopsis thaliana*. Plant Physiol 121:312

Smirnoff N (1993) The role of active oxygen in the response of plants to water deficit and desiccation. New Phytol 125:27-58

Sreenivasulu N, Grimm B, Wobus U, Weschke W (2000) Differential response of antioxidant compounds to salinity stress in salt-tolerant and salt-sensitive seedlings of foxtail millet (*Setaria italica*). Physiol Plant 109:435-42

Srivalli B, Sharma G, Khanna-Chopra R (2003) Antioxidative defense system in an upland rice cultivar subjected to increasing intensity of water stress followed by recovery. Physiol Plant 119:503-12

Strohm M, Jouanin L, Kunert KJ, Pruvost C, Polle A, Foyer CH, Rennenberg H (1995) Regulation of glutathione synthesis in leaves of transgenic poplar (*Populus tremula* × *P. alba*) overexpressing glutathione synthetase. Plant J 7:141-5

Tausz M, ircelj H, Grill D (2004) The glutathione system as a stress marker in plant ecophysiology: is a stress-response concept valid? J Exp Bot 55:1955-62

Tommasini R, Martinoia E, Grill E, Dietz KJ, Amrhein N (1993) Transport of oxidized glutathione into barley vacuoles: evidence for the involvement of the glutathione-S-conjugate ATPase. Z Naturforsch 48:867-71

Tommasini R, Vogt E, Fromenteau M, Hortensteiner S, Matile P, Amrhein N, Martinoia E (1998) An ABC-transporter of *Arabidopsis thaliana* has both glutathione-conjugate and chlorophyll catabolite transport activity. Plant J 13:773-80

Vernoux T, Wilson RC, Seeley KA, Reichheld JP, Muroy S, Brown S, Maughan SC, Cobbett CS, Van Montagu M, Inze D, May MM, Sung ZR (2000) The ROOT MERISTEMLESS1/CADMIUM SENSITIVE2 gene defines a glutathione: dependent pathway involved in initiation and maintenance of cell division during postembryonic root development. Plant Cell 12:97-110

Wachter A, Wolf S, Steininger H, Bogs J, Rausch T (2005) Differential targeting of GSH1 and GSH2 is achieved by multiple transcription initiation: implications for the compartmentation of glutathione biosynthesis in the Brassicaceae. Plant J 41:15-30

Walia H, Wilson C, Condamine P, Liu X, Ismail AM, Zeng L, Wanamaker SI, Mandal J, Xu J, Cui X, Close TJ (2005) Comparative transcriptional profiling of two contrasting rice genotypes under salinity stress during the vegetative growth. Plant Physiol 139:822-35

Walker MA, McKersie BD (1993) Role of the ascorbate-glutathione antioxidant system in chilling resistance of tomato. J Plant Physiol 141:234-9

Wang K (2005) Genetic acclimaton for freezing tolerance. ISB news report

Wang W, Vinocur B, Altman A (2003) Plant responses to drought, salinity and extreme temperatures: towards genetic engineering for stress tolerance. Planta 218:1-14

Wong Y-Y, Ho, C-L, Nguyen PD, Teo S-S, Harikrishna JA, Rahim RA, Wong MCVL (2007) Isolation of salinity tolerant genes from the mangrove plant, *Bruguiera cylindrica* by using suppression subtractive hybridization (SSH) and bacterial functional screening. Aquatic Bot 86:117-22

Xiang C, Oliver, DJ (1998) Glutathione metabolic genes coordinately respond to heavy metals and jasmonic acid in Arabidopsis. Plant Cell 10:1539-50

Xiang C, Werner BL, Christensen EM, Oliver DJ (2001) The biological functions of glutathione revisited in *Arabidopsis* transgenic plants with altered glutathione levels. Plant Physiol 126:564-74

Yadav SK, Singla-Pareek SL, Reddy MK, Sopory SK (2005a) Methylglyoxal levels in plants under salinity stress are dependent on glyoxalase I and glutathione. Biochem Biophys Res Commun 337:61-7

Yadav SK, Singla-Pareek SL, Reddy MK, Sopory SK (2005b) Transgenic tobacco plants overexpressing glyoxalase enzymes resist an increase in methylglyoxal and maintain higher reduced glutathione levels under salinity stress. FEBS Lett 579:6265-71

Yan S-P, Zhang Q-Y, Tang Z-C, Su W-A, Sun W-N (2006) Comparative proteomic analysis provides new insights into chilling stress responses in rice. Mol Cell Proteomics 5:484-96

Zhang M-Y, Bourbouloux A, Cagnac O, Srikanth CV, Rentsch D, Bachhawat AK, Delrot S (2004) A novel family of transporters mediating the transport of glutathione derivatives in plants. Plant Physiol 134:482-91

Zhao S, Blumwald E (1998) Changes in oxidationreduction state and antioxidant enzymes in the roots of jack pine seedlings during cold acclimation. Physiol Plant 104:134-142

Zhu YL, Pilon-Smits EAH, Jouanin L, Terry N (1999) Overexpression of glutathione synthetase in indian mustard enhances cadmium accumulation and tolerance. Plant Physiol 119:73-80

Chapter 11
Recent Advances in Understanding of Plant Responses to Excess Metals: Exposure, Accumulation, and Tolerance

Marjana Regvar(✉) and Katarina Vogel-Mikuš

Abstract Toxicity has been the primary driver of research on excess metals; but in recent decades, the biology of metal accumulation/hyperaccumulation became the main focus of research. The main aim is to develop phytoremedial techniques for diminishing the environmental impact of metal pollution. From the toxicological point of view, only the bioavailable soil metal fractions can affect morphology and/or physiology; therefore methods for quantification of metal stress in plants are continuously evolving. Phenotypic plasticity encoded by fixed genetic components in different species determines a plant's responses to excess metals. Metal exclusion, accumulation, and hyperaccumulation are major plant strategies when facing excess metals, with metal tolerance mechanisms as a prerequisite for coping with the metal(s) in question. Considerable research efforts are being directed into the field of biotechnology of metal hyperaccumulation, but the complexity of plant responses to excess metals makes this task extremely difficult.

1 Introduction

Plant functioning and its regulation were first studied using chemical, physiological, and genetic approaches, dramatic progress has, however, been made as a result of molecular biology techniques. Structural genomic approaches yield new insights on genome mappings, sequencing, genome organization and comparative genome data, while functional genomic approaches (transcriptomics, proteomics, and metabolomics) are needed to link the structural data to plant functioning (Hesse and Höfgen 2001). Model plants such as *Arabidopsis thaliana* and *Thlaspi caerulescens* (Cobbett 2003, Assunção et al. 2003) are used in many laboratories to improve the ratio between the knowledge gained and the working load input. Comparative studies on nonmodel organisms are, on the other hand, used to test the universal

Marjana Regvar
Department of Biology, Biotechnical Faculty, University of Ljubljana, Večna pot 111, SI-1000 Ljubljana, Slovenia
marjana.regvar@bf.uni-lj.si

N.A. Khan et al. (eds.), *Sulfur Assimilation and Abiotic Stress in Plants.*

applicability of the proposed models. Both, powerful analytical techniques and a multidisciplinary approach by experts from different fields of research have contributed significantly to our knowledge of plant responses to excess metals. This chapter focuses on the latest discoveries in metal bioavailability, accumulation, distribution, and detoxification in plants.

2 Toxic Metal Exposure

Major sources of atmospheric emissions are mining, different types of metal production, and combustion of fossil fuels. The principal sources of toxic metals in soils are the disposal of ash residues from coal combustion, disposal of commercial products on land, and contributions from the atmosphere, whereas emissions into aquatic ecosystems originate mainly from the atmosphere, metal smelters, coal burning, and the dumping of sewage sludge. Soil and water metal contents are particularly relevant for plants and algae, and only a fraction of the total soil metals is available for plant uptake (Clemens 2006). Considerable controversy exists in the literature on the precise definition of *bioavailability*, since it has been linked to metal ion activity in soil solution, the exchangeable soil metal fraction, and more recently, to the concentrations of metals that cause ecotoxicity, but the scientific community universally accepts none of these definitions. From the toxicological point of view, only the *bioavailable soil metal fractions* are the ones that can affect the morphology and/or physiology of an organism. The availability of metal ions in soils depends on the soil (pH, surface charge, organic matter content, clay content, oxide minerals, redox potential, soil solution composition), plant (plant species or cultivar, plant parts, age of the plant, interactions with microorganisms), and environmental factors (climatic conditions, management practices, irrigation practices, topography) (Naidu et al. 2003, Moore 2003, Tsao 2003). Biological availability, strictly speaking, means available to living organisms. Thus, although chemical extractions involving dilute salt solutions have been used to measure the bioavailable pool of metals in soils, the availability of metals for plants can only be assessed using in situ techniques that involve growing the organisms of interest in contaminated materials and quantifying the uptake of metal into the organism and assessing the toxicological response. Therefore bioavailability as determined by plant studies should be more correctly termed *phytoavailability* (Naidu et al. 2003).

2.1 *Bioavailability at the Soil-Plant Interface*

Plant roots, soil microbes, and their interactions can increase metal bioavailability through secretion of protons, organic acids, phytosiderophores (PSs), amino acids, and enzymes (see Yang et al. 2005). Secretion of protons by roots acidifies the rhizosphere and increases metal dissolution. Proton extrusion of the roots is operated by plasma membrane H^+-ATPases and other H^+ pumps. In the lupin, P deficiency induces citrate exudation by enhancing the activity of plasma membrane H^+-ATPase and H^+ export (Ligaba et al. 2004). Al-induced exudation of malate, as the basis for the mechanism of Al tolerance in wheat, was found to be accompanied by changes in plasma membrane surface potential and the activation of H^+-ATPase (Ahn et al. 2004). Acetic acid and succinate were found in the rhizosphere of a Cd-accumulating genotype of wheat (Kyle), but not in the nonaccumulating genotype (Arcola) (Cieslinski et al. 1998). Phytosiderophores can be released under Fe deficiency from cereals to increase the mobilization of Fe, Zn, Cu, and Mn in the soil (Römheld 1991). In some dicots root reductases, such as *A. thaliana* NAOH-dependent Fe^{3+}-chelate reductase (NFR) can reduce Fe^{3+} or Cu^{2+} under low Fe and Cu supply to increase plant uptake (Welch et al. 1993, Bagnaresi et al. 1999). No direct evidence for a relationship between root exudation and metal hyperaccumulation in *T. caerulescens* was found (McGrath et al. 1997, 2001, Zhao et al. 2001). Changed soil conditions in the rhizosphere and increased solubility of the retained Zn in the rhizosphere soil of the Zn hyperaccumulator *T. caerulescens* by root-microbe interactions were, however, reported (Whiting et al. 2001). Bacteria have been shown to catalyse redox transformations leading to a decrease in soil metal bioavailability (Yang et al. 2005). For example, a strain of *Xanthomonas maltophyla* was shown to catalyze the reduction and precipitation of highly mobile Cr^{6+} to Cr^{3+}, a significantly less mobile and environmentally less hazardous species, and to induce the transformation of other toxic metal ions including Pb^{2+}, Hg^{2+}, Au^{3+}, Te^{4+}, Ag^+ and oxy-anions such as SeO_4^- (Lasat 2002). Similarly, the bulk of evidence tends to indicate that mycorrhizal fungi inhibit metal uptake by binding metals to components of the mycelium (Joner and Leyval 1997, Joner et al. 2000). Studies of genes of arbuscular mycorrhizal fungi that are putatively expressed under heavy metal stress have recently been reported (González-Guerrero et al. 2005, Hildebrandt et al. 2007).

3 Metal Accumulation and Tolerance in Plants

Besides bioavailability, uptake and translocation efficiencies determine metal accumulation and distribution in plants (Clemens 2006). Roots are the plant organs in closest contact with metal-contaminated soils; therefore they are the most affected by metals. Resistance to excess metals can either be achieved by *avoidance*, when the plant is able to restrict metal uptake into the cells, or *tolerance*, when the plant is able to survive the presence of excess metals. Only a limited number (approximately 420) of plant species have developed the ability to accumulate more than 10,000 $\mu g\ g^{-1}$ of Mn, or Zn; 1,000 $\mu g\ g^{-1}$ Ni, Cu, Pb or Se; and 100 $\mu g\ g^{-1}$ of Cd, far in excess of normal physiological requirements (if any), and far in excess of the levels found in the majority of other species tolerant of metalliferous soils. These so-called *hyperaccumulating plants* possess specific genetic determinants that enable efficient metal uptake and enhanced transport from the roots to the shoots (Baker 1981, 1987, Reeves and Baker 2000, Reeves 2006). Transporters of ligands for metal ions contributing to metal uptake, transport, and detoxification have recently been reviewed (Haydon and Cobbett 2007).

3.1 Metal Transporters

Plant genomes encode large families of metal transporters that vary in their substrate specificities, expression pattern, and cellular localization in governing metal translocation throughout the plant (Colangelo and Guerinot 2006, Haydon and Cobbett 2007). This section focuses on the recently most intensively studied metal transporters in plants.

3.1.1 Metal Uptake Proteins

Metal-uptake transporters are either acting at the plasma membrane (PM) to move metals into the cytoplasm from the apoplast or from intracellular compartments (Fig. 11.1) and involve the yellow stripe-like (YSL) family, the natural resistance associated macrophage protein (NRAMP) family, the zinc-regulated and iron-regulated transporter-like protein (ZIP) families, and the high-affinity Cu uptake proteins (COPT) family (Table 11.1), as extensively reviewed in Yang et al. (2005) and Colangelo and Guerinot (2006).

3.1.1.1 YSL Family

Yellow stripe-like proteins (Table 11.1) are believed to mediate the uptake of metals that are complexed with plant-derived phytosiderophores (PS) or nicotianamine (NA), a non-proteinogenic amino acid found throughout the plant kingdom that serves as a precursor for PS synthesis in grasses (Curie 2001). At the biochemical level, the best-studied member of this plant-specific family is YS1 from maize (ZmYS1), which accumulates in the roots and leaves of Fe-starved plants and functions as a proton-coupled symporter for Fe–PS transport. Analogy with yeast and oocyte transport studies indicates that ZmYS1 might also have other substrates, including Zn, Ni, and Cu. On the basis of sequence similarity to ZmYS1, *A. thaliana* has eight predicted YSL proteins. Considering that nongrasses do not produce or use PS, AtYSLs most probably transport metal–NA complexes (Roberts et al 2004, Schaaf et al. 2004, Colangelo and Guerinot 2006). The gene expression of two family members has recently been studied in detail. *AtYSL2* transcript accumulation increases under conditions of Fe sufficiency or Fe resupply and in response to Cu and Zn. Its mRNA expression pattern and its protein localization in lateral membranes suggest that AtYSL2 might function in the lateral transport of metals in veins (DiDonato et al. 2004, Schaaf et al. 2005, Colangelo and Guerinot 2006). *AtYSL1* is a shoot-specific gene whose transcript levels increase in response to high-Fe conditions. Its expression in young siliquas and the chalazal endosperm, together with data from NA- and Fe-distribution studies, support its role in Fe loading of seeds. *ysl1* shoots contain elevated NA levels, whereas *ysl1* seeds contain two to fourfold less NA (and less Fe) than wild-type plants. The germination of *ysl1* seeds under Fe starvation is slower than that of wild-type plants, and the defect can be overcome by Fe supplementation during germination (Le Jean et al. 2005). Thus, it is apparent that Fe and possibly Zn, Mn, and Cu homeostasis are dependent on YSLs. Further examination of the localization and substrate specificity of various YSLs, and genetic analysis

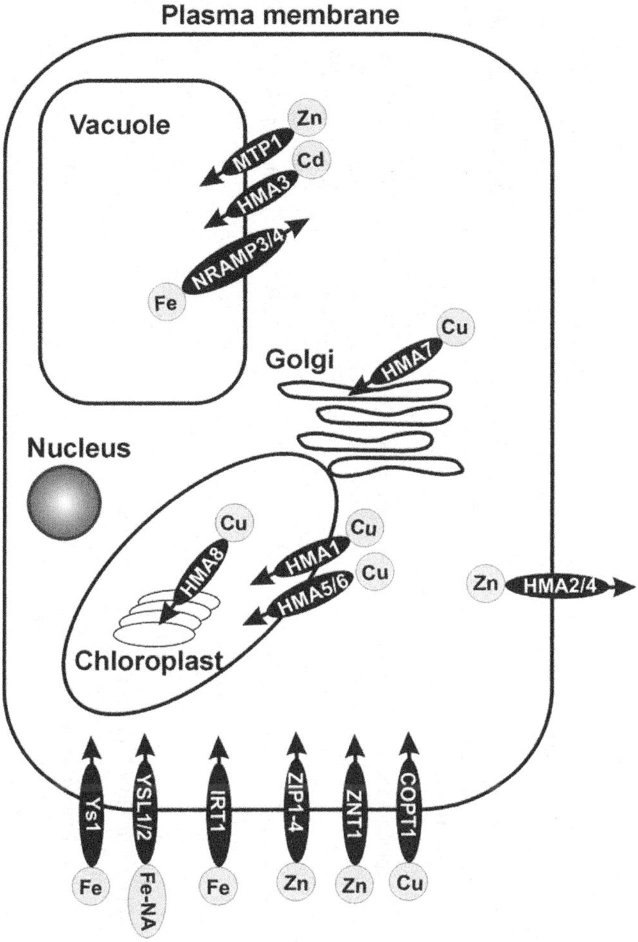

Fig. 11.1 Extra- and intracellular metal transport in plants. Metal delivery to and remobilization from intracellular compartments are important considerations in cellular ion homeostasis and tolerance. The localization of metal transporters with their preferential substrates are summarized in a generic cell, with arrows indicating the direction of transport (adopted after Clemens 2001, Colangelo and Guerinot 2006)

of double (or higher order) mutant plants will help to assign functions to this recently identified plant-specific family of transporters (Colangelo and Guerinot 2006).

3.1.1.2 ZIP Family

Zinc-regulated and iron-regulated transporter-like proteins (Table 11.1) generally contribute to metal-ion homeostasis through the transport of cations into the cytoplasm. More than 100 ZIP proteins have been identified in bacteria, fungi, animals, and plants

Table 11.1 Metal transporters listed by gene family, their tissue expression, cellular localization, function, inducing conditions and proposed/known substrates*

Metal uptake proteins	Tissue expression	Cellular localization and Function	Inducing conditions	Proposed/known substrate	Reference(s)
YSL					
AtYSL1	Siliquae, leaves (xylem parenchyma), flowers, chalazal endosperm of the seeds	Long-distance circulation of Fe-NA, delivery to seeds	+Fe	Fe-NA	Le Jean et al. 2005
AtYSL2	Roots endoderm, pericycle, shoots, reproductive organs, particularly in xylem parenchyma cells	PM, lateral movement of metals in the vasculature, Fe and Zn homeostasis	+Fe, downregulated by –Zn	-	DiDonato et al. 2004
					Schaaf et al. 2005
OsYSL2	Leaves (phloem), roots, seeds	PM, phloem transport of Fe	-Fe in shoots	Fe^{2+}-, Mn^{2+}-NA	Kolke et al. 2004
TcYSL3	Roots, shoots, flowers	-	-	Fe/Ni-NA	Gendre et al. 2007
TcYSL5	Shoots	-	-	Fe/Ni-NA	Gendre et al. 2007
TcYSL7	Flowers	-	-	Fe/Ni-NA	Gendre et al. 2007
ZmYS1	Roots, shoots	Fe^{3+}-PS, Fe^{2+}-NA translocation	-Fe	Fe^{3+}-PS, Fe^{3+}-, Fe^{2+}-, Ni^{2+}-, Cu^{2+}-NA	Schaaf et al. 2004
					Roberts et al. 2004
NRAMP					
AtNRAMP1/ AtNRAMP2	Roots, shoots	Plastid envelope?, Fe distribution in the cell, Fe resistance	-Fe in roots	Fe^{2+}	Curie et al. 2000
AtNRAMP3	Vascular bundles of roots, stems, leaves	VM, Fe-starvation dependent accumulation of Mn-, Zn and Fe acquisition	-	Fe^{2+}	Thomine et al. 2003
AtNRAMP3/ NRAMP4	seeds	VM, Fe remobilization from the vacuole during seed germination	-Fe	Fe^{2+}; Cd^{2+}	Lanquar et al. 2005
OsNRAMP1-2	Roots, leaves	-	-	-	Belouchi et al. 1997
TjNRAMP4				Ni^{2+}	Mizuno et al. 2005

ZIP					
AtIRT1	External cell layers of the roots	PM, Fe^{2+} uptake from soil	-Fe	Fe^{2+}, Mn^{2+}, Zn^{2+}, Co^{2+}, Cd^{2+}	Vert et al. 2002
OsIRT1	Roots	PM, Fe^{2+} uptake from soil	-Fe, -Cu	Fe^{2+}, Mn^{2+}, Zn^{2+}, Co^{2+}, Cd^{2+}	Bughio et al. 2002
OsZIP4	Leaf vascular bundles and mesophyll cells, root vascular bundles, shoot and root meristems	PM, translocation of Zn^{2+} within the plant	-Zn	Zn^{2+}	Ishimaru et al. 2005
OsZIP1	Roots	-	-Cu	-	Ishimaru et al. 2005
MtZIP1	Root, leaf	-	-	Zn^{2+}	López-Millán et al. 2004
MtZIP3	Root, leaf	-	Downreg. by –Mn and –Fe, slightly upreg. by –Zn	Fe^{2+}	López-Millán et al. 2004
MtZIP4	Roots of – Zn plants, leaves	-	Downreg. by –Mn and –Fe, upreg. by -Zn in leaves,	Mn^{2+}	López-Millán et al. 2004
MtZIP5	Leaves	-	Upreg. by –Zn leaves	Zn^{2+}, Fe^{2+}	López-Millán et al. 2004
MtZIP6	Roots of – Fe plants, leaves	-	-	Zn^{2+}, Fe^{2+}	López-Millán et al. 2004
MtZIP7	Leaves	-	-	Mn^{2+}	López-Millán et al. 2004
TcZNT1	Roots, shoots	PM, high affinity Zn^{2+} and low affinity Cd^{2+} uptake	Downreg. by Zn	Zn^{2+}	Pence et al. 2000
TjZNT1	-	-	-	Ni^{2+}, Mn^{2+}, Zn^{2+}, Cd^{2+}	Mizuno et al. 2005
TjZNT2	-	-	-	Ni^{2+}, Mn^{2+}, Cd^{2+}	Mizuno et al. 2005

(continued)

Table 11.1 (continued)

Metal uptake proteins	Tissue expression	Cellular localization and Function	Inducing conditions	Proposed/known substrate	Reference(s)
COPT					
AtCOPT1	Root tips, trichomes, stomata, pollen, embryos	Cu acquisition and accumulation	Downreg. by Cu	Cu	Sancenón et al. 2004
Metal efflux proteins					
P_{1B}-ATPases					
Zn/Cd/Pb/Co cluster					
AtHMA1	Green tissues	Chloroplast envelope, Cu^{2+} import and homeostasis	Cu^{2+}	Cu^{2+}	Seigneurin-Berny et al. 2006
AtHMA2	All plant organs, especially vascular tissues	PM, Zn^{2+} efflux, Zn^{2+} phloem loading, Zn^{2+} homeostasis	Zn^{2+}, Cd^{2+}, less by Pb^{2+}, Ni^{2+}, Cu^{2+} and Co^{2+}	Zn^{2+}, Cd^{2+}	Eren and Argüello 2004, Hussain et al. 2004
AtHMA3	Roots, old rosette and cauline leaves	VM, vacuolar uptake of Cd^{2+} and Pb^{2+} and cellular detoxification	Cd^{2+}	Cd^{2+}, Pb^{2+}	Gravot et al. 2004
AhHMA3	Shoots, roots	Detoxification of Zn	Zn^{2+}	Zn^{2+}	Becher et al. 2004
AtHMA4	Tissues surrounding the root vascular tissues	PM, Zn^{2+} xylem loading in roots, root to shoot translocation, increased Zn^{2+}, Cd^{2+}, Co^{2+} tolerance, detoxification by enhanced efflux in root and shoot apoplast	Zn^{2+}, Cd^{2+}, Pb^{2+}	Zn^{2+}, Cd^{2+}	Verret et al. 2004, 2005 Hussain et al. 2004
Cu/Ag cluster					
AtHMA5	Roots (especially pericycle cells), flowers	Plastid membrane, Cu^{2+} compartmentation and detoxification	Cu^{2+} excess	Cu^{2+}	Andrés-Colás et al. 2005
AtHMA7 (RAN1)	-	Post Golgi compartment, transport of Cu^{2+} into the secretory pathway, and delivering it to secreted or membrane-bound proteins	-	Cu^{2+}	Hirayama et al. 1999, Woeste and Kieber, 2000, Williams and Mills 2005

AtHMA6 (PAA1)	Roots, shoots	Plastid envelope, Cu^{2+} transport into stroma	-	Cu^{2+}	Shikanai et al. 2003, Abdel-Ghany et al. 2005
AtHMA8 (PAA2)	Shoots	Chloroplast tylacoid memenbranes	-	Cu^{2+}	Abdel-Ghany et al. 2005
CDF					
AtMTP1	Roots, shoots, flowers	VM, sequestration of Zn^{2+} into vacuoles, to maintain Zn homeostasis	-	Zn	Kobae et al. 2004, Desbrosses-Fonrouge et al. 2005
AhMTP1	Shoots, roots under high Zn^{2+}	VM, transport of Zn^{2+} into the vacuole	Zn excess in roots	Zn	Dräger et al. 2004
TgMTP1	Roots	PM, VM, Zn^{2+} efflux from the cells	-	Zn	Kim et al. 2004
other					
LCT1	-	-	-	Cd^{2+}, Ca^{2+}	Clemens et al. 1998
NtCBP4	-	PM	-	Pb^{2+}	Arazi et al. 1999

VM, vacuolar membrane; PM, plasma membrane, PS, phytosyderophores; NA- nicotianamine. At-*Arabidopsis thaliana*, Os-*Orhyza sativa*, Zm- *Zea mays*, Tc – *Thlaspi caerulescens*, Tg- *Thlaspi goesingense*, Tj- *Thlaspi japonicum*, Mt- *Medicago truncatula*, Ah – *Arabidopsis halleri*, Nt- *Nicotiana tabaccum*

(Colangelo and Guerinot 2006). *A. thaliana* iron regulated transporter1 (AtIRT1), the founding member of the ZIP family, encodes the major Fe transporter expressed in the external cell layers of the roots, specifically in response to iron starvation, whereas the closely related OsIRT1 appears to play a similar role under Fe limiting conditions in rice (Vert et al. 2002, Bughio et al. 2002). Characterization of OsIRT1 and recent identification of OsIRT2 as an Fe^{2+} transporter reveal that grasses utilize two strategies for Fe uptake: the assimilation of Fe^{2+} and the uptake of Fe^{3+}–PS. ZIP proteins contribute to Zn homeostasis in *A. thaliana* and rice. OsZIP4, which shares a greater than 50% identity with AtZIP1 and OsIRT1 and was able to overcome the growth defect of the Zn-uptake-defective *zrt1 zrt2* yeast strain, but no similar functional complementation was observed in the Fe-, Mn- or Cu-uptake mutants, indicating that OsZIP4 is selective for Zn transport. Its transcript accumulates under Zn deficiency and has been detected in the vascular bundles and apical meristems of roots and shoots, suggesting it may be responsible for Zn transport within plants (Ishimaru et al. 2005, 2006, Colangelo and Guerinot 2006). Six cDNAs encoding ZIP family members were recently identified in the model legume *Medicago truncatula* (Table 11.1), and their ability to complement yeast Zn-, Mn-, and/or Fe-uptake mutants was demonstrated, indicating a function in metal transport. In addition, the metal responsiveness of *MtZIP1* steady-state mRNA levels supports its role in metal homeostasis. MtZIP1 is 59.6% identical to AtZIP1 and appears to function as a Zn transporter (Lopez-Millan et al. 2004, Colangelo and Guerinot 2006).

3.1.1.3 COPT Family

The copper transporter family of high-affinity Cu uptake proteins is found throughout the eukaryotes, including five members in *A. thaliana* named COPT1–COPT5. To date, only COPT1 has been characterized in detail. Plants that express antisense *COPT1* exhibit a decrease in Cu uptake and reduced Cu accumulation in leaves. Consistent with reports on *COPT1* expression in pollen and root tips, *COPT1* antisense plants display pollen development defects and increased root length, which can be reversed by Cu supplementation, pointing to a role for COPT1 in plant growth and development (Sancenón et al. 2003, 2004).

3.1.1.4 NRAMP Family

Plant natural resistance associated macrophage protein (NRAMP) family members (Table 11.1) have been implicated in the transport of several divalent cations, including Fe (Colangelo and Guerinot 2006). AtNRAMP1 can functionally complement an Fe uptake mutant of yeast and increase resistance to toxic Fe levels in plants that overexpress *AtNRAMP1* (Curie et al. 2000). *AtNRAMP3* is expressed in the vascular bundles of leaves and in the central cylinder of roots, irrespective of the Fe-nutritional conditions. By analyzing the metal content of *AtNRAMP3* disruption mutants and the *AtNRAMP3* overexpressing plants it was demonstrated that *AtNRAMP3* diminishes Mn and Zn accumulation upon Fe deficiency. Furthermore, strong ectopic expression of *AtNRAMP3* downregulates expression of the primary Fe uptake transporter IRT1 and root ferric chelate reductase (FRO2), indicating complex regulation among these mechanisms. Using transient expression of the GFP (green fluorescence protein)-tagged AtNRAMP3 protein, it was shown that AtNRAMP3 resides on the vacuolar membrane and was proposed to control Fe-acquisition genes, Mn and Zn accumulation upon Fe starvation and Cd sensitivity by mobilizing Fe, Cd, and other metals from the vacuolar compartment (Thomine et al. 2003). The closely related genes *AtNRAMP3* and *AtNRAMP4* share similar tissue-specific expression patterns, transcriptional regulation by Fe, and subcellular localization at the vacuolar membrane. *nramp3 nramp4* mutant seeds store Fe properly, but the metal is retained in the vacuole globoids and is not released during seed germination as in wild-type seeds, indicating the role of the AtNRAMP3 and AtNRAMP4 in retrieval of globoid-associated Fe during seed germination for seedling development (Lanquar et al. 2005). Both transporters were also shown to transport toxic metals such as Cd^{2+} and disruption of *AtNRAMP3* gene leads to slightly enhanced Cd root growth resistance in *Arabidopsis*, while its overexpression results in root growth hypersensitivity and increased accumulation of Fe on Cd^{2+} treatment (Thomine et al. 2000).

3.1.1.5 Transporters Involved in the Uptake of Nonessential Elements

With the exception of the recently described Cd carbonic anhydrase of marine diatoms, no biological function is known to date for the potentially highly toxic metals Cd and Pb (Lane and Morel 2000, Clemens 2006). It is therefore highly likely that these nonessential metals enter cells through cation transporter(s) with broad specificity, which are induced especially under low-iron conditions. The transporters IRT1, ZNT1, and AtNRAMP3 were shown to transport Cd^{2+} (Pence et al. 2000, Thomine et al. 2000). A search among putative cation transporter cDNAs for effects on growth in the presence of Cd^{2+} upon expression in yeast led to the observation that wheat low-affinity cation transporter 1 (LCT1) renders yeast cells more Cd^{2+} sensitive. LCT1 was originally cloned by complementation of the K^+ high-affinity uptake-deficient yeast mutant CY162 and shown to also mediate Na^+ influx. Its expression leads to increased Cd^{2+} and Ca^{2+} uptake activity in *S. cerevisiae* (Clemens et al. 1998). However, no *LCT1* homologues have been found in *Arabidopsis* or other nonplant species genomes (Mäser et al. 2001).

The first example of a plant transporter possibly mediating Pb^{2+} uptake was the calmodulin-binding cyclic nucleotide-gated channel (NtCBP4) localized in the plasma membrane of tobacco. Overexpression of *NtCBP4* resulted in increased sensitivity toward Pb^{2+} and correlated with enhanced Pb^{2+} accumulation (Arazi et al. 1999).

3.1.2 Metal Efflux Proteins

Transporters that are involved in *metal efflux from the cytoplasm* (Fig. 11.1), either by movement across the plasma membrane (PM) or into organelles, include the P_{1B}-ATPase family and the cation diffusion facilitator (CDF) family (Table 11.1) (Yang et al. 2005, Colangelo and Guerinot 2006). They play significant roles in metal detoxification and tolerance by compartmentation of metals or by loading them into the root or shoot apoplast.

3.1.2.1 P-type ATPase Family

The family of P-type ATP-ases uses energy from ATP hydrolysis to translocate cations across biological membranes and can be divided into several subfamilies, including the heavy metal transporting P_{1B}-ATPases. The eight P_{1B}-ATPases in *A. thaliana* are designated heavy metal transporting P-type ATP-ase HMA1 through HMA8. Four family members (HMA1–HMA4) group with the Zn/Cd/Pb/Co divalent cation transporter class of P_{1B}-ATPases, whereas HMA5–HMA8 encode Cu/Ag monovalent cation transporters (Table 11.1) (Williams and Mills 2005, Baxter et al. 2003). Metal transport studies of HMA2 and HMA4 in yeast showed that HMA2 is indeed a Zn^{2+} ATPase, which can also be activated by $Cd^{2+}$to a similar extent, and to a lesser extent by other divalent cations, while HMA4 transports Zn, Cd, and Pb (Eren and Argüello 2004, Verret et al. 2005). Reverse genetic approaches have been successful in revealing the functions of HMA family members *in planta*. HMA2 and HMA4 play a role in Zn translocation from roots to shoots, since *hma2 hma4* double mutant plants were chlorotic, stunted, and failed to set seeds. They could be resuscitated by watering with high levels of Zn (1 or 3 mM) but not with Co or Cu. Zn levels of *hma2 hma4* plants were two to fourfold lower than those of wild-type plants when grown under 10 mM Zn (+Zn conditions), whereas 35SHMA4 plants (in which HMA4 is overexpressed) accumulated elevated levels of Zn and Cd in leaves. A metal transport assay in yeast showed that HMA2 transports metal substrates out of the cytoplasm, which is in line with the *in planta* localization of HMA2 and HMA4 to the PM, suggesting that these proteins translocate heavy metals out of cells. The vascular expression of HMA2 and HMA4 points to a possible role of these mentioned transport proteins in xylem loading or unloading (Verret et al. 2004, Hussain et al. 2004). A similar root expression pattern was previously observed for the boron transporter BOR1 and for the phosphate transporter PHO1, which have both been implicated in xylem loading. The role of AtHMA2 and AtHMA4 in the translocation of heavy metals from root to shoot tissues makes these transporters potential targets for the generation of transgenic plants that are designed to clean up soils that have been contaminated with toxic metals, such as Cd and Pb (Krämer 2005, Colangelo and Guerinot 2006).

Members of the Cu/Ag-transporting class of P_{1B}-ATPases include HMA7 (responsive to antagonist1 [RAN1]), which is important for Cu delivery to hormone receptors in post-Golgi compartments (Hirayama et al. 1999, Woeste and Kieber

2000). Recently, *HMA5* was characterized as a root-enhanced, Cu-induced gene. The amino-terminal MxCxxC Cu- binding motifs of HMA5 interact with *A. thaliana* ATX1-like Cu chaperones in yeast two-hybrid screens. Considering the Cu hypersensitivity of *hma5* plants, it appears that this HMA family member is involved in Cu detoxification of roots in response to high Cu levels (Andres-Colas et al. 2006). Three P_{1B}-ATPases are involved in Cu transport in the chloroplast. HMA1 and HMA6 (P-type ATPase of *Arabidopsis1* [PAA1]) are localized on the plastid envelope, where they deliver Cu to the stroma. A complete loss of Cu accumulation in chloroplasts might be expected in *hma1 hma6* plants. HMA8 (PAA2) resides at the thylakoid membrane and functions in Cu delivery to the thylakoid lumen. Consistent with its localization *hma8* plants display greatly reduced Cu levels in thylakoids compared to whole chloroplasts, supporting its role in Cu transport to the thylakoid lumen (Abdel-Ghany et al. 2005, Shikanai et al. 2003).

3.1.2.2 CDF Family

The ubiquitous CDF family of metal transporters encodes proton antiporters that efflux heavy metals out of the cytoplasm (Table 11.1) (Colangelo and Guerinot 2006). The first CDF gene characterized in *A. thaliana* was the zinc transporter gene ZAT1, later renamed METAL TOLERANCE PROTEIN1 (MTP1). Its overexpression confers Zn tolerance *in planta* (van der Zaal et al. 1999, Delhaize et al. 2003). Transient expression of AtMTP1::green fluorescent protein (GFP) in *A. thaliana* protoplasts indicates that AtMTP1 is localized in the vacuolar membranes of leaf and root cells, suggesting a role in Zn sequestration in the vacuole. Evidence that *mtp1* knockdown plants accumulate less Zn in various tissues indicates that the proposed defect in vacuolar Zn storage affects Zn uptake or distribution (Kobae et al. 2004, Desbrosses-Fonrouge et al. 2005).

3.2 Long-Distance Transport of Metals

Once taken up by roots, metals are loaded in the xylem sap and translocated to the aerial parts of the plant by the transpiration stream. Only a small fraction of a metal in plants is thought to exist as free ions. A number of small, organic molecules have been implicated in metal ion homeostasis as metal ion ligands facilitating their uptake and transport (Haydon and Cobbett 2007). Metal transporters such as AtHMA2 or AtHMA4 play a role in metal xylem or phloem loading, while AtHMA4, AtNRAMP3, OsZIP4, ZmYSL1, AtISL1, OsISL2 (Table 11.1) may serve for long-distance transport of metals or metal complexes with either phytosiderophores (PS) or nicotianamine (NA) (Hussain et al. 2004, Thomine et al. 2003, Schaaf et al. 2004, Roberts et al. 2004, Koike et al. 2004, Le Jean et al. 2005). In *A. thaliana* expressing *TaPCS1* (phytochelatin synthase from *Triticum aestivum*) gene, significantly decreased Cd^{2+} accumulation in roots and enhanced long-distance Cd^{2+} transport into stems and rosette leaves was found, indicating phytochelatins (PCs) may play an essential role in Cd transport from roots to shoots (Gong et al. 2003). In addition, organic acids, especially citrate, are known as the main metal chelators in the xylem, especially of iron (Briat and Lebrun 1999). Amino acids in the xylem sap are also potential metal chelators of nickel, since increasing free histidine in the xylem sap enhanced translocation of this metal to the shoots, and this could explain nickel hyperaccumulation in some plants (Krämer et al. 1996). Also NA, a methionine derivative involved in sensing Fe status within the plant and translocation of Cu from roots to aerial parts, was shown to bind various metals (Briat and Lebrun 1999, Pianelli et al. 2005).

The mobility of metals from source (leaves) to sink (seeds, roots) tissues via the phloem sap is less understood. It is known that phloem sap contains Fe, Cu, Zn, and Mn arising from their mobilization in source organs. The two principal unknowns regarding phloem metal-loading are the valence state of metals as they enter or transit the phloem system and the identity of the ligands involved in phloem loading. The sole molecule identified as a potential phloem metal transporter is NA, which has been found in a stoichiometry of 1.25 with four metal ions (Fe, Cu, Zn, and Mn) in the phloem sap, leading the authors to speculate on united transport (Stephan et al. 1994). NA could therefore transport metals on long distance, once they have been loaded in the phloem, but this loading would probably require another chelator. Indeed, unidentified compounds of apparent molecular weight much higher than NA that are able to chelate various metals have been found. In *Ricinus communis*, phloem metal chelates of 1,000-1,500 Da for Zn, 1,000-5,000 Da for Mn, and 2,400 Da for iron have been reported. A 4,000 Da Zn-binding peptide has also been identified in *Citrus*. The chemical identity and the role of these metal complexes in the phloem metal-transport processes, if any, remain to be clarified (Briat and Lebrun 1999).

3.3 Metal Uptake and Transport in Metal Hyperaccumulating Plants

Transport mechanisms on the plasma membranes contribute decisively to symplastic metal loading, necessary for further metal translocation to the shoots, a trait that is highly enhanced in metal hyperaccumulating plants (Flowers and Yeo 1992). Molecular studies of metal hyperaccumulating species have revealed that the upregulation of metal transporters from a variety of families is one way in which some hyperaccumulating species achieve increased metal translocation to aerial tissues (Colangelo and Guerinot 2006).

3.3.1 Metal Uptake Proteins

3.3.1.1 YSL and ZIP Family

The role of *YSL* genes (Table 11.1) was demonstrated in metal accumulation by metal hyperaccumulating plant species. In *T. caerulescens YSL* genes are expressed at higher rates and with distinct patterns when compared with their *A. thaliana* homologues. While *TcYSL7* was highly expressed in the flowers, *TcYSL5* was more highly expressed in the shoots, and the expression of *TcYSL3* was equivalent in all the organs tested. In situ hybridizations have shown that *TcYSL7* and *TcYSL5* are expressed around the vasculature of the shoots and in the central cylinder in the roots. Exposure to heavy metals (Zn, Cd, Ni) does not affect the high and constitutive expression of the *TcYSL* genes. By mutant yeast complementation and uptake measurements it was demonstrated that TcYSL3 is an Fe/Ni-NA influx transporter mediating NA-metal, particularly NA-Ni, circulation in metal hyperaccumulating plants (Gendre et al. 2006).

ZIP genes (Table 11.1) were also shown to be an important feature in metal uptake by metal hyperaccumulating plant species. High endogenous *TcZNT1* expression in roots and shoots of the Zn/Cd hyperaccumlator *T. caerulescens* was associated with increased Zn^{2+} uptake in the roots, and the same mechanism likely underlay the enhanced Zn^{2+} uptake into leaf cells. *TcZNT1* encodes a high-affinity Zn transporter that can also mediate low-affinity Cd transport. It would be important to elucidate the regulatory components linking plant Zn status to *ZNT1* gene expression in order to understand how alterations in this pathway contribute to metal hyperaccumulation (Pence et al. 2000, Lasat et al. 2000). The Zn hyperaccumulator *Arabidopsis halleri* has elevated levels (23–24-fold) of the *ZIP6* transcript in roots and shoots, implying its involvement in build-up of this potentially toxic metal (Becher et al. 2004). Two ZIP genes, *TjZNT1* and *TjZNT2*, have been cloned from the Ni hyperaccumulator *Thlaspi japonicum*. Yeast expressing either *TjZNT1* or *TjZNT2* shows increased resistance to Ni^{2+} (Mizuno et al. 2005), highlighting a potential role for these genes in Ni tolerance. Further studies, however, will be needed to determine their functions implicated in Ni hyperaccumulation *in planta* (Colangelo and Guerinot 2006).

3.3.2 Metal Efflux Proteins

3.3.2.1 P-type ATPase and CDF Family

The Zn/Cd hyperaccumulator *A. halleri* safely accumulates 100-fold more Zn than nonaccumulating species. Increased expression of *MTP1* in shoots and *AhHMA3* in shoots and roots exposed to high Zn^{2+} (Table 11.1) were proven to be involved in Zn accumulation and detoxification. *A. halleri* contains three independently segregating and differentially regulated *MTP1* genes, and in a backcross between the hyperaccumulator *A. halleri* and the nonaccumulator *Arabidopsis lyrata*, two *MTP1* loci cosegregate with Zn tolerance. Evidence for AhMTP1-3 mediating Zn transport is provided by functional complementation of the Zn-sensitive phenotype of *zrc1 cot1*, a Zn-hypersensitive strain that lacks the vacuolar CDF family members ZINC RESISTANCE CONFERRING1 (ZRC1) and COBALT TRANSPORTER1 (COT1). Like AtMTP1, AhMTP1-3 localizes on the vacuolar membrane (Dräger et al. 2004, Becher et al. 2004). Expression of *MTP1* or related genes is constitutively high, specifically in the shoots of metal-tolerant metal hyperaccumulating plants, while in roots the transcript levels depend on metal concentrations. Under exposure to low Zn concentrations, root *MTP1* transcript levels in *A. halleri* are comparable to those in *A. thaliana*, whereas at high Zn supply, enhanced Zn sequestration in the root cells of *A. halleri* may participate in protecting the shoot from accumulating excess Zn. The regulation of *MTP1* expression, as observed in *A. halleri*, could thus serve as a model to improve the performance of transgenic plants in the phytoremediation of metal-contaminated soils (Dräger et al. 2004). *Thlaspi goesingense* is an Ni/Zn hyperaccumulator that also relies on enhanced expression of CDF transporters for metal tolerance (Persans et al. 2001). When fused to an epitope of the *Haemophilus influenzae* haemagglutinin protein (TgMTP1b::HA), TgMTP1b localized on vacuolar and plasma membranes. When expressed in *zrc1 cot1* plants, they showed a two-fold reduction in Zn levels, and the level of expression was found to correlate positively with the rate of Zn efflux. These results indicate that TgMTP1b does not confer resistance to high Zn by reversing the defect in vacuolar Zn compartmentalization of *zrc1 cot1*, rather, long- and short-term uptake studies support its role in Zn efflux. Together with its localization at the PM in *A. thaliana* protoplasts, evidence supports a role for MTP1 in Zn efflux at the PM in *T. goesingense*, possibly leading to reduced Zn vacuolar compartmentation in roots and enhanced xylem loading (Kim et al. 2004).

4 Metal Partitioning and Detoxification as Tolerance Strategies in Plants

4.1 Partitioning of Metals at Organ and Tissue Levels

Metal *exclusion from the roots* is achieved by several mechanisms. Enhanced lysis of epidermal and root cap cells with enhanced mucilage secretion were demonstrated to contribute to plant metal tolerance (Llugany et al. 2003, Delisle et al. 2001). Rhizodeposition includes the passive exudation of low molecular weight molecules (sugars, amino acids, organic acids, and hormones), active secretion of high molecular weight molecules (complex carbohydrates, enzymes), lysates, gases, and mucigel (Marschner 1995). Among these, organic acids (malate and citrate) were demonstrated to confer metal tolerance by chelation, although they do not seem to be universally applicable (Ernst et al. 1992, Salt 2001). Nevertheless, in the case of aluminum tolerance, exposure to Al^{3+} triggers the exudation of organic acids, thus preventing its toxicity, which seems to be the result of the constitutive expression of all the necessary metabolic machinery involved. In addition, Cu tolerance in *A. thaliana* was connected to a rapid release of citrate and Pb tolerance in rice to the stimulation of oxalate efflux (Ryan et al. 2001). Symbiosis with arbuscular mycorrhizal fungi was demonstrated to contribute to metal tolerance in plants (Joner and Leyval 1997, Vogel-Mikuš et al. 2006), presumably by root exclusion due to the enhanced adsorption and immobilization of metals on the fungal cell walls and/or compartmentization with complexation mainly in the vacuoles (Joner and Leyval 1997, Joner et al. 2000, Leyval and Joner 2001), thus contributing to the stabilization and succession of plant communities (Van der Heijden et al. 1998, Regvar et al. 2006). Its importance for phytoremediation of metal-polluted sites was recently reviewed elsewhere (Vogel-Mikuš and Regvar 2006).

4.1.1 Distribution of Metals in Roots

Shoot exclusion, on the other hand, indicates that metals have accumulated in the roots but are excluded from further transport. Low metal accumulation with *bioaccumulation factors* (BAF = C_{shoot}/C_{soil}) <1 and minute *translocation factors* (TF = C_{root}/C_{shoot})<<1, indicate *metal exclusion* from the shoots. The majority of plants use this so-called excluder strategy (Salt 2001, Baker 1981, Dahmani-Muller et al. 2000) to cope with excess metals. Thus, when roots absorb metals, they accumulate primarily in the rhizodermis and cortex, and a notable amount of metals are found in the root hairs (Seregin and Ivanov 2001). Metals that reach root cortical cells via apoplastic transport are retained on the walls bound to polygalacturonic acids, with the affinity for metal ions decreasing in the order Pb>Cr>Cu>Ca>Zn, or sequestered in the root vacuoles (Ernst et al. 1992, Clemens 2006). At sublethal concentrations, metals are not found in the stelar parenchyma, while at levels approaching lethal concentrations, when the endodermal barrier is broken, Cd and Pb pass over through the cortex and endodermis and are found in considerable amounts in the cells of the vascular cylinder (Seregin and Ivanov 2001).

4.1.2 Distribution and Partitioning of Metals in Shoots

In nonaccumulators, only low concentrations are found in the stem and leaf tissues, depending primarily on the activity of metal-sequestering pathways in root cells, that probably play a key role in determining the rate of translocation to the aerial parts. A second factor then would be the degree of accessibility and mobilization of the sequestered metal. Thirdly, the efficiency of radial symplastic passage through the root and across the endodermis is important, and finally, xylem loading, i.e., the efflux from xylem parenchyma cells into the xylem (Seregin and Ivanov 2001, Clemens et al. 2002). Metal distribution in plant foliage is species-specific, and its visualization is needed to understand the patterns of accumulation and tolerance of metal-treated plants (Cosio et al. 2005). In young sprouts of soybean, Cd was located in the stems and leaf veins together with leaf age-driven gradients (Cunningham et al. 1975). Uneven Cd accumulation with correspondence between the Cd spots observed on autoradiographs and the necrotic dots observed on the margins of the leaves was found on the leaf surface of *Salix viminalis* and *T. caerulescens* ecotype Prayon (Cosio et al. 2005). At the leaf tissue level the highest Ni concentrations were observed in epidermal cells of the Ni hyperaccumulator *Stackhousia tryonii*, double that recorded in palisade cells (Bhatia et al. 2004).

Similarly, in the Zn/Cd hyperaccumulator *T. caerulescens* preferential localization of metals was observed in the lower and upper epidermis (Vazquez et al. 1994, Küpper et al 1999, Cosio et al. 2005). In addition, Zn was preferentially stored in large vacoularized epidermal cells, while absent from the cells of the stomatal complex (Frey et al. 2000). However, Salt et al. (1995) found an even distribution of points of higher concentration of Cd at the leaf surface of Indian mustard and clearly showed the preferential accumulation of Cd in trichomes. Sequestration of metals in trichomes has been reported in a number of annual plants (Ager et al. 2002, Küupper et al. 1999). They are ideal for allocation of heavy metals because of their localization at the leaf periphery (Cosio et al. 2005), whereas in plants lacking trichomes, metals were found to be located in vessels and at the sites of the transpiration flow pathway, with the epidermis playing an important role in sequestration and metal detoxification (Seregin and Ivanov 2001).

4.1.3 Distribution and Partitioning of Metals in Seeds

The seed coat presents the first barrier to metal absorption in germinating seeds. Cd, Pb, Zn, and Ni were found mainly in the cell walls of seed coats and did not enter the embryos, even at lethal concentrations (Seregin and Ivanov 2001). However, in Zn/Cd hyperaccumulating *Thlaspi praecox* a significant amount of Cd (up to $1350\,\mu g^{-1}$) was accumulated in seeds. Elemental localization within seeds using micro-PIXE showed Cd was mainly accumulated in embryonic tissues, indicating high Cd mobility within the plant tissues. In addition, Zn also preferentially accumulated in embryonic tissues compared to seed coats, although in much lower concentrations (up to $200\,\mu g^{-1}$), while the highest Pb concentrations were observed in the seed coats (Vogel-Mikuš et al. 2007). The rate of metal accumulation in seeds seems to be species-specific, depending mainly on the metal in question; the ability of the plant to sequester metal(s) in roots and leaves, thus lowering free symplastic metal concentrations; and on the transport mechanisms involved in seed metal loading.

4.2 Metal Detoxification at Cellular and Subcellular Levels

Binding metals to the cell walls represents an important detoxification mechanism in both nonaccumulating and metal hyperaccumulating plants, especially at low metal concentrations and short time of exposure. Most studies have shown that in roots of metal hyperaccumulating *T. caerulescens* the majority of Cd was bound to the apoplast, but at higher concentrations Cd was also detected in the root symplast (Wojcik et al. 2005, Vazquez et al. 1992). In metal hyperaccumulating plants, accumulation of large amounts of metals requires the presence of intracellular high-capacity detoxification mechanisms, such as binding to organic acids, amino acids, phytochelatins and metallothioneins, and sequestration in the vacuoles acting as a central storage for ions (Clemens et al. 2002, Briat and Lebrun 1999). Increased expression of *NAS* (nicotianamine syntase) genes was found in *A. halleri* roots when compared to *A. thaliana*, corresponding to 3.5 times higher levels of NA. The higher rate of NA synthesis found for *A. halleri* roots at the transcript, protein, and metabolite levels suggests NA plays an important role in Zn homeostasis in dicots and might well represent one of the molecular factors conferring Zn tolerance and hyperaccumulation on *A. halleri* (Weber et al. 2004). *A. thaliana* plants overexpressing the *TcNAS1* gene overaccumulated NA, and increased NA levels were quantitatively correlated with increased Ni tolerance, indicating the great importance of NA in metal homeostasis of hyperaccumulating plants. These results indicate that it may now be possible to consider transferring NA synthase genes to fast growing plants of high biomass for use in phytoremediation (Pianelli et al. 2005). The function of transport of NA-metal chelates, required for the loading and unloading of vessels, has been assigned to the YSL-like family of proteins. Rapid induction of synthesis and accumulation of phytochelatins in response to Cd, Ag, Pb, Cu, Hg, Zn, Sn, Au, and As was demonstrated (Grill et al. 1987, Schat et al. 2002), with subsequent storage of the metals into vacuoles, providing efficient metal tolerance (Salt and Krämer 2000). The tendency of metals to induce phytochelatins in cell suspension cultures of *Rauvolfia sepentina* decreased in the order Hg >> Cd, As, Fe > Cu, Ni > Sb, Au > Sn, Se, Bi > Pb, Zn (Grill et al. 1987). The latest results, however, indicate that phytochelatin synthesis and metal complexation may contribute to Cd, Zn, and Cu tolerance in metal nontolerant plant populations, whereas they are not responsible for tolerance in metal hyperaccumulating plants, possibly due to the high energy demands required for phytochelatin synthesis, which may be too costly (Schat et al. 2002). Instead of phytochelatins, Cys and other low molecular weight thiols were proposed to be involved in Cd tolerance in *T. caerulescens* (Hernández-Allica et al. 2006). X-ray absorption spectroscopy showed that in mature and senescent leaves of *T. caerulescens* Cd and Zn were mainly complexed to oxygen-ligands, while in young leaves, Cd was mainly bound to S- ligands, and Zn was mainly bound to histidine (Küpper et al. 2004), which also plays an important role in the transport and detoxification of Ni in the Ni hyperaccumulator *T. goesingense* (Krämer et al. 2000).

5 Conclusion

Recent data from structural and functional genomics on uptake and sequestration of metals at plant organ, tissue, and cell levels shed new light on the understanding of the frequently observed exclusion, accumulation, or even hyperaccumulaton of metals in plants. General mechanisms for detoxification of metals in plants at the cellular level are the binding of metals to the apoplast (cell walls and trichomes) and chelation of the metals by ligands, followed by sequestration of the metal-ligand complex into the vacuole (Yang et al. 2005). Intense metal accumulation in the above-ground parts provides a promising approach for both cleaning up anthropogenically contaminated soils (phytoremediation) and for commercial extraction (phytomining) of metals from naturally metal-rich (serpentine) soils (McGrath et al. 1993), or stabilization of highly polluted and eroded sites (Regvar et al. 2006). Still, a deeper understanding of the metal uptake, transport, and detoxification mechanisms of plants is needed before these phytoremedial biotechnologies can brought effectively from theory into practice.

References

Abdel-Ghany SE, Muller-Moule P, Niyogi KK, Pilon M, Shikanai T (2005) Two P-type ATPases are required for copper delivery in *Arabidopsis thaliana* chloroplasts. Plant Cell 17:1233-51

Ahn SJ, Rengel Z, Matsumoto H (2004) Aluminum-induced plasma membrane surface potential and H^+-ATPase activity in near-isogenic wheat lines differing in tolerance to aluminum. New Phytol 162:71-9

Ager FJ, Ynsa MD, Dominguez-Solis JR, Gotor C, Respaldiza MA, Romero LC (2002) Cadmium localization and quantification in the plant *Arabidopsis thaliana* using micro-PIXE. Nuclear Instrum Meth B 189:494-8

Andres-Colas N, Sancenon V, Rodriguez-Navarro S, Mayo S, Thiele DJ, Ecker JR, Puig S, Penarrubia L (2006) The *Arabidopsis* heavy metal P-type ATPase HMA5 interacts with metallochaperones and functions in copper detoxification of roots. Plant J 45:225-36

Arazi T, Sunkar R, Kaplan B, Fromm H (1999) A tobacco plasma membrane calmodulin-binding transporter confers Ni^{2+} tolerance and Pb^{2+} hypersensitivity in transgenic plants. Plant J 20:171-82

Assunção AGL, Schat H, Arts MGM (2003) *Thlaspi caerulescens*, an attractive model species to study heavy metal hyperaccumulation in plants. New Phytol 159:351-60

Bagnaresi P, Thoiron S, Mansion M, Rosignol M, Pupillo P, Briat JF (1999) Cloning and characterization of a maize cytochrome-b(5) reductase with Fe^{3+}- chelate reduction capability. Biochem J 338:499-505

Baker AJM (1981) Accumulators and excluders – strategies in the response of plants to heavy metals. J Plant Nutr 3:643-54

Baker, AJM (1987) Metal tolerance. New Phytol 106:93-111

Baxter I, Tchieu J, Sussman MR, Boutry M, Palmgren MG, Gribskov M, Harper JF, Axelsen KB (2003) Genomic comparison of P-type ATPase ion pumps in *Arabidopsis* and rice. Plant Physiol 132:618-28

Becher M, Talke IN, Krall L, Kramer U (2004) Cross-species microarray transcript profiling reveals high constitutive expression of metal homeostasis genes in shoots of the zinc hyperaccumulator *Arabidopsis halleri*. Plant J 37:251-68

Bhatia NP, Orlic I, Siegele R, Ashwath N, Baker AJM, Walsh KB (2004) Quantitative cellular localisation of nickel in leaves and stems of the hyperaccumulator plant *Stackhousia tryonii* Bailey using nuclear-microprobe (Micro-PIXE) and energy dispersive X-ray microanalysis (EDXMA) techniques. Funct Plant Biol 31:1-14

Briat JF, Lebrun M (1999) Plant responses to metal toxicity. CR Acad Sci Paris 322:43-54

Bughio N, Yamaguchi H, Nishizawa NK, Nakanishi H, Mori S (2002) Cloning an iron-regulated metal transporter from rice. J Exp Bot 53:1677-82

Cieslinski G, Van Rees KC, Szmigielska AM, Huang PM (1998) Low-molecular-weight organic acids in rhizosphere soils of durum wheat and their effect on cadmium bioaccumulation. Plant Soil 203:109-17

Clemens S, Antosiewicz DM, Ward JM, Schachtman DP, Schroeder JI (1998) The plant cDNA LCT1 mediates the uptake of calcium and cadmium in yeast. Proc Natl Acad Sci USA 95:12043-8

Clemens S, Palmgren MG, Krämer U (2002) A long way ahead: understanding and engineering plant metal accumulation. Trends Plant Sci 7:309-15

Clemens S (2006) Toxic metal accumulation, responses to exposure and mechanisms of tolerance in plants. Biochemie 88:1707-19

Cobbett C (2003) Heavy metals and plants – model systems and hyperaccumulators. New Phytol 159:289-93

Colangelo EP, Guerinot ML (2006) Put the metal to the petal: metal uptake and transport through plants. Curr Opin Plant Biol 9:322-30

Cosio C, DeSantis L, Frey B, Diallo S, Keller C (2005) Distribution of cadmium in leaves of *Thlaspi caerulescens*. J Exp Bot 56: 765-75

Cunningham LM, Collings FW, Hutchinson TC (1975) Physiological and biochemical aspects of cadmium toxicity in soybean. I. Toxicity symptoms and autoradiographic distribution of Cd in roots, stems and leaves. In: International conference on heavy metals in the environment, Symposium Proceedings, Institute for Environmental Studies, Universityof Toronto, Ontario, Canada 2: 97-120

Curie C, Alonso JM, Le Jean M, Ecker JR, Briat JF (2000) Involvement of NRAMP1 from *Arabidopsis thaliana* in iron transport. Biochem J 347:749-55

Curie C, Panaviene Z, Loulergue C, Dellaporta SL, Briat JF, Walker EL (2001) Maize yellow stripe1 encodes a membrane protein directly involved in Fe(III) uptake. Nature 409:346-9

Dahmani-Muller HF, van Oort B, Gelie M, Balabane (2000) Strategies of heavy metal uptake by three plant species growing near a metal smelter. Environ Poll 109:231-8

Delhaize E, Kataoka T, Hebb DM, White RG, Ryan PR (2003) Genes encoding proteins of the cation diffusion facilitator family that confer manganese tolerance. Plant Cell 15: 1131-42

Delisle G, Champoux M, Houde M, (2001) Characterization of oxidase and cell death in Al-sensitive and tolerant wheat roots. Plant Cell Physiol 42:324-33

Desbrosses-Fonrouge AG, Voigt K, Schroder A, Arrivault S, Thomine S, Kramer U (2005) *Arabidopsis thaliana* MTP1 is a Zn transporter in the vacuolar membrane which mediates Zn detoxification and drives leaf Zn accumulation. FEBS Lett 579:4165-74

DiDonato RJ Jr, Roberts LA, Sanderson T, Eisley RB, Walker EL (2004) *Arabidopsis* YELLOW STRIPE-LIKE2 (YSL2): a metal-regulated gene encoding a plasma membrane transporter of nicotianamine-metal complexes. Plant J 39:403-14

Dräger DB, Desbrosses-Fonrouge AG, Krach C, Chardonnens AN, Meyer RC, Saumitou-Laprade P, Kramer U (2004) Two genes encoding *Arabidopsis halleri* MTP1 metal transport proteins cosegregate with zinc tolerance and account for high MTP1 transcript levels. Plant J 39:425-39

Eren E, Argüello JM (2004) Arabidopsis HMA2, a divalent heavy metaltransporting P(IB)-type ATPase, is involved in cytoplasmic Zn^{2+} homeostasis. Plant Physiol 136:3712-23

Ernst WHO, Verkleij JAC, Schat H (1992) Metal tolerance in plants. Acta Bot Neerl 41:229-48

Flowers TJ, Yeo AR (1992) Solute transport in plants. Chapman & Hall, Edmundsbury

Frey B, Keller C, Zierold K, Schulin R (2000) Distribution of Zn in functionally different leaf epidermal cells of the hyperaccumulator *Thlaspi caerulescens*. Plant Cell Environ 23:675-87

Gendre D, Czernic P, Conéjéro G, Pianelli K, Briat J- F, Lebrun M, Mari S (2006) TcYSL3, a member of the YSL gene family from the hyperaccumulator *Thlaspi caerulescens*, encodes a nicotianamine-Ni/Fe transporter. Plant J 49:1-15

Gong JM, Lee DA, Schroeder JI (2003) Long-distance root-to-shoot transport of phytochelatins and cadmium in *Arabidopsis*. Proc Natl Acad Sci USA 100:10118-23

González-Guerrero M, Azcón-Aguillar C, Mooney M, Valderas A, MacDiarmid CW, Eide DJ, Ferrol, N (2005) Characterization of a *Glomus intraradices* gene encoding a putative Zn transporter of the cation diffusion facilitator family. Fungal Genet Biol 42:130-40

Gravot A, Lieutaud A, Verret F, Auroy P, Vavasseur A, Richaud P (2004) AtHMA3, a plant P_{1B}-ATPase, functions as a Cd/Pb transporter in yeast. FEBS Lett 561:22-8

Grill E, Winnacker E-L, Zenk MH 1987. Phytochelatins, a class of heavy-metal-binding peptides from plants are functionally analogous to metallothioneins. Proc Natl Acad Sci USA 84: 439-43

Haydon MJ, Cobbett CS (2007) Transporters of ligands for essential metal ions in plants. New Phytol 174:499-506

Van der Heijden MGA, Klironomos JN, Ursic M, Moutoglis P, Streitwolf-Engel R, Boller T, Wiemken A, Sanders IR (1998) Mycorrhizal fungal diversity determines plant biodiversity and productivity. Nature 396:69-72

Hernández-Allica J, Garbisu C, Becerril JM, Barrutia O, Garcia-Plazola JI, Zhao FJ, McGrath SP (2006) Synthesis of low molecular weight thiols in response to Cd exposure in *Thlaspi caerulescens*. Plant Cell Environ 29:1422-9

Hildebrandt U, Regvar M, Bothe H (2007) Arbuscular mycorrhiza and heavy metal tolerance. Phytochemistry 68:139-46

Hirayama T, Kieber JJ, Hirayama N, Kogan M, Guzman P, Nourizadeh S, Alonso JM, Dailey WP, Dancis A, Ecker JR (1999) Responsive-to-antagonist1, a Menkes/Wilson diseaserelated copper transporter, is required for ethylene signaling in *Arabidopsis*. Cell 97:383-93

Hussain D, Haydon MJ, Wang Y, Wong E, Sherson SM, Young J, Camakaris J, Harper JF, Cobbett CS (2004) P-type ATPase heavy metal transporters with roles in essential zinc homeostasis in *Arabidopsis*. Plant Cell 16:1327-39

Ishimaru Y, Suzuki M, Kobayashi T, Takahashi M, Nakanishi H, Mori S, Nishizawa NK (2005) OsZIP4, a novel zinc-regulated zinc transporter in rice. J Exp Bot 56:3207-14

Ishimaru Y, Suzuki M, Tsukamoto T, Suzuki K, Nakazono M, Kobayashi T, Wada Y, Watanabe S, Matsuhashi S, Takahashi M, Nakanishi H, Mori S, Nishizawa NK (2006) Rice plants take up iron as an Fe^{3+}- phytosiderophore and as Fe^{2+}. Plant J 45:335-46

Joner EJ, Leycal C (1997) Uptake of ^{109}Cd by roots and hyphae of a *Glomus mosseae/Trifolium subterraneum* mycorrhiza from soil amended with high and low concentrations of cadmium. New Phytol 135:353-60

Joner EJ, Briones R, Leyval C (2000) Metal-binding capacity of arbuscular mycorrhizal mycelium. Plant Soil 226:227-34

Kim D, Gustin JL, Lahner B, Persans MW, Baek D, Yun DJ, Salt DE (2004) The plant CDF family member TgMTP1 from the Ni/Zn hyperaccumulator *Thlaspi goesingense* acts to enhance efflux of Zn at the plasma membrane when expressed in *Saccharomyces cerevisiae*. Plant J 39:237-51

Kobae Y, Uemura T, Sato MH, Ohnishi M, Mimura T, Nakagawa T, Maeshima M (2004) Zinc transporter of Arabidopsis thaliana AtMTP1 is localized to vacuolar membranes and implicated in zinc homeostasis. Plant Cell Physiol 45:1749-58

Koike S, Inoue H, Mizuno D, Takahashi M, Nakanishi H, Mori S, Nishizawa NK (2004) OsYSL2 is a rice metal-nicotianamine transporter that is regulated by iron and expressed in the phloem. Plant J 39:415-24

Krämer U, Cotter-Howells JD, Charnock JM, Baker AJM, Smith AC (1996) Free histidine as a metal chelator in plants that accumulate nickel. Nature (London) 379:635-8

Krämer U, Pickering IJ, Prince RC, Raskin I, Salt DE (2000) Subcellular localization and speciation of nickel in hyperaccumulator and non-accumulator *Thlaspi* species. Plant Physiol 122:1343-53

Krämer U (2005) Phytoremediation: novel approaches to cleaning up polluted soils. Curr Opin Biotech 16:133-41

Küpper H, Zhao FJ, McGrath SP (1999) Cellular compartmentation of zinc in leaves of the hyperaccumulator *Thlaspi caerulescens*. Plant Physiol 119:305-11

Küpper H, Mijovilovich A, Klaucke-Mayer W, Kroneck PHM (2004) Tissue and age-dependent differences in the complexation of cadmium and zinc in the cadmium/zinc hyperaccumulator *Thlaspi caerulescens* Ganges ecotype revealed by X-ray absorption spectroscopy. Plant Physiol 134:748-57

Lane T, Morel F (2000) A biological function for cadmium in marine diatoms. P Natl Acad Sci USA 97:4627-31

Lanquar V, Lelievre F, Bolte S, Hames C, Alcon C, Neumann D, Vansuyt G, Curie C, Schroder A, Kramer U (2005) Mobilization of vacuolar iron by AtNramp3 and AtNramp4 is essential for seed germination on low iron. EMBO J 24:4041-51

Lasat MM, Pence NS, Garvin DF, Ebbs SD, Kochian LV (2000) Molecular physiology of zinc transport in the Zn hyperaccumulator *Thlaspi caerulescens*. J Exp Bot 51:71-9

Lasat MM (2002) Phytoextraction of toxic metals: a review of biological mechanisms. J Environ Qual 31:109-20

Llugany M, Lombini A, Poschenrieder C, Barcelo J (2003) Different mechanisms account for enhanced copper resistance in *Silene armeria* from mine spoil and serpentine sites. Plant Soil 251:55-63

Leyval C, Joner EJ (2001) Bioavailability of heavy metals in the mycorrhizosphere. In: Gobran RG, Wenzel WW, Lombi E (eds) Trace metals in the rhizosphere, CRC Press, Florida, pp 165-85

Le Jean M, Schikora A, Mari S, Briat JF, Curie C (2005) A loss-of function mutation in AtYSL1 reveals its role in iron and nicotianamine seed loading. Plant J 44:769-82

Ligaba A, Yamaguchi M, Shen H, Sasaki T, Yamamoto Y, Matsumoto H (2004) Phosphorus deficiency enhances plasma membrane H^+-ATPase activity and citrate exudation in greater purple lupin (*Lupinus pilosus*). Funct Plant Biol 31:1075-83

Lopez-Millan AF, Ellis DR, Grusak MA (2004) Identification and characterization of several new members of the ZIP family of metal ion transporters in *Medicago truncatula*. Plant Mol Biol 54:583-96

Marschner H, 1995. Mineral nutrition of higher plants, 2nd ed. Academic Press, London

Mäser P, Thomine S, Schroeder I, Ward JM, Hirschi K, Sze H, Talke IN, Amtmann A, Maathuis FJM, Sanders D, Harper JF, Tchieu J, Gribskov M, Persans MW, Salt DE, Kim SA, Guerinot ML (2001) Phylogenetic Relationships within cation transporter families of *Arabidopsis*. Plant Physiol 126:1646-67

McGrath SP, Sidoli CMD, Baker AJM and Reeves RD (1993) The potential for the use of metal-accumulating plants for the in situ decontamination of metal-polluted soils. In: Eijsackers HJP, Hamers T (eds) Integrated soil and sediment research: a basis for proper protection, Kluwer Academic Publishers, Dordrecht, pp 673-6

McGrath SP, Shen ZG, Zhao FJ (1997) Heavy metal uptake and chemical changes in the rhizosphere of *Thlaspi caerulescens* and *Thlaspi ochroleucum* grown in contaminated soils. Plant Soil 180:153-9

McGrath SP, Zhao FJ, Lombi E (2001) Plant and rhizosphere processes involved in phytoremediation of metal-contaminated soils. Plant Soil 232:207-14

Mizuno T, Usui K, Horie K, Nosaka S, Mizuno N, Obata H (2005) Cloning of three ZIP/Nramp transporter genes from a Ni hyperaccumulator plant *Thlaspi japonicum* and their Ni^{2+}-transport abilities. Plant Physiol Biochem 43:793-801

Moore MR (2003) Risk assessment in environmental contamination and environmental health. In: Naidu R, Gupta VVSR, Rogers S, Kookana RS, Bolan NS, Adriano DC (eds) Bioavailability, toxicity and risk relationships in ecosystems, Science Publishers, Inc., Enfield, pp. 3-19

Naidu R, Rogers S, Gupta VVSR, Kookana RS, Bolan NS, Adriano DC (2003) Bioavailability of metals in the soil plant environment and its potential role in risk assessment. In: Naidu R, Gupta VVSR, Rogers S, Kookana RS, Bolan NS, Adriano DC (eds) Bioavailability, toxicity and risk relationships in ecosystems, Science Publishers, Inc., Enfield, pp. 21-57

Pence NS, Larsen PB, Ebbs SD, Letham DL, Lasat MM, Garvin DF, Eide D, Kochian LV (2000) The molecular physiology of heavy metal transport in the Zn/Cd hyperaccumulator *Thlaspi caerulescens*. Proc Natl Acad Sci USA 97:4956-60

Persans MW, Nieman K, Salt DE (2001) Functional activity and role of cation-efflux family members in Ni hyperaccumulation in *Thlaspi goesingense*. Proc Natl Acad Sci USA 98: 9995-10000

Pianelli K, Mari S, Marque's L, Vacchina V, Lobinski R, Lebrun M, Czernic P (2005) Nicotianamine over-accumulation confers resistance to nickel in *Arabidopsis thaliana*. Trans Res 14:739-48

Reeves RD, Baker AJM (2000) Metal-accumulating plants. In: Raskin I, Ensley BD (eds) Phytoremediation of toxic metals. Using plants to clean up the environment, John Whiley & Sons, Inc., NY, pp. 193-229

Reeves RD (2006) Hyperaccumulation of trace elements by plants. In Morel JL, Echevarria G, Goncharova N (eds) Phytoremediation of metal contaminates soils (Nato Science Series: IV: Earth and Environmental Sciences), Springer-Verlag, pp 25-52

Regvar M, Vogel-Mikuš K, Kugonič N, Turk B, Batič F (2006) Vegetational and mycorrhizal successions at a metal polluted site: Indications for the direction of phytostabilisation? Environ Pollut 144:976-84

Roberts LA, Pierson AJ, Panaviene Z, Walker EL (2004) Yellow stripe1, Expanded roles for the maize iron-phytosiderophore transporter. Plant Physiol 135:112-20

Römheld V (1991) The role of phytosiderophores in acquisition of iron and other micronutrients in graminaceous species: an ecological approach. Plant Soil 130:127-34

Ryan PR, Delhaize E, Jones DL (2001) Function and mechanism of organic anion exudation from plant roots. Ann Rev Plant Physiol Plant Mol Biol 52:527-60

Salt DE, Prince RC, Pickering IJ, Raskin I (1995) Mechanisms of cadmium mobility and accumulation in Indian Mustard. Plant Physiol 109:1427-33

Salt DE, Krämer U (2000) Mechanisms of metal hyperaccumulation in plants. In: Raskin I, Ensley BD (eds) Phytoremediation of toxic metals: using plants to clean up the environment. John Wiley & Sons Inc, New York, pp. 193-229

Salt D (2001) Responses and adaptations of plants to metal stress. In: Hawkesford MJ, Buchner P (eds) Molecular analysis of plant adaptation to the environment. Kluwer Academic Publishers, Dodrecht, pp. 159-79

Sancenón V, Puig S, Mira H, Thiele DJ, Penarrubia L (2003) Identification of a copper transporter family in *Arabidopsis thaliana*. Plant Mol Biol 51:577-87

Sancenón V, Puig S, Mateu-Andres I, Dorcey E, Thiele DJ, Penarrubia L (2004) The *Arabidopsis* copper transporter COPT1 functions in root elongation and pollen development. J Biol Chem 279:15348-55

Schaaf G, Ludewig U, Erenoglu BE, Mori S, Kitahara T, von Wiren N (2004) ZmYS1 functions as a proton-coupled symporter for phytosiderophore- and nicotianamine-chelated metals. J Biol Chem 279:9091-6

Schaaf G, Schikora A, Haberle J, Vert G, Ludewig U, Briat JF, Curie C, von Wiren N (2005) A putative function for the *Arabidopsis* Fe-phytosiderophore transporter homolog AtYSL2 in Fe and Zn homeostasis. Plant Cell Physiol 46:762-74

Schat H, Llugany M, Vooijs R, Hartley-Whitaker J, Bleeker P (2002) The role of phytochelatins in constitutive and adaptive heavy metal tolerances in hyperaccumulator and non-hyperaccumulator metallophytes. J Exp Bot 53:2381-92

Seigneurin-Berny D, Gravot A, Auroy P, Mazard C, Kraut A, Finazzi G, Grunwald D, Rappaport F, Vavasseur A, Joyard J et al (2006) HMA1, a new Cu-ATPase of the chloroplast envelope, is essential for growth under adverse light conditions. J Biol Chem 281:2882-92

Seregin IV, Ivanov VB (2001) Physiological aspects of cadmium and lead toxic effects on higher plants. Russ J Plant Physiol 48:523-44

Shikanai T, Muller-Moule P, Munekage Y, Niyogi KK, Pilon M (2003) PAA1, a P-type ATPase of *Arabidopsis*, functions in copper transport in chloroplasts. Plant Cell 15:1333-46

Stephan UW, Schmidke I, Pich A (1994) Phloem translocation of Fe, Cu, Mn, and Zn in *Ricinus* seedlings in relation to the concentrations of nicotianamine, an endogenous chelator of divalent metal ions, in different seedling parts. Plant Soil 165:181-8

Thomine S, Wang R, Ward JM, Crawford NM, Schroeder JI (2000) Cadmium and iron transport by members of a plant metal transporter family in *Arabidopsis* with homology to Nramp genes. Proc Natl Acad Sci USA 97:4991-6

Thomine S, Lelievre F, Debarbieux E, Schroeder JI, Barbier-Brygoo H (2003) AtNramp3, a multispecific vacuolar metal transporter involved in plant responses to iron deficiency. Plant J 34:685-95

Tsao DT (2003) Phytoremediation. Adv Biochem Eng Biot 78. Springer-Verlag Berlin

Vázquez MD, Barceló J, Poschenreider C, Mádico J, Hatton P, Baker AJM, Cope GH (1992) Localization of zinc and cadmium in *Thlaspi caerulescens* (Brassicaceae), a metallophyte that can hyperaccumulate both metals. J Plant Physiol 140:350-55

Vázquez MD, Poschenreider C, Barceló J, Baker AJM, Hatton P, Cope GH (1994) Compartmentation of zinc in roots and leaves of the zinc hyperaccumulator *Thlaspi caerulescens* J & C Presl. Bot Acta 107:243-50

Verret F, Gravot A, Auroy P, Leonhardt N, David P, Nussaume L, Vavasseur A, Richaud P (2004) Overexpression of AtHMA4 enhances root-to-shoot translocation of zinc and cadmium and plant metal tolerance. FEBS Lett 576:306-12

Verret F, Gravot A, Auroy P, Preveral S, Forestier C, Vavasseur A, Richaud P (2005) Heavy metal transport by AtHMA4 involves the N-terminal degenerated metal binding domain and the C-terminal His11 stretch. FEBS Lett 579:1515-22

Vert G, Grotz N, Dedaldechamp F, Gaymard F, Guerinot ML, Briat JF, Curie C (2002) IRT1, an *Arabidopsis* transporter essential for iron uptake from the soil and for plant growth. Plant Cell 14:1223-33

Vogel-Mikuš K, Regvar M (2006) Arbuscular mycorrhiza as a tolerance strategy in metal contaminated soils-prospects in phytoremediation. Nova Science Publishers, pp. 37-56

Vogel-Mikuš K, Pongrac P, Kump P, Ne čemer M, Regvar M (2006) Colonisation of a Zn, Cd and Pb hyperaccumulator *Thlaspi praecox* Wulfen with indigenous arbuscular mycorrhizal fungal mixture induces changes in heavy metal and nutrient uptake. Environ Poll 139:362-71

Vogel-Mikuš K, Pongrac P, Kump P, Ne čemer M, Sim ǎ čJ, Pelicon J, Budnar M, Povh B, Regvar M (2007) Localisation and quantification of elements within seeds of Cd/Zn hyperaccumulator *Thlaspi praecox* by micro-PIXE. Environ Poll 147:50-9

Weber M, Harada E, Vess C, Roepenack-Lahaye E, Clemens S (2004) Comparative microarray analysis of *Arabidopsis thaliana* and *Arabidopsis halleri* roots identifies nicotianamine synthase, a ZIP transporter and other genes as potential metal hyperaccumulation factors. Plant J 37:269-81

Welch RM, Norwell WA, Schafer SC, Shaff Je, Kochian LV (1993) Introduction of iron (III) and copper (II) reduction in pea (*Pisum sativum* L.) roots by Fe and Cu status: does the root-cell plasmalema Fe (III)-chelate reductase perform a general role in regulating cation uptake? Planta 190:555-61

Whiting SN, de Souza MP, Terry N (2001) Rhizosphere bacteria mobilize Zn for hyperaccumulation by *Thlaspi caerulescens*. Eviron Sci Technol 15:3144-50

Williams LE, Mills RF (2005) P(1B)-ATPases-an ancient family of transition metal pumps with diverse functions in plants. Trends Plant Sci 10:491-502

Woeste KE, Kieber JJ (2000) A strong loss-of-function mutation in RAN1 results in constitutive activation of the ethylene response pathway as well as a rosette-lethal phenotype. Plant Cell 12:443-55

Wójcik M, Vangronsveld J, Haen JD, Tukiendorf A (2005) Cadmium tolerance in *Thlaspi caerulescens* II. Localization of cadmium in *Thlaspi caerulescens*. Environ Exp Bot 53:163-71

Yang X, Feng Y, He Z, Stoffella PJ (2005) Molecular mechanisms of heavy metal hyperaccumulation and phytoremediation. J Trace Elem Med Biol 18:339-53

van der Zaal BJ, Neuteboom LW, Pinas JE, Chardonnens AN, Schat H, Verkleij JA, Hooykaas PJ (1999) Overexpression of a novel *Arabidopsis* gene related to putative zinc-transporter genes from animals can lead to enhanced zinc resistance and accumulation. Plant Physiol 119:1047-55

Zhao FJ, Hamon RE, McLaughlin MJ (2001) Root exudates of the hyperaccumulator *Thlaspi caerulescens* do not enhance metal mobilization. New Phytol 151:613-20

Chapter 12
Role of Sulfate and S-Rich Compounds in Heavy Metal Tolerance and Accumulation

Michela Schiavon and Mario Malagoli(✉)

Abstract Plants can withstand the potentially toxic effects exerted by a number of heavy metals through exclusion, which consists in restricted metal transport into plant tissues, or accumulation of metals, accompanied by the development of concomitant internal tolerance mechanisms. Plant tolerance and accumulation to heavy metals are known to be related to sulfur assimilation. The presence of metals can differently modulate genes involved in sulfate uptake and induce the sulfate assimilatory pathway. The regulation of S assimilation may be necessary to ensure an adequate supply of sulfur compounds required for heavy metal detoxification. The individuation of limiting steps of the sulfate assimilatory pathway in heavy metal tolerance and accumulation allows the possibility of genetic engineering approaches in order to develop plants with augmented phytoremediation capacity.

1 Introduction

Plants can adapt to a wide range of environmental stress conditions using appropriate physiological responses. In the case of heavy metals stress, these mainly consist in modulation of the activity of plasma membrane and/or vacuolar transporters, secretion of metal chelating organic acids such as malate or citrate in the rhizosphere, and biosynthesis of intracellular metal chelators (Clemens 2006).

In this context, sulfur (S) plays a pivotal role because sulfate transporters can mediate the entry of sulfate-analogues into the cells (Maruyama-Nakashita et al. 2007), and S-containing compounds like glutathione (GSH), phytochelatins (PCs), and metallothioneins (MTs) can improve the tolerance of plants to several metals and metalloids through complexation and/or further sequestration of toxic forms inside cellular vacuoles (Xiang and Oliver 1998, Cobbett and Goldsbrough 2002, Hall 2002). Because of this, many studies have been focused on heavy metals

Mario Malagoli
Department of Agricultural Biotechnology, University of Padua, Agripolis, 35020 Legnaro PD, Italy
mario.malagoli@unipd.it

N.A. Khan et al. (eds.), *Sulfur Assimilation and Abiotic Stress in Plants.*

uptake, accumulation, and tolerance in relation to sulfur availability to plants, and efforts have been made and others are underway to increase the phytoremediation potential of such species with traits that make them highly suitable for the clean up of heavy-metal-contaminated sites (Pilon-Smits and Pilon 2002, Cherian and Oliveira 2005, Pilon-Smits 2005, Pilon-Smits and Freeman 2006).

One promising approach to enhance plant metal accumulation and tolerance is genetic engineering, which allows the insertion of foreign genes of interest into plants. Such genes can be overexpressed by modulation of specific promoters, which leads to overproduction of desirable molecules involved in heavy metal tolerance and accumulation, like GSH, PCs, and MTs. Several transgenics developed in this way have been successful with respect to one or more metal(loids) in a number of trials.

2 Sulfate's Role in Heavy Metal Uptake

Heavy metal(loid)s like selenium (Se), chromium (Cr), and molybdenum (Mo) may cause sulfur deficiency in plants by inhibiting the uptake of sulfate from the environment when they are furnished to plants in the forms of selenate, chromate, and molybdate, respectively (Hopper and Parker 1999, Alhendawi et al. 2005, Maruyama-Nakashita et al. 2007, Schiavon et al. 2007, 2008). On the other hand, such elements as cadmium (Cd), zinc (Zn), and copper (Cu) have been reported to induce the absorption of sulfate for sustaining greater sulfur demand during the biosynthesis of the sulfur-containing compounds, glutathione and phytochelatins, notably involved in metal tolerance (Nocito et al. 2002, 2006).

On account of this, the interactions between sulfate and heavy metals must be considered when plants are going to be employed for the remediation of metal-contaminated sites, as the accumulation of contaminants in plant tissues might be altered by the sulfate concentration in the substrate where plants grow. Also, the use of sulfur-containing fertilizers during agronomic practices should be tightly managed to avoid excessive accumulation of nonessential elements and trace nutrients in crops.

2.1 Plant Availability of Chromate, Selenate, and Molybdate as Influenced by the Competing Ion Sulfate

The uptake of heavy metals and metalloids by plants is governed by many soil factors, which often include the presence of competitive ions (Hopper and Parker 1999). In particular, the group VI elements selenium and chromium are known to function as toxic analogues of sulfur (Mikkelsen and Wan 1990, Maruyama-Nakashita et al. 2007).

Selenium is an essential trace element in human and animal diets (Rayman 2000), as it is required for the production of selenoproteins involved in scavenging injurious free radicals (Zhang et al. 2003, Sors et al. 2005, Galeas et al. 2006). Although higher plants do not need Se for their metabolism (Novoselov et al. 2002, Sors et al. 2005), they import Se from the environment mainly in the forms of selenate (SeO_4^{2-}), selenite (SeO_3^{2-}), or organic Se compounds (Ellis and Salt 2003, Sors et al. 2005). The rate and form of Se uptake is strictly dependent on the concentration and chemical form of Se in the soil solution, as well as on the rhizosphere conditions (Sors et al. 2005). Generally, when selenate and selenite are present at equimolar concentrations, the uptake of selenate is greater than that of selenite (Mikkelsen et al. 1987, Zayed et al. 1998, Pilon-Smits et al. 1999a, Zhang et al. 2003).

The antagonistic interaction between sulfate and selenate for active transport into roots, as well as interference during their assimilation in plants, has long been investigated (Ferrari and Renosto 1972, Mikkelsen and Wan 1990, Bailey et al. 1995, Barak and Goldman 1997, Cherest et al. 1997, Shibagaki et al. 2002, White et al. 2004). Because of its close similarity in properties to sulfate, selenate is rapidly transported over membranes by the activity of sulfate transporters. Hence, it competitively inhibits the influx of sulfate and accesses the sulfur metabolic pathway for insertion into the Se nonprotein amino acid analogues selenocysteine (SeCys) and selenomethionine (SeMet) (Ferrari and Renosto 1972, Mikkelsen and Wan 1990, Hopper and Parker 1999, Ellis and Salt 2003). This competition is generally considered to cause the toxic reaction of plants to selenate, which can substitute sulfur in proteins and other sulfur compounds, leading to the reduced synthesis of important sulfur-molecules such as cysteine (Cys), methionine (Met), and glutathione (GSH), and to the disruption of several biochemical processes (Anderson and Scarf 1983, Bailey et al. 1995, Galeas et al. 2006, Kassis et al. 2007).

A number of mutants lacking sulfate transporter genes have been isolated from yeasts and plants by using the toxic effects of selenate as sulfate analogue (Cherest et al. 1997, Shibagaki et al. 2002). The finding that *A. thaliana sel* mutants were tolerant to selenate because they lacked root high-affinity sulfate transporter genes *SULTR1;1* or *SULTR1;2*, undoubtedly indicated a role for these transporters in taking up both sulfate and selenate from the soil solution into the root cells (Shibagaki et al. 2002, Yoshimoto et al. 2002, Kassis et al. 2007), and suggested that root growth and potentially root tip activity could represent a specific target of selenate toxicity in plants. Indeed, the selenate resistance phenotype of the *A. thaliana sel* mutants containing a lesion in the *SULTR1;2* gene was accompanied by a significant increase of root growth and root sulfate to selenate ratio under selenate treatment, while the sulfate uptake capacity of roots was clearly reduced, as well as the contents of sulfate and selenate in plant tissues (Kassis et al. 2007). The mechanism conferring the selenate resistance phenotype to the *A. thaliana sel* mutants is thought to be root-specific, as differences in sulfate to selenate ratio between mutants and wild-type were observed only in roots. The mechanism most likely consists of the combined effects of a restricted selenate uptake and a defensive role of sulfate against the toxicity of selenate induced on root growth.

In order to better exploit the use of plants in Se phytoremediation additional studies focused on Se uptake and accumulation versus sulfur availability in various plant species (Mikkelsen and Wan 1990, Bailey et al. 1995, Barak and Goldman 1997, Hopper and Parker 1999, White et al. 2004). In general, the provision of high sulfate concentrations to plants leads to a significant decline of selenate uptake and its content in the aerial plant tissues (Bailey et al. 1995, White et al. 2004); in contrast, increasing selenate concentrations promoted the accumulation of both sulfur and selenium in leaves of *A. thaliana* (White et al. 2004), and of *Hordeum vulgare* and *Oryza sativa* at low sulfate concentration (Mikkelsen and Wan 1990).

The other analogue of sulfur, chromium, is regarded as one of the most noxious pollutants thought to pose a threat to animal and human health. In nature Cr is largely widespread and most frequently found in the trivalent (+3) and hexavalent (+6) oxidation states. Such forms, which can interchange depending upon several physical, biological, and chemical processes occurring in waters and soils (Kotas and Stasicka 2000), differ significantly in toxicity and environmental hazard. In particular, Cr^{6+} compounds (chromates and dichromates) are considered extremely toxic and harmful for living organisms due to their powerful oxidizing and carcinogenic properties. They can presumably react with cellular reducing cofactors to produce Cr^{3+} ions and valence intermediate-Cr ion species, which are most likely involved in chromium genotoxicity (Maruyama-Nakashita et al. 2007). Since a role for Cr in plant metabolism has not been evidenced so far, it seems reasonable to suppose that plants have not evolved specific mechanisms to take up this metal ion from external substrates (Zayed and Terry 2003).

Similar to selenate, the transport of chromate into the cells is active and appears to be mediated by sulfate carriers (Shewry and Peterson 1974, Skeffington et al. 1976, Cherest et al. 1997, Kleiman and Cogliatti 1997, Kim et al. 2005, Maruyama-Nakashita et al. 2007, Schiavon et al. 2007). The influx of chromate showed the Michaelis–Menten kinetics when the external concentration of Cr varied from 10 to 160 μM, and was competitively inhibited by 10 μM sulfate (Skeffington et al. 1976). The drastic inhibition (up to 75%) of chromate uptake by 250 μM sulfate or other Group VI anions (selenate, molybdate, and arsenate) was also observed by Shewry and Peterson (1974) in barley seedlings, except when chromate and sulfate were provided to plants at equimolar concentrations.

Further investigation on the uptake of chromate, in wheat reported the enhancement of chromate absorption in sulfate-deprived plants within 5 hours of exposure to either 1 or 5 μg ml^{-1} chromate, most due to likely due to the lack of sulfate-competition (Kleiman and Cogliatti 1997). The increase of chromate uptake was also dependent upon the duration of the sulfate deprivation pretreatment, as 5 days rather than 2 days of sulfur absence in the nutrient solution could cause a stronger de-repression of the sulfate transport system, which in turn induced the activity of the plasma membrane transporters implied in the uptake of sulfate and sulfate-chemical analogues like chromate, as already observed for selenate (Lee 1982).

A recent work suggests that chromate may affect the sulfate uptake capacity in *Zea mays* plants by competing for the active binding site of sulfate transporters and/or repressing the high-affinity sulfate transport system at transcriptional level (Schiavon et al. 2007). Indeed, short-term exposure to chromate at concentrations ranging within 0.05-1 mM was found to specifically inhibit the rate of sulfate influx in maize-pretreated S-deprived plants, and 2 days of 200 μM chromate treatment strongly repressed the gene expression of the maize root high-affinity sulfate transporter *ZmST1;1*. Sustaining a role of chromate as competitor of sulfate for transport into the cells, the maximum chromium accumulation in maize plant tissues was recorded under S limitation as a result of the higher rate of the chromate uptake observed in S-starved plants following few hours of Cr exposure. Concomitantly, chromate lowered the concentration of sulfur and sulfate in S-provisioned plants to the basal level of S-starved, thus suggesting that maize plants may respond to chromate stress by reducing S accumulation (Schiavon et al. 2007). Similar results were also obtained in *B. juncea*, where the inhibition of the root low-affinity sulfate transporter and the reduction of the sulfate uptake capacity were observed concomitantly with the enhancement of sulfur assimilation as well as of GSH biosynthesis (Schiavon et al. 2008).

The involvement of sulfate transporters in the chromate uptake has been also corroborated by investigation on transgenics. The constitutive expression of the root high-affinity sulfate transporter encoded by the SHST1 gene from *Stylosanthes hamata*, resulted in enhanced capacity to take up chromate in *Brassica juncea* seedlings and mature plants (Lindblom et al. 2006). Transgenic tobacco plants overexpressing a putative yeast transcriptional activator MSN1 involved in chromium accumulation (Chang et al. 2003) showed greater accumulation of both Cr and S and higher tolerance to chromate. At the same time, the overexpression of MSN1 in transgenics increased the transcript level of the *NtST1* gene (*Nicotiana tabacum* high affinity sulfate transporter 1), therefore finally indicating that chromate and sulfate share a common transport system to enter into cells. In support of this, the expression of *NtST1* in *Saccharomyces cerevisiae* increased the yeast ability to accumulate Cr and S (Kim et al. 2005).

Unlike Cr and Se, molybdenum (Mo) is an essential micronutrient for plant metabolism (Marschner 1995, Kaiser et al. 2005) functioning as constituent of the mononuclear Mo enzymes involved in detoxifying excess sulfite (sulfite oxidase), purine catabolism (xanthine dehydrogenase), nitrate assimilation (nitrate reductase), and phytohormones biosynthesis (aldehyde oxidase[s]) (Hale et al. 2001, Mendel and Hänsch 2002, Mendel and Florian 2006). Inside cells the metal itself is biologically inactive if not complexed by a special cofactor (Mendel and Schwarz 1999). With the exception of the bacterial nitrogenase, where Mo is included in the FeMo-cofactor, Mo is bound to a pterin, thereby forming the molybdenum cofactor (Moco), which is the active compound at the catalytic site of all other Mo-enzymes (Mendel and Florian 2006).

To date, Mo transport systems and mechanisms of homeostasis have been identified and characterized in bacteria and some lower-order eukaryotes (Mendel and Häensch 2002), while in plants they are still weakly understood (Zimmer and Mendel 1999, Kaiser et al. 2005). Solely in the model plant *Arabidopsis thaliana* has a Mo-specific metal transporter, AMA1, been cloned and characterized (Palmgren and Harper 1999).

However, similarities in physiological responses to Mo between prokaryotes and eukaryotes exist (Kaiser et al. 2005), and evidences from bacteria studies indicate that Mo is actively transported across the plasma membranes of cells in the form of molybdate (Kannan and Ramani 1978), most likely competing with the similar-sized anion sulfate for the binding transport sites in uptake at root surface (Stout and Meagher 1948, Stout et al. 1951, Self et al. 2001). Sustaining this, the provision of sulfate to sulfur-starved plants altered the transport of Mo in the xylem sap of tomato (Alhendawi et al. 2005), lowering the Mo concentration of the sap and increasing sulfur levels. The analyses of Mo transport as measured by root pressure exudates also revealed that the absence of sulfate in the nutrient-growing medium could enhance Mo uptake by plants.

The effect of sulfate on molybdate uptake is not only detectable at the root/soil interface. Soybean plants showed decreased molybdenum levels in aerial plant tissues as the sulfate supply increased, even though molybdenum was applied as a foliar spray (Kannan and Ramani 1978). Following the first studies by Stout and Meagher (1948), sulfate has been proved to act as an effective regulator of molybdenum uptake in many plant species under a wide range of growing and environmental conditions (Macleod et al. 1997).

Recently, efforts to conceive plant-based tests for quantifying environmental concentration of heavy metals have been made due to their low cost and lack of requirement for technical means for analysis. From this standpoint, the employment of sulfur-responsive promoters of sulfate transporter genes may represent a promising tool for monitoring sulfate analogues. Maruyama-Nakashita et al. (2007) used a fusion gene construct consisting of a sulfur-responsive promoter region of the *A. thaliana* root high-affinity sulfate/selenate transporter SULTR1;2 and green fluorescent protein (GFP; $P_{SULTR1;2}$-GFP) to quantify the external levels of selenate and chromate via GFP accumulation. The $P_{SULTR1;2}$-GFP transgenics significantly increased in GFP following selenate or chromate treatment. The increase in GFP was linearly dependent upon the amount of Se and Cr in the growing medium, suggesting the potential use of $P_{SULTR1;2}$-GFP plants as indicators in detecting environmental selenate and chromate concentrations. However, for practical applications the sensitivity of $P_{SULTR1;2}$-GFP plants for selenate and chromate is still too low within the environmental limitations of Se and Cr, but more sensitive indicators could be designed through arranging the promoter regions and reporter proteins.

2.2 Enhanced Sulfate Uptake in Response to Cadmium, Zinc, or Copper Excess

Genes encoding sulfate transporters are involved in Cd-detoxification mechanisms, which imply the synthesis of phytochelatins (Nocito et al. 2002, 2006); they are strictly regulated by the accumulation of intermediates generated along the sulfate reductive assimilatory and GSH biosynthesis pathways, which may function as signals for the modulation of sulfate uptake (Hawkesford and De Kok 2006).

The induction of the sulfate uptake capacity in *Zea mays* plants exposed to toxic concentrations of either Cd, Zn, or Cu is closely dependent on the upregulation of the high-affinity sulfate transport system (HATS), as well as on the external metal concentration (Nocito et al. 2002, 2006). In particular, the enhancement in transcript accumulation of the maize root high-affinity sulfate transport gene, ZmST1;1, was observed following all Cd, Zn, and Cu treatments. It occurred earlier and was more pronounced under 10 μM Cd exposure than that occurring upon sulfur starvation, which was compared during the same time courses (Nocito et al. 2002, 2006). Concomitantly with the induced sulfate uptake activity and sulfate transporters transcription, the content of nonprotein thiols (NPTs) significantly increased following Cd or Zn treatments, and was accompanied by a transient depletion or no variation of the GSH pools in plus Cd and Zn, respectively. Neither NPTs nor total GSH were affected by excess Cu; however, the ratio GSH to glutathione oxidized strongly varied, thus suggesting that Cu could induce sulfate uptake in response to a reduction in the root GSH pool.

In *Brassica juncea* the downregulation of a low-affinity sulfate transporter most likely involved in the translocation of sulfate, *BjST1*, was observed in response to Cd exposure (Heiss et al. 1999). The authors interpreted this effect with the need of plants to retain high sulfate amounts in roots, which may be required for filling the greater demand of GSH and S-chelating compounds to cope Cd stress.

3 Cellular Signaling of Heavy Metals: Activation of Internal S-Related Chelating Mechanisms

Several transition metal ions are essential for plants and algae but, together with nonessential heavy metal ions, they can be toxic to living cells when present in excess. Excess metal ions activate redox mechanisms, leading to the formation of hydroxyl radicals through Haber-Weiss and Fenton reactions (Halliwell and Gutteridge 1990).

In plants, the response to the uptake and accumulation of heavy metals involves a variety of potential mechanisms, including the biosynthesis of two classes of metal-chelating compounds, phytochelatins (PCs) and metallothioneins (MTs), which bind metals in thiolate complexes (Cobbett and Goldsbrough 2002).

As the accumulation of metals rises, plants must cope with the enhanced requirement for amino acids, in particular cysteine, needed for the formation of MTs and PCs. Their synthesis is energy expensive and requires increasing amounts of sulfur and nitrogen. Therefore, the increasing demand for the synthesis of S-containing metal ligands might affect plant growth and consequently limit the use of certain plant species as possible phytoremediators (Tong et al. 2004). On metal-rich soil certain plant species and varieties have adapted a very pronounced hypertolerance to toxic metal concentration. More than 400 plant species are known to hyperaccumulate metals, including Cu, Ni, Zn, Cd, As, and Se (Baker and Brooks 1989)

3.1 Gluthatione and Phytochelatins

The tripeptide (γ-GluCysGly) glutathione (GSH) plays a central role in several physiological processes, including regulation of sulfate transport, signal transduction, conjugation of metabolites, detoxification of xenobiotics (Xiang et al. 2001), and the expression of stress-responsive genes (Mullineaux and Rausch 2005). Indeed, GSH acts as an antioxidant, quenching the reactive oxygen species (ROS) generated in response to stress before they cause damages to cells (Xiang et al. 2001). When present in excess, several heavy metals may induce variation in plant GSH content (Rauser 2001). In plants exposed to Cd the cellular GSH concentration transiently declined (Steffens 1990), inducing an increased activity of the two enzymes of GSH synthesis, γ-glutamylcysteine synthetase (γ-ECS) and glutathione synthetase (GS) (Rüegsegger et al. 1990, Rüegsegger and Brunold 1992).

The availability of cysteine may restrict the rate of GSH synthesis (Noctor et al. 1996), and therefore an increased assimilatory flux for sulfur should be required. Upstream of GSH synthesis is the assimilation of sulfate as well as the biosynthesis of cysteine. The transcriptional upregulation of several steps in these pathways has been thoroughly described. These include the uptake of sulfate (Nocito et al 2002, 2006), the sulfate activation catalyzed by ATP sulfurylase, and the following reduction to sulfite by adenosine 5-phosphosulfate reductase (Heiss et al 1999), the synthesis of cysteine promoted by O-acetylserine(thiol)lyase (Dominguez-Solis et al 2001), and the biosynthesis of GSH involving two enzymes, γ-glumatyl synthetase and GSH synthetase (Xiang and Oliver 1998). However, the mechanisms at the basis of the upregulation are still poorly understood. In one study, under metal excess jasmonate was found to be involved in the transcriptional control of GSH biosynthesis genes (Xiang and Oliver 1998), with the upregulation of the genes encoding γ-ECS, GS and GSH reductase (GR). It remains unclear whether the upregulation is directly triggered by metal sensing or by the increased requirement for sulfur caused by GSH enhanced synthesis.

Although heavy metal hyperaccumulators have been widely studied in the last years with the aim of better understanding the biochemical and molecular basis of metal tolerance, the molecular signaling pathways that control these mechanisms are not fully understood. In plants of *Thlaspi* Ni-hyperaccumulators, the elevated levels GSH found to be involved in Ni-tolerance were due to the constitutively increased activity of Ser acetyltransferase (SAT) (Freeman et al. 2004).

GSH serves an additional function in plant responses to heavy metal stress as a precursor of metal-chelating oligopeptides, the phytochelatins (PCs), which are synthesized by a reaction of transpeptidation of γ-glutamyl-cysteinyl units catalyzed by phytochelatin synthase (PCS) (Grill et al. 1989). Plants produce phytochelatins (PCs) in response to excess of several metal ions, including Ni, As, Cd, Zn, Ag, Sb, Cu, Hg, Pb, and Te (Cobbett 2003). PCs are small peptides of the general structure (-Glu-Cys)nGly, where n = 2–11, although (γ-Glu-Cys)2 -Gly (PC2) and (γ-Glu-Cys)3-Gly (PC3) are the most common (Cobbett and Goldsbrough 2002). No other environmental factors are known that would induce PC accumulation.

The vacuole is generally the main storage site for metals in yeast and plant cells, and there is evidence that phytochelatin-metal complexes are pumped into the plant vacuole (Salt and Rauser 1995).

PCS is activated through metal ions and/or metal-GS complexes (Vatamaniuk et al. 2000), and no signaling cascade seems to be involved in the activation of the enzyme since only the substrate is needed for the functioning of the enzyme in vitro (Clemens 2006).

However, the precise mechanism of enzyme activation by either free heavy metal ions or metal-thiolate complexes is still a matter of debate (Ducruix et al. 2006).

The most convincing evidence of the role of PCs in heavy metal detoxification was obtained by Ha et al. (1999) by comparing the relative sensitivity of PC synthase-deficient mutants of *Arabidopsis* and *Schizosaccharomyces pombe* to different heavy metals. PC-deficient mutants of both organisms were sensitive to Cd and arsenate, while for many other metals (Zn, Cu, Hg, Ni, Ag) and selenite, little or no sensitivity was observed. Vatamaniuk et al. (2000) showed that purified tagged AtPCS1 catalyzed PC synthesis in the presence of Cu, Zn, Mg, Ni, or Co, while Oven et al. (2002) found that purified untagged AtPCS1 did not show any activity with Mg, Ni, or Co. The regulation of PCS has been extensively studied. In *Thlaspi* Ni-hyperaccumulators Freeman et al. (2005) showed that salicylic acid activated Ser acetyltransferase (SAT), enhancing GSH accumulation and GR activity, while inhibiting PCS activity. This finding was explained as a need for hyperaccumulators to maintain high levels of GSH acting as antioxidant. The paradoxical observation that overexpression of *Arabidopsis* phytochelatin synthase provides Cd sensitivity is a confirmation that depletion of GSH pools for phytochelatin synthesis can reduce metal tolerance (Lee et al. 2003). In the same plant species, the cellular pool of available glycine constituted a limiting factor for the synthesis of PCs and iso-PCs under Cd stress (Ducruix et al. 2006).

3.2 *Metallothioneins*

Metallothioneins (MTs) are a superfamily of low molecular mass cysteine-rich, metal-binding proteins. Their ability to bind such metal ions as Cd, Zn, Cu, Co, Ag, and Hg is attributable to the arrangement of the Cys residues. According to Cobbett and Goldsbrough (2002) MTs can be classified into three classes, based on the arrangement of Cys residues: Class-I MTs are monomers with two Cys-rich clusters separated by a spacing region; Class-II MTs are translational monomers in which Cys residues are scattered throughout the entire sequence, and are further subdivided into four types (MT1, MT2, MT3, MT4) based on amino acid sequence; and Class-III MTs consist of peptide chains of variable length.

Many genes and cDNAs encoding MTs have been isolated in plants (Cobbett and Goldsbrough 2002), and functional complementation has been useful to show a relation between plant MT genes and plant metal tolerance (Giritch et al. 1998, Ma et al. 2003, Kohler et al. 2004, Zhang et al. 2004, Castiglione el al. 2006). Although the possible involvement of MTs in plant metal detoxification and homeostasis has been thoroughly studied and a number of metal responsive elements (MRE) have been identified in the promoter regions of the MTs genes and characterized in yeasts and mammalian (Cobbett and Goldsbrough 2002), little is known about the transcriptional regulation of plant MTs genes in response to heavy metals. To date the promoter

PvSR2 isolated in tobacco (Qi et al. 2007) represents the only example of plant MRE sequences-containing promoters that is heavy-metal-specific-responsive. No evidences of other MRE or MRE-like sequences conferring heavy-metal-specific-responsiveness to MT genes in plants have yet been provided. Therefore, alternative mechanisms of sensing, such as signal transduction pathways that involve mitogen-activated protein kinases (MAPKs), may be hypothesized to induce the transcription of defense genes, including MT genes. Indeed, several distinct MAPK pathways have been recently corroborated to be activated in response to copper and cadmium stress (Jonak et al. 2004).

4 Enhancement of Heavy Metal Accumulation Through Overexpression of Enzymes Involved in S-Assimilation and GSH/PCs Biosynthesis

Naturally occurring metal hyperaccumulators are promising candidates for the phytoextraction of heavy metals from contaminated sites. However, factors such as slow growth, shallow root system, and small biomass production often limit their use (Chaney et al. 1997, Pilon-Smits 2005).

Genetic engineering approaches represent a powerful tool to increase the ability of plants to remediate environmental pollutants (Pilon-Smits and Pilon 2002, Tong et al. 2004, Pilon-Smits 2005). On account of this, two possible strategies can be used: enhancing the biomass productivity of hyperaccumulators or improving the tolerance and/or the accumulation of heavy metals in high biomass producing and fast-growing plants (Pilon-Smits and Pilon 2002, Pilon-Smits and Freeman 2006). The latter is more easily achievable, as plant productivity is under the control of several genes and is fairly hard to obtain through a single gene insertion (Pilon-Smits and Pilon 2002).

The enhancement of plant metal tolerance and/or accumulation can be carried out by the acceleration of an existing plant process that is limiting for metal remediation or by the transfer and overexpression of a new pathway from other organisms, ranging from bacteria to mammalian, into plants (Clemens et al 2002, Pilon-Smits and Pilon 2002, Cherian and Oliveira 2005). The foreign genes introduced are usually integrated in the nuclear genome or targeted to the chloroplasts, and appropriate promoters are used to modulate their expression (Pilon-Smits and Freeman 2006).

The understanding of the mechanisms involved in metal tolerance and accumulation is required to develop transgenics that may be successfully employed in phytoremediation. To date, a number of metal detoxification systems have been genetically and functionally characterized at the molecular level in plants, yeasts, and bacteria, and several works have utilized the overexpression approach to dissect the involvement of enzymes in alleviating heavy metal stress in plants. In particular, some transgenics with enhanced metal tolerance and accumulation have been developed by overexpression of sulfate transporters (Lindblom et al. 2006), enzymes involved in the sulfate assimilation (Pilon-Smits et al. 1999b, Dominiguez-Solis et al. 2001, Dominiguez-Solis et al. 2004, Kawashima et al. 2004, Van Huysen et al. 2004, Wangeline et al. 2004), and biosynthesis of GSH, PCs, and MTs (Hasegawa et al. 1997, Zhu et al. 1999ab, Lee et al. 2003, Li et al. 2004).

The overexpression of the *Escherichia coli gshI* and *gshII* genes encoding γ-ECS and GS respectively, conferred to the metal accumulator *B. juncea* increased Cd tolerance and accumulation (Zhu et al. 1999ab). Specifically, overexpression of *gshII* in the cytosol increased Cd concentrations in the shoot up to 25% and total Cd accumulation per shoot up to threefold compared with the wild-type, and promoted the synthesis of GSH and PCs. The overexpression of *E. coli gshI* targeted to the plastids resulted in transgenic plants that well tolerated and accumulated shoot Cd concentration 40%–90% higher than in the wild-type. This was most likely because of greater production of GSH (1.5- to 2.5-fold) and PCs. Overexpression of γ-ECS also led to greater accumulation of total sulfur in the shoot of *B. juncea* as already found in poplar, thus indicating an added benefit of enhanced sulfur metabolism (Arisi et al. 1997, Zhu et al. 1999a).

The induced synthesis of thiols in response to Cd was also observed in *A. thaliana* plants overexpressing OASTL and, together with the increased rate of cysteine biosynthesis, is thought to be responsible for the augmented Cd tolerance and accumulation (Dominiguez-Solis et al. 2001, Dominguez-Solis et al. 2004).

The overexpression of adenosine triphosphate sulfurylase (APS) led to increased selenate uptake, reduction and tolerance in *B. juncea* with rates of Se accumulation in shoots 2- to 3-fold and 1.5-fold higher in shoots and roots, respectively, compared to wild-type (Pilon-Smits et al. 1999b). Similarly, transgenic *B. juncea* plants overexpressing the enzymes cystathionine-γ-synthase and selenocysteine methyltransferase promoting selenium (Se) organication and volatilization showed increased tolerance to Se and higher rates of Se volatilization (LeDuc et al. 2004, Van Huysen et al. 2004). Also, glutathione reductase (GR), an enzyme that regenerates GSH, was overexpressed in *B. juncea* and was found to be most effective in providing chloroplastic tolerance. Indeed, GR transgenic plants had increased Cd tolerance of the plastids when the gene was targeted there, while the whole plant Cd tolerance was not affected (Pilon-Smits et al. 2000).

In another study, transgenic *B. juncea* plants overexpressing γ-ECS, GS, and APS showed improved phytoremediation ability when tested on soil contaminated with mixtures of metals (Bennett et al. 2003). The ECS and GS transgenics accumulated more Cd (+50%) and Zn (+45%for GS and +93%for ECS) in shoot than wild-type. Additionally, the ECS transgenics contained greater amounts of Cr (+170%), Cu (+140%), and Pb (+200%) relative to wild-type plants.

Other efforts to increase heavy metal tolerance and accumulation in plants have focused on the enhancement of the plant ability to sequester metals in nontoxic forms inside cells through the overexpression of genes directly encoding metal-chelating compounds like PCs or MTs. Paradoxically, the overexpression of the phytochelatins synthase *AtPCS1* in *A. thaliana* did not enable plants to be more tolerant and/or to accumulate more Cd and Zn; rather it induced hypersensitivity to these metals, which the authors explain as a result of the toxicity of supraoptimal levels of PCs generated when compared with GSH levels (Lee et al. 2003). Conversely, *AtPCS1* overexpressing *A. thaliana* plants showed more tolerance to arsenate and high expression of many unknown thiol products (Li et al. 2004). However, no significant accumulation of As occurred in the above-ground plant tissues. Gisbert et al. (2003) reported that the overexpression in *Nicotiana glauca* of a phytochelatins synthase encoding gene from wheat, TaPCS1, greatly improved its tolerance to such metals as lead (Pb) and Cd. The significant increase of Pb concentrations in up to 50% in the shoot and 85% in roots indicates that *Nicotiana glauca* represents a promising candidate for Pb and Cd phytoextraction from soil.

The overexpression of MTs often led to enhanced plant tolerance to the metals tested. Transgenic tobacco and rapeseed overexpressing the *MT2* human gene resulted in greater tolerance to Cd (Misra and Gedamu 1989), as well as the overexpression of the *CUP1* yeast gene in cauliflower (Hasegawa et al. 1997). In a related study, the overexpression of *CUP1* enabled tobacco plants to taken up Cu at higher rates, up to threefold relative to wild-type (Thomas et al. 2003). Greater Cu accumulation was also observed in *A. thaliana* transformants overexpressing the pea MT gene *PsMTA* (Evans et al. 1992).

While most of the above-mentioned investigations using transgenics have been realized in the laboratory and greenhouse, only one trial has been done in the field (Banuelos et al. 2005). This study successfully confirmed the results obtained in laboratory experiments with transgenics overexpressing the enzyme involved in sulfate/selenate reduction, which showed fivefold higher accumulation of Se in the field when grown on soil polluted with Se, boron (B), and other salts. Other field experiments with transgenics are presently underway (Pilon-Smits and Freeman 2006).

5 Concluding Remarks

Much progress in understanding the role of sulfur in heavy metals uptake and accumulation in plants has been made in the last years. However, a comprehensive picture of the signaling network necessary for a coordinated cellular response to heavy metals is still lacking, mainly because of the complex cross talk existing between heavy metals and other stress-signaling pathways, which involve redox mechanisms and antioxidant molecules.

Future research should be aimed at identifying new metal-responsive genes and elucidating the regulatory mechanisms of plant responses to heavy metal stress. The comprehension of these mechanisms will be helpful for engineering plants with enhanced ability for metal tolerance and accumulation, considering the success obtained in the field trials by using transgenics in metal remediation.

References

Alhendawi R, Kirkby EA, Pilbeam DJ (2005) Evidence that sulfur deficiency enhances molybdenum transport in xylem sap of tomato plants. J Plant Nutr 28:1347-53

Anderson JW, Scarf AR (1983) Selenium and plant metabolism. In: Robb DA, Pierpoint WS (eds), Metal and micronutrients: uptake and utilization by plants, Academic Press, New York, pp. 241-75

Arisi ACM, Noctor G, Foyer CH, Jouanin L (1997) Modification of thiol contents in poplars (*Populus tremula* x *P. alba*) overexpressing enzymes involved in glutathione synthesis. Planta 203:362-73

Bailey FC, Knight AW, Ogle RS, Klaine SJ (1995) Effect of sulfate level on selenium uptake by *Ruppia maritima*. Chemosphere 30:579-91

Baker AJM, Brooks RR (1989) Terrestrial higher plants which hyperaccumulate metallic elements – a review of their distribution, ecology and phytochemistry. Biorecovery 1:81-126

Banuelos G, Terry N, LeDuc DL, Pilon-Smits EAH, Mackey B (2005) Field trial of transgenic Indian mustard plants shows enhanced phytoremediation of selenium-contaminated sediment. Environ Sci Technol 39:1771-7

Barak P, Goldman IL (1997) Antagonistic relationship between selenate and sulfate uptake in onion (*Allium cepa*): implications for the production of organosulfur and organoselenium compounds in plants. J Agric Food Chem 45:1290-4

Bennet LS, Burkhead JL, Hale KL, Terry N, Pilon M, Pilon-Smits EAH (2003) Analysis of transgenic Indian mustard plants for phytoremediation of metal-contaminated mine tailings. J Environ Qual 32:432-40

Castiglione S, Franchin C, Fossati T, Lingua C, Torrigiani P, Biondi S (2006) High zinc concentrations reduce rooting capacity and alter metallothionein gene expression in white poplar (*Populus alba* L. cv. Villafranca). Chemosphere 67:1117-26

Chaney RL, Malik M, Li YM, Brown SL, Brewer EP, Angle JS, Baker AJM (1997) Phytoremediation of soil metals. Curr Opin Biotechnol 8:279-84

Chang KS, Won JI, Lee MR, Lee CE, Kim KH, Park KY, Kim S-K, Lee JS, Hwang, S (2003) The putative transcriptional activator MSN1 promotes chromium accumulation in *Saccharomyces cerevisiae*. Mol Cells 16:291-6

Cherest H, Davidian J-C, Thomas D, Benes V, Ansorge W, SurdinKerjan Y (1997) Molecular characterization of two high affinity sulfate transporters in *Saccharomyces cerevisiae*. Genetics 145:627-35

Cherian S, Oliveira MM (2005) Transgenic plants in phytoremediation: recent advances and new possibilities. Environ Sci Technol 24:9377-90

Clemens S (2006) Toxic metal accumulation, responses to exposure and mechanisms of tolerance in plants Biochimie 88:1707-19

Clemens S, Palmgren M, Krämer U (2002) A long way ahead: understanding and engineering plant metal accumulation. Trends Plant Sci 7:309-15

Cobbett CS (2003) Genetic and Molecular analysis of phytochelatin biosynthesis, regulation and function. In: Davidian JC et al (eds) Sulfur transport and assimilation in plants, Backhuys Publishers, Leiden, The Netherlands, pp. 69-77

Cobbett CS, Goldsbrough P (2002) Phytochelatins and metallothioneins: roles in heavy metal detoxification and homeostasis. Annu Rev Plant Biol 53:159-82

Dominguez-Solis JR, Gutierrez-Alcala G, Vega JM, Romero LC, Gotor C (2001) The cytosolic O-acetylserine(thiol)lyase gene is regulated by heavy metals and can function in cadmium tolerance. J Biol Chem 276 (33):9297-02

Dominguez-Solis JR, Lopez-Martin MC, Ager FJ, Ynsa MD, Romero LC, Gotor C (2004) Increased cysteine availability is essential for cadmium tolerance and accumulation in *Arabidopsis thaliana*. Plant Biotechnol J 2:469-76

Ducruix C, Junot C, Fiévet JB, Villiers F, EzanE, Bourguignon J (2006) New insights into the regulation of phytochelatin biosynthesis in *A. thaliana* cells from metabolite profiling analyses. Biochimie 88:1733-42

Ellis DR, Salt DE (2003) Plant, selenium and human health. Curr Opin Plant Biol 6:273-9

Evans KM, Gatehouse JA, Lindsay WP, Shi J, Tommey AM, Robinson NJ (1992) Expression of pea metallothionein-like gene PsMTA function. Plant Mol Biol 20:1019-28

Ferrari G, Renosto F (1972) Regulation of sulphate by excised barley root in the presence of selenate. Plant Physiol 49:114-16

Freeman JL, Persans MW, Nieman K, Albrecht C, Peer W, Pickering IJ, Salt DE (2004) Increased glutathione biosynthesis plays a role in nickel tolerance in *Thlaspi* nickel hyperaccumulators. Plant Cell 16:2176-91

Freeman JL, Garcia D, Kim D, Hopf A, Salt DE (2005) Constitutively elevated salicylic acid signals glutathione-mediated nickel tolerance in *Thlaspi* nickel hyperaccumulators. Plant Physiol 137:1082-91

Galeas ML, Zhang L, Freeman JL, Wegner M, Pilon-Smits EAH (2006) Seasonal fluctuations of selenium and sulfur accumulation in selenium hyperaccumulators and related nonaccumulators. New Phytol 173:517-25

Giritch A, Ganal M, Stephan UW, Bäumlein H (1998) Structure, expression and chromosomal localisation of the metallothionein-like gene family of tomato. Plant Mol Biol 37:701-14

Gisbert C, Ros R, Haro AD, Walker DJ, Bernal MP, Serrano R, Avino JN (2003) A plant genetically modified that accumulates Pb is especially promising for phytoremediation. Biochem Biophys Res Commun 303:440-5

Grill E, Löeffler S, Winnacker E-L, Zenk MH (1989) Phytochelatins, the heavy-metal-binding peptides of plants, are synthesized from glutathione by a specific γ-glutamylcysteine dipeptidyl transpeptidase (phytochelatin synthase). Proc Natl Acad Sci USA 86:6838-42

Ha S-B, Smith AP, Howden R, Dietrich WM, Bugg S, O'Connell MJ, Goldsbrough P B, Cobbett C S (1999) Phytochelatin synthase genes from *Arabidopsis* and the yeast, *Schizosaccharomyces pombe*. Plant Cell 11:1153-64

Hale KL, McGrawth SP, Lombi E, Stack SM, Terry N, Pickering IJ, George GN, Pilon Smits EAH (2001) Molybdenum sequestration in *Brassica* species. A role for anthocyanins? Plant Physiol 126:1391-1402

Hall JL (2002) Cellular mechanisms for heavy metal detoxification and tolerance. J Exp Bot 366:1-11

Halliwell B, Gutteridge J M C (1990) Role of free radicals and catalytic metal ions in human disease-an overview. Methods Enzymol 186:1-85

Hasegawa I, Terada E, Sunairi M, Wakita H, Shinmachi F, Noguchi A, Nakajima M, Yazaki J (1997) Genetic improvement of heavy metal tolerance in plants by transfer of the yeast metallothionein gene (CUP1). Plant Soil 196:277-81

Hawkesford MJ, De Kok LJ (2006) Managing sulphur metabolism in plants. Plant Cell Environ 29:382-95

Heiss S, Schäfer H J, Haag-Kerwer A, Rausch T (1999) Cloning sulfur assimilation genes of *Brassica juncea* L.: cadmium differentially affects the expression of a putative low-affinity sulfate transporter and isoforms of ATP sulfurylase and APS reductase. Plant Mol Biol 39:847-57

Hopper JL, Parker DR (1999) Plant availability of selenate and selenate as influenced by the competing ions phosphate and sulfate. Plant Soil 210:199-207

Jonak C, Nakagami H, Hirt H (2004) Heavy metal stress. activation of distinct mitogen-activated protein kinase pathways by copper and cadmium. Plant Physiol 136:3276-83

Kaiser BN, Gridley KL, Ngaire BJ, Phillips T, Tyerman SD (2005) The role of molybdenum in agricultural plant production. Ann Bot 96:745-54

Kannan S, Ramani S (1978) Studies on molybdenum absorption and transport in bean and rice. Plant Physiol 62:179-81

Kassis EE, Cathala N, Rouached H, Fourcroy P, Berthomieu P, Terry N, Davidian JC (2007) Characterization of a selenate-resistant *Arabidopsis* mutant. root growth as a potential target for selenate toxicity. Plant Physiol 143:1231-41

Kawashima CG, Noji M, Nakamura M, Ogra Y, Suzuki KT, Saito K (2004) Heavy metal tolerance of transgenic tobacco plants over-expressing cysteine synthase. Biotechnol Lett 26(2):153-7

Kim YJ, Kim JH, Lee CE, Mok YG, Choi JS, Shin HS, Hwanga S (2005) Expression of yeast transcriptional activator MSN1 promotes accumulation of chromium and sulfur by enhancing sulfate transporter level in plants. FEBS Lett 580:206-10

Kleiman ID, Cogliatti DH (1997) Uptake of chromate in sulfate deprived wheat plants. Environ Pollut 97:131-5

Kohler A, Blaudez D, Chalot M, Martin F (2004) Cloning and expression of multiple metallothioneins from hybrid poplar. New Phytol 164:83-93

Kotas J, Stasicka Z (2000) Chromium occurrence in the environment and methods of its speciation. Environ Pollut 107:263-83

Lee RB (1982) Selectivity and kinetics of ion uptake by barley plants following nutrient deficiency. Ann Bot 50:429-49

LeDuc DL, Tarun AS, Montes-Bayon M, Meija J, Malit MF, Wu CP, Abdel Samie M, Chiang CY, Tagmount A, de Souza M, Neuhierl B, Bock A, Caruso J, Terry N (2004) Overexpression of selenocysteine methyltransferase in *Arabidopsis* and indian mustard increases selenium tolerance and accumulation. Plant Physiol 135(1):377-83

Lee S, Moon JS, Ko TS, Petros D, Goldsbrough PB, Korban SS (2003) Overexpression of *Arabidopsis* phytochelatin synthase paradoxically leads to hypersensitivity to cadmium stress. Plant Physiol 131(2):656-63

Li Y, Dhankher OP, Carreira L, Lee D, Chen A, Schroeder JI, Balish RS, Meagher RB (2004) Overexpression of phytochelatin synthase in *Arabidopsis* leads to enhanced arsenic tolerance and cadmiumhypersensitivity. Plant Cell Physiol 45(12):1787-97

Lindblom SD, Abdel-Ghany SE, Hanson BR, Hwang S, Terry N, Pilon-Smits EAH (2006) Constitutive expression of a high-affinity sulfate transporter in *Brassica juncea* affects metal tolerance and accumulation. J Environ Qual 35:726-33

Ma M, Lau P-S, Jia Y-T, Tsang W-K, Lam SKS, Tam NFY, Wong Y-S (2003) The isolation and characterization of type 1 metallothionein (MT) cDNA from a heavy metal-tolerant plant, *Festuca rubra* cv. Merlin. Plant Sci 164:51-60

Macleod JA, Gupta UC, Stanfiled B (1997) Molybdenum and sulfur relationships in plants. In: Gupta UC (ed) Molybdenum in agriculture. Cambridge: Cambridge University Press. Proc Natl Acad Sci USA 92:9373-7

Marschner H (1995) Mineral nutrition of higher plants. Academic Press, San Diego

Maruyama-Nakashita A, Inoue E, Saito K, Takahashi H (2007) Sulfur-responsive promoter of sulfate transporter gene is potentially useful to detect and quantify selenate and chromate. Plant Biotechnol 24:261-3

Mendel RR, Schwarz GR (1999) Molybdoenzymes and molybdenum cofactor in plants.
Crit Rev Plant Sci 18(1):33-69

Mendel RR, Hänsch R (2002) Molybdoenzymes and molybdenum cofactor in plants. J Exp Bot 53(375):1689-98

Mendel RR, Florian B (2006) Cell biology of molybdenum. Biochim Biophys Acta 1763(7):621-35

Mikkelsen RL, Haghnia GH, Page AL (1987) Effects of pH and selenium oxidation state on the selenium accumulation and yield of alfalfa. J Plant Nutr 10:937-50

Mikkelsen RL, Wan HF (1990) The effect of selenium on sulfur uptake by barley and rice. Plant Soil 121:151-3

Misra S, Gedamu L (1989) Heavy metal tolerant transgenic *Brassica napus* L. and *Nicotiana tabacum* L. plants. Theor Appl Genet 78:161-8

Mullineaux P M, Rausch T (2005) Glutathione, photosynthesis and the redox regulation of stress-responsive gene expression. Photosynth Res 86:459-74

Nocito FF, Pirovano L, Cocucci M, Sacchi GA (2002) Cadmium induced sulfate uptake in maize roots. Plant Physiol 129:1872-9

Nocito FF, Lancilli C, Crema B, Fourcoy P, Davidian J-C, Sacchi GA (2006) Heavy metal stress and sulfate uptake in maize roots. Plant Physiol 141:1138-48

Noctor G, Strohm M, Jouanin L, Kunert KJ, Foyer CH, Rennenberg H (1996) Synthesis of glutathione in leaves of transgenic poplar overexpressing γ-glutamylcysteine synthetase. Plant Physiol 112:1071-8

Novoselov SV, Rao M, Onoshko NV, Zhi H, Kryukov GV, Xiang Y, Weeks DP, Hatfield DK, Gladyshev NV (2002) Selenoproteins and selenocysteine insertion system in the model plant system, *Chlamydomonas reinhardtii*. EMBO J 21:3681-93

Oven M, Page JE, Zenk MH, Kutchan TM (2002) Molecular characterization of the homo-phytochelatin synthase of soybean *Glycine max*. J Biol Chem 277:4747-54

Palmgren MG, Harper JF (1999) Pumping with plant P-type ATPases. J Exp Bot 50:883-93

Pilon-Smits EAH (2005). Phytoremediation. Annu Rev Plant Biol 56:15-39

Pilon-Smits EAH, Pilon M (2002) Phytoremediation of metals using trangenic plants. Crit Rev Plant Sci 21(5):439-56

Pilon-Smits EAH, Freeman JL (2006) Environmental cleanup using plants: biotechnological advances and ecological consideration. Rev Front Ecol Environ 4(4):203-10

Pilon-Smits EAH, de Souza MP, Hong G, Amini A, Bravo RC, Payabyab ST, Terry N (1999a) Selenium volatalization and accumulation by twenty aquatic plant species. J Environ Qual 28:1011-18

Pilon-Smits EAH, Hwang S, Mel Lytle C, Zhu Y, Tai JC, Bravo RC, Chen Y, Leustek T, Terry N (1999b) Overexpression of ATP sulfurylase in Indian mustard leads to increased selenate uptake, reduction, and tolerance. Plant Physiol 119:123-32

Pilon-Smits EAH, Zhu YL, Sears T, Terry N (2000) Overexpression of glutathione reductase in *Brassica juncea*: effects on cadmium accumulation and tolerance. Physiol Plant 110:455-60

Qi X, Zhang Y, Chai T (2007) Characterization of a novel plant promoter specifically induced by heavy metal and identification of the promoter regions conferring heavy metal responsiveness. Plant Physiol 143:50-9

Rauser WE (2001) The role of glutathione in plant reaction and adaptation to excess metals. In: Grill D, Tausz M, De Kok LJ (eds) Significance of glutathione in plant adaptation to the environment, Kluwer Academic Publisher, Dordrecht, The Netherlands, pp. 123-54

Rayman MP (2000) The importance of selenium to human health. Lancet 356:233-41

Rüegsegger A, Schmutz D, Brunold C (1990) Regulation of glutathione synthesis by cadmium in *Pisum sativum* L. Plant Physiol 93:1579-84

Rüegsegger A, Brunold C (1992) Effect of cadmium on γ-glutamylcysteine synthesis in maize seedlings. Plant Physiol 99:428-33

Salt DE, Rauser WE (1995) MgATP-dependent transport of phytochelatins across the tonoplast of oat roots Plant Physiol 107:1293-1301

Schiavon M, Wirtz M, Borsa P, Quaggiotti S, Hell R, Malagoli M (2007) Chromate differentially affects the expression of a high affinity sulfate transporter and isoforms of components of the sulfate assimilatory pathway in *Zea mays* (L.). Plant Biol 9:662-671

Schiavon M, Pilon-Smits E, Wirtz M, Hell R, Malagoli M (2008) Interactions between chromium and sulfur metabolism in *Brassica juncea*. J Environ Qual (In press)

Self WT, Grunden AM, Hasona A, Shanmugam KT (2001) Molybdate transport. Res Microbiol 152:311-21

Shewry PR, Peterson PJ (1974) The uptake and transport of chromium by barley seedlings (*Hordeum vulgare* L.) J Exp Bot 25:785-97

Shibagaki N, Rose A, McDermott JP, Fujiwara T, Hayashi H, Yoneyama T, Davies JP (2002) Selenate-resistant mutants of *Arabidopsis thaliana* identify Sultr1,2 a sulfate transporter required for efficient transport of sulfate into roots. Plant J 29:475-86

Skeffington RA, Shewry PR, Peterson PJ (1976) Chromium uptake and transport in barley seedlings (*Hordeum vulgare* L.) Planta 132:209-14

Sors TG, Ellis DR, Salt DE (2005) Selenium uptake, translocation, assimilation and metabolic fate in plants. Photosynth Res 86:373-89

Steffens JC (1990) The heavy metal-binding peptides of plants. Annu Rev Plant Physiol Plant Mol Biol 41:553-75

Stout PR, Meagher WR (1948) Studies of the molybdenum nutrition of plants with radioactive molybdenum. Science 108:471-3

Stout PR, Meagher WR, Pearson GA, Johnson CM (1951) Molybdenum nutrition of plant crops: I. The influence of phosphate and sulfate on the absorption of molybdenum from soil and solution cultures. Plant Soil 3:51-87

Thomas JC, Davies EC, Malick FK, Enreszl C, Williams CR, Abbas M, Petrella S, Swisher K, Perron M, Edwards R, Ostenkowski P, Urbanczyk N, Wiesend WN, Murray KS (2003) Yeast metallothionein in transgenic tobacco promotes copper uptake from contaminated soils. Biotechnol Prog 19:273-80

Tong YP, Kneer R, Zhu YG (2004) Vacuolar compartmentalization: a second-generation approach to engineering plants for phytoremediation Trends Plant Sci:7-9

Van Huysen T, Terry N, Pilon-Smits EAH (2004) Exploring the selenium phytoremediation potential of transgenic indian mustard overexpressing ATP sulfurylase or cystathionine-gamma- synthase Int J Phytoremediation 6(2):1-8

Vatamaniuk OK, Mari S, Lu Y, Rea PA (2000) Mechanism of heavy metal ion activation of phytochelatin (PC) synthase. J Biol Chem 275:31451-9

Wangeline AL, Burkhead JL, Hale KL, Lindblom SD, Terry N, Pilon M, Pilon-Smits EAH (2004) Overexpression of ATP sulfurylase in indian mustard: effects on tolerance and accumulation of twelve metals. J Environ Qual 33 (1):54-60

White PJ, Bowen HC, Parmaguru P, Fritz M, Spracklen WP, Spiby RE, Meacham MC, Mead A, Harriman M, Trueman LJ, Smith BM, Thomas B, Broadley MR (2004) Interactions between selenium and sulphur nutrition in *Arabidopsis thaliana*. J Exp Bot 55(404):1927-37

Xiang C, Oliver D J (1998) Glutathione metabolic genes coordinately respond to heavy metals and jasmonic acid in *Arabidopsis*, Plant Cell 10:1539-50

Xiang C, Werner BL, Christensen EM, Oliver DJ (2001) The biological functions of glutathione revisited in *Arabidopsis* transgenic plants with altered glutathione levels. Plant Physiol 126:564-74

Zayed A, Lytle CM, Terry N (1998) Accumulation and volatilization of different chemical species of selenium by plants. Planta 206:284-92

Zayed AM, Terry N (2003) Chromium in the environment: factors affecting biological remediation. Plant Soil 249:139-56

Zhang Y, Pan G, Chen J, Hu Q (2003) Uptake and transport of selenite and selenate by soybean seedlings of two genotypes. Plant Soil 253:437-43

Zhang Y-W, Tam NFY, Wong YS (2004) Cloning and characterization of type 2 metallothionein-like gene from a wetland plant, *Typha latifolia*. Plant Sci 167:869-77

Zhu YL, Pilon-Smits EAH., Tarun AS, Weber SU, Jouanin L, Terry N (1999a) Cadmium tolerance and accumulation in Indian mustard is enhanced by overexpressing glutamylcysteine synthetase. Plant Physiol 121:1169-77

Zhu Y, Pilon-Smits EAH, Jouanin L, Terry N (1999b) Overexpression of glutathione synthetase in *Brassica juncea* enhances cadmium tolerance and accumulation. Plant Physiol 119:73-9

Zimmer W, Mendel RR (1999) Molybdenum metabolism in plants. Plant Biol 1:160-8

Yoshimoto N, Takahashi H, Smith FW, Yamaya Y, Saito K (2002) Two distinct high-affinity sulfate transporters with different inducibilities mediate uptake of sulfate in *Arabidopsis* roots. Plant J 29(4):465-73

Chapter 13
Sulfur Assimilation and Cadmium Tolerance in Plants

N.A. Anjum, S. Umar(✉), S. Singh, R. Nazar and N.A. Khan

Abstract Sulfur (S) is an essential element for growth and physiological functioning of plants. S uptake and assimilation in higher plants are crucial factors determining crop yield, quality, and even resistance to various biotic and abiotic stresses. The sulfur assimilation pathway, which leads to cysteine (Cys) biosynthesis, involves high- and low-affinity sulfate transporters and several enzymes. The biochemical and genetic regulation of these pathways is affected by oxidative stress, sulfur deficiency, and heavy metal exposure. In fact thiols are the main form of reduced sulfur in plants to cope with heavy metal stress through enhanced synthesis of heavy metal chelating molecules, glutathione, and phytochelatins (PCs). Cadmium is the most potent activator of phytochelatins (PCs). The present chapter summarizes the available data and information on various aspects and control of enzymatic and nonenzymatic pathways in relation to sulfate assimilation-reduction and metabolism of organic reduced S-containing compounds when plants are challenged by cadmium.

1 Introduction

Sulfur (S) is an essential nutrient, both for plants and animals. It is fourth in importance after nitrogen (N), phosphorus (P), and potassium (K), and is considered vital for proper plant growth and development (Syers et al. 1987). The importance of S as a plant nutrient has been recognized for a long time, but active research started in the second half of the 20th century, when widespread S deficiencies were observed (Duke and Reisenauer 1986). S uptake and assimilation in higher plants are crucial factors in determining crop yield, quality, and even resistance to various biotic and abiotic stresses (Anjum 2006). In fact inorganic S is converted to nutritionally and functionally important S-containing compounds like Cys, Met, several co-enzymes, thioredoxins, sulpholipids, and vitamins, i.e.,. biotin, thiamine, and ferredoxin, through a cascade of enzymatic steps (Hell 1997, Saito 2000). Sulfur is

S. Umar
Department of Botany, Faculty of Science, Hamdard University, New Delhi, 110062, India
s_umar9@hotmail.com

N.A. Khan et al. (eds.), *Sulfur Assimilation and Abiotic Stress in Plants.*

important in the formation of sulfhydryl (S-H) and disulphide bonds (S-S). A good part of the sulfur incorporated into organic molecules in plants is located in thiol (-SH) groups in proteins (cys-residues) or nonprotein thiols. Due to their particular redox properties, thiol groups can be oxidized, thereby forming disulphides (S-S groups). These bonds are important for stabilization of protein structures, and in many enzymes thiol groups form the active centers (Noji and Saito 2003). As a part of the Cys molecule, the sulfur group, called a thiol, is strongly nucleophilic, making it ideally suited for biological redox processes. Redox control regulates enzymes and protects against oxidative damage.

2 Sulfur Assimilation and Importance in Plant Development

Plants are the major food source for humans and other animals, providing carbohydrates, protein, lipids, and vitamins (Imsande 2003). The importance of S as a plant nutrient is becoming more imminent due to its effect on crop productivity and quality. Agronomic responses of crop growth and yield to the addition of S fertilizer are well documented. Several greenhouse and field experiments with various crop plants as test plants have indicated that effect of S is primarily on the number of grains per pod/siliqua or spike, indicating that S deficiency either increases the mortality of flowers/florets or reduces the initiation of florets (Archer 1974, Islam et al. 1999, Ahmad et al. 2005, Anjum 2006). Other yield components such as number of pods/siliqua and tillers and 100(0)-grain weight are affected to a lesser extent by S availability. A large number of studies have reported a marked influence of applied S on the yields of several cereals, pulses, oilseeds, vegetables, forages, and other crops (Pasricha et al. 1987, Tandon 1991, Aulakh and Chhibba 1992, Ahmad et al. 1998, Aulakh and Pasricha 1998). Oilseed rape has a high requirement for sulfur and is particularly sensitive to any shortfall in sulfur supply (Ahmad et al. 2005, Anjum 2006). Yield responses of oilseed rape to sulfur supply have been reported in many countries (Walker and Booth 1992). In fact sulfur is an important nutrient for oilseed rape due to its association with yield and also a range of quality factors. It is required by Brassicas for the synthesis of the sulfur-bearing compounds glucosinolates. Seeds from many plants, especially of legume crops, contain low concentrations of small 2S proteins that are relatively rich in cysteine and methionine (Shewry and Pandya 1999). Soybean also produces low molecular weight polypeptides that contain disproportionately high methionine content (George and de Lumen 1991, Paek et al. 2000). In fact the availability of reduced sulfur (i.e., cysteine and methionine) is the rate-limiting factor for the regulation of β-conglycinin chains that are usually synthesized only during late seed development (Meinke et al. 1981).

2.1 Enzymes of Sulfur Assimilation

Sulfur is found in soil in the form of sulfate, and through a set of reaction is converted to sulfide and into an N/C-skeleton form cysteine (Cys) or its homologues

(Droux 2004). The assimilation of sulfate could be summarized in four steps: (1) uptake of sulfate; (2) activation of sulfate; (3) reduction of sulfate; (4) synthesis of cysteine. Sulfate uptake is facilitated by sulfate transporters; once sulfate is within cells, it can be stored or can enter the metabolic stream. Metabolism of sulfate is initiated by its activation by the reaction of adenylation catalyzed by ATP-sulfurylase. The reaction product adenosine 5′-phosphosulfate (APS) is a branch point intermediate, which can be channeled toward reduction or sulfation (Leustek et al. 2000). Activation of sulfate reduction is the dominant route for assimilation and is carried out in plastids (Brunold and Suter 1989, Rotte 1998, Leustek et al. 2000, Saito 2000). APS is reduced to sulfite by APS-reductase (APR) (Leustek and Saito 1999, Kopriva and Koprivova 2003), and finally sulfite is reduced to sulfide by sulfite reductase (SiR). Sulfide is then transferred to activated serine (Ser) by O-acetylserine(thiol)lyase (OAS-TL) to form cysteine (Cys). The formation of cysteine is a direct coupling step between sulfur and nitrogen assimilation in plants (Brunold 1990, 1993, Brunold et al. 2003). Cysteine is the precursor or sulfur donor for most other organic S-compounds in plants. In addition, cysteine is the precursor of glutathione (GSH), a low molecular weight, water-soluble nonprotein thiol compound which functions in protection of plants against varied environmental stresses (De Kok et al. 2005).

2.1.1 ATP-Sulfurylase (ATPS; EC 2.7.7.4)

Activation of the relatively inert sulfate occurs through binding to ATP catalyzed by ATP sulfurylase (EC 2.7.7.4, ATPS). In fact ATPS catalyzes the first step in sulfate assimilation, the adenylation of sulfate to adenosine-5′-phosphosulfate (APS) from ATP and sulfate.

The formation of APS is an energetically unfavorable process, which is driven forward by the consumption of APS by subsequent reactions, reduction to sulfite by APS reductase, or phosphorylation to PAPS by APS kinase (APK). Because of the thermodynamic balance, ATPS enzyme activity is routinely measured either in the back reaction by measurement of ATP synthesized from APS and pyrophosphate or indirectly by the molybdolysis assay which determines the rate of AMP production (Segel et al. 1987). *Arabidopsis thaliana* contains a three-member, highly homologous, expressed gene family encoding plastid localized forms of ATP sulfurylase; APS1 (Leustek et al 1994), APS2, and APS3 (Murillo and Leustek, 1995). All three cDNA clones functionally complement a *met3* (ATP sulfurylase) mutant strain of *Saccharomyces cerevisiae* (Murillo and Leustek 1995). APS1 is the most highly expressed member of this gene family. The APS polypeptides share homology with ATP-sulfurylases from fungi, a marine worm, and a chemoautotrophic bacterium, but not from *Escherichia coli* or *Rhizobium meliloti* (Murillo and Leustek 1995). Analysis of recombinant APS3 indicates that the protein is structurally and kinetically similar to fungal ATP-sulfurylase, but very different from the *E. coli* enzyme. The APS3 polypeptide is a homotetramer. Despite the sequence, structural, and kinetic differences between higher plant and *E. coli*

ATP-sulfurylases, *APS2* and *APS*, are able to functionally complement *E. coli cysD* and *cysN* (ATP-sulfurylase) mutant strains (Murillo and Leustek 1995). Plant ATPS is a homotetramer of 52-54 kDa polypeptides (Murillo and Leustek 1995). In plants, ATPS activity was detected in chloroplasts and in the cytosol of spinach leaves (Lunn et al. 1990, Renosto et al. 1993) and in proplastids of pea roots (Brunold and Suter 1989). However, ATPS was found in the cytosol and mitochondria in *Euglena gracilis* (Li et al. 1991). Low ATPS activity was measured in etiolated pea seedlings, which increased after transfer into the light, but decreased again in the leaves during further incubation (von Arb and Brunold 1986). Klonus et al. (1994) reported an ATPS-mRNA in potato leaves, stems, and roots, but not in tubers. There are reports of both decreasing and increasing trends of ATPS activity in various plants. In poplars, ATPS activity diminished slowly with the leaf age (Hartmann et al. 2000). In *Arabidopsis*, the foliar ATPS activity continually declined during the plant growth. During this time the more abundant chloroplastic ATPS activity was found to decrease, while cytosolic activity increased (Rotte and Leustek 2000). This observation indicates different functions of ATPS in the two compartments: sulfate reduction in the plastids and activation of sulfate for synthesis of sulfonated compounds in the cytosol (Rotte and Leustek 2000). ATPS activity may be localized to bundle sheath cells of chloroplasts. There are reports that approximately 80%-100% of total leaf ATPS activity in maize may be confined to bundle sheath cells (Gerwick et al. 1980, Passera and Ghisi 1982, Schmutz and Brunold 1984). Furthermore, it was revealed that mRNA of ATPS is present exclusively in RNA isolated from bundle-sheath cells of maize, revealing that in maize, the intercellular distribution of ATPS is regulated on the transcriptional level (Kopriva et al. 2001).

2.1.2 APS Reductase (APR; EC 1.8.4.9)

Sulfate reduction is the dominant route for assimilation and is carried out in plastids (Brunold and Suter 1989, Rotte 1998). In fact the sulfate reduction is carried out in two steps, i.e., (1) APS reductase transfers two electrons to APS to produce sulfite and (2) sulfite reductase transfers 6 electrons from ferredoxin to produce sulfide.

APS sulfotransferase, 5 adenylylphosphosulfate reductase, adenosine 5′-phosphosulfate reductase, are synonyms of APS reductase (Setya et al. 1996, Suter et al. 2000). APS reductase reduces the sulfate residue of APS into sulfite. APS reductase is a unique enzyme. It is now known that APS reductase possesses a transit peptide that allows translocation of the mature protein to plastids. Kopriva and Koprivova (2004) observed in an in vivo study that APS reductase is present as a homodimer most probably linked by a disulfide bond of the conserved Cys residue. Bick et al. (1998) reported that the mature APS reductase consists of two distinct domains viz., N- and C-domains and in fact, the N-terminal domain of APS reductase resembles PAPS reductase while the C-terminal domain exhibits homology to thioredoxin and acts as a glutaredoxin using reduced GSH as the electron donor. APS reductase catalyzes a thiol-dependent two-electron reduction of APS to sulfite. There is a great deal of evidence indicating that APS sulfotransferase is a prime regulation point in SO_4^{2-} assimilation (Brunold and Rennenberg 1997). In fact vast literature is available regarding situation-specific changes in the activity of this enzyme in a variety of plant species after sulfur starvation (Gutierrez-Marcos et al. 1996, Takahashi et al. 1997), exposure to reduced sulfur compounds, heavy-metal stress (Heiss et al. 1999, Lee and Leustek 1999), or other stresses. Heavy metals induce the synthesis of phytochelatins, and high concentrations of metal ions significantly increase the demand for Cys. Recent studies indicate that one potential mechanism for regulating APS sulfotransferase activity may involve changes in the steady-state mRNA level (Leustek and Saito 1999). The importance of APS sulfotranferase in sulfate assimilation is due to the fact that this enzyme involves changes in the steady-state mRNA level, while on the other hand SO_3^{2+} reductase does not appear to be appreciably regulated at the mRNA level (Bork et al. 1998). However, further studies are required to know the extent of the abundance of mRNA that regulates the changes in APS sulfotransferase activity in plants under normal and stress conditions.

Detailed information regarding APS reductase came through genetic studies. The amino acid sequence of plant APS reductase revealed a multidomain composition (Leustek et al. 2000, Suter et al. 2000). It is synthesized as a precursor with an amino terminal plastid transit peptide. The amino terminal domain of the mature protein is homologous to PAPS reductase, and the C-terminal domain is homologous to thioredoxin, a redox enzyme. APS reductase is able to use GSH or DTT (dithiothreitol) as an electron source. APS reductase is thought to be one of the key regulators of the sulfate reduction pathway. Its activity and steady-state mRNA level increased markedly and coordinately in response to sulfate starvation (Gutierrez-Marcos et al. 1996, Takahashi et al. 1997, Yamaguchi et al. 1999), oxidative stress, and/or exposure to heavy metals (Leustek et al. 2000, Heiss et al. 1999). The oxidative stress and heavy metal exposure have been shown to increase the demand for glutathione and, hence, the cysteine necessary for glutathione synthesis (Hesse et al. 2004). However, the other sulfate assimilatory enzymes are also regulated but to a lesser degree (ATP sulfurylase) or are constitutively expressed (sulphite reductase) (Bork et al. 1998). Vauclare et al. (2002) reported that excess feeding of plants with Cys and GSH decreases the activity of APS reductase and the level of transcripts; implying that increased internal cysteine and GSH levels might control sulfate assimilation. Further, it can be concluded from the split-root experiments that GSH, and not cysteine, is acting as a signal (Lappartient and Touraine 1996, Hesse et al. 2004).

2.1.3 SO_3^{2+} Reductase (SiR; EC 1.8.7.1)

Sulfite reductase catalyzes the transfer of six electrons from ferredoxin to sulfite to produce sulfide, S^{2-}. The sulfite reductase found in plant cells consists of a homo-oligomer containing a siroheme and an iron-sulfur cluster per subunit. Sulfite reductase is localized in plastids of both photosynthetic and nonphotosynthetic tissues. Electrons are supplied to ferredoxin from PSI in photosynthetic cells and from NADPH in nonphotosynthetic cells. The proper combination of different isoforms of ferredoxin, ferredoxin-NADP$^+$ reductase, and sulfite reductase is critical for efficient sulfite reduction (Yonekura-Sakakibara et al. 2000). In fact SO_3^{2+} reductase completes the reduction of sulfur with using electrons donated from reduced Fd. The formed sulfide is incorporated in to Cys, catalyzed by O-ace tyleserine(thiol)lyase, with O-acetylserine as substrate.

2.1.4 O-acetylserine (THIOL) lyase (OAS-TL) or Cys Synthase (OAS-TL; EC 2.5.1.47)

O-acetylserine (thiol) lyase (OASTL), a key enzyme of plant sulfur metabolism, catalyzes the formation of Cys from sulfide and O-acetylserine. In fact the Cys biosynthetic pathway involves several enzymatic reactions (Brunold and Rennenberg 1997, Leustek and Saito 1999). The SO_4^{2-} is reduced to SO_3^{2-} and then sulfide (S^{2-}) through the sulfate reduction pathway. The final step of Cys biosynthesis is the incorporation of S^{2-} into Cys. The reaction is catalyzed by Cys synthase [O-acetyl-L-Ser (thiol)-lyase, EC 4.2.99.8; CSase], which uses S^{2-} and O-acetyl-L-Ser as the substrates. This final step of Cys biosynthesis seems to exist necessarily in three major compartments of plant cells, i.e., cytosol, chloroplasts, and mitochondria, since the presence of CSase has been demonstrated in these three compartments from several plants (Brunold and Suter 1989, Lunn et al. 1990). In fact the metabolic pathways involved in the biosynthesis of Cys are regulated with a high degree of complexity. Availability of OAS, synthesized by Ser acetyl transferase (SAT), is generally regarded as a limiting factor and a positive signal for sulfur assimilation and Cys biosynthesis (Rennenberg 1983). A further level of control is also provided by the formation of a bi-enzyme complex between OASTL and SAT, in which the properties of the two enzymes are drastically modified and whose stability is dependent on the availability of the OASTL substrates, OAS and sulfide (Droux et al. 1998, Leustek and Saito 1999). Under nonstressed conditions, overproduction of OASTL in plants seems to have less significant effects on the level of the noncellular thiols than overproduction of SAT. It is not surprising, considering the fact that in pea in vivo, OASTL is present in a huge molar excess over SAT in all compartments (Ruffet et al. 1995, Droux 2003). In fact, it is unclear how to explain the observed increase of thiol levels in tobacco OASTL-transformants without modification of the current models. One possibility is that the OASTL/SAT ratio in tobacco might be much lower than in pea. Nevertheless, under stress conditions, OASTL-overproducing transformants contain more thiols and are more tolerant to stress than the control plants, suggesting that the increased potential for cysteine synthesis in such conditions still gives them an important advantage.

2.1.5 Serine Acetyltransferase (SAT; EC 2.3.1.30)

This enzyme catalyzes the formation of OAS from Ser and acetyl CoA. In fact SAT is responsible for the entry step from Ser-metabolism to Cys biosynthesis. A large number of reports are available in the literature regarding isolation of cDNA clones encoding SAT from plant species viz., watermelon (Saito et al. 1995), spinach (Noji et al. 2001), *Arabidopsis thaliana* (Bogdanova et al. 1995, Hell and Bogdanova 1995, Ruffet et al. 1995, Roberts and Wray 1996, Howarth et al. 1997), and Chinese chive (*Allium tuberosum* (Urano et al., 2000). Depending upon the subcellular localization, three isoforms of SAT enzyme have been reported. These SAT-isoforms were designated as SAT-c (cytoplasmic isoform), SAT-p (plastidic isoform), and SAT-m (mitochondrial isoform). cDNAs of these three SAT-isoforms have been cloned from *Arabidopsis thaliana* (Noji et al. 1998).

SAT is one of the major regulatory factors in the biosynthesis of Cys in plants. In fact the feedback inhibition of SAT activity by various Cys concentration has been reported to regulate the biosynthesis of Cys in plants (Hell and Bogdanova 1995, Saito et al. 1995, Roberts and Wray 1996, Noji et al. 1998). However, the inhibition of SAT activity depends upon the subcellular isoforms of SAT and plant-specific SAT (Saito et al. 1995, Urano et al. 2000, Noji et al. 2001).

Noji et al. (1998) reported in plants that there are two types of Ser acetyltransferase that differ in their sensitivity to the Cys inhibition. Difference of sensitivity to Cys means that Ser acetyltransferase has a regulatory role through the feedback inhibition in Cys biosynthesis and it depends on the subcellular compartmentation (Noji et al. 1998, Saito 2000).

2.2 SAT – OAS-TL Bi-enzyme Complex and Cysteine

Two pathways complete the Cys biosynthesis in plants: the pathway of transport, activation, and reduction of sulfate into sulfide, and the pathway supplying amino acid moiety, which is derived from serine (Ser) through *O*-acetyl-L-serine (OAS), and then yielding Cys by the reaction of incorporating sulfide moiety into β-position of alanine (Leustek and Saito 1999, Saito 1999, Hawkesford and Wray 2000, Leustek et al. 2000, Saito 2000). Ser acetyltransferase (SAT) and Cys synthase (OAS (thiol)-lyase) are the enzymes committing to the final step of this pathway (Noji and Saito 2003).

Distinct isorforms are localized in plastids, the cytosol, and in mitochondria (Hesse et al. 1999, Saito 1999). However, the multimeric property of serine acetyltransferase (SAT) is known, further, there are reports that SAT may form a complex in association with Cys synthase (OAS-TL) i.e. SAT-OAS-TL bi-enzyme complex (Nakamura et al. 1988, Nakamura and Tamura 1990, Bogdanova and Hell 1997, Droux et al. 1998, Noji and Saito 2003). Within the bi-enzyme complex, SAT is enzymatically active, whereas OAS-TL is not, but the excess amount of OAS-TL has been shown to catalyze the incorporation of sulfide to form cysteine. The findings suggest that free OAS-TL is responsible for cysteine synthesis and that it also functions as a regulatory subunit of SAT (Leustek et al. 2000). Droux et al. (1998) reported that the ratio of OAS-TL to SAT in chloroplasts is 300:1, so the majority of OAS-TL is in free form. Berkowitz et al. (2002) reported that OAS gets accumulated and allosterically disrupts the bi-enzyme complex under sulfide-limitation and as a result SAT gets inactivated. However, when the sulfur assimilation activity is induced, the OAS level decreases and the bi-enzyme complex is resumed (Rausch and Wachter 2005).

One of the factors that regulate GSH biosynthesis is cysteine availability, because exogenous addition of Cys has been shown to increase the GSH content (Farago and Brunold 1994). In addition, thiol group of Cys is of great importance. It is highly reactive, and the Cys residue present in active center plays critical roles in the catalytic function of some proteins, so-called SH proteins.

3 Thiols, Sulfur-Containing Proteins and Peptides: Role in Cadmium Tolerance

Cadmium (Cd^{2+}), severely toxic to normal plant growth and development, is a widespread pollutant with a long biological half-life (Chien et al. 2002). Its addition to the agricultural soil is generally due to continuous application of Cd^{2+}-rich phosphatic fertilizers (McLaughlin et al. 1999). Cd is easily taken up by plants and then enters the food chain, resulting in a serious health issue for humans when food is grown in Cd-supplemented/contaminated soils or growth medium (Hall 2002). Cadmium has been shown to disturb photosynthetic activity (Siedlecka et al. 1997, Khan et al. 2006, Mobin and Khan 2007, Samiullah et al. 2007, Singh et al. 2007), and the process of uptake and translocation of mineral nutrients lead to significant alterations in the normal plant growth (Wu and Zhang 2002a,b, Zhang et al. 2002, Khan et al. 2006, Samiullah et al. 2007, Khan et al. 2007). The causes of these dysfunctions include irreversible changes to protein conformation by forming metal thiolate bonds and alteration of the cell wall and membrane permeability by binding to nucleophilic groups (Ramos et al. 2002) through generation of reactive oxygen species (ROS). Plants adopt different strategies to reduce Cd-induced oxidative damage. Increasing the activities of antioxidative enzymes increases tolerance of plants to stress (Gratao et al. 2006).

Exposure of higher plants to heavy metals also induces a significant alteration in S metabolism (Tukendorf and Rauser 1990). Cadmium exposure in wheat plants induces activities of ATP-sulfurylase and antioxidative enzymes leading to tolerance (Khan et al. 2007). Sulfur is an essential macronutrient that plays a vital role in the regulation of plant growth and development (Ernst 1998). S metabolism tightly regulates the biosynthesis of phytochelatins in plants. The monothiols, Cys and GSH, are actively involved in PC synthesis and also in metal sequestration in plants (Thangavel et al. 2007). The additional S ions have been shown to enhance the stability of the PC-Cd^{2+} complex (Oritz et al. 1992).

A major part of sulfur incorporated into various organic molecules is located in thiols –SH) groups in proteins (Cys-residues) or nonprotein thiols (Tausz et al. 2003). Thiol groups have redox properties and hence play important roles in the stress response of plants. The –SH/S-S status controls the three-dimensional molecular structure of proteins (Saito 1999). On the other hand sulfur-containing proteins and peptides play a crucial role in the survival of plants under heavy metal stress, including cadmium. Among S-containing proteins and peptides the phytochelatins (PCs) (γ-glutamylcysteine polymers and enzymatically synthesized peptides from GSH), metallothionins (MTs) (cysteine-rich polypeptides and short, gene-encoded polypeptides), and glutathione (GSH) (cysteine-containing tripeptide) are important (Cobbett 2003).

3.1 Thiols and Stress Adaptation

Thiols (-SH groups) are the main form of reduced sulfur in plants. Thiols may be found either as protein (such as cys-residues; thioredoxins, glutaredoxins) or non-protein thiols (such as CoA, GSH and its homologues) (Schurman and Jacquot 2000, Tausz et al. 2003).

Various stress factors are nearly associated with an enhanced production of active/reactive oxygen species (AOS/ROS). ROS comprises several potentially toxic compounds such as hydrogen peroxide, superoxide anion free radical, and hydroxyl free radical of singlet oxygen. All of these radicals have been shown to damage proteins, lipids, nucleic acids, and pigments in various plant species (Foyer and Noctor 2000). Thiols have strong redox properties and have role in stress response in plants. Glutathione (γ-glutamyl-cysteinyl-glycine) is an abundant and ubiquitous thiol with unique structural properties (May et al. 1998). The presence of γ-glutamyl linkage found in GSH makes it stable, while the strong nucleophilic nature of central cysteine makes GSH a powerful cellular antioxidant.

L-Cysteine (Cys) and GSH are called monothiols and are involved actively in the systhesis of phytochelatins (PCs) and also in metal sequestration in plants. Cys is one of the sulfur-containing amino acids of the 20 standard amino acids found in proteins. In fact the inorganic sulfur is first fixed into Cys by the Cys biosynthetic pathway (Saito 1999). The thiol group of Cys is of great importance. It is highly reactive, and the Cys residue present in active cites plays critical roles in the catalytic function of some proteins, so-called SH proteins. Beside, two -SH groups can form a covalent S-S bond by oxidation. These disulphide bonds are known for their importance in the establishment of tertiary and, in some cases, quaternary structures of proteins (Saito 1999). Moreover, the Cys residues found in some proteins are hence necessary to retain their structure by S-S bonds. The redox cycle in most cells have been shown to be governed by the oxidation-reduction-lead sulfide-disulphide residues-interconversion process.

Thioredoxin (Trx) is a small protein (12–14 kDa) with two cysteines in its active site that can form a disulfide in the oxidized form [Trx-(S-S)]. Together with the enzyme thioredoxin reductase (EC 1.6.4.5), which uses NADPH to convert Trx-(S-S) into reduced thioredoxin [Trx-$(SH)_2$], thioredoxin is involved in the regulation of enzyme activities as well as in scavenging hydroperoxides and H_2O_2 (Halliwell and Gutteridge 1989, Meyer et al. 1999). Thioredoxin has long been known to regulate the activities of a number of the Calvin cycle enzymes in the chloroplast stroma (Buchanan 1980). Thioredoxins are reduced enzymatically by an NADP-dependent or by a ferredoxin (light)-dependent reductase and transmit the regulatory signal to a selected target enzymes through disulfide/dithiol interchange reactions (Schurmann and Jacquot 2000). Recently, another group of small thiol proteins, the glutaredoxins, which were well known as redox components in bacteria and mammal systems, have been characterized in plants (Rouhier et al. 2002). The chemistry of disulfide thiol interchange has been extensively used to mediate redox reactions, including the proteins thioredoxin, glutaredoxin, and protein disulfide isomerases. These proteins are nearly ubiquitous and play fundamental roles in different types of regulation (Fig. 13. 1). Buchanan (1991) was perhaps among the pioneers who reported the function(s) of thioredoxin in plants. The light reactions of the process of photosynthesis are strictly coordinated with the dark reactions of CO_2 fixation. In fact, the coordination mechanism relies on the reductive activation of specific enzymes by thioredoxin, which is reduced by photosynthetically reduced ferredoxin (Fd). Fig. 13.1 shows how thioredoxin functions as a regulatory factor through reduction of a disulfide on a target enzyme. In this example of carbon assimilation, the source of electrons for thioredoxin reduction is Fd, which is reduced via the light reactions of photosynthesis. Thioredoxins also exist in the cytoplasm of plants where NADPH + H^+ serves as an electron source. Recent evidence shows that thioredoxin has the potential to act as an oxidant mediating the formation of a disulfide bond on a target enzyme (Stewart et al. 1998). It is speculated that this activity could be important for activation of antioxidant enzymes during oxidative stress.

3.1.1 Biosynthesis and Transport of Glutathione and Its Regulatory Enzymes

Glutathione is a tripeptide (γ-glutamyl-cysteinyl Gly) synthesized both in the cytosol and in the chloroplasts of plant cells, through the sequential action of γ-glutamyl Cys synthetase (γ-ECS; EC 6.3.2.2) and glutathione synthetase (GSHS; EC 6.3.2.3). It plays numerous roles, including storage and transport of reduced

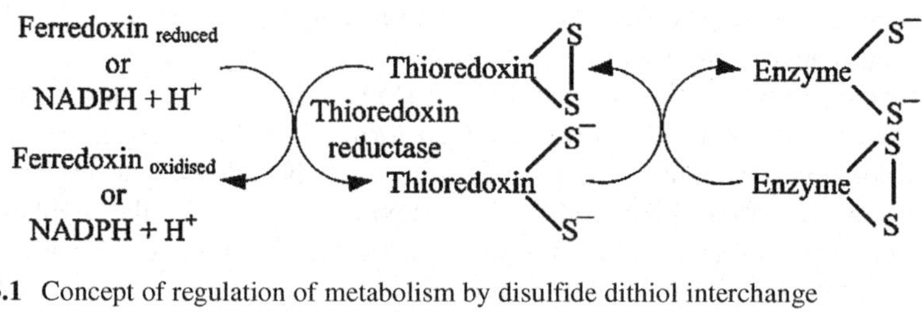

Fig. 13.1 Concept of regulation of metabolism by disulfide dithiol interchange

sulfur, control of sulfur assimilation, control of redox status, protection against biotic and abiotic stresses, protein folding, and in the cell cycle (May et al. 1998, Foyer et al. 2001).

GSH has been shown to scavenge several peroxides efficiently. The intracellular medium is buffered in the reduced state by GSH. Upon oxidation, one GSH can react with another to produce the disulfide form (GSSG). It is interesting to note here that the process of reductive inactivation of peroxides depends mostly on how efficiently the two forms of glutathione (GSH and GSSG) are interconverted (May et al. 1998). Participation of glutathione in the Halliwell-Asada cycle allows the destruction of hydrogen peroxide produced by oxidative stress (Kunert and Foyer 1993). A number of environmental stress conditions have been shown to trigger the biosynthesis of GSH and also regulate its level and redox status in various plant species.

As discussed in previous section, stress-induced AOS accumulation triggers the GSH level and its biosynthetic pathway (Smith et al. 1984), and thus there are indications of stress-induced regulation of the activity of enzymes of GSH biosynthetic pathway (Ruesegger et al. 1990, Ruesegger and Brunold 1992, Chen and Goldsbrough 1994) in several plant species. The characterization and identification of factors controlling the activities of γ-ECS and GSHS enzymes are of great importance and also contribute to the understanding of how the synthesis of GSH is controlled in plants.

3.1.2 Transport of GSH

Glutathione is found both in the cytosol and the chloroplasts of plant cells (Foyer and Halliwell 1976, Bielawski and Joy 1986, Klapheck et al. 1987), but the precise determination of compartment concentrations is confounded by the possible exchange between the different compartments during subcellular fractationation (Klapheck et al. 1987). Information regarding the transport or movement of GSH between compartments by specific amino acid or peptide transporters is meager. Jamai et al. (1996) performed an experiment to characterize the GSH uptake system in broad bean (*Vicia faba*) protoplast. They reported that GSSG was taken up at about twice the rate of GSH. GSH uptake was inhibited by GSSG and GS conjugates. Conversely, GSSG uptake was inhibited by GSH and GS conjugates. Various amino acids and peptides affected the transport of neither GSH nor GSSG. Altogether, the data suggested that GSH, GSSG, and GS conjugates may be absorbed by a common uptake system that differed from transporters for amino acids and for di- and tripeptides. In addition they performed electrophysiological and pH measurement studies and concluded that proton cotransport may mediate the uptake of glutathione in leaf tissues. Another report regarding GSH transport came from studies in tobacco (Schneider et al. 1992). Although possible GSH transporters have been described in literature (Frommer et al. 1994; Steiner et al., 1994), specific studies have been confined only to plasmalemma (Schneider et al. 1992, Jamai et al. 1996). Several high- and low-affinity transporters have been identified in plants (Frommer et al. 1994, Song et al. 1996). Much information regarding GSH transporters came into light through studies in yeast (*Saccharaomyces cerevisiae*) (Bourbouloux et al. 2000). Researchers grew yeast on growth media containing labeled GSH and performed uptake assays and gene disruption studies and concluded that the YJL212c open reading frame (ORF) encodes the GSH transporter, namely HGT1. Furthermore, it was shown that HGT1 GSH-transporter exhibits high affinity for GSH, GSSG, and the glutathione-N-ethylmaleimide conjugate (GS-NEM) (Miyake et al. 2002).

3.1.2a and 3.1.2b γ-Glu-Cys Synthase/ γ-Glu-Cys Synthetase (γ-ECS, EC 6.3.2.2)

This is the first enzyme that catalyses the ATP-dependent ligation of cysteine and glutamate form c-EC in a reaction mechanism analogous to that catalyzed by GS, except that it is the c-carboxyl group of glutamate which condenses with the a-amino group of cysteine. Increased γ-ECS activities have been measured in several Cd-exposed plants, such as pea (Ruegsegger and Brunold 1992), tobacco, maize (Ruegsegger et al. 1990), and tomato (Chen and Goldsbrough, 1994). In fact the demand for increased GSH synthesis in Cd-exposed plants is accommodated by the activation of γ-ECS activity and in parallel, substantial increases in the activity of key enzymes responsible for S-assimilation (May et al. 1998).

Ample literature is available regarding rate-limiting role of γ-ECS in the GSH biosynthetic pathway (Hell and Bergmann 1990, Kovari et al. 1997). Enzyme γ-ECS, along with GSHS, has been found in both cytosol and plastid of roots and leaves (Hell and Bergmann 1990). It is proposed that the cytosol and plastid of roots and leaves possess the GSH uptake transporters. Most of the information regarding γ-ECS and GSHS-induced potentials to enhance GSH levels came into light through genetic studies. Hell and Bergmann (1988, 90) were the pioneers that first purified γ-ECS and GSHS and found their homogeneity in tobacco. Little is known about the biochemistry of GSH biosynthesis in plants, which is a serious obstacle to conventional cloning procedures. Much information regarding complementary DNAs encoding γ-ECS and GSHS came into light from studies in *Arabidopsis thaliana* (May and Leaver 1994, Rawlins et al. 1995, Ullman et al. 1996, May et al. 1998). High sequence homology of cDNAs encoding γ-ECS has been studied in tomato (Kovari et al. 1997) and *B. juncea* (Schafer et al. 1998). GSH level can be engineered to a certain extent so as improve protection against various biotic and abiotic stresses in plant species. γ-ECS-constructs (encoding cytosolic or plastidic γ-ECS) and GR-constructs, all driven from the 35S-promotor, have yielded transgenic plants (*Arabidopsis*) with moderately increased GSH content (Foyer et al. 1995, Xiang et al. 2001). Mittler et al. (2004) confirmed the fact that a link exists between GSH and other antioxidants and that GSH is indeed a part of a regulated anti-ROS network. Chen et al. (2003) studied increasing vitamin C content of plants through enhanced ascorbate recycling. They found that plants overexpressing DHAR enzyme also exhibited an increased GSH content.

GSH pool has been measured in several plant species exposed to a number of abiotic stresses (May et al. 1996, Anjum 2006), and it is now known that several environmental factors can change GSH pool size. In vitro synthesis of GSH can continue even when its level exceeds the *Ki* value of γ-ECS (Hell and Bergmann 1990). Overexpression of γ-ECS may increase plant tolerance to a number of environmental stresses, including Cd (Kovari et al. 1997). Data from transgenic and mutation studies reveal that γ-ECS gene mutated in *cad2-1* mutant of *Arabidopsis* was reduced by 60% compared to wild-type, while the transformations of *cad2-1* with wild-type γ-ECS gene restored Cd-tolerance, along with restoring the GSH level and the activity of γ-ECS enzyme. This assumption is supported by studies in which cysteine was supplied to leaf discs of poplar plants. Irrespective of the level of activity of γ-ECS and GSHS, incubation with cysteine enhanced foliar GSH contents (Strohm et al. 1995, Noctor et al. 1996). This observation has been explained by cellular cysteine concentrations being low compared with K_M-values of γ-ECS for cysteine or close to these values (Bergmann and Rennenberg 1993). Higher total GSH levels have also been achieved through overexpression of γ-glutamylcysteine synthetase (γ-ECS, Noctor et al. 1996, Zhu et al. 1999b) or glutathione synthetase (GSHS) (Zhu et al. 1999a).

The regulation and compartmentalization of GSH synthesis may orchestrate plant defense in varied environmental conditions. Wachter et al. (2005) studied the localization of γ-ECS enzyme in *Arabidopsis thaliana* and revealed that this enzyme is confined exclusively to plastids, while another enzyme, GSHS, is confined to cytosol, as evidenced by the presence of GSHS-transcripts encoding cytosolic proteins. It may be concluded from above data that γ-ECS (plastid-derived) is the precursor of the synthesis of GSH in cytosol and that this cytosolic GSH must be transported to other GSH-containing compartments (Wachter et al. 2005). The regulation of γ-ECS enzyme expression has revealed the existence of multiple controls, including transcriptional activation in response to various biotic and abiotic stresses, including Cd (Xiang and Oliver 1998, Vanacker et al. 2000, Noctor et al. 2002, Rusch and Wachter 2005), and translational control via a 5′ UTR binding complex (Xiang and Bertrand 2000, Wachter et al. 2005). Jez et al. (2004) observed that γ-ECS activity can also be regulated through posttranslation. As stated above GSH is found both in reduced (GSH) as well in oxidized (GSSG) forms; this is possible via GR activity, and hence it maintains the cellular redox status (Russo et al. 1995, Shaul et al. 1996, Sanchez-Fernandez et al. 1997).

3.1.2c GSH Synthase (GSHS, EC 6.3.2.3)/GS/Glutathione Synthetase

This enzyme catalyzes the ATP-dependent formation of a peptide bond between the a-carboxyl group of cysteine in γ-EC and the a-amino group of glycine to form GSH. The catalytic mechanism involves an acylphosphate intermediate resulting from transfer of the γ -phosphate of ATP to the cysteinyl carboxyl group. In fact the a-amino group of glycine reacts with the acylphosphate group, forming a peptide bond and releasing inorganic phosphate. Information regarding the overexpression of GSH synthetase enzyme in transformants is meager. Much information came from the studies on poplar transformants (Foyer et al. 1995, Strohm et al. 1995, Noctor et al. 1998a, b). Foyer et al. (1995) succeeded in overexpressing GSHS enzyme in the cytosol of poplar transformants relative to untransformed poplars; they obtained a 300-fold enhancement in the activities of extractable foliar GSH. However, no significant effect on foliar thiol content was noticed (Foyer et al. 1995, Strohm et al. 1995). In an exogenous supply of γ-EC, Strohm et al. (1995) observed a higher rate of GSH synthesis in leaf discs isolated from these poplars. Arisi et al. (1997), through immunoblotting of leaf extracts of transformed poplar, confirmed the presence of a polypeptide of approximately 35.6 kDa. In model plant *A. thaliana*, GSHS has been encoded by GSH2 gene (At5g27380). Despite the presence of a unique gene the GSHS activity was associated with plastids and cytosol of plant cells (Hell and Bergman 1990).

3.2 Regulation of Level of GSH under Cd Stress

GSH and its metabolism play an important role in plant responses and adaptation to various natural stress conditions. Apart from metal chelation, GSH is involved in multiple metabolic roles such as intracellular redox state regulation, scavenging of reactive oxygen species, transport of GSH-conjugated amino acids, storage of sulfur, and other functions related to the cell cycle, plant growth, and cell death (Noctor et al. 1998a, b). The concentration and redox state of intracellular glutathione pools depends on the complex interplay of numerous factors. Glutathione redox state is remarkably constant, but stress-induced oxidative stress leads to the oxidation of the pool (Noctor et al. 2002). A decrease in GSH level in Cd-exposed moongbean and rapeseed plants has been reported (Anjum 2006). The level of GSH depends upon the availability of the substrates Cys and, to some extent, glycine. Hence, it is noteworthy to observe the regulation of these amino acids in GSH biosynthesis and also in metal detoxification mechanisms. Several authors have reported a significant decrease in GSH levels in seedlings and cell cultures of different plants treated with varying heavy metals including Cd (Scheller et al. 1987, Tukendorf and Rauser1990, Di Baccio et al. 2005, Le Faucheur et al. 2005, Wojcik et al. 2005, Lima et al. 2006, Thangavel et al. 2007). Thangavel et al. (2007) noted a decreasing trend in GSH with all concentrations of Cd. Howden et al. (1995) reported that intracellular GSH acts as a “first line defense” against Cd toxicity in *A. thaliana.* A possible reason for the decrease in GSH in red spruce cells may be that initially GSH acts as a first line of defense against Cd toxicity by complexing the internal metal ions before the induction of PC synthesis becomes effective. GSH is also actively involved in secondary defensive mechanisms such as antioxidants, by scavenging free radicals in Cd induced oxidative stress in *P. abies* (Schroder et al. 2003) and other plants such as *T. aestivum* L. (Ranieri et al. 2005) and *Helianthus annus* L. (Gallego et al. 2005). Anjum (2006) observed significant increases in GSH level on sulfur supply to Cd-exposed moongbean and rapeseed plants. Contrarily, a more detailed study with canola roots indicated that interactions between GSH synthesis and sulfate assimilation differ from those between GSH synthesis and oxidative stress (Lappartient and Touraine 1996). However, unlike studies in which glutathione was shown to accumulate (Smith et al. 1984, Smith 1985, Willekens et al. 1997), exposure of roots to H_2O_2 caused GSH levels to decrease (Lappartient and Touraine 1997).

3.3 Ascorbic Acid (AsA) and Glutathione (GSH) Pools–Interaction and Stress Tolerance

Limited literature is available regarding the importance of AsA and GSH pools size in stress tolerance in various plant species (Sanchez-Fernandez et al. 1997, May et al. 1998). Sanchez-Fernandez et al. (1997) were able to show that a reduction in the meristematic activity of *Arabidopsis* roots caused by depletion of GSH can be alleviated by an exogenous supply of AsA. However, Grant et al. (1996) showed that an exogenous supply of AsA cannot rescue GSH-deficient yeast. In fact, for the proper functioning of the ascorbate-glutathione cycle (AGC), there are three independent redox couples namely, AsA/DHA, GSH/GSSG, and NADPH/NADP. There are indications that these redox couples may act as sensors and effectors, thus providing a direct link between environmental stress and morphological adaptations of plants through alterations in the patterns of cell division in the primary root apical meristem (Kerk and Feldman 1995, Sanchez-Fernandez et al. 1997). AsA has been implicated in the regulation of cell cycle progression from G_1 to S phase (Smirnoff 1996, Liso et al. 1988) and cell elongation (De Tullio et al. 1999). In addition, AsA and GSH levels may be necessary for G_1/S transition in the cell cycle. The G_1 phase cell cycle is capable of responding to extracellular stimuli, which determines G_1/S transition and whether the cell will enter quiescence, differentiation, or death (May et al. 1998). The cells of quiescent center (QC) have been shown to be stimulated by the levels of AsA and GSH. Kerk and Feldman (1995) reported a notable depletion of AsA in the QC of *Zea mays* and that AsA could stimulate cells of QC to enter into the cell cycle. It was shown in another study that, while high levels of GSH were associated with the epidermal and cortical initials, GSH levels were found markedly lower in the cells of the QC, which have an extended G_1(May et al. 1998). In addition, there are reports stating that AsA regeneration can occur independently of GSH and that an AsA/DHA redox couple may function in the absence of GSH pool in *cad2* in *Arabidopsis* (May et al. 1998). Further, GSH independent DHAR, nonenzymatic reduction of DHA, direct reduction of MDHA by MDHAR, or direct regeneration of AsA by electrons from PS I support the argument that AsA-regeneration can be independent of GSH (Asada 1994, May et al. 1998). The characterization of the AsA biosynthetic pathway in plants would certainly help resolve the regulatory and functional interactions between the pools of AsA and GSH (May et al. 1998).

3.4 Phytochelatins

Plants respond to metal toxicity by initiating a wide range of cellular defense mechanisms. These include immobilization, exclusion, and compartmentalization of metals and synthesis of phytochelatins [PCs, (γ-GluCys)n-Gly], stress proteins, and ethylene (Sanita' di Toppi and Gabbrielli 1999). Information available in the literature mostly concerns chelating mechanisms. Potential metal-binding ligands include amino acids and organic acids and two classes of cysteine-rich peptides, the PCs and the metallothioneins (MTs). MTs are small, gene-encoded, cysteine-rich polypeptides and the PCs, which, in contrast, are enzymatically synthesized, cysteine-rich peptides (Rauser 1999, Cobbett 2000a,b, Clemens 2001). Mendoza-Cozatl et al. (2005) showed that the enhanced synthesis of PCs and their sulfur-containing metabolic precursors, a reduced form of glutathione (GSH), γ-glutamylcysteine (γ-EC), cysteine (Cys), and sulfide, are mainly involved against Cd stress from yeast to plants.

Further observations also reveal that genes in both the sulfur assimilation and GSH biosynthetic pathways are regulated in response to Cd exposure. Studies in *Brassica juncea* (Schafer et al. 1998, Lee and Leustek 1999) and in *Arabidopsis* (Xiang and Oliver 1998) have demonstrated the coordinated transcriptional regulation of genes involved in sulfur transport and assimilation and in GSH biosynthesis and that the Cd exposure consequently induces PC synthesis from GSH.

PCs form a family of structures with increasing repetitions of the γ-EC (γ-Glu-Cys) dipeptide followed by a terminal Gly; $(\gamma\text{-Glu-Cys})_n$-Gly. The number of c-EC moiety varies (n = 2–11) depending upon the PC derivatives. In addition, a number of structural variants, for example, $(\gamma\text{-Glu-Cys})_n$-βAla, $(\gamma\text{-Glu-Cys})_n$-Ser, and $(\gamma\text{-Glu-Cys})_n$-Glu, known also as *iso-PCs*, have been identified in some plant species (Rauser 1995, 1999, Zenk 1996, Cobbett 2000a,b). The phytochelatins (PCs) are synthesized nontranslationally using glutathione as a substrate by phytochelatin synthase (PCS) or γ-EC dipeptidyl transpeptidase (EC 2.3.2.15), a constitutive cytoplasmic enzyme that is activated upon exposure to several heavy metals including Cd (Grill et al. 1989, Rauser 1995, Cobbett 2000a,b, Cobbett and Goldsbrough 2002). The regulation of PCS activity is the key factor in PC synthesis. The genes for PCS have now been identified in *Arabidopsis* and yeast (Clemens et al. 1999, Ha et al. 1999, Vatamaniuk et al. 1999). Howden et al. (1995) isolated a series of Cd-sensitive mutants of *Arabidopsis* that varied in their ability to accumulate PCs; the amount of PCs accumulated by the mutants correlated with the degree of sensitivity to Cd (Howden et al. 1995). Using *Brassica juncea*, it has been shown that Cd accumulation is accompanied by a rapid induction of PC biosynthesis and that the PC content was theoretically sufficient to chelate all Cd taken up. The cultured cells of azuki beans that were Cd hypersensitive were shown to lack in PCS activity (Inouhe et al. 2000). Using *Arabidopsis*, Xiang and Oliver (1998) showed that treatment with Cd and Cu resulted in increased transcription of the genes for glutathione synthesis, and the response was specific for those metals thought to be detoxified by PCs. Interestingly, jasmonic acid (JA) treatment activated the same set of genes, although jasmonic acid production was not stimulated by heavy metals in plant cell cultures (Blechert et al. 1995). Zhu et al. (1999a,b) overexpressed the γ-glutamylcysteine synthetase gene from *E. coli* in *Brassica juncea*, resulting in increased biosynthesis of glutathione and PCs and an increased tolerance to Cd. A similar approach was taken with *Arabidopsis*; γ-glutamylcysteine synthetase was expressed in both sense and antisense orientations, resulting in plants with a wide range of glutathione levels (Xiang et al. 2001). Plants with low glutathione levels were hypersensitive to Cd, although elevating the levels above wild-type did not increase metal resistance. Recently, Mendoza-Cozatl and Moreno-Sanchez (2006) proposed a kinetic modeling pathway for GSH and PC synthesis in plants. They demonstrated in unstressed conditions that the rate-limiting step in GSH synthesis is controlled by γ-glutamylcysteine synthetase (γ-ECS), while under Cd stress at least two enzymes (γ-ECS and PCS) play a major role in increasing PC synthesis and/or alternatively diminishing the GSH-S-transferases to maintain the GSH demand. The type and amount of PCs synthesized in response to Cd depends on the concentration of Cd. Thangavel et al. (2007) reported higher PC2 levels (109%–270%) in Cd-exposed red spruce cells compared to control and suggested that tolerance to higher Cd toxicity levels can be correlated with an elevated PC synthesis. Similar results have been reported in *Betula pendula* (Gussarsson 1994), *Datura innoxia* (Delhaize et al. 1989), *Zea mays* (Tukendorf and Rauser 1990), *Pisum sativum* (Lima et al. 2006), and *Thlaspi caerulescens* (Wojcik et al., 2006) plants exposed to Cd concentrations.

The synthesis of PC has been known in the whole plant kingdom (Gekeler et al. 1989). In fact plants tested so far for PC synthesis, including the majority of algae and several fungi, are capable of PC synthesis (Mehra et al. 1988, Gekeler et al. 1989, Kneer et al. 1992). Cd stands second in terms of its potential to induce PC synthesis after Hg (Ernst 1996). PCs are produced from GSH, hGSH, hydrooxymethyl glutathione, or γ-glutamylcysteine by a transpeptidase, the constitutive PC synthase enzyme (Grill et al. 1989, Chen and Huerta 1997).

PC synthase (PCS), the enzyme responsible for synthesis of PCs, requires a metal–glutathione (GSH) complex as one of its substrates (Vatamaniuk et al. 1999). PCs are therefore synthesized when the cytoplasmic concentration of heavy metal ions is able to provide sufficient levels of this substrate. In some cases, PCs can be detected in plant tissues prior to an increase in metal concentration, and PC synthesis increases with exposure to higher levels of cadmium (Cobbett and Goldsbrough 2002). High correlations between shoot PC concentrations and the degree of Cd-induced growth inhibition in maize and wheat have been reported by Keltjens and van Beusichem (1998). Therefore, analysis of PCs in plants could be a useful biochemical indicator of Cd stress.

Phytochelatin synthase is a constitutive enzyme and requires posttranslational activation by heavy metals (Grill et al. 1989, De Knecht et al. 1995, Klapheck et al. 1995). This enzyme catalyzes the conversion of GSH to PCs and has been characterized as a specific γ-glutamyl cystein dipeptidyl transpeptidase (EC. 2.3.2.15) (Vatamaniuk et al. 2000, 2001). Detailed information concerning PC synthase enzyme is now largely known through genetic studies in various plant species. Howden et al. (1995) were among the pioneers who identified PC synthase gene *(cad1)* in *Arabidopsis thaliana*. They isolated the PC synthase gene from Cd-sensitive and PC-deficient mutants. Some of the important PC synthase genes have been noted in the Table 13.1, along with other details.

It has been shown that above mentioned genes [AtPCS1 (CAD1), SpPCS and TaPCS1] encode 40%-50% sequences similar to the 50-55 kDa polpeptides active in the synthesis of PC from GSH (Cobbett 2000a,b, Takagi et al. 2002). AtPCS1, a gene isolated from *Arabidopsis*, encodes an important protein. It has been revealed that this protein shows heterologous expression in *S. cerevisae* and confers heavy-metal tolerance by promoting the Cd-dependent accumulation of PCs. In fact AtPCS1 has been shown to encode a single polypeptide species that, when purified to homogeneity, was found sufficient for Cd activated PC synthesis from GSH (Cobbett et al. 1998).

3.4.1 PC-Assisted Cd Detoxification

A large number of studies have demonstrated the critical role of PCs in Cd detoxification and tolerance of plants to Cd (Howden et al. 1995, Inouhe et al. 2000). In contrast, studies with plant species that exhibit unusual hypertolerance to Cd, such as *Silene vulgaris* and *Thlaspi caerulescens*, indicate that PCs are not responsible

Table 13.1 Genes involved in PC biosynthetic pathway

Organism	PC biosynthesis genes	Activity	Reference
Arabidopsis thaliana	CAD1 (AtPCS1)	PC synthase	Howden et al. 1995
		PC biosynthesis	Clemens et al. 1999
Arabidopsis thaliana	CAD2 (AtPCS2)	GCS/GSH biosynthesis	Howden et al. 1995, Vatamaniuk et al. 1999
Saccharomyces prombe	Gsh1	GCS/GSH biosynthesis	Clemens et al. 1999, Ha et al. 1999
Saccharomyces prombe	Gsh2	GCS/GSH biosynthesis	Clemens et al. 1999, Ha et al. 1999
Saccharomyces prombe	PCS1 (SpPCS1)	PC synthase/PC biosynthesis	Clemens et al. 1999, Ha et al. 1999
Triticum aestivum	TaPCS1	PC synthase/PC biosynthesis	Clemens et al. 1999
Arabidopsis thaliana	CAD2/RML1	γ-glu-cys-synthatase/ GSH biosynthesis	Cobbett, 2000a b, Vernoux et al. 2001
Candida elegans	Pcs1	PC synthase/PC biosynthesis	Vatamaniuk et al. 2000
Saccharomyces prombe	Gsh1	γ-glu-cys-synthatase/ GSH biosynthesis glutathioine synthetase/GSH biosynthesis	Glaeser et al. 1991, Mutoh and Hayashi 1988, Glaeser et al. 1991
Saccharomyces prombe	Gsh2		Mutoh and Hayashi 1988

for the observed metal tolerance phenotypes (De Knecht et al. 1992, Ebbs et al. 2002). However, studies with PC-deficient, *cad1-3* mutant of *Arabidopsis* and the PC synthase targeted deletion mutant of *S. pombe* revealed that PCs are essential for the detoxification of a number of heavy metals, including Cd (Ha et al. 1999). Howden et al. (1995) demonstrated in experiments with CAD1 mutants of *A. thaliana* that PCs can inactivate various heavy metals. It was found that these CAD1 mutants, which are unable to synthesize PCs, exhibited extreme sensitivity to Cd (Piechalak et al. 2002). As indicated by the hypersensitivity of PC-deficient *Arabidopsis* CAD1 mutants to Cd (Howden et al. 1995), PC synthase genes conribute most markedly to Cd detoxification *in planta* (Cobbett et al. 1998, Vatamaniuk et al. 1999). Several studies were completed taking *A. thaliana*, PC-deficient mutant CAD1-3, and wild-type plants to test the potential effect of prior exposure to different Cd concentrations on Cd uptake and accumulation. These results suggest that the PC-deficient mutant CAD1-3 accumulates less Cd than the wild-type (Larsson et al. 2002). The possibility that the differences in Cd accumulation in mutant and wild-type lines may be due to the cytosolic Cd regulation, which is inhibited by the complexation of Cd by PCs (Speiser et al. 1992). In some plants and in the yeasts *S. pombe* and *Candida glabrata*, sulfide ions play an important role in the efficacy of Cd detoxification by PCs. HMW PC-Cd complexes contain both Cd and acid-labile sulfide. The analysis of Cd-sensitive mutants of *S. pombe* deficient in PC-Cd complexes has provided evidence for the importance of sulfide in the function of PCs (Speiser et al. 1992, Juang et al. 1993). Juang et al. (1993) have shown that cysteine sulfinate could also be utilized to form different S-containing compounds, which may be intermediates or carriers in the pathway of sulfide incorporation into HMW complexes (Fig. 13. 2). However, this is only true with Cd ions; whether sulfide is involved in the detoxification of other metal ions by PCs is unknown (Cobbett and Goldsbrough 2002).

The mechanism of Cd detoxification is a complex process. Exposure to Cd stimulates the synthesis of PCs (Srivastava et al. 2004), which rapidly form a low molecular weight (LMW) complex with Cd (Fig. 13. 2) (Sanita di Toppi 1999). These complexes acquire acid labile sulfur (S^{2-}) at the tonoplast, and form a high molecular weight (HMW) complex with a higher affinity for Cd ions (Hu et al. 2001). Free Cd ions can also enter the vacuole by means of a $Cd^{2+}/2H^+$ antiport system (Gries and Wagner 1998). Because of the acidic pH of the vacuole, these complexes dissociate, and Cd can be complexed by vacuolar organic acids like oxalate, citrate, and malate (Memon et al. 2001) and possibly by amino acids. In fact, incorporation of labile S in PC-Cd complexes was observed in *Silene vulgaris* and *B. juncea* and could be, therefore, of importance in the general Cd detoxification mechanism (Speiser et al. 1992, Verkleij et al. 2003). On the other hand, formation of GSH-PC complex may provide protection from Cd-induced damage(s) within the cytosol (Howden et al. 1995). GSH was able to complex all Cd present in Cd-treated leaves of tobacco plants (Vogeli-Lange and Wagner 1996). Two sulphydryl (SH) groups from GSH and/or PCs should complex one Cd^{2+} ion. However, the formation of Cd-PC complexes and their transportation to vacuoles is certainly an important tool in Cd-detoxifcation and hence may allow recycling of

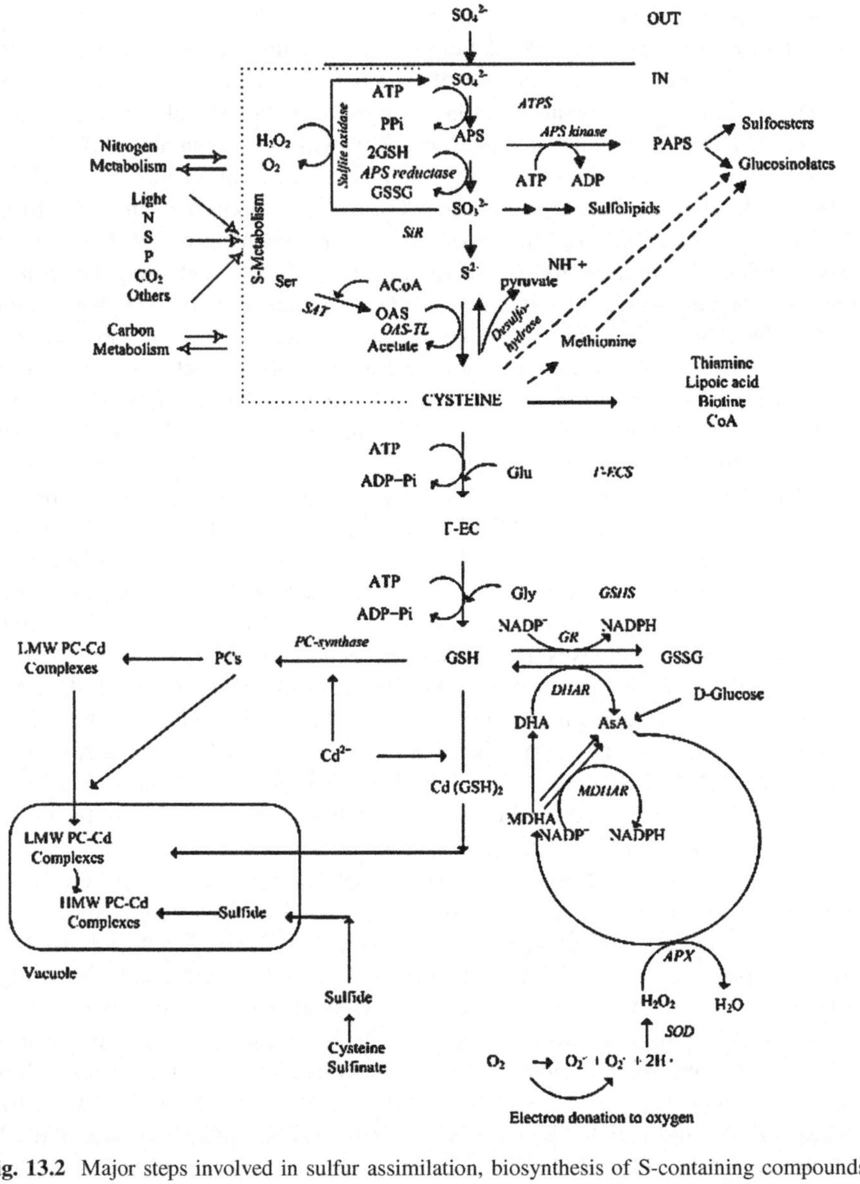

Fig. 13.2 Major steps involved in sulfur assimilation, biosynthesis of S-containing compounds and their involvement in detoxification of superoxide radicals and Cd^{2+}, and the steps involved in the compartmentalization of Cd^{2+} in plants. The drawing also shows the interconnection and dependence of plant metabolism on the external supply of various factors *viz*., nutrients, light, and water; the scheme also depicts factors affecting S-assimilation and interrelation with nitrogen and carbon metabolism in plants. *Summary*: External sulfate is taken up through members of a multigene family of sulfate transporters; activation of sulfate of shoot takes place in leaves by ATP via ATPS. The product APS is reduced by APS-reductase, using GSH as reductant. Alternatively, APS is further activated by APS-kinase to form PAPS, which is required for various sulfation reactions, including the biosynthesis of glucosinolates. Excess sulfite is converted in to back to

some PCs as well as the accumulation of Cd in tissues without a concomitant increase in PC concentrations (Maier et al. 2003).

4 Critical Appraisal and Future Prospects

As sulfur constitutes one of the macronutrients necessary for the plant life cycle, the two processes the uptake and assimilation of sulfur in higher plants are the crucial factors determining plant growth and vigor, crop yield, and resistance to biotic and abiotic stresses. The inorganic SO_4^{2-} assimilated and reduced by plants appears ultimately in cysteine and methionine, which constitute about 80%-90% of the total sulfur in most plants. Nearly all of cysteine and methionine are in protein. In the nonprotein fraction, reduced glutathione (GSH) is ubiquitous and is commonly a major plant constituent of storage S. In coenzyme A (CoA), the active site of the molecule is the SH-group. The SH-group reacts with organic acids and becomes esterified with an acyl group of an organic acid. Coenzyme A thus serves as a carrier of acyl groups. Acetyl CoA is formed when CoA reacts with acetic acid. This is an important example of an activated acid of this type and plays a very significant role in fatty acid and lipid metabolism. Some other sulfur-containing compounds are cysteine, cystathione, homocysteine and methionine, S-methyl cysteine and S-methyl methionine, sulfate esters, and sulpholipids. Sulfur is also present in certain other constituents in the members of cruciferae family, which give these plants their characteristic taste and smell. These are glucosides and isothiocyanic esters containing both S and N. Sulfate assimilation is an essential pathway being a source of reduced sulfur for various cellular processes and for the synthesis of glutathione, a major factor in plant stress-defense. Reports are available on the metabolic control of sulfate uptake and the mediation of further steps by Cys and GHSH and/or low levels of N. In addition many reports have shown that sulfate assimilation is well coordinated with the assimilation of nitrate and carbon. Deficiency of nitrate has been shown to affect the sulfate assimilatory steps in plants. During nitrate deficiency, sulfate assimilation is reduced, and the capacity to reduce nitrate is diminished in plants starved for sulfate. The availability of Cys is a crucial factor in GSH synthesis, but an adequate

Fig. 13.2 (continued) sulfate by enzyme sulfite oxidase using O_2 as an electron acceptor, releasing H_2O_2. Sulfite is reduced by SiR to sulfide. Ser, which comes from N-metabolism, is added with Acetyl CoA via enzyme SAT. Furthermore, sulfide is incorporated to OAS via OAS-TL to form cysteine (Cys). Cys is the primary product of S-assimilation, and acts as SH donor for glucosinolates, GSH, etc. Sulfide can be released from cysteine via the action of desulfohydrases. After being synthesized from Glu-Cys-Gly in two ATP-dependent reactions catalyzed by γ-ECS and GSHS, glutathione (GSH) acts as precursor of phytochelatins (PCs), a reaction catalyzed by PC-synthase. Furthermore, GSH acts as a key role player in ascorbate-glutathione cycle (AGC). AGC is a major ROS-scavenging antioxidant pathway that operates both in chloroplasts and cytosol, where cycling of redox molecules (AsA and GSH) takes place with effective utilization of AGC-enzymes (DHAR, MDHAR, APX and GR)

supply of glutamate and glycine are also of much importance. It has also been proposed that various environmental and nutritional conditions and compounds act as molecular signals in the regulatory processes of both GSH synthesis and sulfur assimilation in plants. It has been revealed through transcriptome analyses that interactions of S, N, and C metabolism are very complex. A new perspective in this regard must be pointed out so as to dissect the mechanism(s) of regulation of sulfate assimilation at the molecular level. The information regarding signaling molecules and pathways is still meager. Although much has been achieved in finding the majority of transcription factor genes responding to sulfur deficiency in plants, identification of some other trans elements, molecular signals, and sulfur-responsive elements like β–conglycinin promoter, would certainly expand our knowledge. Furthermore, sulfate uptake and reduction and Cys biosynthesis are highly regulated in plants at the gene expression and enzyme activity levels; in addition these processes are also regulated by other factors, such as subcellular localization of enzymes and developmental and spatial activity patterns and differences between plant species. The validation and verification of the respective regulatory steps should be metabolically engineered with genetic tools such as forward and reverse genetic processes so as to exploit the results in order to manipulate the fluxes of metabolites in plants. Molecular approaches should be narrowed to manipulate sulfate assimilatory enzymes such as ATP-sulfyrylase, APS-reductase, and sulfite reductase. Strategies should also aim to manipulate steps of pathways leading to the production of thiols and their products in plants through overexpressing Ser acetyl transferase (SAT), γ-glutamyl-cysteine synthase, and GSHS enzymes in respect to stress physiology. We have much information on PC biosynthesis, but through advanced molecular genetic approaches we must keep in mind the numerous other aspects of PC biosynthesis and function and the ways in which they, too, are regulated at a cellular and physiological level in response to heavy-metal exposure. These include aspects of sulfur assimilation, GSH and sulphide biosynthesis, PC compartmentalization, and the signal pathways through which metal toxicity leads to gene regulation. Glutathione, one of the reduced S-compounds, is one promising research area. Plants overproducing GSH may prove useful for the purposes of ecodetoxification. Similarly, plants with enhanced capacities for synthesis of PCs due to increased rates of GSH formation could be applied to bioremediate poor soils through the removal of heavy metals. However, the sulfur assimilatory pathway and the role of sulfur and S-containing proteins in plant-cell metabolism are still to be studied in detail and much remains to be discovered.

References

Ahmad A, Khan I, Anjum NA, Diva I, Abdin MZ, Iqbal M (2005) Effect of timing of sulfur fertilizer application on growth and yield of rapeseed. J Plant Nutr 28:1049-59

Anjum NA (2006) Effect of abiotic stresses on growth and yield of *Brassica campestris* L. and [*Vigna radiata* (L.) Wilczek] under different sulfur regimes. Ph. D. thesis. Jamia Hamdard, New Delhi

Archer MJ (1974) A sand culture experiment to compare the effects of sulphur on five wheat cultivars (*Triticum aestivum* L.). Aust J Agric Res 25:369-80

Arisi AC, Noctor G, Foyer CH, Jouanin L (1997) Modification of thiol contents in poplars (*Populus tremulaxP. alba*) overexpressing enzymes involved in glutathione synthesis. Planta 203:362-72

Asada K (1994) Production and action of active oxygen in photosynthetic tissue. In: Foyer CH, Mullineaux PM (eds) Causes of photooxidative stress and amelioration of defense system in plants, CRC Press, Boca Raton, pp. 77-104

Aulakh MS, Chhibba IM (1992) Sulphur in soils and responses of crops to its application in Punjab. Fert News 37:33-45

Aulakh MS, Pasricha NS (1998) The effect of green manuring and fertilizer N application on enhancing crop productivity in mustard-rice rotation in semiarid subtropical regions. Euro J Agron 8:51-8

Bergmann L, Rennenberg H (1993) Glutathione metabolism in plants. In: De Kok LJ, Stulen I, Rennenberg H, Brunold C, Rauser WE (eds) Sulphur nutrition and assimilation in higher plants: regulatory, agricultural and environmental aspects, SPB Academic Publishing, The Hague, pp. 102-23

Berkowitz O, Wirtz M, Wolf A, Kuhlmann J, Hell R (2002) Use of biomolecular interaction analysis to elucidate the regulatory mechanism of the cysteine synthase complex from *Arabidopsis thaliana*. J Biol Chem 277:30629-34

Bick JA, Åslund F, Chen Y, Leustek T (1998) Glutaredoxin function for the carboxyl terminal domain of the plant-type 5′-adenylylsulfate (APS) reductase. Proc Natl Acad Sci USA 95:8404-9

Bielawski W, Joy KW (1986) Reduced and oxidized glutathione and glutathione-reductase activity in tissues of *Pisum sativum*. Planta 169:267-72

Blechert S, Brodschelm W, Hölder S, Kammerer L, Kutchan TM, Mueller MJ, Xia Z-Q, Zenk MH (1995) The octadecanoic pathway: signal molecules for the regulation of secondary pathways. Proc Natl Acad Sci USA 92:4099-4105

Bogdanova N, Bork C, Hell R (1995) Cysteine biosynthesis in plants: isolation and functional identification of a cDNA encoding a serine acetyltransferase from *Arabidopsis thaliana*. FEBS Lett 358:43-7

Bogdanova N, Hell R (1997) Cysteine synthesis in plants: protein-protein interactions of serine acetyltransferase from *Arabidopsis thaliana*. Plant J 11:251-62

Bork C, Schwenn JD, Hell R (1998) Isolation and characterization of a gene for assimilatory sulfite reductase from *Arabidopsis thaliana*. Gene 212:147-53

Bourbouloux A, Shahi P, Chakladar A, Delrot S, Bachhawat AK (2000) Hgt1p, a high affinity glutathione transporter from the yeast *Saccharomyces cerevisiae*. J Biol Chem 275:13259-65

Brunold C (1990) Regulation of sulfate sulfide. In: Renenberg H, Brunold C, De Kok LJ, Stulen I (eds) Sulfur nutrition and sulfur assimilation in higher plants: fundamental, environmental and agricultural aspects, SPB Academic Publishing, The Hague, pp. 13-31

Brunold C (1993) Regulatory interactions between sulfate and nitrate assimilation. In: De Kok LJ, Stulen I, Renenberg H, Brunold C, Rauser W (eds) Sulfur nutrition and sulfur assimilation in higher plants: regulatory, agricultural and environmental aspects, SPB Academic Publishing, The Hague, pp. 125-38

Brunold C, Rennenberg H (1997) Regulation of sulfur metabolism in plants: first molecular approaches. Prog Bot 58:164-86

Brunold C, Suter M (1989) Localization of enzymes of assimilatory sulfate reduction in pea roots. Planta 179:228-34

Brunold C, Von Ballmoss P, Hesse H, Fell D, Kopriva S (2003) Interactions between sulfur, nitrogen and carbon metabolism. In: Davidian J-C, Grill D, De Kok LJ, Stulen I, Hawkesford MJ, Schnug E, Rennenberg H (eds) Sulfur transport and assimilation in plants: regulation, interaction and signaling, Backhuys Publishers, Leiden, pp. 45-56

Buchanan BB (1980) The role of light in the regulation of chloroplast enzymes. Annu Rev Plant Physiol 31:341-74

Buchanan BB (1991) Regulation of CO_2 assimilation in oxygenic photosynthesis: the ferredoxin/thioredoxin system. Perspectives on its discovery, present status, and future development. Arch Biochem Biophys 288:1-9

Chen J, Goldsbrough PB (1994) Increased activity of γ-glutamylcysteine synthetase in tomato cells selected for cadmium tolerance. Plant Physiol 106:233-9

Chen Y, Huerta AJ (1997) Effects of sulfur nutrition on photosynthesis in cadmium–treated barley seedlings. J Plant Nutr 20:845-56

Chen Z, Young TE, Ling J, Chang S, Gallie D R (2003) Increasing vitamin C content of plants through enhanced ascorbate recycling. Proc Natl Acad Sci USA 100:3525-30

Chien HF, Lin CC, Wang JW, Chen CT, Kao CH (2002) Changes in ammonium ion content and glutamine synthetase activity in rice leaves caused by excess cadmium are a consequence of oxidative damage. Plant Growth Regul 36:41-7

Clemens S (2001) Molecular mechanisms of plant metal tolerance and homeostasis. Planta 212:475-86

Clemens S, Kim EJ, Neumann D, Schroeder JI (1999) Tolerance to toxic metals by a gene family of phytochelatin synthases from plants and yeast. EMBO J 18:3325-33

Cobbett CS (2000a) Phytochelatins and their roles in heavy metal detoxification. Plant Physiol 123:825-32

Cobbett CS (2000b) Phytochelatin biosynthesis and function in heavy metal detoxification. Curr Opin Plant Biol 3:211-16

Cobbett CS (2003) Metalothioneins and phytochelatins: Molecular aspects. In: Abrol YP, Ahmad A (eds) Sulphur in plants, Kluwer Academic Publishers, The Netherlands, pp. 177-88

Cobbett CS, Goldsbrough P (2002) Phytochelatins and metallothioneins: roles in heavy metal detoxification and homeostasis. Annu Rev Plant Biol 53:159-82

Cobbett CS, May MJ, Howden R, Rolls B (1998) The glutathione-deficient, cadmiumsensitive mutant, *cad2-1*, of *Arabidopsis thaliana* is deficient in γ-glutamylcysteine synthetase. Plant J 16:73-8

De Knecht JA, Koevoets PLM, Verkleij JAC, Ernst WHO (1992) Evidence against a role for phytochelatins in naturally selected increased cadmium tolerance in *Silene vulgaris* (Moench) Garcke. New Phytol 122:681-8

De Knecht JA, Van Baren N, ten Bookum WM, Wong Fong Sang HW, Koevoets PLM, Schat H, Verkleij JAC (1995) Synthesis and degradation of phytochelatins in cadmium-sensitive and cadmium-tolerant *Silene vulgaris*. Plant Sci 106:9-18

De Kok LJ, Castro A, Durenkamp M, Kralewska A, Posthumus FS, Elisabeth, Stuiver E, Yang L, Stulen I (2005) Pathways of plant sulphur uptake and metabolism-an overview. Landbauforschung Volkenrode 283:5-13

De Tullio MC, Paciolla C, Dalla Vecchia F, Rascio N, D'Emerico S, De Gara L, Liso R, Arrigoni O (1999) Changes in onion root development induced by the inhibition of peptidyl-pronyl hydroxylase and influence of the ascorbtae system on cell division and elongation. Planta 209:424-34

Delhaize E, Jackson PJ, Lujan LD, Robinson NJ (1989) Poly (γ-glutamylcysteinyl) glycine synthesis in *Datura innoxia* and binding with cadmium. Plant Physiol 89:700-6

Di Baccio D, Kopriva S, Sebastiani L, Rennenberg H (2005) Does glutathione metabolism have a role in the defence of poplar against zinc excess? New Phytol 167:73-80

Droux M (2003) Plant serine acetyltransferase: new insights for regulation of sulfur metabolism in plant cells. Plant Physiol Biochem 41:619-27

Droux M (2004) Sulfur assimilation and the role of sulphur in plant metabolism: a survey. Photosynth Res 79:331-48

Droux M, Ruffet M-L, Douce R, Job D (1998) Interactions between serine acetyltransferase and O-acetylserine (thiol) lyase in higher plants: structural and kinetic properties of the free and bound enzymes. Eur J Biochem 225:235-45

Duke SH, Reisenauer HM (1986) Roles and requirements of sulfur in plant nutrition. In: Tabatabai MA (ed) Sulfur in agriculture, Agron Mongor, ASA, CSSA, SSA, Madison, WI, pp. 123-68

Ebbs S, Lau I, Ahner B, Kochian L (2002) Phytochelatin synthesis is not responsible for Cd tolerance in the Zn/Cd hyperaccumulator *Thlaspi caerulescens* (J & C Presl.). Planta 214:635-40

Ernst WHO (1996) Schermetalle. In: Brunold CH, Ruegsegger A, Braendle R (eds) Stress bei pflanzen, Verlag Paul Haupt, Berlin, pp. 191-219

Ernst WHO (1998) Sulfur metabolism in higher plants: Potential for phytoremediation. Biodegradation 9:311-18

Farago S, Brunold C (1994) Regulation of thiol contents in maize roots by intermediates and effectors of glutathione synthesis. J Plant Physiol 144:433-7

Foyer CH, Halliwell B (1976) The presence of glutathione and glutathione reductase in chloroplasts: a proposed role in ascorbic acid metabolism. Planta 133:21-5

Foyer CH, Noctor G (2000) Oxygen processing in photosynthesis: regulation and signaling. New Phytol 146:359-88

Foyer CH, Souriau N, Perret S, Lelandais M, Kunert KJ, Pruvost C, Jouanin L (1995) Overexpression of glutathione reductase but not glutathione synthetase leads to increases in antioxidant capacity and resistance to photoinhibition in poplar trees. Plant Physiol 109:1047-57

Foyer CH, Theodoulou FL, Delrot S (2001) The functions of inter- and intracellular glutathione transport systems in plants. Trends Plant Sci 6:486-92

Frommer WB, Hummel S, Rentsch D (1994) Cloning of an Arabidopsis transporting protein related to nitrate and peptide transporters. FEBS letters 347:185-9

Gekeler W, Grill E, Winnacker EL, Zenk MH (1989) Survey of the plant kingdom for the ability to bind heavy metals through phytochelatins. Z Naturforsch 44:361-9

Gallego M, Kogan MJ, Azpilicueta CE, Pena C, Tomaro ML (2005) Glutathione-mediated antioxidative mechanisms in sunflower (Helianthus annus L.) cells in response to cadmium stress. Plant Growth Regul 46:267-76

George AA, de Lumen BO (1991) A novel methionine-rich protein in soybean seed: Identification, amino acid composition, and N-terminal sequence. J Agric Food Chem 39:224-7

Gerwick BC, Ku SB, Black CC (1980) Initiation of sulfate activation: a variation in C_4 photosynthesis plants. Science 209:513-15

Glaeser H, Coblenz A, Kruczek R, Ruttke I, Ebert-Jung A, Wolf K (1991) Glutathione metabolism and heavy metal detoxification in *Schizosaccharomyces pombe*. Isolation and characterization of glutathione-deficient, cadmium-sensitive mutants. Curr Genet 19:207-13

Grant CM, MacIver FH, Dawes IW (1996) Glutathione is an essential metabolite required for resistance to oxidative stress in the yeast *Saccharomyces cerevisiae*. Curr Genet 29:511-15

Gratao PL, Gomes-Junior RA, Delite FS, Lea PJ, Azevedo RA (2006) Antioxidant stress responses of plants to cadmium. In: Khan NA, Samiullah (eds) Cadmium toxicity and tolerance in plants, Narosa Publishers, New Delhi, pp. 1-36

Gries GE, Wagner GJ (1998) Association of nickel versus transport of cadmium and calcium in tonoplast vesicles of oat roots. Planta 204:390-6

Grill E, Löffler S, Winnacker E-L, Zenk MH (1989) Phytochelatins, the heavy-metal-binding peptides of plants, are synthesized from glutathione by a specific γ-glutamylcysteine dipeptidyl transpeptidase (phytochelatin synthase). Proc Natl Acad Sci USA 86:6838-42

Gussarsson M (1994) Cadmium induced alterations in nutrient composition and growth of *Betula pendula* seedlings: The significance of fine roots as a primary target for cadmium toxicity. J Plant Nutr 17:2151-63

Gutierrez-Marcos JF, Roberts MA, Campbell EI, Wray JL (1996) Three members of a novel small gene family from *Arabidopsis thaliana* able to complement functionally an *Escherichia coli* mutant defective in PAPS reductase activity encode proteins with a thioredoxin-like domain and "APS reductase" activity. Proc Natl Acad Sci USA 93:13377-82

Ha S-B, Smith AP, Howden R, Dietrich WM, Bugg S, O'Connell MJ, Goldsbrough PB, Cobbett CS (1999) Phytochelatin synthase genes from *Arabidopsis* and the yeast, *Schizosaccharomyces pombe*. Plant Cell 11:1153-64

Hall JL (2002) Cellular mechanisms for heavy metal detoxification and tolerance. J Exp Bot 53:1-11

Halliwell B, Gutteridge JMC (1989) Free radicals in biology and medicine, 2nd edn. Oxford: Clarendon Press

Hartmann T, Mult S, Suter M, Rennenberg H, Herschbach C (2000) Leaf age-dependent differences in sulfur assimilation and allocation in poplar (*Populus tremula*x*P. alba*) leaves. J Exp Bot 51:1077-88

Hawkesford MJ, Wray JL (2000) Molecular genetics of sulfate assimilation. Adv Bot Res 33:159-223

Heiss S, Schäfer H, Haag-Kerwer A, Rausch T (1999) Cloning sulfur assimilation genes of *Brassica juncea* L.: cadmium differentially affects the expression of a putative low affinity sulfate transporter and isoforms of ATP sulfurylase and APS reductase. Plant Mol Biol 39:847-57

Hell R (1997) Molecular physiology of plant sulfur metabolism. Planta 202:138-48

Hell R, Bergmann L (1988) Glutathione synthetase in tobacco suspension cultures: catalytic properties and localization. Physiol Plant 72:70-6

Hell R, Bergmann L (1990) γ-Glutamylcysteine synthetase in higher plants: catalytic properties and subcellular localization. Planta 180:603-12

Hell R, Bogdanova N (1995) Characterization of a full-length cDNA encoding a serine acetyltransferase from *Arabidopsis thaliana*. Plant Physiol 109:1498

Hesse H, Lipke J, Altmann T, Hoefgen R (1999) Molecular cloning and expression analyses of mitochondrial and plastidic isoforms of cysteine synthase (*O*-acetylserine(thiol)lyase) from *Arabidopsis thaliana*. Amino Acids 16:113-31

Hesse H, Nikiforova V, Gskieare B, Rainer Hoefgen R (2004) Molecular analysis and control of cysteine biosynthesis: integration of nitrogen and sulfur metabolism. J Exp Bot 55:1283-92

Howarth JR, Roberts MA, Wray JL (1997) Cysteine biosynthesis in higher plants: A new member of the *Arabidopsis thaliana* serine acetyltransferase small gene family obtained by functional complementation of an *Escherichia coli* cysteine auxotroph. Biochem Biophys Acta 1350:123-7

Howden R, Goldsbrough PB, Andersen CR, Cobbett CS (1995) Cadmium-sensitive, cad1 mutants of *Arabidopsis thaliana* are phytochelatin deficient. Plant Physiol 107:1059-66

Hu SX, Lau KWK, Wu M (2001) Cadmium sequestration in *Chalamydomonas reinhardtii*. Plant Sci 161:987-96

Imsande J (2003) Sulphur nutrition and legume seed quality. In: Abrol YP, Ahmad A (eds) Sulphur in plants, Kluwer Academic Publishers, The Netherlands, pp. 295-304

Inouhe M, Ito R, Ito S, Sasada N, Tohoyama H, Joho M (2000) Azuki bean cells are hypersensitive to cadmium and do not synthesize phytochelatins. Plant Physiol 123:1029-36

Islam MR, Islam MS, Jahiruddin M, Hoque MS (1999) Effects of sulfur, zinc and boron on yield, yield components and nutrient uptake of wheat. Pak J Sci Ind Res 42:137-40

Jamai A, Tommasini R, Martinoia E, Delrot S (1996) Characterization of glutathione uptake in broad bean leaf protoplasts. Plant Physiol 111:1145-52

Jez JM, Cahoon RE, Chen S (2004) *Arabidopsis thaliana* glutamate-cysteine ligase: functional properties, kinetic mechanism and regulation of activity. J Biol Chem 279:33463-70

Juang R-H, MacCue KF, Ow DW (1993) Two purine biosynthetic enzymes that are required for cadmium tolerance in *Schizosaccharomyces pombe* utilize cysteine sulfinate *in vitro*. Arch Biochem Biophys 304:392-401

Keltjens WG, van Beusichem ML (1998) Phytochelatins as biomarkers for heavy metal stress in maize (*Zea mays* L.) and wheat (*Triticum aestivum* L.): combined effects of copper and cadmium. Plant Soil 203:119-26

Kerk NM, Feldman LJ (1995) A biochemical model for the initiation and maintenance of the quiescent center: implications for organization of root meristems. Development 121:2825-33

Khan NA, Ahmad I, Singh S, Nazar R (2006) Variation in growth, photosynthesis and yield of five wheat cultivars exposed to cadmium stress. World J Agric Sci 2:223-6

Khan NA, Samiullah, Singh S, Nazar R (2007) Activities of antioxidative enzymes, sulphur assimilation, photosynthetic activity and growth of wheat (*Triticum aestivum*) cultivars differing in yield potential under cadmium stress. J Agron Crop Sci 193:435-44

Klapheck S, Latus C, Bergmann L (1987) Localization of glutathione synthetase and distribution of glutathione in leaf of *Pisum sativum* L. J Plant Physiol 131:123-31

Klapheck S, Schlunz S, Bergmann L (1995) Synthesis of phytochelatins and homo-phytochelatins in Pisum sativum L. Plant Physiol 107:515-21

Klonus D, Hofgen R, Willmitzer L, Riesmeier JW (1994) Isolation and characterization of two cDNA clones encoding ATP-sulfurylases from potato by complementation of a yeast mutant. Plant J 6:105-12

Kneer R, Kutchan RM, Hochberger A, Zenk MH (1992) *Saccharomyces cerevisiae* and *Neurospora crassa* contain heavy metal sequestering phytochelatin. Arch Microbiol 157:305-10

Kopriva S, Buchert T, Fritz G, Suter M, Weber M, Benda R, Schaller J, Feller U, Schurmann P, Scurmann V, Trautwein AX, Kroneck PM, Brunold C (2001) Plant adenosine 5′-phosphosulfate reductase is a novel iron-sulfur protein. J Biol Chem 276:42881-6

Kopriva S, Koprivova A (2003) Sulfate assimilation: a pathway which likes to surprise. In: Abrol YP, Ahmad A (eds) Sulphur in plants, Kluwer Academic Publishers, The Netherlands, pp. 87-112

Kopriva S, Koprivova A (2004) Plant adenosine 5′-phosphosulfate reductase: the past, the present, and the future. J Exp Bot 55:1775-83

Kovari IA, Cobbett CS, Goldsbrough PB (1997) Expression of tomato γ-Glu-Cys synthetase in the *Arabidopsis* cad2 mutant restores cadmium tolerance. Plant Physiol 114:126

Kunert KJ, Foyer CH (1993) Thiol/disulphide exchange in plants. In: De Kok LJ (ed) Sulfur nutrition and assimilation in higher plants, SPB Academic Publishing The Hague, The Netherlands, pp. 139-51

Lappartient AG, Touraine B (1996) Demand-driven control of root ATP sulphurylase activity and SO_4^{2-} uptake in intact canola. Role of phloem-translocated glutathione. Plant Physiol 111:147-57

Larsson EH, Asp H, Bornman JF (2002) Influence of prior Cd^{2+} exposure on the uptake of Cd^{2+} and other elements in the phytochelatin-deficient mutant, cad1-3, of *Arabidopsis thaliana*. J Exp Bot 53:447-53

Le Faucheur S, Behra R, Sigg L (2005) Thiol and metal contents in periphyton exposed to elevated copper and zinc concentrations: a field and microcosm study. Environ Sci Technol 39:8099-8107

Lee S, Leustek T (1999) The affect of cadmium on sulfate assimilation enzymes in *Brassica juncea*. Plant Sci 141:201-7

Leustek T, Martin MN, Bick J-A, Davies JP (2000) Pathways and regulation of sulfur metabolism revealed through molecular and genetic studies. Annu Rev Plant Physiol Plant Mol Biol 51:141-65

Leustek T, Murillo M, Cervantes M (1994). Cloning of a cDNA in coding ATP sulfurylase from *Arabidopsis thaliana* by functional expression in *Saccharomyces cerevisiae*. Plant Physiol 105: 897-902

Leustek T, Saito K (1999) Sulfate transport and assimilation in plants. Plant Physiol 120:637-43

Li JJ, Saidha T, Schiff JA (1991) Purification and properties of two forms of ATP sulfurylase from Euglena. Biochim Biophys Acta 1078:68-76

Lima AIS, Pereira SIA, Figueira EMAP, Caldeira GCN, Caldeira HDQM (2006) Cadmium detoxification in roots of *Pisum sativum* seedlings: relationship between toxicity levels, thiol pool alterations and growth. Environ Exp Bot 55:149-62

Liso R, Innocenti AM, Bitoni MB, Arrigoni O (1988) Ascorbic acid induced pregrssion of quiescent centre cells from G1 to S phase. New Phytol 110:469-71

Lunn J, Droux M, Martin J, Douce R (1990) Localization of ATP sulfurylase and O-acetylserine (thiol)lyase in spinach leaves. Plant Physiol 94:1345-52

Maier EA, Matthews RD, McDowell JA, Walden RR, Ahner BA (2003) Environmental cadmium levels increase phytochelatin and glutathione in lettuce grown in a chelator-buffered nutrient solution. J Environ Qual 32:1356-64

May MJ, Leaver CJ (1994) *Arabidopsis thaliana* γ-glutamylcysteine synthetase is structurally unrelated to mammalian, yeast and *E. coli* homologs. Proc Natl Acad Sci USA 91:10059-63

May MJ, Parker JE, Daniels MJ, Leaver CJ, Cobbett CS (1996) An *Arabidopsis* mutant depleted in glutathione shows unaltered responses to fungal and bacterial pathogens. Mol Plant-Microbe Interact 9:349-56

May MJ, Vernoux T, Leaver C, Van Montagu M, Inze D (1998) Glutathione homeostasis in plants: implications for environmental sensing and plant development. J Exp Bot 49:649-67

McLaughlin, MJ, Parker DR, Clarke JM (1999) Metals and micronutrients: Food safety issues. Field Crops Res 60:143-63

Mehra RK, Tarbet EB, Gray WR, Winge DR (1988) Metal-specific synthesis of 2 metallthioneins and γ-glutamyl-transferase peptides in *Candida glabrata*. Proc Natl Acad Sci USA 85:8815-19

Meinke DW, Chen J, Beachy RN (1981) Expression of storage-protein genes during soybean seed development. Planta 153:130-9

Memon AR, Ozdemir A, Aktoprakligi D (2001) Heavy metal accumulation in plants. Biotech Biotechnol Equip 15:44-8

Mendoza-Cozatl D, Loza-Tavera H, Hernandez-Navarro A, Moreno-Sanchez R (2005) Sulfur assimilation and glutathione metabolism under cadmium stress in yeast, protists and plants. FEMS Microbiol Rev 29:653-71

Mendoza-Cozatl DG, Moreno-Sanchez R (2006) Control of glutathione and phytochelatin synthesis under cadmium stress. Pathway modeling for plants. J Theor Biol 238:919-36

Meyer Y, Verdoucq L, Vignols F (1999) Plant thioredoxins and glutaredoxins: identity and putative roles. Trends Plant Sci 4:388-94

Mittler R, Vanderauwera S, Gollery M, Breusegem FV (2004) Reactive oxygen gene network of plants. Trends Plant Sci 9:490-8

Miyake T, Kanayama M, Sammoto H, Ono B (2002) A novel *cis*-acting cysteine-responsive regulatory element of the gene for the high-affinity glutathione transporter of *Saccharomyces cerevisiae*. Mol Genet Gen 266:1004-11

Mobin M and Khan NA (2007) Photosynthetic activity, pigment composition and antioxidative response of two mustard (*Brassica juncea*) cultivars differing in photosynthetic capacity subjected to cadmium stress. J Plant Physiol 164:601-10

Murillo M, Leustek T (1995) ATP sulfurylase from *Arabidopsis thaliana* and *Escherichia coli* are functionally equivalent but structurally and kinetically divergent. Nucleotide sequence of two ATP sulfurylase cDNAs from *Arabidopsis thaliana* and analysis of a recombinant enzyme. Arch Biochem Biophys 323:195-204

Mutoh N, Hayashi Y (1988) Isolation of mutants of *Schizosaccharomyces pombe* unable to synthesize cadystin, small cadmium-binding peptides. Biochem Biophys Res Commun 151:32-9

Nakamura K, Hayama A, Masada M, Fukushima K, Tamura G (1988) Purification and some properties of plant serine acetyltransferase. Plant Cell Physiol 29:689-93

Nakamura K, Tamura G (1990) Isolation of serine acetyltransferase complexed with cysteine synthase from *Allium tuberosum*. Agric Biol Chem 53:2537-8

Noctor G, Arisi A-CM, Jouanin L, Foyer CH (1998b) Manipulation of glutathione and amino acid biosynthesis in the chloroplast. Plant Physiol118:471-82

Noctor G, Arisi A-CM, Jouanin L, Kunert KJ, Rennenberg H, Foyer CH (1998a) Glutathione: biosynthesis, metabolism, and relationship to stress tolerance explored in transformed plants. J Exp Bot 49:623-47

Noctor G, Gomez LA, Vanacker H, Foyer CH (2002) Interactions between biosynthesis, comparmentation and transport in the control of glutathione homeostasis and signaling. J Exp Bot 53:1283-3304

Noctor G, Strohm M, Jouanin L, Kunert K-J, Foyer CH, Rennenberg H (1996) Synthesis of glutathione in leaves of transgenic poplar overexpressing γ-glutamylcysteine synthetase. Plant Physiol 112:1071-8

Noji M, Inoue K, Kimura N, Gouda A, Saito K (1998) Isoform-dependent differences in feedback regulation and subcellular localization of serine acetyltransferase involved in cysteine biosynthesis from *Arabidopsis thaliana*. J Biol Chem 273:32739-45

Noji M, Saito K (2003) Sulfur amino acids: biosynthesis of cysteine and methionine. In: Abrol YP, Ahmad A (eds) Sulphur in plants, Kluwer Academic Publishers, The Netherlands, pp. 135-44

Noji M, Takagi Y, Kimura N, Inoue K, Saito M, Horikoshi M, Saito F, Takahashi H, Saito K (2001) Serine acetyltransferase involved in cysteine biosynthesis from spinach: Molecular cloning, characterization and expression analysis of cDNA encoding a plasticic isoform. Plant Cell Physiol 42:627-34

Oritz DF, Kreppel L, Speiser DM, Scheel G, McDonald G, Ow DW (1992) Heavy metal tolerance in the fission yeast requires an ATP-binding cassette-type vacuolar membrane transporter. EMBO J 11:3491-9

Paek NC, Sexton PJ, Naeve SL, Shibles R (2000) Differential accumulation of soybean seed storage protein subunits in response to sulfur and nitrogen nutritional sources. Plant Prod Sci 3:268-74

Pasricha NS, Aulakh MS, Bahl GS, Baddesha HS (1987) Nutritional Requirements of Oilseed and Pulse Crops in Punjab. Res Bull 15. PAU, Ludhiana

Passera C, Ghisi R (1982) ATP sulphurylase and O-acetylserine sulphydrylase in isolated mesophyll protoplasts and bundle sheath strands of S-deprived maize leaves. J Exp Bot 33:432-8

Piechalak A, Tomaszewska B, Baralkiewicz D, Malecka A (2002) Accumulation and detoxification of lead ions in legumes. Phytochemistry 60:53-162

Ramos I, Esteban E, Lucena JJ, Garate A (2002) Cadmium uptake and subcellular distribution in plants of Lactica sp. Cd-Mn interaction. Plant Sci 162:761-7

Ranieri A, Castagna A, Scebba F, Careri M, Zagnoni I, Predieri G, Pagliari M, Sanita di Toppi L (2005) Oxidative stress and phytochelatin characterisation in bread wheat exposed to cadmium excess. Plant Physiol Biochem 43:45-54

Rausch T, Wachter A (2005)) Sulfur metabolism: a versatile platform for launching defence operations. Trends Plant Sci 10:503-9

Rauser WE (1995) Phytochelatins and related peptides: structure, biosynthesis, and function. Plant Physiol 109:1141-9

Rauser WE (1999) Structure and function of metal chelators produced by plants: The case for organic acids, amino acids, phytin and metallothioneins. Cell Biochem Biophys 31:19-48

Rawlins MR, Leaver CJ, May MJ (1995) Characterisation of an *Arabidopsis thaliana* cDNA encoding glutathione synthetase. FEBS Letters 376:81-6

Rennenberg H (1983) Role of O-acetylserine in hydrogen sulfide emission from pumpkin leaves in response to sulfate. Plant Physiol 73:560-5

Renosto F, Patel HC, Martin RL, Thomassian C, Zimmerman G, Segel IH (1993) ATP sulfurylase from higher plants: kinetic and structural characterization of the chloroplast and cytosol enzymes from spinach leaf. Arch Biochem Biophys 307:272-85

Roberts MA, Wray JL (1996) Cloning and characterization of an *Arabidopsis* cDNA clone encoding an organellar isoform of serine acetyltransferase. Plant Mol Biol 30:1041-9

Rotte C (1998) Subcellular localization of sulphur assimilation enzymes in *Arabidopsis thaliana* (L.) *HEYNH*. Diplomarbeit thesis. Carl von Ossietzky Univ. Oldenburg, Germany. pp. 87

Rotte C, Leustek T (2000) Differential subcellular localization and expression of ATP sulfurylase and 5′-adenylylsulfate reductase during ontogenesis of *Arabidopsis* leaves indicates that cytosolic and plastid forms of ATP sulfurylase may have specialized functions. Plant Physiol 124:715-24

Rouhier N, Gelhaye E, Jacquot J-P (2002) Exploring the active site of plant glutaredoxin by site-directed mutagenesis. FEBS letters 511:145-9

Ruegsegger A, Brunold C (1992) Effect of cadmium on γ- glutamylcysteine synthesis in maize seedlings. Plant Physiol 99:428-33

Ruegsegger A, Schmutz D, Brunold C (1990) Regulation of glutathione synthesis by cadmium in *Pisum sativum* L. Plant Physiol 93:1579-84

Ruffet M-L, Lebrum M, Droux M, Douce R (1995) Subcellular distribution of serine actetyl transferase from Pidum sativum and characterization of an *Arabidopsis thaliana* putative cytosolic isoform. Eur J Biochem 227:500-9

Rusch T, Wachter A (2005) Sulfur metabolism: a versatile platform for launching defence operations. Trends Plant Sci 10:503-9

Russo T, Zambrano N, Esposito F, Ammendola R, Cimino F, Fiscella M, Jackman J, O'Connor M, Anderson CW, Apella E (1995) A p53-independent pathway for activation of WAF1/C1P1 expression following oxidative stress. J Biol Chem 270:29386-91

Saito K (1999) Biosynthesis of cysteine. In: Singh BK (ed) Plant amino acids: biochemistry and biotechnology, Marcel Dekker Inc, New York, pp. 267-91

Saito K (2000) Regulation of sulfate transport and synthesis of sulfur-containing amino acids. Curr Opin Plant Biol 3:188-95

Saito K, Yokoyama H, Noji M, Murakoshi I (1995) Molecular cloning and characterization of a plant serine acetyltransferase playing a regulatory role in cysteine biosynthesis from watermelon. J Biol Chem 270:16321-6

Samiullah, Khan NA, Nazar R, Ahmad I (2007) Physiological basis for reduced photosynthesis and growth of cadmium-treated wheat cultivars differing in yield potential. J Food Agric Environ 5:375-7

Sanchez-Fernandez R, Fricker M, Corben LB, White NS, Sheard N, Leaver CJ, Van Montagu M, Inze D, May MJ (1997) Cell proliferation and hair tip growth in the *Arabidopsis* root are under mechanistically different forms of redox control. Proc Natl Acad Sci USA 94:2745-50

Sanita di Toppi L, Gabbrielli R (1999) Response to cadmium in higher plants. Environ Exp Bot 41:105-30

Schafer HJ, Haag-Kerwer A, Rausch T (1998) cDNA cloning and expression analysis of genes encoding GSH synthesis in roots of the heavy metal accumulator *Brassica juncea* L.: evidence or Cd-induction of a putative mitochondrial γ-glutamylcysteine synthetase isoform. Plant Mol Biol 37: 87-97

Scheller HB, Huang B, Hatch E, Goldsbrough PB (1987) Phytochelatin synthesis and glutathione levels in response to heavy metals in tomato cells. Plant Physiol 85:1031-5

Schmutz D, Brunold C (1984) Intercellular localization of assimilatory sulfate reduction in leaves of *Zea mays* and *Triticum aestivum*. Plant Physiol 74:866-70

Schneider A, Martini N, Rennenberg H (1992) Reduced glutathione (GSH) transport in cultured tobacco cells. Plant Physiol Biochem 30:29-38

Schroder P, Fischer C, Debus R, Wenzel A (2003) Reaction of detoxification mechanisms in suspension cultured spruce cells (*Picea abies* L. Karst.) to heavy metals in pure mixture and soil eluates. Environ Sci Pollut Res 10:225-34

Schurmann P, Jacquot J-P (2000) Plant thioredoxin systems revisited. Annu Rev Plant Physiol Plant Mol Biol 51:371-400

Segel IH, Renosto F, Seubert PA (1987) Sulfate-activating enzymes. Methods Enzymol 143:334-49

Setya A, Murillo M, Leustek T (1996) Sulfate reduction in higher plants: molecular evidence for a novel 5-adenylylphosphosulfate (APS) reductase. Proc Natl Acad Sci USA 93:13383-8

Shaul O, Mironov V, Burssen S, Van Montagu MV, Inze D (1996) Two Arabidopsis cyclin promoters mediate distinctive transcriptional osscilation in synchronized tobacco 3Y-2 cells. Proc Natl Acad Sci USA 93:4868-72

Shewry PR, Pandya MJ (1999) The 2S albumin storage proteins. In: Shewry PR, Casey R (eds) Seed proteins, Kluwer Academic Publishers, Dordrecht, pp. 563-86

Siedlecka A, Krupa Z, Samuelsson G, Oquist G, Gardestrom P (1997) Primary carbon metabolism in *Phaseolus vulgaris* plants under Cd(II)/Fe interaction. Plant Physiol Biochem 35:951-7

Singh S, Khan NA, Nazar R, Anjum NA (2007) Photosynthetic traits and activities of antioxidant enzymes in blackgram (*Vigna mungo* L. Hepper) under cadmium stress. Am J Plant Physiol 3:25-32

Smirnoff N (1996) The Function and metabolism of ascorbic acid in plants. Ann Bot 78:661-9

Smith IK (1985) Stimulation of glutathione synthesis in photorespiring plants by catalase inhibitors. Plant Physiol 79:1044-7

Smith IK, Kendall AC, Keys AJ, Turner JC, Lea PJ (1984) Increased levels of glutathione in a catalase deficient mutant of barley (*Hordeum vulgare* L.) Plant Sci Letters 37:29-33

Song W, Steiner HY, Zang L, Naider F, Becker JM, Stacey G (1996) Cloning of a second *Arabidopsis* peptide transport gene. Plant Physiol 110:171-8

Speiser DM, Abrahamson SL, Banuelos G, Ow DW (1992) *Brassica juncea* produces a phytochelatin-cadmiumsulfide complex. Plant Physiol 99:817-21

Srivastava S, Tripathi RD, Awivedi UN (2004) Synthesis of phytocheltins and modulation of antioxidants in responses to cadmium stress in *Cuscuta reflexa* – an angiospermic parasite. J Plant Physiol 161:665-74

Steiner HY, Song W, Zhang L, Naider F, Becker JM, Stacey G (1994) An *Arabidopsis* peptide transporter is a member of a new class of membrane transport proteins. Plant Cell 6:1289-99

Stewart EJ, Aslund F, Beckwith J (1998) Disulfide bond formation in the *Escherichia coli* cytoplasm: an in vivo role reversal for the thioredoxins. EMBO J 17:5543-50

Strohm M, Jouanin L, Kunert KJ, Pruvost C, Polle A, Foyer CH, Rennenberg H (1995) Regulation of glutathione synthesis in leaves of transgenic poplar (*Populus tremula x P. alba*) overexpressing glutathione synthetase. Plant J 7:141-5

Suter M, Von Ballmoos P, Kupriva S, Opdencamp R, Schaller J, Kuhlemeier C, Scurmann P, Brunold C (2000) Adenosine 5-sulphosulfate sulpho transferease and adenosine 5-phospho sulpho reductase are identical enzymes. J Biol Chem 275:930-6

Syers JK, Skinner RJ, Curtin D (1987) Soil and fertilizer sulphur in UK agriculture. Proc Fertilizer Society No. 379, The Fertilizer Society, Peterborough

Takagi M, Satofuka H, Amano S, Mizuno H, Eguchi Y, Kazumasa H, Kazuhisa M, Fukui K, Imanaka T (2002) Cellular toxicity of cadmium ions and their detoxification by heavy metal spesific plant peptides, phytochelatins, expressed in mammalian cells. J Biochem. 131:233-9

Takahashi H, Yamazaki M, Sasakura N, Watanabe A, Leustek T, de Almeida Engler J, Engler G, Van Montagu M, Saito K (1997) Regulation of sulfur assimilation in higher plants: A sulfate transporter induced in sulfate-starved roots plays a central role in *Arabidopsis thaliana*. Proc Natl Acad Sci USA 94:11102-7

Tandon HLS (1991) Sulphur Research and Agricultural Production in India, 3rd Edition. The Sulphur Institute, Washington DC

Tausz M, Gullner G, Komives T, Grill D (2003) The role of thiols in plant adaptation to environmental stress. In: Abrol YP, Ahmad A (eds) Sulphur in plants, Kluwer Academic Publishers, The Netherlands, pp. 221-44

Thangavel P, Long S, Minocha R (2007) Changes in phytochelatins and their biosynthetic intermediates in red spruce (*Picea rubens* Sarg.) cell suspension culture under cadmium and zinc stress. Plant Cell Tiss Organ Cult 88:201-16

Tukendorf A, Rauser WE (1990) Changes in glutathione and phytochelatins in roots of maize seedlings exposed to cadmium. Plant Sci 70:155-66

Ullman P, Gondet L, Potier S, Bach TJ (1996) Clonning of *Arabidopsis thaliana* glutathione synthetase (*GSH2*) by functional complementation of a yeast *gsh2* mutatnt. Eur J Biochem 236:662-9

Urano Y, Manabe T, Noji M, Saito K (2000) Molecular cloning and functional characterization of cDNAs encoding cysteine synthase and serine acetyltransferase that may be responsible for high cellular cysteine content in *Allium tuberosum*. Gene 257:269-77

Vanacker H, Carver TLW, Foyer CH (2000) Early H_2O_2 accumulation in mesophyll cells leads to induction of glutathione during the hyper-sensitive response in the barley-powdery mildew interaction. Plant Physiol 123:1289-1300

Vatamaniuk OK, Bucher EA, Ward JT, Rea PA (2001) A new pathway for heavy metal detoxification in animals - Phytochelatin synthase is required for cadmium tolerance in *Caenorhabditis elegans*. J Biol Chem 276:20817-20

Vatamaniuk OK, Mari S, Lu Y-P, Rea PA (1999) AtPCS1, a phytochelatin synthase from *Arabidopsis*: isolation and *in vitro* reconstitution. Proc Natl Acad Sci USA 96:7110-15

Vatamaniuk OK, Mari S, Lu Y-P, Rea PA (2000) Mechanism of heavy metal ion activation of phytochelatin (PC) synthase - Blocked thiols are sufficient for PC synthase-catalyzed transpeptidation of glutathione and related thiol peptides. J Biol Chem 275:31451-9

Vauclare P, Kopriva S, Fell D, Suter M, Sticher L, von Ballmoos P, KraÈhenbuÈhl U, Op den Camp R, Brunold C (2002) Flux control of sulfate assimilation in *Arabidopsis thaliana*: adenosine 5-phosphosulfate reductase is more susceptible to negative control by thiols than ATP sulphurylase. Plant J 31:729-40

Verkleij JAC, Sneller FEC, Schat H (2003) Metallothioneins and Phytochelatins: Ecophysiological aspects. In: Abrol YP, Ahmad A (eds) Sulphur in plants, Kluwer Academic Publishers, The Netherlands, pp. 163-76

Vernoux T, Wilson RC, Seeley KA, Reichheld JP, Muroy S, Brown S, Maughan SC, Cobbett CS, Van Montagu M, Inze D, May MJ, Sung ZR (2001) The *ROOT MERISTEMLESS1/CADMIUM SENSITIVE2* gene defines a glutathione-dependent pathway involved in initiation and maintenance of cell division during postembyronic root development. Plant Cell 12:97-109

Vogeli-Lange R, Wagner GJ (1996) Relationship between cadmium, glutathione and cadmium-binding peptides (phytochelatins) in leaves of intact tobacco seedlings. Plant Sci 114:11-18

von Arb C, Brunold C (1986) Enzymes of assimilatory sulfate reduction in leaves of *Pisum sativum*: activity changes during ontogeny and in vivo regulation by H_2S and cysteine. Physiol Plant 67:81-6

Wachter A, Wolf S, Steininger H, Bogs J, Rausch T (2005) Differential targeting of GSH1 and GSH2 is achieved by multiple transcription initiation: implications for the compartmentation of glutathione biosynthesis in the *Brassicaceae*. Plant J 41:15-30

Walker KC, Booth EJ (1992) Sulfur research on oilseed rape in Scotland. Sulfur Agric 16:15-19

Willekens H, Chamnongpol S, Davey M, Schraudner M, Langebartels C, Van Montagu M, Inze D, Van Camp W (1997) Catalase is a sink for H_2O_2 and is indispensable for stress defence in C_3 plants. EMBO J 16:4806-16

Wojcik M, Skorzynska-Polit E, Tukiendorf A (2006) Organic acids accumulation and antioxidant enzyme activities in *Thlaspi caerulescens* under Zn and Cd stress. Plant Growth Regul 48:145-55

Wojcik M, Vangronsveld J, Tukiendorf A (2005) Cadmium tolerance in *Thlaspi caerulescens* I. Growth parameters, metal accumulation and phytochelatin synthesis in response to cadmium. Environ Exp Bot 53:151-61

Wu FB, Zhang G-P (2002a) Genotypic differences in effect of Cd on growth and mineral concentrations in barley seedlings. Bull Environ Contam Toxicol 69:219-27

Wu FB, Zhang G-P (2002b) Alleviation of cadmium-toxicity by application of zinc and ascorbic acid in barley. J Plant Nutr 25:2745-61

Xiang C, Bertrand D (2000) Glutathione synthesis in *Arabidopsis*: Multilevel controils coordinate responses to stress. In: Brunold C, et al. (eds) Sulfur nutrition and sulfur assimilation in higher plants, Paul Haupt, pp. 409-12

Xiang C, Oliver DJ (1998) Glutathione metabolic genes co-ordinately respond to heavy metals and jasmonic acid in *Arabidopsis*. The Plant Cell 10:1539-50

Xiang C, Werner BL, Christensen EM, Oliver DJ (2001) The biological functions of glutathione revisited in *Arabidopsis* transgenic plants with altered glutathione levels. Plant Physiol 126:564-74

Yamaguchi Y, Nakamura T, Harada E, Koizumi N, Sano H (1999) Differential accumulation of transcripts encoding sulfur assimilation enzymes upon sulfur and/or nitrogen deprivation in *Arabidopsis thaliana*. Biosci Biotechnol Biochem 63:762-66

Yonekura-Sakakibara K, Onda Y, Ashikari T, Tanaka Y, Kusumi T, Hase T (2000) Analysis of reductant supply systems for ferredoxin-dependent sulfite reductase in photosynthetic and nonphotosynthetic organs of maize. Plant Physiol 122:887-94

Zenk MH (1996) Heavy metal detoxification in higher plants-a review. Gene 179:21-30

Zhang G-P, Motohiro F, Hitoshi S (2002) Influence of cadmium on mineral concentrations and yield components in wheat genotypes differing in Cd tolerance at seedling stage. Field Crops Res 77:93-9

Zhu Y, Pilon-Smits EAH, Jouanin L, Terry N (1999a) Overexpression of glutathione synthetase in *Brassica juncea* enhances cadmium tolerance and accumulation. Plant Physiol 119:73-9

Zhu Y, Pilon-Smits EAH, Tarun A, Weber SU, Jouanin L, Terry N (1999b) Cadmium tolerance and accumulation in Indian mustard is enhanced by overexpressing γ-glutamylcysteine synthetase. Plant Physiol 121:1169-77

Chapter 14
Glutathione Metabolism in Bryophytes under Abiotic Stress

David J. Burritt

Abstract Plants living in natural communities rely entirely on numerous defense mechanisms to cope with abiotic stress. The mechanisms that enable plants to perceive the environmental conditions in which they grow and to respond to abiotic stressors are complex, consisting of cascades of multiple reactions. In recent years it has become apparent that sulfur metabolism, and in particular the tripeptide thiol glutathione, play critical roles in protecting the cells of plants under abiotic stress. Although numerous studies have investigated the impact of abiotic stress on the cells of higher plants, far fewer studies have investigated the impact on the cells of bryophytes. In this chapter the importance of glutathione for the survival of bryophytes exposed to selected abiotic stressors is discussed.

1 Introduction

In living organisms stress caused by nonliving environmental factors, such as a lack of water, extreme temperatures, high light, ultraviolet radiation, or environmental pollutants, such as heavy metals, is referred to as abiotic stress. As normal plant growth and morphogenesis depend upon environmental factors, for example, photosynthesis and the various environmental cues used to control development, abiotic stresses can be particularly harmful to plants.

Abiotic stress is a major factor influencing crop and food production in much of the world, and many agricultural and agronomic practices (e.g., irrigation) have been designed to reduce or avoid the risk of abiotic stress, so as to optimize crop growth and hence productivity. Plants living in natural communities are also exposed to abiotic stress, and these rely entirely on the numerous defense mechanisms that plants have evolved to cope with such stress.

In an ever more industrialized world and with the likelihood of significant climate change in the foreseeable future, abiotic stress is becoming an increasing

David J. Burritt
Department of Botany, University of Otago, P.O. Box 56, Dunedin, New Zealand
david@planta.otago.ac.nz

N.A. Khan et al. (eds.), *Sulfur Assimilation and Abiotic Stress in Plants.*

threat not only to crop and food production, but also to fragile ecosystems, many of which have already been subjected to substantial anthropological disturbance. Although numerous studies have been conducted investigating the influence of abiotic stressors on the growth, development, biochemistry, physiology, and genetics of higher plants (Vij and Tyagi 2007), in particular crop plants, relatively few studies have been conducted on the lower plants and in particular the bryophytes. Of the studies conducted on byrophytes most have concentrated on gross physiological responses or changes in community dynamics (Davey 1997, Wasley et al. 2006). Few studies have investigated the biochemical, cellular, and molecular responses of bryophytes to abiotic stress.

2 The Bryophytes and Their Ecological Significance

The bryophytes, typically including the mosses, liverworts, and hornworts, is a grouping of plants that recent research has shown may not be monophyletic (Mishler et al. 1994, Kenrick and Crane 1997). With the exception of the flowering plants, the bryophytes are the most diverse group of land plants, with an estimated 20,000 or more species worldwide. Among the land plants the bryophytes have a basal phylogenetic position, with today's bryophytes being survivors of the radiation of land plants, in the Devonian Period, over 400 million years ago (Mishler et al. 1994, Kenrick and Crane 1997).

Most bryophytes are small green plants, only a few centimeters high, although some mosses can grow to 50 cm or more in diameter. Bryophytes reproduce vegetatively, or by means of spores instead of seeds, and have no roots, instead being anchored to a substrate by rhizoids. Although rhizoids may play a limited role in the absorption of water and nutrients, most bryophytes appear to absorb water and nutrients directly into their leaves, stems, or thallus tissues.

Bryophytes can be found over a wide range of environments, except in salt water. No entirely marine bryophytes are known, but a few can be found growing on coastline rocks, where they are able to tolerate salt spray (Ackerman 1998). The majority of bryophytes are found growing on shaded or exposed soil or rocks, on the bark of living or dead trees, or on litter in humid forests. Bryophytes are also found in abundance in swamps, bogs, and fens, with some growing submerged or emergent in streams or ponds. When present in abundance, for example in the rainforests of the Pacific Northwest, where bryophytes make up a significant proportion of the biomass, they play important roles in the cycling of water and nutrients. They also play important roles in relationships with many other organisms, including higher plants, fungi, and animals such as protozoa, nematodes, and insects (Gerson 1982). Bryophytes can also be found in extreme environments such as in the arctic tundra, the Antarctic, and in many arid and desert areas. Relatively few bryophytes, found in arid or desert areas, are highly physiologically and biochemically adapted to such harsh environments (Merrill 1991).

Although the mechanisms that enable higher plants to perceive the environmental conditions in which they grow and to respond to abiotic stress are complex, consisting of cascades of multiple reactions, it has become apparent in recent years that sulfur metabolism, and in particular the tripeptide thiol glutathione, play critical roles (Noctor et al. 1998). Studies are now beginning to show an equally important role for glutathione in bryophytes under abiotic stress.

3 A Brief Overview of Glutathione Functions and Synthesis in Plants

Glutathione (L-γ-glutamyl-L-cysteinyl-glycine) is a small tripeptide, not encoded by a gene, but instead synthesized enzymatically. It is the most abundant nonprotein thiol in plants, existing in two interchangeable forms, a reduced form (GSH) and an oxidized form (GSSG) (Noctor and Foyer 1998). As well as being a mobile pool of reduced sulfur, glutathione has numerous cellular and physiological functions. It acts as an important cellular antioxidant in many stress responses, and has recently been implicated in the regulation of plant growth and development (Espunya et al. 2006). Glutathione also plays an important role in enabling plants to grow in polluted environments, being involved in the localization of toxic molecules and xenobiotics to the vacuole.

Glutathione is a major end product of the sulfate assimilation (reduction) pathway. The reduced form is synthesized in two ATP-dependent steps. In the first step, γ-glutamylcysteine (γ -EC) is synthesized from L-glutamate and cysteine by the enzyme γ -glutamylcysteine synthetase (γ–ECS (GSH1), which catalyzes the formation of a peptide bond between the amino group of cysteine and the carboxyl group of glutamate. In the second step, the enzyme glutathione synthetase (GS (GSH2) catalyzes the formation of a peptide bond between the α-amino group of glycine and the cysteinyl carboxyl group of γ -EC. The γ -ECS-catalyzed reaction is the rate-limiting step of GSH biosynthesis, and photosynthetic production of ATP can restrict the rate of GSH biosynthesis. An important step in GSH production is cysteine biosynthesis, which takes place in the chloroplasts, mitochondria, and in the cytosol of *Arabidopsis* cells (Rausch and Wachter 2005). Cysteine biosynthesis and hence GSH biosynthesis are linked to the pathway of reductive sulfate assimilation, a pathway regulated not only in response to the availability of sulfur, but also to environmental conditions that influence plant growth and development and hence that alter the demand for reduced sulfur (Rausch and Wachter 2005).

Sulfate uptake in plants is mediated by high- and low-affinity sulfate transport proteins (STPs), located in the plasma membrane (Hawkesford 2003). Once taken up by plant cells, sulfate molecules are activated by the enzyme ATP sulphurylase (ATPS), which catalyzes its adenylylation, producing 5′-adenylylsulphate, commonly referred to as adenosine 5′-phosphosulphate (APS), which contains a high-energy phosphoric acid – sulfuric acid anhydride bond required for subsequent metabolic reactions. APS is then reduced via two consecutive enzyme-catalyzed reactions. The first is carried out by APS reductase (APR) and results in the transfer of two electrons to APS, producing sulfite, and the second by sulfite reductase (SiR), which reduces sulfite by six electrons to sulfide. Sulfide is then incorporated into O-acetylserine (OAS), produced by a reaction that is catalyzed by the enzyme serine acetyltransferase (SAT), and by O-acetylserine (thiol) lyase (OAS-TL) to form cysteine (Kopriva and Rennenberg 2004).

Unlike in higher plants, where APS is directly reduced to sulfite via APR (Kopriva et al. 2002), in some bacteria and fungi, APS is first phosphorylated by APS kinase (APK) to phosphoadenosine 5′ -phosphosulphate (PAPS) before being reduced by PAPS reductase (PAPSR) to sulfite (Koprivova et al. 2002). Recently studies of sulfate metabolism in the moss *Physcomitrella patens* found both mechanisms of sulfite production present (Koprivova et al. 2002, Kopriva and Koprivova 2004). This is possibly an important difference between bryophytes and higher plants, as studies on bryophytes under abiotic stress have shown very high levels of glutathione and hence there would be a high demand for reduced sulfur (Rother et al. 2006).

Examples of the genes encoding the enzymes required for glutathione synthesis have been isolated from examples of both higher and lower plants. In the *Arabidopsis thaliana* genome, γ –ECS is encoded by a single gene (*gsh1*), as is GS (*gsh2*). A recent study of the 5′-ends of the GSH1 and GSH2 mRNAs revealed two mRNA populations for each gene, populations of transcripts with short 5′-UTRs, and less abundant populations with long 5′-UTRs (Wachter et al. 2005). Further analysis of the GSH1 transcripts showed that both short and long mRNA populations contained an entire putative plastidic transit peptide (TP) sequence. Analysis of the GSH2 transcripts showed that the long transcript population contained a complete TP sequence, but the more abundant short transcript lacked a complete TP sequence. In vivo targeting studies further revealed that in cells of both *A. thaliana* and *Brassica juncea* GSH1 was exclusively localized to the plastids; GSH2, while also present in chloroplasts, was largely located in the cytosol (Wachter et al. 2005).

Recent studies investigating the presence and expression of genes involved in glutathione biosynthesis in the moss *Physcomitrella patens* suggest that *gsh1* and *gsh2* are also likely be present as single copies, as they are present as single copies in *A. thaliana* and *B. juncea*, and the moss has a lower level of gene redundancy when compared to higher plants (Rother et al. 2006). Genes involved in the sulfate reduction pathway have also been isolated from *P. patens* with the *apr* and *papr* being present as single-copy genes (Koprivova et al. 2002) and the two copies of each of the genes encoding OAS-TL and SAT (Rother et al. 2006).

4 Abiotic Stress and Bryophytes

Bryophytes, like most other plants, are often subjected to a wide range of abiotic stressors. However, unlike for higher plants, where the impacts of theses stressors on plant physiology and biochemistry have been extensively investigated, much less is known about the bryophytes. The three abiotic stresses most extensively investigated in bryophytes are desiccation and drought, metal toxicity, and organic pollutants, and it is these three stressors that will be addressed in the following pages.

5 Glutathione and Desiccation Tolerance in Bryophytes

Many bryophytes have the ability to switch from a fully hydrated metabolically active state to a dry metabolically inactive state, quickly and in a fully reversible manner (Proctor 2001). Although some liverworts have been shown to be desiccation tolerant, most research on desiccation tolerance in bryophytes has concentrated on desiccation-tolerant mosses. Although it is clear from the literature that the responses of bryophytes to desiccation are complex and multifaceted, a common factor associated with desiccation in many organisms, including bryophytes, is oxidative stress. For example, in a recent study on bryophytes and lichens by Minibayeva and Beckett (2001), desiccation was found to greatly enhance the oxidative burst, measured as superoxide production, in a number of liverworts and mosses.

Several studies have addressed the importance of reactive oxygen species (ROS) production and antioxidant metabolism in desiccation-tolerant bryophytes, but the importance of glutathione remains inconclusive (Dhindsa 1987 and 1991, Seel et al. 1992). Dhindsa (1987) examined the glutathione status and its relationship to protein synthesis in the desiccation-tolerant moss, *Tortula ruralis*. In this study a small increase in GSSG was observed during rapid desiccation; however, in contrast a small decrease in total glutathione, with an increase in the percentage of oxidized glutathione, was observed in slowly desiccated specimens. Upon rehydration of the slowly desiccated specimens, GSSG rapidly declined to normal levels, but when rapidly desiccated specimens were rehydrated, there was an immediate, sharp increase in GSSG as a percentage of total glutathione. When GSSG (5 mM) was supplied exogenously, during rehydration, protein synthesis was strongly inhibited in both rapidly and slowly desiccated specimens. Seel et al. (1992) compared the levels of activated oxygen processing enzymes and antioxidant levels in the desiccation-sensitive moss *Dicranella palustris* and the desiccation-tolerant moss *Tortula ruraliformis*. They found that the total glutathione concentration in *T. ruraliformis* was greater than that in *D. palustris* and that desiccation did not cause a decrease the glutathione concentration in *T. ruraliformis*, however, high irradiation in the hydrated state did. In contrast, high irradiance caused a decrease in glutathione concentration in both hydrated and desiccated *D. palustris* plants to a level below detection.

Although glutathione is known to be an important cellular antioxidant, involved in ROS scavenging, there is increasing evidence that its role as a regulator of cell redox state may be more important. Changes in cellular redox status can be brought about by altered glutathione levels and/or redox state, and there is considerable evidence that glutathione can strongly influence redox-sensitive reactions by adjusting the redox environment of the cell. Under some abiotic stresses, such as desiccation, changes in glutathione levels and/or redox state may indirectly regulate the functions of other proteins, possibly via redox-active protein co-factors such as the glutaredoxins (Rouhier et al. 2004). In addition, changes in glutathione levels and/or redox state could be involved in the regulation of genes associated with stress tolerance in bryophytes. In a recent study on the aquatic liverwort *Riccia fluitans*, glutathione was shown to be able to regulate the expression of a peroxide-detoxifying chloroplast 2-Cys peroxiredoxin gene, possibly via its role in ascorbate metabolism (Horling et al. 2001)

6 Glutathione and Bryophytes Exposed to Metal Pollutants

Although micronutrients such as Cu, Fe, Mn, or Zn are essential for plant growth and development, these elements become toxic if taken up at higher concentrations than required. In addition, plants are often exposed to nonessential elements, having no known metabolic function, many of which are grouped together into one major category termed heavy metals. A category that includes Cd, Hg, and Pb which are toxic to plants even at very low concentrations.

Metal pollutants can negatively influence many aspects of plant physiology, including alterations in chlorophyll biosynthesis and electron transport and the activities of essential enzymes, including those involved in photosynthesis (Cobbett 2002). Uptake of metallic pollutants can also interfere with plant growth and metabolism by triggering the production of (ROS) such as the OH^{-} and O^{2-} radicals, which can cause the disintegration of biomembranes by lipid peroxidation and oxidative damage to macromolecules such as proteins and nucleic acids.

Higher plants generally cope with metal pollutants, either by avoidance or tolerance mechanisms. Higher plants avoid or reduce metal uptake by secreting root exudates that complex metals, or by immobilizing these potentially damaging molecules onto or within the cell wall matrix, thus avoiding uptake into the cytosol. Plants can also tolerant toxic and excess essential metals by strict regulation of cellular metal homeostasis, a process in which glutathione plays important roles (Callahan et al. 2006).

Exposure of higher plants, or fungi, to high concentrations of micronutrients such as Cu or toxic heavy metals such as Cd induces the synthesis of phytochelatins (PCs), glutathione (GSH)-derived metal-complexing polymers of various sizes with general structure: $(\gamma\text{-GluCys})_n\text{Gly}$ ($n = 2$ to 11). In the case of cadmium, two phytochelatin molecules coordinate with an ion of cadmium, and the phytochelatin-cadmium complex is then transported into the vacuole via an ATP-binding cassette (ABC) type transporter, essentially removing the toxic metal ion from the cytosol (Clemens et al. 2002, Mendoza-Co'zatl et al. 2005). In addition to forming complexes with phytocelatins, heavy metals and toxins can be directly conjugated to GSH via the action of glutathione S-transferases. The conjugates are then transported from the cytosol into the vacuole by membrane-associated glutathione S-conjugate (GS-X) pumps (Ficker and Meyer 2001). Interestingly, this mechanism is not only thought to function as a mechanism to reduce cellular damage from toxins, but is also thought to be involved in cell pigmentation and the storage of antimicrobial compounds (Marrs 1996).

As mentioned previously, bryophytes do not possess roots, and most appear to absorb water and nutrients directly into their leaves, stems, or thallus tissues. Because of this fact, the bryophytes, and in particular the mosses, tend to accumulate chemical elements more or less indiscriminately and so are very good indicators of soil conditions. This fact also makes them extremely vulnerable to air-borne and other pollutants taken into solution (Longton 1980), and as such they are considered to be very sensitive indicators of air, water, and soil pollution. It has been suggested that analysis of pollutants in bryophyte specimens could be used to monitor changes in pollution levels over time (Siebert et al. 1996).

A number of recent studies have shown that glutathione plays a very important role in protecting bryophytes exposed to toxic metals from cellular damage. Although contradictory evidence can be found in the published literature, with some studies suggesting the presence of PCs in bytophytes (Jackson et al. 1991), recent investigations of Cd-stressed plants suggest that many bryophytes may not synthesize significant amounts of PCs, but instead increase GSH biosynthesis (Bruns et al. 2001, Bleuel et al. 2005). Studies on a number byophytes, both mosses and liverworts, exposed to toxic metals support the view that increased GSH synthesis is an important tolerance mechanism in bryophytes. When the moss *Taxithelium nepalense* was exposed to high concentrations of Cr or Pb, large increases in both glutathione and ascorbate were observed (Choudhury and Panda 2005). The importance of GSH and increased cysteine biosynthesis has also been demonstrated in the moss *Sphagnum squarrosum* exposed to lead (Saxena et al. 2003) and the liverwort *Lunularia cruciata* exposed to Cd (Basile et al. 2005). In a study on the aquatic moss *Fontinalis antipyretica*, Bleuel et al. (2005) found evidence for cytosolic localization of sulphydryl (SH)-group chelated Cd and suggested that Cd may bind directly to GSH in some bryophytes. Bruns et al. (2001) suggested that mosses could respond to Cd stress differently than plants such as *A. thaliana* as, arguably, mosses lack the metal chelating PCs and so must have different mechanisms for metal detoxification. In a more recent study on sulfate assimilation in *P. patens* Rother et al. (2006) found that exposure of this moss to Cd induced the expression of genes involved in cysteine and GSH biosynthesis and increased the activities of the corresponding enzymes, resulting in greater intracellular concentrations of thiol metabolites and GSH. This evidence supports Bruns et al. (2001) hypothesis and suggests that in this moss cytoplasmic metal detoxification is achieved by intracellular chelation and sequestration through GSH.

Our own experiments with the aquatic liverwort *Riccia fluitans* show that when plants are exposed to Cu or Cd there are significant increases in the levels of glutathione (Fig. 14.1) and that treatment of plants with the glutathione biosynthesis L-buthionine sulfoximine (BSO), a nontoxic, highly specific inhibitor of γ-glutamylcysteine synthase in plants (Vernoux et al. 2000, Potters et al. 2004), largely stops the metal-induced increases. The importance of increased glutathione synthesis for the health of *R. fluitans* plants was clearly demonstrated when a Pulse Amplitude Modulation (PAM-2000) portable chlorophyll fluorometer (Walz, Effeltrich, Germany) was used to measure chlorophyll fluorescence. To quantify the efficiency of photon capture by open Photosystem

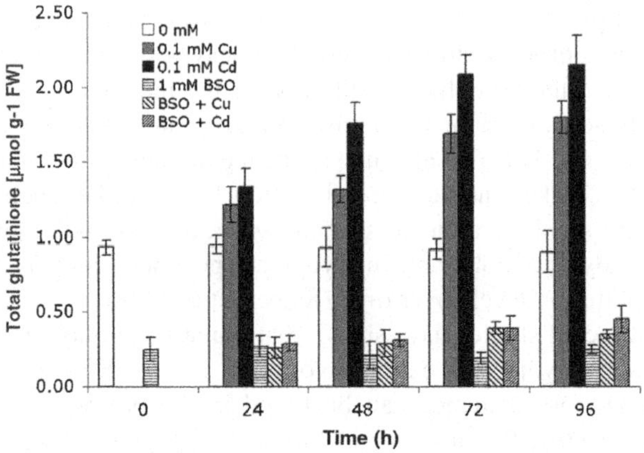

Fig. 14.1 Changes in the glutathione levels in *Riccia Fluitans* plants grown in sterile liquid culture in the presence of 0.1 mM Cu, 0.1 mM Cd, 1 mM of the glutathione biosynthesis inhibitor BSO, or these chemicals in combination. Plants were cultured using the method of Hellwege et al. (1996) and total glutathione extracted and assayed according to Burritt and MacKenzie (2002)

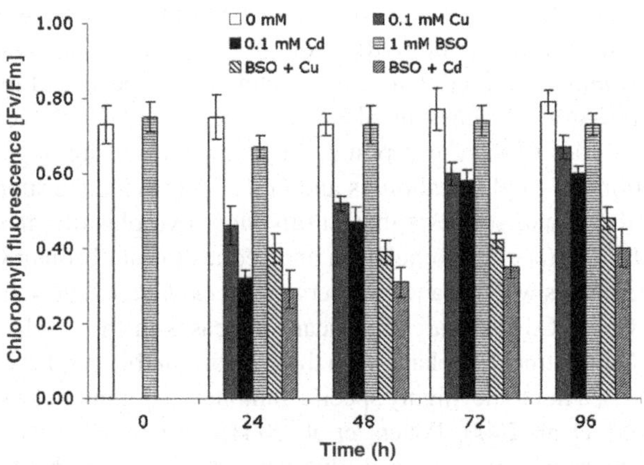

Fig. 14.2 Changes in the Fv/Fm values of *Riccia Fluitans* plants grown in sterile liquid culture in the presence of 0.1 mM Cu, 0.1 mM Cd, 1 mM of the glutathione biosynthesis inhibitor BSO, or these chemicals in combination. Plants were cultured using the method of Hellwege et al. (1996) and Fv/Fm values were determined according to Burritt and MacKenzie (2002)

II (PSII) reaction centers, thallus surfaces were exposed to a saturating pulse of light for determination of the ratio of variable (Fv) to maximal fluorescence (Fm) (Butler and Kitajima 1975). The Fv/Fm ratio is commonly used as an indicator of the health of plants exposed to environmental stress, with plants having higher Fv/Fm ratios considered healthier than plants with lower Fv/Fm ratios (Burritt and MacKenzie 2003).

Exposure of *R. fluitans* plants to Cu or Cd caused a decline in Fv/Fm, with a gradual but incomplete recovery of Fv/Fm occurring over the next 72 hours, along with increased glutathione levels (Fig. 14.2). Treatments of *R. fluitans* with BSO alone caused no decline in Fv/Fm, but when plants were treated with BSO and Cu or Cd no recovery of the Fv/Fm ratio was observed (Fig. 14.2). These results suggest that glutathione may play an important role in protecting *R. fluitans* plants from metal-induced damage such as occurs in other bryophytes and higher plants.

7 Organic Pollutants and Glutathione in Bryophytes

As mentioned previously, mosses have been found to be effective organisms for monitoring environmental pollutants. In addition to metals, mosses have also been found to be useful for monitoring chlorinated hydrocarbons, e.g., benzohexachloride (BHC) and polycyclic aromatic hydrocarbons (PAHs) (Thomas 1984, Knulst et al. 1995). Knulst et al. (1995) studied the concentrations of HCB, PCB, and PAHs in the epiphytic mosses (*Hylocomium splendens* and *Pleurozium schreberi*) in a region in central Sweden and found levels of 0.4–1.7 ng/g (DW) of HCB, 2–28 ng/g DW PCB (the of sum of 7 identified PCB-congeners), and PAH concentrations between 39 and 730 ng/g DW, mainly consisting of nonvolatile PAHs. Moss samples from across Europe have also been analyzed by a variety of research groups in Germany, the Netherlands, Denmark, Norway, and Iceland. *Hypnum cupressiforme* was analyzed at most sites, and a clear PAH concentration gradient was detected, which was high in the industrial centers in middle Europe and low in northern Europe (Thomas 1986). Active monitoring, using bags of sphagnum moss, has also been used to successfully test PCB and organochlorine pesticide levels in Canada (Strachan and Glooschenko 1988) and PAHs levels 1 km from an aluminium production plant in Botlek, the Netherlands (Wegener et al. (1992)). Despite the successful use of these bryophytes as tools to help monitor pollution, little is known about the mechanisms used by bryophytes to cope with exposure to these pollutants and in particular the role of glutathione.

Of the pollutants detailed above, the PAHs are among the most toxic and, as PAHs are made up of two or more benzene rings, highly persistent. PAHs are found in numerous commonly used products, including asphalt, coal tar, fuel oils, and numerous other oil-containing products (Neff 1979). They are often released into the environment as byproducts of petroleum-based industries, through incomplete combustion of fuel oils, or through oil spills (Neff 1979). PAHs commonly found in the environment include benz[a]anthracene, benz[a]pyrene, fluorene, fluoranthene, naphthalene, and phenanthrene.

The impact of PAHs on animals and humans has been extensively investigated, for example, benz[a]pyrene is one of the most toxic and carcinogenic substances known (Nair et al. 1991). Once taken up into animal cells, PAHs are metabolized in a process that results in the formation of reactive electrophilic metabolites and the production of ROS (Burczynski et al. 1999, Burchiel and Luster 2001). As PAHs are highly hydrophobic, in terrestrial environments they are thought to be tightly bound to soil particles and so their bioavailability is considered limited (Pilon-Smits 2005). In aquatic environments PAHs are likely to show much greater bioavailability, with studies on marine and freshwater ecosystems following major and simulated oil spills showing negative impacts on algae and aquatic plants respectively (Marshall and Edgar 2003, Tkalec et al. 1998). Despite these studies, in contrast to information about animals, little is known about the cellular consequences of PAH uptake in plants (Alkio et al. 2005). Studies on higher plants exposed to PAHs have demonstrated the importance of antioxidant metabolism and, in particular, glutathione (Flocco et al 2004, Alkio et al. 2005). Flocco et al. (2004) tested the hypothesis that transgenic Indian mustard (*Brassica juncea*) plants overexpressing ECS or GS, which have two-fold higher levels of glutathione and total nonprotein thiols, would have enhanced tolerance to organic pollutants, including the PAH phenanthrene. Exposure of plants to organics pollutants significantly enhanced total nonprotein thiol levels in both wild-type and transgenic plants (Flocco et al. 2004). Flocco et al. (2004) concluded that glutathione could be important for detoxification, via conjugation to glutathione, of many organic xenobiotics including phenanthrene and that overexpression of enzymes involved in GSH biosynthesis offers a promising approach to create plants with the enhanced capacity to tolerate not only heavy metals, but also certain organic pollutants.

Our own experiments with the liverworts *R. fluitans* have shown that, as with metal pollutants, glutathione may play a very important role in detoxification of PAHs in bryophytes. Exposure of *R. fluitans* to the PAHs fluorene or phenanthrene caused a rapid increase in the levels of glutathione (Fig. 14.3). Exposure to BSO inhibited this PAH-induced increase (Fig. 14.3). The importance of increased glutathione synthesis for the health of *R. fluitans* plants was again clearly demonstrated using chlorophyll fluorescence. Exposure of *R. fluitans* plants to PAHs caused a decline in Fv/Fm within 24 hours of exposure, with a gradual recovery of Fv/Fm occurring over the next 72 hours along with increased glutathione levels (Fig. 14.4). Treatments of *R. fluitans* with BSO alone caused no decline in Fv/Fm, but when plants were treated with BSO and a PAH no recovery of Fv/Fm was observed (Fig. 14.4). These results suggest that glutathione may play an important role in protecting *R. fluitans* plants from PAH-induced damage and that it may facilitate plant recovery. Although the mechanism by which glutathione protects plants from PAHs is not known, in human tissues exposed to PAHs, glutathione S-transferase activities increase significantly upon exposure (Elovaara 2007), suggesting a conjugative mechanism. In addition, exposure of plants to PAHs has been shown to cause oxidative stress (Alkio et al. 2005), hence the antioxidative properties of glutathione may also be important.

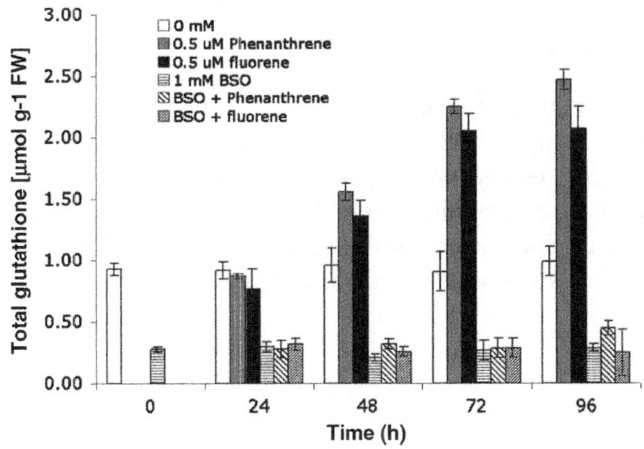

Fig. 14.3 Changes in the glutathione levels of *Riccia Fluitans* plants grown in sterile liquid culture in the presence of the PAHs fluorene or phenanthrene, 1 mM of the glutathione biosynthesis inhibitor BSO, or these chemicals in combination. Plants were cultured using the method of Hellwege et al. (1996) and total glutathione extracted and assayed according to Burritt and MacKenzie (2002)

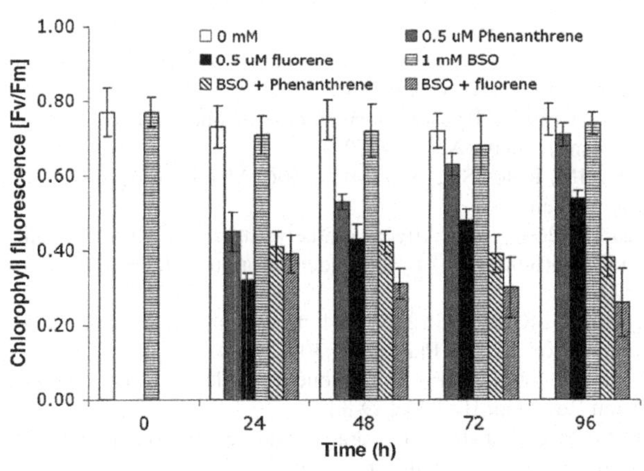

Fig. 14.4 Changes in the Fv/Fm values of *Riccia Fluitans* plants grown in sterile liquid culture in the presence of the PAHs fluorene or phenanthrene, 1 mM of the glutathione biosynthesis inhibitor BSO, or these chemicals in combination. Plants were cultured using the method of Hellwege et al. (1996) and Fv/Fm values were determined according to Burritt and MacKenzie (2002)

8 Conclusion

From the studies conducted to date it is clear that glutathione plays important roles in the protection not only of higher plants, but also of bryophytes. Current efforts to sequence the genomes of bryophyte species and the increasing interest in the roles of bryophytes in plant communities should ensure that in the coming years we will see many new developments that will help us to understand the mechanisms that protect, and the signals that regulate, the responses of bryophytes to abiotic stress.

References

Ackerman JD (1998) Is the limited diversity of higher plants in marine systems the result of biophysical limitations for reproduction or evolutionary and physiological constraints? Func Ecol 12:979-82

Alkio M, Tabuchi TM, Wang XC, Colon-Carmona A (2005) Stress responses to polycyclic aromatic hydrocarbons in Arabidopsis include growth inhibition and hypersensitive response-like symptoms. J Exp Bot 56:2983-94

Basile A, di Nuzzo RA, Capasso C, Sorbo S, Capasso A, Carginale V (2005) Effect of cadmium on gene expression in the liverwort *Lunularia cruciata*. Gene 356:153-9

Bleuel C (2005) The use of the aquatic moss *Fontinalis antipyretica* L. ex Hedw. as a bioindicator for heavy metals: Cd^{2+} accumulation capacities and biochemical stress response of two *Fontinalis* species. Sci Total Environ 345:13

Bruns I (2001) Cadmium lets increase the glutathione pool in bryophytes. J Plant Physiol 158:79

Burchiel SW, Luster MI (2001) Signaling by environmental polycyclic aromatic hydrocarbons in human lymphocytes. Clin Immnu 98:2-10

Burczynski ME, Lin HK, Penning TM (1999) Isoform-specific induction of a human aldo-keto reductase by polycyclic aromatic hydrocarbons (PAHs), electrophiles, and oxidative stress: Implications for the alternative pathway of PAH activation catalyzed by human dihydrodiol dehydrogenase. Cancer Res 59:607-14

Burritt DJ, MacKenzie S (2003) Antioxidant metabolism during acclimation of *Begonia x erythrophylla* to high light levels. Ann Bot 91:783-94

Butler WL, Kitajima M (1975) Energy-transfer between photosystem-2 and photosystem-1 in chloroplasts. Biochim Biophys Acta 396:72-85

Callahan DL, Baker AJM, Kolev SD, Wedd AG (2006) Metal ion ligands in hyperaccumulating plants. J Biol Inorg Chem 11:2-12

Choudhury S, Panda SK (2005) Toxic effects, oxidative stress and ultrastructural changes in moss *Taxithelium nepalense* (Schwaegr.) Broth. under chromium and lead phytotoxicity. Water Air Soil Pollut 167:73-90

Clemens S, Palmgren MG, Kramer U (2002) A long way ahead: understanding and engineering plant metal accumulation. Trends Plant Sci 7:309-15

Cobbett C (2002) Phytochelatins and metallothioneins: Roles in heavy metal detoxification and homeostasis. Annu Rev Plant Biol 53:159-82

Davey MC (1997) Effects of short-term dehydration and rehydration on photosynthesis and respiration by Antarctic bryophytes. Environ Exp Bot 37:187-98

Dhindsa RS (1987) Protein-synthesis during rehydration of rapidly dried *Tortula-ruralis*: evidence for oxidation injury. Plant Physiol 85:1094-8

Dhindsa RS (1991) Drought stress, enzymes of glutathione metabolism, oxidation injury, and protein-synthesis in *Tortula ruralis*. Plant Physiol 95:648-51

Elovaara E, Mikkola J, Stockmann-Juvala H, Luukkanen L, Keski-Hynnila H, Kostiainen R, Pasanen M, Pelkonen O, Vainio H (2007) Polycyclic aromatic hydrocarbon (PAH) metabolizing enzyme activities in human lung, and their inducibility by exposure to naphthalene, phenanthrene, pyrene, chrysene, and benzo(a)pyrene as shown in the rat lung and liver. Arch Toxicol 81:169-82

Espunya MC, Diaz M, Moreno-Romero J, Martinez MC (2006) Modification of intracellular levels of glutathione-dependent formaldehyde dehydrogenase alters glutathione homeostasis and root development. Plant Cell Environ 29:1002-11

Ficker MD, Meyer AJ (2001) Confocal imaging of metabolism in vivo: pitfalls and possibilities. J Exp Bot 52:631-40

Flocco CG, Lindblom SD, Smits EAHP (2004) Overexpression of enzymes involved in glutathione synthesis enhances tolerance to organic pollutants in *Brassica juncea*. Int J Phytorem 6:289-304

Gerson, U (1982) Bryophytes and invertebrates. In: Smith AJE (ed) Bryophyte ecology, Chapman and Hall, New York, pp. 291-332

Hawkesford MJ (2003) Transporter gene families in plants: the sulphate transporter gene family - redundancy or specialization? Physiol Plant 117:155-63

Hellwege EM, Dietz KJ, Hartung W (1996) Abscisic acid causes changes in gene expression involved in the induction of the landform of the liverwort *Riccia fluitans* L. Planta 198:423-32

Horling F, Baier M, Dietz KJ (2001) Redox-regulation of the expression of the peroxide-detoxifying chloroplast 2-Cys peroxiredoxin in the liverwort *Riccia fluitans*. Planta 214:304-13

Jackson PP, Robinson NJ, Whitton BA (1991) Low-molecular-weight metal-complexes in the fresh-water moss rhynchostegium-riparioides exposed to elevated concentrations of Zn, Cu, Cd and Pb in the laboratory and field. Environ Exp Bot 31:359-66

Kenrick P, Crane PR (1997) The origin and early evolution of plants on land. Nature 389:33-9

Knulst JC, Westling HO, Brorstromlunden E (1995) Airborne organic micropollutant concentrations in mosses and humus as indicators for local versus long-range sources. Environ Monit Assess 36:75-91

Kopriva S, Koprivova A (2004) Plant adenosine 5′-phosphosulphate reductase: the past, the present, and the future. J Exp Bot 55:1775-83

Kopriva S and Rennenberg H (2004) Control of sulphate assimilation and glutathione synthesis: interaction with N and C metabolism. J Exp Bot 55:1831-42

Kopriva S, Suter M, von Ballmoos P, Hesse H, Krahenbuhl U, Rennenberg H, Brunold C (2002) Interaction of sulfate assimilation with carbon and nitrogen metabolism in *Lemna minor*. Plant Physiol 130:1406-13

Longton RE (1980) Physiological ecology of mosses In: Taylor RJ and Leviton AE (eds) The mosses of North America Symposium. Pacific Division, American Association for the Advancement of Science, San Francisco, pp. 77-113

Marrs KA (1996) The functions and regulation of glutathione s-transferases in plants. Annu Rev Plant Physiol Plant Mol Biol 47:127-58

Marshall PA, Edgar GJ (2003) The effect of the Jessica grounding on subtidal invertebrate and plant communities at the Galapagos wreck site. Mar Poll Bull 47:284-95

Mendoza-Co'zatl D, Loza-Tavera H, Hernandez-Navarro A, Moreno-Sanchez R (2005) Sulfur assimilation and glutathione metabolism under cadmium stress in yeast, protists and plants. FEMS Microbiol Rev 29:653–671

Merrill GLS (1991) Bryophytes of Konza Prairie Research Natural Area, Kansas. Bryologist 94:383-91

Minibayeva F, Beckett RP (2001) High rates of extracellular superoxide production in bryophytes and lichens, and an oxidative burst in response to rehydration following desiccation. New Phytol 152:333-41

Mishler BD, Lewis LA, Buchheim MA, Renzaglia KS, Garbary DJ, Delwiche CF, Zechman FW, Kantz TS, Chapman RL (1994) Phylogenetic-relationships of the green-algae and bryophytes. Ann Miss Bot Gard 81:451-83

Nair RV, Fisher EP, Safe SH, Cortez C, Harvey RG, Digiovanni J (1991) Novel coumarins as potential anticarcinogenic agents. Carcinogenesis 12:65-9

Neff JM (1979) Polycyclic aromatic hydrocarbons in the aquatic environment. Sources: fate and biological effects. Applied Science Publishers, London, pp. 44-60

Noctor G, Arisi ACM, Jouanin L, Kunert KJ, Rennenberg H, Foyer CH (1998) Glutathione: biosynthesis, metabolism and relationship to stress tolerance explored in transformed plants. J Exp Bot 49:623-47

Noctor G, Foyer CH (1998) Ascorbate and glutathione: Keeping active oxygen under control. Annu Rev Plant Physiol Plant Mol Biol 49:249-79

Pilon-Smits E. (2005) Phytoremediation. Annu Rev Plant Biol 56:15-39

Potters G, Horemans N, Bellone S, Caubergs RJ, Trost P, Guisez Y, Asard H (2004) Dehydroascorbate influences the plant cell cycle through a glutathione-independent reduction mechanism. Plant Physiol 134:1479-87

Proctor M (2001) Patterns of desiccation tolerance and recovery in bryophytes. Plant Growth Regul 35:147-56

Rausch T, Wachter A (2005) Sulfur metabolism: a versatile platform for launching defence operations. Trends Plant Sci 10:503-9

Rother M, Krauss GJ, Grass G, Wesenberg D (2006) Sulphate assimilation under Cd^{2+} stress in *Physcomitrella patens* - combined transcript, enzyme and metabolite profiling. Plant Cell Environ 29:1801-11

Rouhier N, Gelhaye E, Jacquot JP (2004) Plant glutaredoxins: still mysterious reducing systems. Cell Mol Life Sci 61:1266-77

Saxena A, Saxena DK, Srivastava HS (2003) The influence of glutathione on physiological effects of lead and its accumulation in moss *Sphagnum squarrosum*. Water Air Soil Poll 143:351-61

Seel WE, Hendry GAF, Lee JA (1992) Effects of desiccation on some activated oxygen processing enzymes and antioxidants in mosses. J Exp Bot 43:1031-7

Siebert A, Bruns I, Krauss GJ, Miersch J, Markert B (1996) The use of the aquatic moss *Fontinalis antipyretica* L ex Hedw as a bioindicator for heavy metals. Sci Tot Environ 177:137-44

Strachan WNJ and Glooschenko WA (1988) Moss bags as monitors of organic contamination in the atmosphere. Bull Environ Contam Toxicol 40:447-50

Thomas W (1984) Statistical models for the accumulation of PAH, chlorinated hydrocarbons and trace metals in epiphytic *Hypnum Cupressiforme*. Water Air Soil Poll 22:351-71

Thomas W (1986) Representativity of mosses as biomonitor organisms for the accumulation of environmental chemicals in plants and soils. Ecotox Environ Saf 11:339-46

Tkalec M, Vidakovic-Cifrek Z, Regula I (1998) The effect of oil industry "high density brines" on duckweed *Lemna minor* L. Chemosphere 37:2703-15

Vernoux T, Wilson RC, Seeley KA, Reichheld JP, Muroy S, Brown S, Maughan SC, Cobbett CS, Van Montagu M, Inze D, May MJ, Sung ZR (2000) The ROOT MERISTEMLESS1/CADMIUM SENSITIVE2 gene defines a glutathione-dependent pathway involved in initiation and maintenance of cell division during postembryonic root development. Plant Cell 12:97-109

Vij S, Tyagi AK (2007) Emerging trends in the functional genomics of the abiotic stress response in crop plants. Plant Biotech J 5:361-80

Wachter A, Wolf S, Steininger H, Bogs J, Rausch T (2005) Differential targeting of GSH1 and GSH2 is achieved by multiple transcription initiation: implications for the compartmentation of glutathione biosynthesis in the Brassicaceae. Plant J 41:15-30

Wasley J, Robinson SA, Lovelock CE, Popp M (2006) Some like it wet - biological characteristics underpinning tolerance of extreme water stress events in Antarctic bryophytes. Func Plant Biol 33:443-55

Wegener JWM, van Schaik MJM, Aiking H (1992) Active biomonitoring of polycyclic aromatic hydrocarbons by means of mosses. Environ Poll 76:15-18

Chapter 15
Allocation of Sulfur to Sulfonium Compounds in Microalgae

Simona Ratti and Mario Giordano(✉)

Abstract Many algae are able to use S surplus to produce sulfonium compounds with different functions. Among these, dimethylsulfoniopropionate (DMSP) has attracted the attention of the scientific community for its multiple functions. DMSP is produced by some salt-tolerant angiosperms and by many marine algae. Different phytoplankton groups show different abilities to produce this compound: dinoflagellates, diatoms, and coccolithophores usually produce large amounts of DMSP, while Chlorophytes and cyanobacteria tend to produce very little DMSP, if any. DMSP has been proposed to act as thermoprotectant, osmoprotectant, antioxidant, and antigrazing agent. The antigrazing activity of DMSP requires its cleavage by a specific lyase to produce dimethylsulfide (DMS) and acrylate, the latter being the antigrazing agent. DMS has been suggested to play an important role in climate control. This review discusses the allocation of sulfur to different sulfonium compounds, mostly focusing on the role played by DMSP in response to abiotic and biotic stresses and on their implications in the ecology and evolution of phytoplankton.

1 Introduction

Sulfur is an essential element for the growth of photosynthetic organisms. While its metabolism has been studied in depth for flowering plants (Hell 1997), relatively little is known for algae (Giordano et al. 2007). Sulfur is acquired from the environment as sulfate; it is reduced to sulfide and is then incorporated as a thiol group in cysteine (Leustek and Saito 1999). In higher plants and algae, S-containing metabolites deriving from cysteine participate in a variety of cellular processes, such as resistance against diseases (Booth and Walker 1997), tolerance to oxidation (Lappartient and Touraine 1997, Sunda et al. 2002), tolerance to heavy metals (Schafer et al. 1997), and osmotic stress (Storey et al. 1993, Stefels 2000).

Mario Giordano
Laboratorio di Fisiologia delle Alghe, Dipartimento di Scienze del Mare,
Università Politecnica delle Marche, Via Brecce Bianche, 60131 Ancona, Italy
m.giordano@univpm.it

N.A. Khan et al. (eds.), *Sulfur Assimilation and Abiotic Stress in Plants.*

Sulfate abundance in freshwater and marine environments can be very different. In lakes, sulfate concentration is more variable and usually lower (0.01-1 mM; Holmer and Storkholm 2001, Giordano et al. 2007) than in the sea, where sulfate concentration is quite constant and around 29 mM (Pilson 1998). Consequently, S utilization may differ substantially in freshwater and marine species. In general, freshwater algae are more efficient in the use of S per unit of C (Heldal et al. 2003, Ho et al. 2003). Marine algae are often characterized by a lower C:S ratio compared to their freshwater counterparts; nevertheless, the large excess of S in the sea allows them to allocate substantial amounts of S in a number of compounds with different functions, such us sulfated extracellular polysaccharides (Painter 1983, Farias et al. 2000) and sulfonium compounds like S-adenosyl-L-methionine (Adomet, Cantoni 1953), S-methyl-L-methionine (SMM, Maw 1981), phosphotidylsulfocholine (PSC, Bisseret et al. 1984), cis-2-(dimethylsulfonio)cyclopropane carboxylate (Gonyauline, Nakamura et al. 1992), 3S-5-dimethylsulfonio-3-hydroxypentanoate (Gonyol, Nakamura et al. 1993), and 3-dimethylsulfoniopropionate (DMSP). DMSP is certainly the most studied of these molecules, and it is allegedly involved in response mechanisms to thermal (Kirst et al. 1991, Nishiguchi and Somero 1992, Karsten et al. 1992), osmotic (Stefels 2000, Welsh 2000), oxidative (Sunda et al. 2002), and grazing stresses (Dacey and Wakeham 1986, Wolfe et al. 1997, Wolfe 2000, Van Alstyne et al. 2001, Strom et al. 2003a, Strom et al. 2003b, Kasamatsu et al. 2004b). DMSP is produced by some flowering plants, such as *Spartina* sp., *Saccharum* sp., *Wollastonia biflora, Posidonia* sp., *Zostera* sp., and by a lot of marine macro and microalgae (Keller et al. 1989, Otte and Morris 1994, Hanson and Gage 1996, Kocsis et al. 1998, Otte et al., 2004). Among microalgae, dinoflagellates, coccolithophores, and diatoms are the taxa that, on average, produce and accumulate more DMSP. In this review, we shall focus on the role of DMSP produced by marine algae in response mechanisms to biotic and abiotic stresses.

2 Sulfur Acquisition and Assimilation by Algae

Algae acquire S as sulfate. There are some indications that, in algae, sulfate uptake is highly regulated by sulfate availability (Hawkesford and Wray 2000, Giordano et al. 2005). Yildiz and co-workers (1994) demonstrated that *Chlamydomonas reinhardtii* grown under S-deficient conditions rapidly increases both affinity and capacity for sulfate uptake, whereas both these parameters decrease when *C. reinhardtii* is cultured in the presence of nonlimiting sulfate concentrations. In *C. reinhardtii*, S limitation promotes the expression of genes encoding high-affinity sulfate transporters and arylsulfatase (de Hostos et al. 1989, Yildiz et al., 1994, Pollock et al. 2005). Arylsulfatases are released in the extracellular medium and facilitate the utilization of exogenous, esterified sulfate, thanks to their ability to cleave ester bonds between sulfate and organic substances (Fig. 15.1; de Hostos et al. 1988, Ravina et al. 2002). The transport systems work both in the light and in the dark and are constituted by integral membrane proteins with 12 transmembrane domains (Pollock et al. 2005). These transporters promote sulfate/H^+ co-transport (Pollock et al. 2005). A proton gradient generated by an ATPase localized in the plasma membrane drives the transport (Perez-Castineira et al. 1992, Yildiz et al. 1994, Pollock et al. 2005). In *C. reinhardtii*, a sulfate permease in the chloroplast envelope was also found. This transport system is presumably responsible for sulfate import into the plastid, where primary assimilation of sulfate and biosynthesis of cysteine and methionine take place (Fig. 15.1; Leustek et al. 2000, Chen et al. 2003, Melis and Chen 2005). This plastidial sulfate permease is encoded by four nuclear genes, whose transcription and translation are induced under S-limitation (Chen et al. 2003, Chen and Melis 2004, Melis and Chen 2005, Chen et al. 2005, Pollock et al. 2005). The holocomplex is an ABC-type transporter located in the inner chloroplast envelope (Sirko et al. 1990, Melis and Chen 2005). Unfortunately, the relatively detailed understanding of sulfate acquisition in *Chlamydomonas* cannot be transferred *sic et simpliciter* to other microalgae. With respect to marine phytoplankton, it should be considered that sulfate concentration in the oceans is quite high and constant; the presence of a finely regulated system may thus be unnecessary (Giordano et al. 2005).

Sulfate is a chemically very stable compound, and its activation is required prior to assimilation. Sulfate assimilation is initiated by the enzyme ATP-sulfurylase, which catalyzes the adenylation of sulfate to 5′-adenylylsulfate (APS, Fig. 15.1). APS is then reduced to sulfite by an APS-reductase (APR, Hell 1997, Bick and Leustek 1998, Saito 2000). Plant APR is unique in that it is able to use reduced glutathione as the electron source. Gao and collaborators (2000) showed that APR from algae is a highly homologous to APR from flowering plants. Sulfite is reduced to

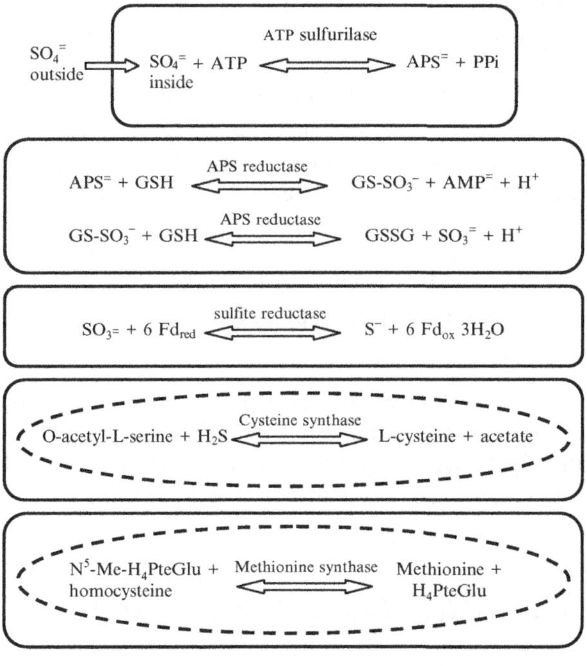

Fig. 15.1 Sulfate metabolism in algae. Fd, Ferredoxin; GSH reduced glutathione; GSSG, oxidized glutathione; H_4PteGlu, tetrahydropteryoglutamine; Me-H_4PteGlu, Methyltetrahydropteryol-L-glutamine. The continuous line represents the plasmalemma; the dashed line represents the chloroplast envelop

sulfide by sulfite reductase; sulfide is then incorporated into cysteine. In higher plants, cysteine formation takes place in the cytosol, in the chloroplast, and in the mitochondrion, while in *C. reinhardtii* it occurs only in the chloroplast (Hell 1997, Ravina et al. 2002). Cysteine is the first stable organic compound containing reduced S; all other S-compounds (among which is the other S-amino acid, methionine) derive from it (Hell 1997, Bork et al. 1998). No methionine synthetases were isolated from plastids, suggesting that cysteine is exported to the cytosol for subsequent modification; cysteine is thus generated in the cytosol and then reimported in the chloroplast (Eichel et al. 1995, Ravina et al. 2002).

3 Biosynthesis of Sulfonium Compounds by Algae

3.1 Synthesis of S-Adenosyl-L-Methionine (AdoMet)

About 80% of the methionine produced from cysteine is used for the synthesis of S-adenosyl-L-methionine (AdoMet). All the information available on AdoMet was obtained from higher plants, and no direct investigations of this pathway have been conducted in algae. AdoMet is the major methyl-group donor in a number of transmethylation reactions and, in plants, it is also an intermediate in the biosynthesis of polyamine and of the phytohormone ethylene (Cantoni 1952, Cantoni 1953, Giovanelli et al. 1985, Zarembinski and Theologis 1994, Nakano et al. 2000).

The synthesis of AdoMet from Met and ATP is catalyzed by cytosolic AdoMet synthetase. The catalysis of this enzyme involves the formation of an enzyme-bound inorganic tripolyphosphate intermediate that is cleaved by a tripolyphosphatase activity of the enzyme (Chou and Talalay 1972, Lee et al. 1997, Ravanel et al. 1998).

$$\text{L-Met} + \text{ATP} \rightarrow \text{AdoMet} + \text{PPP}_i$$
$$\text{PPP}_i + \text{H}_2\text{O} \rightarrow \text{P}_i + \text{PP}_i$$

In higher plants, AdoMet synthase is encoded by a gene family; members of this gene family have been cloned and characterized in different higher plant species, revealing that this synthetase is highly conserved within plants (Ravanel et al. 1998). Whether these genes are also conserved in algae remains to be ascertained.

3.2 Synthesis of S-Methyl-L-Methionine (SMM)

S-methyl-L-methionine (SMM) is produced in many higher plants and in algae (Maw 1981). SMM and S-adenosyl-L-homocysteine are the products of the reaction between AdoMet and L-methionine catalyzed by the enzyme AdoMet:Met S-methyltransferase. SMM can be converted back to L-methionine by the SMM: L-homocysteine S-methyltransferase, or it can be used by SMM hydrolase to produce DMS and L-homoserine (Mudd and Datko 1990, Cooper 1996). SMM metabolism, in combination with AdoMet metabolism, constitutes the so-called "S-methylmethionine cycle" in which AdoMet is converted to L-methionine and adenosine:

1. AdoMet + L-methionine → SMM + S-adenosyl-L-homocysteine (catalyzed by AdoMet:methionine S-methyltransferase)
2. SMM + L-homocysteine → 2 L-methionine + H^+ (catalyzed by SMM:L-homocysteine S-methyltransferase)
3. S-adenosyl-L-homocysteine + H_2O ↔ Adenosine + L-homocysteine (catalyzed by S-adenosyl-L-homocysteine hydrolase)
4. AdoMet + H_2O → Adenosine + L-methionine + H^+

3.3 Synthesis of Phosphatidylsulfocholine (PSC)

Phosphatidylsulfocholine (phosphatidyl-S,S-dimethyl mercaptoethanol, PSC) was initially found in the membranes of the nonphotosynthetic diatom *Nitzschia alba*, where it replaces the membrane phospholipids phosphatidylcholine (PC, Anderson et al. 1978a, b). Later, PSC was also found in several photosynthetic diatoms (Bisseret et al. 1984). In the nonphotosynthetic species, PSC completely replaces PC; in photosynthetic species, the two lipids coexist and PC is the dominant form (Kates and Volcani 1966, Bisseret et al. 1984). In diatoms, PSC is synthesized from serine, via cysteine and methionine, in a cystathionine pathway identical to that described for higher plants (EQ. 1; Kates and Volcani 1996). Methionine is then converted to DMSP (see paragraph 3.4), which serves as a substrate for sulfocholine synthesis (EQ. 2). Sulfocholine is finally incorporated into PSC by phosphocholine transferase in the Kennedy nucleotide pathway. (EQ. 3-4, Kates and Volcani 1996).

$$\text{Ser} \xrightarrow{+\text{HS}^-} \text{Cys} \xrightarrow{+\text{HomoSer}} \text{Cystathionine} \longrightarrow \text{HomoCys} \xrightarrow{\text{ATP}} \text{Met} \qquad \text{(EQ. 1)}$$

$$\text{Met} \longrightarrow \text{DMSP} \longrightarrow \text{Sulfocholine} \qquad \text{(EQ. 2)}$$

$$\text{Sulfocholine} \xrightarrow{\text{ATP}} \text{P-sulfocholine} \xrightarrow{\text{CTP}} \text{CDP-sulfocholine} \qquad \text{(EQ. 3)}$$

$$\text{CDP-sulfocholine} \xrightarrow{\text{1,2-diacylglycerol}} \text{PSC} \qquad \text{(EQ. 4)}$$

3.4 Synthesis of 3-Dimethylsulfonioproprionate (DMSP)

DMSP is a tertiary sulfonium compound produced by some halophytic angiosperm and many marine algae (Keller et al. 1989). In DMSP producers, this compound comprises a high percentage (50%-90%) of total organic S in the cell, with concentrations between 50 and 400 mM (Matrai and Keller 1994, Keller et al. 1999, Wolfe 2000, Yoch 2002). Four different DMSP biosynthetic pathways were identified, two in plants, one in the marine macroalgae *Enteromorpha intestinalis* (Gage et al. 1997), and one was hypothesized for the heterotrophic dinoflagellate *Crypthecodinium cohnii* (Uchida et al. 1996); in all cases, methionine is the point of origin of the pathway. Gage and colleagues (1997), using in vivo isotope labeling, identified the four steps necessary for DMSP production in *E. intestinalis*. Following the fate of [^{35}S]Met they found that, initially, a substrate-specific 2-oxoglutarate-dependent aminotransferase converts L-methionine into 4-methylthio-2-oxobutyrate (MTOB). MTOB is then reduced by an NADPH-linked reductase to 4-methylthio-2-hydroxybutyrate (MTHB) that is subsequently S-methylated to 4-dimethylsulfonio-2-hydroxybutyrate (DMSHB) by a S-adenosylmethionine-dependent methyltransferase. The last reaction yields DMSP via the oxidative decarboxylation of DMSHB (Gage et al. 1997, Summers et al. 1998). Gage et al. (1997) tested the same biosynthetic pathway in three marine phytoplankton species, the Prymnesiophycea *Emiliania huxleyi*, the diatom *Melosira nummuloides*, and the Prasinophycea *Tetraselmis* sp. The authors found that these planktonic species produce DMSP in the same way as *E. intestinalis*. A slightly different pathway was proposed for DMSP biosynthesis in the dinoflagellate *C. cohnii*; the path includes the same steps of deamination, decarboxilation, and S-methylation, but in this last case the synthesis starts with the decarboxylation of methionine instead of deamination, leading to different intermediates. Decarboxlation of methionine, in fact, produces methylthiopropylamine, which is then deaminated to methylmercaptoproprionate (MMPA, Uchida et al. 1996). The final step is S-methylation of MMPA to form DMSP. Nakamura et al. (1997) suggested that a similar biosynthetic path is present also in the photosynthetic dinoflagellates *Gonyaulax polyedra, Amphydinium* sp. and *Pyrocystis lunula*. In higher plants, instead, DMSP biosynthesis involves SMM as an intermediate, and in these organisms SMM is converted to DMSP via 3-dimethylsulfoniopropionaldehyde (DMSP-ald, Hanson and Gage 1996, Kocsis et al. 1998).

Catabolism of DMSP requires enzymatic cleavage to acrylate, dimethylsulfide (DMS), and a proton (Wolf et al. 2000). Acrylate is toxic for many organisms. Its conjugated double bond structure, in fact, makes acrylate a Michael acceptor (Michael 1887) that may react with glutathione, thus causing its depletion (Freidig et al. 1999). Further to this specific mechanism, acrylate toxicity may be due to its hydrophobicity and acidity (Sieburth 1960, 1961, Freidig et al. 1999). DMS is a volatile compound that is oxidized to acidic aerosol sulfates in the atmosphere; the products of DMS oxidation act as cloud condensation nuclei, influencing global albedo and consequently global climate, prior to their deposition as acid rain (Lovelock et al. 1972, Nguyen et al. 1983, Bates et al. 1987, Charlson et al. 1987, Liss et al. 1997).

Biogenic DMS, derived from algal and bacterial degradation of DMSP, is a very important source of atmospheric S (1.5×10^{13} g of S y^{-1} to the atmosphere) and represents one of the major players in the global S cycle (Malin 1996). The biological cleavage of DMSP is catalyzed by DMSP lyases; these enzymes (or their activity) have been found in the marine fungus *Fusarium lateritium* (Basic and Yoch 1998), in bacteria (de Souza and Yoch 1995a,b, Yoch et al. 1997), and in several marine algae (Stefels and van Boekel 1993, de Souza et al. 1996, Stefels and Dijkhuizen 1996, Steinke and Kirst 1996, Niki et al. 2000). The amount of DMSP-lyase activity is species-specific and probably strain-specific (Keller 1991, Stefels and van Boekel 1993, de Souza et al. 1996, Stefels and Dijkhuizen 1996, Steinke and Kirst 1996, Wolfe and Steinke 1996, Steinke et al. 1998, Niki et al. 2000, Wolfe et al. 2002). DMSP lyase and its substrate are physically separated in algal cells, and this may explain why healthy microalgae produce lower amounts of DMS and acrylate than cells subjected to stresses (Stefels and van Boekel 1993, Wolfe and Steinke 1996, Steinke et al. 1998). In fact, the production of DMS and acrylate from DMSP increases during mechanical and chemical stress (Wolfe et al. 2002, Bucciarelli and Sunda 2003), senescence (Nguyen et al. 1988, van Boekel et al. 1992, Stefels and van Boekel 1993, Kwint and Kramer 1995, Laroche et al. 1999), viral lysis, microbial attack (Leck et al. 1990, Bratbak et al. 1995, Hill et al. 1998, Kiene and Linn 2000), and during grazing by zooplankton, when the DMSP and its enzyme come in contact (Dacey and Wakeham 1986, Belviso et al. 1990, Wolfe et al. 1994, Christaki et al. 1996, Levasseur et al. 1996, Liss et al. 1997, Wolfe et al. 1997, Van Alstyne et al. 2001, Van Alstyne and Houser 2003).

DMSP catabolism does not always imply DMS production. Same bacteria are able to demethylate DMSP to 3-mercaptoproprionate. This reaction involves methiolproprionate as an intermediate, which is then demethiolated to yield methanethiol and acrylate (Kiene and Taylor 1988, Taylor and Gilchrist 1991, Taylor and Visscher 1996).

3.5 *Other Sulfonium Compounds of Microalgae*

Two other sulfonium compounds, *cis*-2-(dimethylsulfonio)cyclopropane carboxylate (Gonyauline) and of 3S-5-dimethylsulfonio-3-hydroxypentanoate (Gonyol), have been detected in some dinoflagellates (Nakamura et al. 1993, 1996, 1997). Gonyauline is a zwitterionic S-compound produced by the dinoflagellate *Gonyaulax polyedra*, which is involved in the regulation of the bioluminescent circadian rhythm (Roenneberg et al. 1991, Nakamura et al. 1992). Gonyol is also believed to play a role in circadian rhythm regulation; it was identified in several dinoflagellates, especially when they were grown on a medium rich in methionine and acetate (Nakamura et al. 1993, 1996, 1997); in organisms like *Pyrocystis lunula*, gonyol is often the most abundant sulfonium compound (Nakamura 1997). Nakamura et al. (1992, 1993, 1997) conducted experiments on *G. polyedra* and showed that gonyauline and gonyol receive a methyl group from methionine via DMSP. DMSP thus appears to be also involved in the regulation of circadian rhythms in dinoflagellates (Nakamura 1993, 1997, Miller 2004 and references therein).

3.6 Osmotic Stress

Phytoplankton responses to osmotic stress are initially characterized by rapid water fluxes usually followed by adjustments of osmolyte concentrations to balance the osmotic pressure of the external medium and maintain cell turgor (Kirst 1990, Welsh 2000). A number of osmolytes are used by microalgae, such as ions and low molecular weight organic molecules. Usually, these osmolytes fit the description of a compatible solute: at low water potential, they can be accumulated at very high concentration with no negative effects, have a stabilizing and protective action on membranes and enzymes, and are promptly degraded when osmolarity decreases. A large variety of compatible solutes are produced by algae; often different compounds are used depending on the nutritional conditions and resource availability (Welsh 2000). Organic N-compounds (proline, betaines) and some carbohydrates (e.g., nonreducing sugars, polyols, heterosides) can play this role (Borowitza 1981, Reed 1990, Kirst 1990, Stefels 2000). Microalgae also use DMSP as osmolyte, and there are several reports showing a direct coupling between intracellular DMSP concentration and salinity, especially under long-term stress conditions (Vairavamurthy et al. 1985, Dickson and Kirst 1987ab, Karsten et al. 1992, Kirst 1996, Stefels 2000, Van Bergeijk et al. 2003). Intracellular DMSP concentration increases under N-depleted conditions (Dacey et al. 1987, Gröne and Kirst 1992, Keller and Korjeff-Bellows 1996, Bucciarelli and Sunda 2003). This may be due to the fact that DMSP is a structural analogue of glycine betaine, a quaternary ammonium compound that, like DMSP, acts as a compatible solute (Rhodes and Hanson 1993). DMSP differs from betaine only by the presence of S in the place of N. Under conditions of limited N availability, betaine production by phytoplankton declines, while DMSP production tends to increase. The production of DMSP is likely to contribute to the saving of nitrogen in N-poor and S-rich environments such as most oceanic water (Dacey et al. 1987, Turner et al. 1988, Grone and Kirts 1992, McNeil et al. 1999).

3.7 Hypothermic Stress

DMSP and the products of its degradation have also been proposed to act as a cryoprotectant. Several authors reported that production and accumulation of DMSP increase concomitantly with a temperature decline, both in macro and microalgae (Karsten et al. 1990, 1992, 1996, Sheets and Rhodes 1996, Rijssel and Gieskes 2002). This increase has been reported for phytoplankton of various taxonomy, such as the Prasinophycea *Tetraselmis subcordiformis* (Sheets and Rhodes 1996), diatoms, the Prymnesiophyceae *Emiliania huxleyi* (Meyerdierks 1997), and *Phaeocystis antarctica*

(Baumann et al. 1994). However, Kasamatsu et al. (2004a) conducted a comparison between DMSP production in temperate and polar psychrophilic diatoms and found that there were no significant differences in the intracellular accumulation of DMSP between these two groups of algae. This suggests that DMSP accumulation is not a prerequisite for the maintenance of the psychrophilic properties in the species considered in that study. DMSP accumulation at low temperature may in fact be related to its role as compatible soluble: it has been suggested that DMSP works best as a compatible solute at low temperature (Nishiguchi and Somero 1992, Karsten et al. 1996), when it is especially effective in protecting enzymes. DMSP, for instance, appears to be an excellent cryoprotectant for phosphofructokinase under conditions of cold-induced denaturation, but not at high temperature when, instead, DMSP could destabilize the enzyme (Nishiguchi and Somero 1992, Welsh 2000).

3.8 Oxidative Stress

DMSP and its breakdown products (DMS and acrylate) are powerful scavengers of reactive oxygen species. They thus represent an antioxidant system regulated, at least in part, by the enzymatic cleavage of DMSP itself (Jakob and Heber 1996, Sunda et al. 2002). Cell concentration of DMSP and/or its breakdown increase in the presence of all those factors that induce oxidative stress, especially exposure to UV radiations and H_2O_2, CO_2 limitation, Fe limitation, and high Cu^{++} concentration (Sunda et al. 2002, Bucciarelli and Sunda 2003). Various reactive oxygen species are targeted by the different components of this antioxidant system. DMS is very reactive toward oxygen singlet; both DMS and acrylate are very effective hydroxyl radicals (·OH) quenchers, while DMSP exerts only a modest protective action against ·OH (Wilkinson et al. 1995). Unlike the polar/charged (at physiological pH) species DMSP and acrylate, DMS can diffuse into membranes; this allows DMS to act as a powerful countermeasure to lipid peroxidation (Sunda et al. 2002). The reaction between DMS and ·OH generates dimethylsulphoxide (DMSO) that, being more hydrophylic than DMS, does not go through membranes and accumulates in the cell (Lee and de Mora 1999, Simó et al. 2000). DMSO promptly reacts with ·OH; the reaction of DMSO with ·OH generates the antioxidant methane sulphinic acid (MSNA). MSNA is a water-soluble molecule that reacts with ·OH (Scaduto 1995). Sunda et al. (2002) demonstrated that this elaborate antioxidant system, encompassing both water- and lipid-soluble compounds, is induced by the presence of oxidative stressors and is modulated by both the intracellular accumulation of DMSP and by the activity of DMSP lyase, which, in turn, controls the amount of the other scavengers.

The stimulation of DMSP production in response to N limitation may also be associated with its antioxidant function, since N deficiency increases oxidative stress and makes the synthesis of N-containing antioxidants such as glutathione difficult (Evans and Terashima 1987, Berges and Falkowzski 1998, Logan et al. 1999).

4 Ecological and Evolutionary Impact of DMSP in Algae

What was the original function of DMSP when it appeared and what competitive advantage it gave to the phytoplankters that produced it is an open question. The history of the oceans and the evolutionary trajectory of planktonic communities may provide some clues: Proterozoic oceans were characterized by vast anoxic regions, the consequence of which was a severe shortage of N in the photic oxic zones (Anbar and Knoll 2002). Since most of the major DMSP producers appeared in this period of Earth's history (Katz et al. 2004) it seems plausible that DMSP initially appeared as a substitute for the N-containing analogues with osmoprotectant functions (e.g., glycine betaine); the close structural similarity between DMSP and glycine betaine may have made it relatively easy to recruit portions of the existing pathways to make the new, N-free osmoprotectant.

The prevalence of DMSP producers in the ocean (mostly algae of the "red line" *sensu* Falkowski et al. 2004), however, was only established at the beginning of the Mesozoic, long after their first appearance (Katz et al. 2004). In this eon, oceans were well oxygenated, rich in sulfate, and inhabited by numerous and effective grazers (Bambach 1993, Canfield 1998, Anbar and Knoll 2002, Knoll 2003). It has been proposed (Norici et al. 2005, Giordano et al. 2007) that at this time of transition the DMSP producers, due to the steep increase in grazing pressure, were favored over those organisms with little DMSP, thanks to the repulsive/toxic action that DMSP degradation products exerted on grazers (Sieburth 1960, Sieburth 1961, Freidig et al. 1999).

References

Anbar AD, Knoll AH (2002) Proterozoic ocean chemistry and evolution: a bioinorganic bridge? Science 297:1137-42

Anderson R, Kates M, Volcani BE (1978a) Identification of the sulfolipids in the non-photosynthetic diatom *Nitzschia alba*. Biochim Biophys Acta 528:89-106

Anderson R, Livermore BP, Kates M, Volcani BE (1978b) The lipid composition of the non-photosynthetic diatom *Nitzschia alba*. Biochim Biophys Acta 528:77-88

Bambach RK (1993) Seafood through time: changes in biomass, energetics, and productivity in the marine ecosystem. Paleobiol 19:372-97

Basic MK, Yoch DC (1998) *In vivo* characterization of dimethylsulfonioproprionate lyase in the fungus *Fusarium lateritium*. Appl Environ Microbiol 64:106-11

Bates TS, Charlson RJ, Gammon RH (1987) Evidence for the climatic role of marine biogenic sulfur. Nature 329:319-21

Baumann MEM, Brandini FP, Staubes R (1994) The influence of light and temperature on carbon-specific DMS release by culture of *Phaeocystis aantarctica* and three Antarctic diatoms. Mar Chem 45:129-36

Belviso S, Kim SK, Rassoulzadegan, Krajka B, Nguyen BC, Mihalopoulos N, Buat-Menard P (1990) Production of dimethylsulfonio propionate (DMSP) and dimethylsulfide (DMS) by a microbial food web. Limnol Oceanogr 35:1810-21

Berges JA, Falkowski PG (1998) Physiological stress and cell death in marine phytoplankton: Induction of proteases in response to nitrogen or light limitation. Limnol Oceanogr 43:129-35

Bick JA, Leustek T (1998) Plant sulfur metabolism-the reduction of sulfate to sulfite. Curr Opin Plant Biol 1:240-4

Bisseret P, Ito S, Tremblay PA, Volcani BE, Dessort D, Kates M (1984) Occurrence of phosphatidylsulfocholine, the sulfonium analog of phosphatidylcoline in some diatoms and algae. Biochim Biophys Acta 796:320-7

Booth E, Walker K (1997) The effectiveness of foliar glucosinolate content raised by sulfur application on disease control in oilseed rape. In: Cram WJ, De Kok LJ, Stulen I, Brunold C, Rennenberg H (eds) Sulphur metabolism in higher plants, Backhuys Publisher, Leiden, The Netherlands, pp. 327-9

Bork C, Schwenn J-D, Hell R (1998) Isolation and characterization of a gene for assimilaory sulfite reductase from *Arabidopsis thaliana*. Gene 212:147-53

Borowitzka LJ (1981) Solute accumulation and regulation of cell water activity. In Paleg LG, Aspinall D (eds) Physiology and biochemistry of drought resistance in plants, Academic Press, Melbourne, pp. 97-30,

Bratbak G, Levasseur M, Michaud S, Cantin G, Fernández E, Heimdal BR, Heldal M (1995) Viral activity in relation to *Emiliania huxleyi* blooms: a mechanism of DMSP release? Mar Ecol Prog Ser 128:133-42

Bucciarelli E, Sunda WG (2003) Influence of CO_2, nitrate, phosphate, and silicate limitation on intracellular dimethylsulfoniopropionate in batch cultures of the coastal diatom *Thalassiosira pseudonana*. Limnol Oceanogr 48:2256-65

Canfield DE (1998) A new model for Proterozoic ocean chemistry. Nature 396:450-3

Cantoni GL (1952) The nature of the active methyl donor formed enzymatically form L-methionine and adenosinetriphosphate. J Am Chem Soc 74:2942-3

Cantoni GL (1953) S-Adenosylmethionine: a new intermediate formed enzymatically from L-methioninc and adenosinetriophosphate. J Biol Chem 204:403-6

Charlson RJ, Lovelock JE, Andreae MO, Warren SG (1987) Oceanic phytoplankton, atmospheric sulphur, cloud albedo and climate. Nature 326:655-61

Chen HC, Melis A (2004) Localization and function of SulP, a nuclear-encoded chloroplast sulfate permease in *Chlamydomonas reinhardtii*. Planta 220:198-210

Chen HC, Newton AJ, Melis A (2005) Role of SulP, a nuclear-encoded chloroplast sulfate permease, in sulfate transport and H_2 evolution in *Chlamydomonas reinhardtii*. Photosynth Res 84:289-96

Chen HC, Yokthongwattana K, Newton AJ, Melis A (2003) *SulP*, a nuclear gene encoding a putative chloroplast-targeted sulfate permease in *Clamydomonas reinhardtii*. Planta 218:98-106

Chou T-C, Talalay P (1972) The mechanism of S-adenosyl-L-methionine synthesis by purified preparations of baker's yeast. Biochem 11:1065-73

Cooper AJL (1996) Chemical and biochemical properties of sulfonium compounds. In Kiene RP, Visscher PT, Keller MD, Kirst GO (eds) Biological and environmental chemistry of DMSP and related sulfonium compounds, Plenum Press, New York, pp. 75-86

Dacey JWH, King GM, Wakeham SG (1987) Factors controlling emission of dimethylsulphide from salt marshes. Nature 330:643-5

Dacey JWH, Wakeham SG (1986) Oceanic dimethylsulfide: production during zooplankton grazing on phytoplankton. Science 233:1314-15

de Hostos E, Schilling J, Grossman AR (1989) Structure and expression of the gene encoding the periplasmic arylsulfatase in *Chlamydomonas reinhardtii*. Mol Gen Genet 218:229-39

de Hostos E, Togasaki RK, Grossman Ar (1988) Purification and biosynthesis of derepressible periplasmic arylsulfatase from *Chlamydomonas reinhardtii*. J Cell Biol 106:29-37

de Souza MP, Chen YP, Yoch DC (1996) Dimethylsulfoniopropionate lyase from the marine macroalga *Ulva curvata:* purification and characterization of the enzyme. Planta 199:433-8

de Souza MP, Yoch DC (1995a) Comparative physiology of dimethyl sulphide production by dimethylsulfoniopropionate lyase in *Pseudomonas doudoroffii* and *Alcaligenes* sp. strain M3A. Appl Environ Microbiol 61:3986-91 de Souza MP, Yoch DC (1995b) Purification and characterization of dimethylsulfoniopropionate lyase from an *Alcaligenes*-like dimethyl sulphide-producing marine isolate. Appl Environ Microbiol 61:21-6

Dickson DJJ, Kirst GO (1986) The role of dimethylsulphonioproprionate, glycine betaine and homarine in the osmoacclimation of *Platymonas subcordiformis*. Planta 167:536-43

Dickson DM, Kirst GO (1987a) Osmotic adjustment in marine eukaryotic algae, the role of inorganic ions, quaternary ammonium, tertiary sulphonium and carbohydrate solutes. I. Diatoms and Rhodophyte. New Phytol 106:645-55

Dickson DM, Kirst GO (1987b) Osmotic adjustment in marine eukaryotic algae, the role of inorganic ions, quaternary ammonium, tertiary sulphonium and carbohydrate solutes. II. Prasinophytes and Haptophytes. New Phytol 106:657-66

Eichel J, Gonzales JC, Hotze M, Matthews RG, Schröder J (1995) Vitamin-B12-independent methionine synthase from a higher plant (*Catharantus roseus*). Molecular characterization, regulation, heterologous expression, and enzyme properties. Eur J Biochem 230:1053-8

Evans JR, Terashima I (1987) Effects of nitrogen nutrition on electron transport components and photosynthesis in spinach. Aust J Plant Physiol 14:59-68

Falkowski PG, Katz ME, Knoll A, Quigg A, Raven JA, Schofield O, Taylor FJR (2004) The evolution of modern eukaryotic phytoplankton. Science 305:354-60

Farias WRL, Valente AP, Pereira MS, Mourão PAS (2000) Structure and anticoagulant activity of sulfated galactans. Isolation of a unique sulfated galactan from the red algae *Botryocladia occidentalis* and comparison of its anticoagulant action with that of sulfated galactans from invertebrates. J Biol Chem 275:29299-397

Freidig AP, Verhaar HJM, Hermens JLM (1999) Comparing the potency of chemicals with multiple modes of action in aquatic toxicology: acute toxicity due to narcosis versus reactive toxicity of acrylic compounds. Environ Sci Technol 33:3038-43

Gage DA, Rhodes D, Nolte KD, Hicks WA, Leustek T, Cooper AJL, Hanson AD (1997) A new route for synthesis of dimethylsulphonioproprionate in marine algae. Nature 387:891-4

Gao Y, Schofield OME, Leustek T (2000) Characterization of sulfate assimilation in marine algae focusing on the enzyme 5′-adenylylsulfate reductase. Plant Physiol 123:1087-96

Giordano M, Norici A, Hell R (2005) Sulfur and phytoplankton: acquisition, metabolism and impact on the environment. New Phytol 166:371-82

Giordano M, Norici A, Ratti S, Raven JAR. (2008) Role of sulfur for phytoplankton: acquisition, metabolism and ecology. In: Hell R, Leustek T, Knaff D, Dahl C (eds.), Advances in Photosynthesis Research Vol. 27: Sulfur metabolism in phototrophic organisms, Govindjee (series ed.), Springer, Berlin, Germany, pp. 405–423

Giovanelli J, Mudd SH, Datko AH (1985) Quantitative analysis of pathways of methionine metabolism and their regulation in *Lemna*. Plant Physiol 78:555-60

Gröne T, Kirst GO (1991) Aspects of dimethylsulfoniopropionate effects on enzymes isolated from the marine phytoplankter *Tetraselmis subcordiformis* (Stein). J Plant Physiol 138:85-91

Gröne T, Kirst GO (1992) The effect of nitrogen deficiency, methionine and inhibitors of methionine metabolism on the DMSPp contents of *Tetraselmis subcordiformis* (Stein). Mar Bio 112:497-503

Hanson AD, Gage DA (1996) 3-dimethylsulfonioproprionate biosynthesis and use by flowering plants. In Kiene RP, Visscher PT, Keller MD, Kirst GO (eds) Biological and environmental chemistry of DMSP and related sulfonium compounds, Plenum Press, New York, pp. 75-86,

Hawkesford MJ, Wray JL (2000) Molecular genetics of sulfur assimilation. Adv Bot Res 33:159-23

Heldal M, Scanlan DJ, Norland S, Thingstad F and Mann NH (2003) Elemental composition of single cells of various strains of marine *Prochlorococcus* and *Synechococcus* using X-ray microanalysis. Limnol Oceanogr 48:1732-43

Hell R (1997) Molecular physiology of plant sulfur metabolism. Planta 202:138-48

Hill RW, White BA, Cottrell M, Dacey JWH (1998) Virus-mediated total release of dimethylsulfoniopropionate from marine phytoplankton: a potential climate process. Aquat Microb Ecol 14:1-6

Ho T-Y, Quigg A, Finkel ZV, Milligan AJ, Wyman K, Falkowski PG and Morel FMM (2003) The elemental composition of some marine phytoplankton. J Phycol 39:145-59

Holmer M, Storkholm P (2001) Sulfate reduction and sulfur cycling in lake sediments: a review. Freshwater Biol 46:431-51

Jakob B, Heber U (1996) Photoproduction and detoxification of hydroxyl radicals in chloroplasts and leaves and (its) relation to photoinactivation of photosystem I and II. Plant Cell Physiol 37:625-9

Karsten U, Kuck K, Vogt C, Kirst GO (1996) Dimethylsulfoniopropionate production in phototrophic organisms and its physiological function as a cryoprotectant. In Kiene RP, Visscher PT, Keller MD, Kirst GO (eds) Biological and environmental chemistry of DMSP and related sulfonium compounds, Plenum Press, New York, pp. 143-53

Karsten U, Wiencke C, Kirst GO (1990) The effect of light intensity and daylenght on the β-dimethylsulphonio-propionate (DMSP) content of green macroalgae at different irradiances. Plant Cell Environ 13:989-93

Karsten U, Wiencke C, Kirst GO (1992) Dimethylsulfioniopropionate (DMSP) accumulation in green macroalgae from polar to temperate regions: interactive effect of light versus salinity and light versus temperature. Polar Biol 12:603-60

Kasamatsu N, Hirano T, Kudoh S, Odate T, Fukuchi M (2004a) Dimethylsulfonioproprionate production by psychrophylic diatom isolates. J Phycol 40:874-8

Kasamatsu N, Kawaguchi S, Watanabe S, Odate T, Fukuchi M (2004b) Possible impacts of zooplankton grazing on dimethylsulfide production in the Antarctic Ocean. Can J Fish Aquat Sci 61:736-43

Kates and Volcani (1966) Lipids of diatoms. Biochim Biophys Acta 116:264-76

Kates M, Volcani BE (1996) Biosynthetic patways for phosphatidylsulfocholine, the sulfonium analogue of phoaphatidylcholine, in diatoms. In Kiene RP, Visscher PT, Keller MD, Kirst GO (eds) Biological and environmental chemistry of DMSP and related sulfonium compounds, Plenum Press, New York, pp. 109-19,

Katz ME, Finkel ZV, Grzebyk D, Knoll AH, Falkowski PG (2004) Evolutionary trajectories and biogeochemical impacts of marine eukaryotic phytoplankton. Annu Rev Ecol Evol Syst 35:523-56

Keller MD (1991) Dimethylsulphide production and marine phytoplankton: the importance of species composition and cell size. Biol Oceanogr 6:375-82

Keller MD, Bellows WK, Guillard RRL (1989) Dimethyl sulfide production in marine phytoplantkon. In Saltzman ES, Cooper WJ (eds) Biogenic sulfur in the environment, American Chemical Society, Washington DC, pp. 167-82,

Keller MD, Kiene RP, Matrai PA, Bellows WK (1999) Production of glycine betaine and dimethy lsulfoniopropionate in marine phytoplankton. I. Batch cultures. Mar Biol 135:237-48

Keller MD, Korjeff-Bellows W (1996) Physiological aspects of the production of dimethylsulfoniopropionate (DMSP) by marine phytoplankton. In: Kiene RP, Visscher PT, Keller MD, Kirst GO (eds) Biological and environmental chemistry of DMSP and related sulfonium compounds, Plenum, New York, pp. 131-42

Kiene RP, Linn LJ (2000) The fate of dissolved dimethylsulfoniopropionate (DMSP) in seawater: tracer studies using ^{35}S-DMSP. Geochim Cosmochim Acta 64:2797-2810

Kiene RP, Taylor BF (1988) Demethylation of dimethylsulfonioproprionate and production of thiols in anoxic marine sediments. Appl Environ Microbiol 54:2208-12

Kirst GO (1990) Salinity tolerance in eukaryotic marine algae. Annu Rev Plant Physiol Mol Biol 41:21-53

Kirst GO (1996) Osmotic adjustment in phytoplankton and macroalgae. In: Kiene RP, Visscher PT, Keller MD, Kirst GO (eds) Biological and environmental chemistry of DMSP and related sulfonium compounds, Plenum, New York, pp. 121-9

Kirst GO, Thiel C, Wolff H, Nothnagel J, Wanzek M, Ulmke R (1991) Dimethylsulfoniopropiona te (DMSP) in ice-algae and its possible biological role. Mar Chem 35:381-8

Knoll AH (2003) Biomineralization and evolutionary history. Rev Mineral Geochem 54:329-56

Kocsis MG, Nolte KD, Rhodes D, Shen T, Gage DA, Hanson AD (1998) Dimethylsulfoniopropi onate biosynthesis in *Spartina alterniflora*. Evidence that S-Methylmethionine and dimethylsulfoniopropylamine are intermediates. Plant Physiol 117:273-81

Kwint RLJ, Kramer KJM (1995) DMS production by plankton communities. Mar Ecol Prog Ser 121:227-37

Lappartient AG, Touraine B (1997) Comparison between demand-driven regulation of ATP sulfuphurylase activity and responses to oxidation stress. In: Cram WJ, De Kok LJ, Stulen I, Brunold C, Rennenberg H (eds) Sulphur metabolism in higher plants, Backhuys Publisher, Leiden, The Netherlands, pp. 203-5

Laroche D, Vezina AF, Levasseur M, Gosselin M, Stefels J, Keller MD, Matrai PA, Kwint RLJ (1999) DMSP synthesis and exudation in phytoplankton: a modelling approach. Mar Ecol Prog Ser 180:37-49

Leck C, Larsson U, Bagander LE, Johansson S, Hajdu S (1990) Dimethyl sulphide in the Baltic sea: annual variability in relation to biological activity. J Geophys Res 95:3353-63

Lee JH, Chae HS, Lee JH, Hwang B, Hahn KW, Kang BG, Kim WT (1997) Structure and expression of two cDNAs encoding S-adenosyl-L-methionine synthetase of rice (*Oryza sativa* L.) Biochim Biophys Acta 1354:13-18

Lee PA, de Mora SJ (1999) Intracellular dimethylsulfoxide (DMSO) in unicellular marine algae: Speculations on its origin and possible biological role. J Phycol 35:8-18

Leustek T, Saito K (1999) Sulfate transport and assimilation in plants. Plant Physiol 120:637-43

Leustek T, Martin MN, Bick J-A and Davies JP (2000) Pathways and regulation of sulfur metabolism revealed through molecular and genetic studies. Annu Rev Plant Physiol 51:141-65

Levasseur M, Michaud S, Egge J, Cantin G, Nejstgaard JC, Sanders R, Fernadez E, Solberg PT, Heimdal B, Gosselin M (1996) Production of DMSP and DMS during a mesocosm study of an *Emiliania huxleyi* bloom: influence of bacteria and *Calanus finmarchicus* grazing. Mar Biol 126:609-18

Liss PS, Hatton AD, Malin G, Nightingale PD, Turner SM (1997) Marine sulphur emissions. Phil Trans R Soc Lond B 352:159-69

Logan BA, Demmig-Adams B, Rosenstiel TN, Adams III WW (1999) Effect of nitrogen limitation on foliar antioxidants in relationship to other metabolic characteristics. Planta 209:213-20

Lovelock JE, Maggs RJ, Rasmussen RA (1972) Atmospheric dimethyl sulphide and the natural sulphur cycle Nature 237:452-3

Malin G (1996) The role of DMSP and DMS in the global sulfur cycle and climate regulation. In: Kiene RP, Visscher PT, Keller MD, Kirst GO (eds) Biological and environmental chemistry of DMSP and related sulfonium compounds, Plenum, New York, pp. 177-89

Matrai PA, Keller MD (1994) Total organic sulfur and dimethylsulfoniopropionate in marine phytoplankton: intracellular variation. Mar Biol 119:61-8

Maw GA (1981) The biochemistry of sulphonium salts. In: Stirling CJM (ed) The chemistry of the sulfonium group. Part 2,Wiley, New York, pp 703-70

McNeil SD, Nuccio ML, Hanson AD (1999) Betaines and related osmoprotectants. Targets for metabolic engineering of stress resistance. Plant Physiol 120:945-9

Melis A, Chen HC (2005) Chloroplast sulfate transport in green algae – genes, proteins and effects. Photosynth Res 86:299-307

Meyerdierks D (1997) Ecophysiology of the dimethylsulfonioproprionate (DMSP) content of temperate and polar phytoplankton communities in comparison with cultures of the coccolithophore *Emiliania huxleyi* and the Antarctic diatom *Nitszchia lecointei*. Ber Polarforsch p. 233

Michael, A (1887) Ueber die addition von natriumacetessig- und natriummalon- säureäthern zu den aethern ungesättigter säuren. J Prakt Chem35:349-56

Miller TR (2004) Swimming for sulfur: analysis of the *Roseobacter*- dinoflagellate interaction. PhD dissertation, University of Maryland

Mudd SH, Datko AH (1990) The S-methylmethiomine cycle in *Lemna paucicostata*. Plant Physiol 93:623-30

Nakamura H, Fujimaka K, Smapai O, Murai A (1993) Gonyol: Methionine-induced sulfonium accumulation in a dinoflagellate *Gonyaulax polyedra*. Tet Lett 34:8481-4

Nakamura H, Ohtoshi M, Sampei O, Akashi Y, Murai A (1992) Synthesis and absolute configuration of (+)-gonyauline: A modulating substance of bioluminescent circadian rhythm in the unicellular alga *Gonyaulax polyedra*. Tet Lett 33:2821-2

Nakamura H, Jin T, Funahashi M, Murai A (1996) Intergovernmental Oceanographic Commission of UNESCO. In: Yasumoto T, Oshima Y, Fukuyo Y (eds) Harmful and Toxic Algal Blooms, Intergovernmental Oceanographic Commission of UNESCO, Paris, pp. 515-18

Nakamura H, Jin T, Funahashi M, Fujimaki K, Sampei O, Murai A, Roenneberg T, Hastings J W (1997) Biogenesis of sulfonium compounds in a dinoflagellate; methionine cascade. Tetrahedron 53:9067-74

Nakano Y, Koizumi N, Kusano T, Sano H (2000) Isolation and properties of an S-adenosyl-L-methionine binding protein from the green alga, *Chlamydomonas reinhardtii*. J Plant Physiol 157:707-11

Nguyen BC, Belviso S, Mihalopoulos N, Gostan J, Nival P (1988) Dimethyl sulphide production during natural phytoplankton blooms. Mar Chem 24:133-41

Nguyen BC, Bonsang B, Gaudry A (1983) The role of the ocean in the global atmospheric sulphur cycle. J Geophys Resear 88:10903-14

Niki T, Kunugi M, Otsuki A (2000) DMSP-lyase activity in five marine phytoplankton species: its potential importance in DMS production. Mar Biol 136:759-64

Nishiguchi MK, Somero GN (1992) Temperature- and concentration- dependence of compatibility of the organic osmolyte β-dimethylsulfoniopropionate. Cryobiol 29:118-24

Norici A, Hell R, Giordano M (2005) Sulfur and primary production in aquatic environments: an ecological perspective. Photosynth Res 86:409-17

Otte ML, Morris JT (1994) Dimethylsulphoniopropionate (DMSP) in *Spartina alterniflora* Loisel Aquat Bot 48:239-59

Otte ML, Wilson G, Morris JT, Moran BM (2004) Dimethylsulphoniopropionate (DMSP) and related compounds in higher plants. J Exp Bot 55:1919-25

Painter TJ (1983) Algal polysaccharides. In: Aspinall GO (ed) The polysaccharides, Academic Press, New York, pp. 195-285:

Perez-Castineira JR, Prieto JL, Vega JM (1992) Sulfate uptake in *Chlamydomonas reinhardtii* Phyton 32:91-3

Pilson MEQ (1998) An introduction to the chemistry of the sea. Prentice Hall, Upper Saddle River, NJ, USA

Pollock SV, Pootakham W, Shibagaki N, Moseley JL, Grossman AR (2005) Insights into the acclimation of *Chlamydomonas reinhardtii* to sulfur deprivation. Photosynth Res 86:475-89

Ravanel S, Gakière B, Job D, Douce R (1998) The specific feature of methionine biosynthesis and metabolism in plants. Proc Natl Acad Sci USA 95:7805-12

Ravina CG, Chang CI, Tsakraklides GP, McDermott JP, Vega JM, Leustek T, Gotor C, Davies JP (2002) The sac mutants of *Chlamydomonas reinhardtii* reveal transcriptional and posttranscriptional control of cysteine biosynthesis. Plant Physiol 130:2076-84

Reed RH (1990) Solute accumulation and osmotic adjustment. In: Cole KM, Sheath RG (eds) Biology of the red algae, Cambridge University Press, pp. 147-70

Rhodes D, Hanson AD (1993) Quaternary ammonium and tertiary sulfonium compounds in higher plants. Annu Rev Physiol Plant Mol Biol 44:357-84

Roenneberg T, Nakamura H, Cranmer LD, Ryan K, Kishi Y, Hastings JW (1991) Gonyauline: a novel endogenous substance shortening the period of the circadian clock of a unicellular algae. Experientia 47:103-6

Saito K (2000) Regulation of sulfate transport and synthesis of sulfur-containing amino acids. Curr Opin Plant Biol 3:188-95

Scaduto RC (1995) Oxidation of DMSO and methane sulfinic acid by the hydroxyl radical. Free Radic Biol Med 18:271-7

Schafer HJ, Greiner S, Rausch T, Haag-Derwer A (1997) In seedlings of the heavy metal accumulator *Brassica juncea* Cu^{2+} differentially affects transcript amounts for γ-glutamylcysteine synthetase (γ-ECS) and metallothionein (MT2) FEBS Lett 404:216-20

Sheets EB, Rhodes D (1996) Determination of DMSP and other sulfonium compounds in *Tetraselmis subcordiformis* by plasma desorption mass spectrometry. In: Kiene RP, Visscher PT, Keller MD, Kirst GO (eds) Biological and environmental chemistry of DMSP and related sulfonium compounds, Plenum Press, New York, pp. 55-63

Sieburth JM (1960) Acrylic acid, an "antibiotic" principle in *Phaeocystis* blooms in Antartic waters. Science 132:676-7

Sieburth JM (1961) Antibiotic properties of acrylic acid, a factor in gastrointestinal antibiosis of polar marine animals. J Bact 82:72-9

Simó R, Pedrós-Alló C, Malin G, Grimalt JO (2000) Biological turnover of DMS, DMSP, and DMSO in contrasting open-sea waters. Mar Ecol Prog Ser 203:1-11

Sirko A, Hryniewicz MM, Hulanicka DM, Boeck A (1990) Sulfate and thiosulfate transport in *Escherichia coli* K-12: nucleotide sequence and expression of the cysTWAM gene cluster. J Bacteriol 172:3351-7

Stefels J (2000) Physiological aspects of the production and conversion of DMSP in marine algae and higher plants. J Sea Resear 43:183-97

Stefels J, Dijkhuizen L (1996) Characteristics of DMSP-lyase in *Phaeocystis* sp. (Primnesiophyceae) Mar Ecol Prog Ser 131:307-13

Stefels J, van Boekel WHM (1993) Production of DMS from dissolved DMSP in axenic cultures of the marine phytoplankton species *Phaeocystis* sp. Mar Ecol Prog Ser 97:11-18

Steinke M, Kirst GO (1996) Enzymatic cleavage of dimethylsulfoniopropionate (DMSP) in cell-free extracts of the marine macroalga *Enteromorpha clathrata* (Roth) Grev. (Ulvales, Chlorophyta) J Exp Mar Biol Ecol 201:73-85

Steinke M, Wolfe GV, Kirst GO (1998) Partial characterisation of dimethylsulfoniopropionate (DMSP) lyase isozymes in 6 strains of *Emiliania huxleyi*. Mar Ecol Prog Ser 175:215-25

Storey R, Gorham J, Pitman MG, Hanson AD, Gage D (1993) Response of *Melanthera biflora* to salinity and water stress. J Exp Bot 44:1551-60

Strom S, Wolfe G, Holmes J, Stecher H, Shimeneck C, Lambert S, Moreno E (2003a) Chemical defense in the microplankton I: Feeding and growth rates of heterotrophic protists on the DMS-producing phytoplankter *Emiliania huxleyi*. Limnol Oceanogr 48:217-29

Strom S, Wolfe GV, Slajer A, Lambert S, Clough J (2003b) Chemical defence in the microplankton II: inhibition of protist feeding by β-dimethylsulfoniopropionate (DMSP) Limnol Oceanogr 48:230-7

Summers PS, Nolte KD, Cooper AJL, Leustek T, Rhodes D, Hanson AD (1998) Identification and stereospecificity of the first three enzymes of 3-dimethylsulfoniopropionate biosynthesis in a chlorophyte alga. Plant Physiol 116:369-78

Sunda W, Kieber DJ, Kiene RP, Huntsman S (2002) An antioxidant function for DMSP and DMS in marine algae. Nature 418:317-20

Taylor BF, Gilchrist DC (1991) New routes of aerobic biodegradation of dimethylsulfoniopropri-onate. Appl Environ Microbiol 57:3581-4

Taylor BF, Visscher PT (1996) Metabolic pathways involved in DMSP degradation. In Kiene RP, Visscher PT, Keller MD, Kirst GO (eds) Biological and environmental chemistry of DMSP and related sulfonium compounds, Plenum, New York, pp. 265-76

Turner SM, Maline G, Liss PS, Harbour DS, Holligan PM (1988) The seasonal variation of dimethyl sulphide and dimethylsulfonioproprionate concentrations in nearshore waters. Limnol Oceanogr 33:364-75

Uchida A, Ooguri T, Ishida T, Kitaguchi H, Ishida Y (1996) Biosynthesis of dimethylsulfoniopro-prionate in *Crypthecodinium conhii* (Dinophyceae). In Kiene RP, Visscher PT, Keller MD, Kirst GO (eds) Biological and environmental chemistry of DMSP and related sulfonium compounds, Plenum, New York, pp. 97-107

Vairavamurthy A, Andreae MO, Iverson RL (1985) Biosynthesis of dimethylsulfide and dimethy lpropiothetin by *Hymenomonas carterae* in relation to sulfur source and salinity variations. Limnol Oceanogr 30:59-70

Van Alstyne KL, Houser LT (2003) Dimethylsulfide release during macroinvertebrate grazing and its role as an activated chemical defense. Mar Ecol Prog Ser 250:175-81

Van Alstyne KL, Wolfe GV, Freidenburg TL, Neill A, Hicken C (2001) Activated defence systems in marine macroalgae: evidence for an ecological role for DMSP cleavage. Mar Ecol Prog Ser 213:53-65

Van Bergeijk SA, Van der Zee C, Stal LJ (2003) Uptake and excretion of dimethylsulfoniopropionate is driven by salinity changes in marine benthic diatom *Cylindrotheca closterium*. Eur J Phycol 38:341-9

van Boekel WHM, Hansen FC, Riegman R, Bak RPM (1992) Lysis-induced decline of a *Phaeocystis* spring bloom and coupling with the microbial foodweb. Mar Ecol Prog Ser 81:269-76

Van Rijssel M, Gieskes WWC (2002) Temperature, light, and dimethylsulfonioproprionate (DMSP) content of *Emiliania huxleyi* (Prymnesiophyceae) J Sea Res 48:17-27

Welsh DT (2000) Ecological significance of compatible solute accumulation by micro-organisms: from single cells to global climate. FEMS Microbiol Rev 24:263-90

Wilkinson F, Helman WP, Ross AB (1995) Rate constants for the decay and reactions of the lowest electronically excites singlet state of molecular oxygen in solution. An expanded and revised compilation J Phys Chem Ref Data 24:663-1021

Wolfe GV (2000) The chemical defence ecology of marine unicellular plankton: constraints, mechanisms, and impacts. Biol Bull 198:225-44

Wolfe GV, Sherr EB, Sherr BF (1994) Release and consumption of DMSP from *Emiliania huxleyi* during grazing by *Oxyrrhis marina*. Mar Ecol Prog Ser 111:111-19

Wolfe GV, Steinke M (1996) Grazing-activated production of dimethyl sulphide (DMS) by two clones of *Emiliania huxleyi*. Limnol Oceanogr 41:1151-60

Wolfe GV, Steinke M, Kirst GO (1997) Grazing-activated chemical defence in a unicellular marine alga. Nature 387:894-7

Wolfe GV, Strom SL, Holmes JL, Radzio T, Olson MB (2002) Dimethylsulfoniopropionare cleavage by marine phytoplankton in response to mechanical, chemical, or dark stress. J Phycol 34:948-60

Yildiz FH, Davies JP, Grossman AR (1994) The regulation of photosynthetic electron transport in *Chlamydomonas reinhardtii* during sulfur-limited and sulfur-sufficient growth. Plant Physiol 104:981-7

Yoch DC (2002) Dimethylsulfoniopropionate: its sources, role in the marine food web, and biological degradation to dimethylsulfide. Appl Environm Microbiol 68:5804-15

Yoch DC, Ansede JH, Rabinowitz KS (1997) Evidence for intracellular and extracellular dimethy lsulfoniopropionate (DMSP) lyases and DMSP uptake sites in two species of marine bacteria. Appl Environ Microbiol 63:3182-8

Zarembinski TI, Theologis A (1994) Ethylene biosynthesis and action: a case of conservation. Plant Mol Biol 26:1579-97

Chapter 16
Accumulation and Transformation of Sulfonated Aromatic Compounds by Higher Plants – Toward the Phytotreatment of Wastewater from Dye and Textile Industries

Jean-Paul Schwitzguébel(✉), Stéphanie Braillard, Valérie Page and Sylvie Aubert

Abstract Sulfonated anthraquinones are precursors of many synthetic dyes and pigments, recalcitrant to biodegradation and thus not eliminated by classical wastewater treatments. In the development of a phytoremediation process to remove sulfonated aromatic compounds from industrial effluents, the most promising results have been obtained with *Rheum rabarbarum* (rhubarb), a plant species producing natural anthraquinones. Rhubarb is not only able to accumulate, but also to transform, these xenobiotic chemicals. Even if the biochemical mechanisms involved in the detoxification of sulfonated anthraquinones are not yet fully understood, they probably have cross talks with secondary metabolism, redox processes, and plant energy metabolism. Therefore, key problems of xenobiotic detoxification are addressed, as well as the potential role of the apoplast, often neglected, and the possible links with plant sulfur metabolism.

1 Sulfonated Aromatic Compounds in Wastewater from Dye and Textile Industries

Annually more than 1 million tons of at least 10,000 synthetic dyes and pigments are produced worldwide. Of this amount, approximately 280,000 tons are discharged every year into the environment, mainly via industrial effluents (Willetts and Ashbolt 2000). Synthetic sulfonated anthraquinones (Fig. 16.1) are the parent compounds for a large palette of dyes and an important starting material in their production. Effluents from dye, textile, and detergent industries, but also leachates from landfills, are thus often contaminated with sulfonated aromatics; as a result, these chemicals have an actual impact on the environment, especially fresh water (Young and Yu 1997, Riediker et al. 2000, Schwitzguébel et al. 2002). The pollution of many rivers by sulfur-organic xenobiotics is thus largely due to this class of

Jean-Paul Schwitzguébel
Laboratory for Environmental Biotechnology (LBE), Swiss Federal Institute of Technology Lausanne (EPFL), Station 6, CH-1015 Lausanne, Switzerland
jean-paul.schwitzguebel@epfl.ch

N.A. Khan et al. (eds.), *Sulfur Assimilation and Abiotic Stress in Plants.*

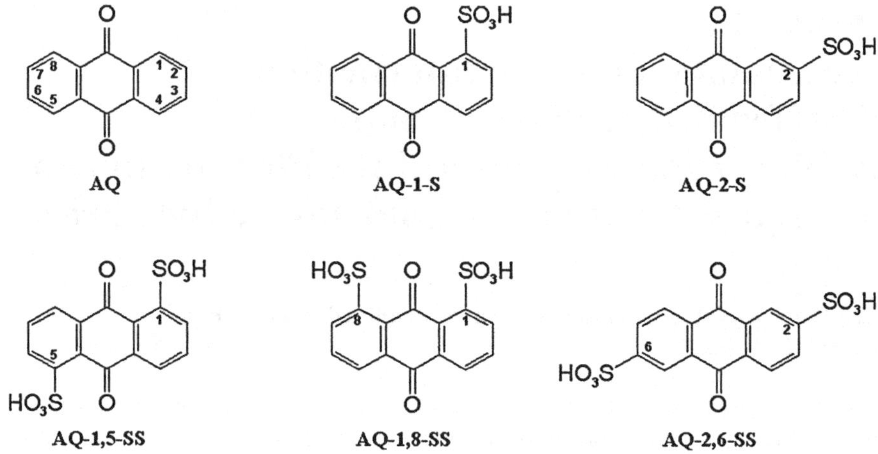

Fig. 16.1 Chemical structure of anthraquinone and sulfonated derivatives. IUPAC names of the compounds:
AQ-1-S: 9,10-dioxo-9,10-dihydro-1-anthracenesulfonic acid
AQ-2-S: 9,10-dioxo-9,10-dihydro-2-anthracenesulfonic acid
AQ-1,8-SS: 10-dioxo-9,10-dihydro-1,8-anthracenedisulfonic acid
AQ-1,5-SS: 10-dioxo-9,10-dihydro-1,5-anthracenedisulfonic acid
AQ-2,6-SS: 10-dioxo-9,10-dihydro-2,6-anthracenedisulfonic acid

chemicals, and many of them have acute and/or chronic effects on aquatic organisms (Greim et al. 1994).

Removal of sulfonated aromatic compounds is thus a major challenge for the dye and textile industries, not only because of the color of these chemicals, but also because of their recalcitrance and toxicity (Banat et al. 1996). During the last decade, several physical techniques have been tested, such as adsorption, precipitation, ion exchange, or filtration (Nigam et al. 2000, Robinson et al. 2001). Chemical treatment possibilities based on electrochemistry, different oxidative processes, and enzymatic degradation have also been investigated. However, these physicochemical treatments have major disadvantages, including high cost, low efficiency, and inapplicability to a wide variety of dyes, as well as the formation of byproducts, which create disposal problems associated with contaminated wastes (Vandevivere et al. 1998, Robinson et al. 2001).

Anionic dyes are intentionally designed to be recalcitrant under typical usage conditions, and it is this property that makes treatment difficult. Because they contain at least one sulfonic group and often also varying substitutions such as nitro groups, these chemicals are not uniformly susceptible to bio-decolorization and biodegradation (McMullan et al. 2001).

Bacterial degradation of dyes and byproducts often requires unusual catabolic activities rarely found in a single species (Cook et al. 1999). An important step appears to be catalyzed by dioxygenases adding oxygen across the double bond bearing the sulfonate group, leading to its elimination (McMullan et al. 2001). Unfortunately, a rather limited substrate range has been observed for bacterial isolates containing these enzymes, and the accumulation of dead-end products often occurs (Schwitzguébel et al. 2002).

The decolorization of synthetic dyes including azo and anthraquinone derivatives has also been examined in white-rot fungal cultures, known to produce powerful laccases and other peroxidases (Young and Yu 1997, Claus et al. 2002, Nyanhongo et al. 2002). However, it appears that concentrations above 10 to 125 mg l^{-1}, depending on the individual dye structure, cause slower decolorization (Young and Yu 1997). Fungal production of the key enzyme responsible for dye decolorization, lignin peroxidase, is also often dependent on nutrient carbon or nitrogen limitation, not always present in industrial effluents (Wong and Yu 1999). The limited ability of microorganisms to degrade sulfonoaromatic compounds, and thus to cope with various mixtures of these xenobiotics, limits the efficiency and, therefore, the use of conventional wastewater treatment plants. In this context, the development of alternative biological treatments to eliminate these pollutants from industrial effluents is a requirement (Robinson et al. 2001). More precisely, constructed wetlands and hydroponic systems are able to remove organic pollutants from industrial wastewater (Biddlestone et al. 1991, Furukawa and Fujita 1993, Davies and Cottingham 1994, Haberl et al. 2003). Both processes are based on the use of appropriate plant species and offer a potentially low-cost, low-maintenance approach to treat effluents containing recalcitrant xenobiotics.

2 Plants Are Able to Accumulate and Detoxify Many Xenobiotic Chemicals

Because plants are static and live in a competitive and sometimes hostile environment, they have evolved mechanisms like heat-shock proteins and molecular chaperones that protect them from environmental abiotic stress (Sandermann 2004, Wang et al. 2004, Kochhar and Kochhar 2005, Saidi et al. 2007). The defense process includes the detoxification of xenobiotic compounds and the synthesis and excretion of simple or complex molecules to protect them from excessive competition or pathogen attack. Higher plants thus produce a large number of secondary metabolites, which are chemical substances produced for purposes other than primary physiological functions and life-sustaining processes. Secondary metabolites often play a major role in the interactions of the plant with its environment or serve as signal molecules (Singer et al. 2003, Wink 2003). It is likely that the metabolism of xenobiotics uses at least partially secondary metabolic pathways and has an impact on the fundamental metabolism of the plant. Consequently there must be a limit on the amount of pollutant that can be accumulated and detoxified without disrupting the normal metabolic processes.

Plant metabolism is extremely diverse and can be exploited to treat recalcitrant pollutants not degradable by bacteria. Actually, higher plants are able to metabolize a great number of foreign compounds, originating from other living organisms or human industrial processes and defined as xenobiotics, and may therefore be considered as "green livers," acting as an important global sink for environmental pollutants, in many cases detoxifying them (Macek et al. 2000, Meagher 2000, Zaalishvili et al. 2000, Sandermann 2004, Schwitzguébel 2004).

Phytotechnologies are ecotechnologies relating to the use of vegetation to resolve environmental problems in watershed management by prevention of landscape degradation and by remediation and restoration of degraded ecosystems, control of environmental processes, and monitoring and assessment of environmental quality. Phytotechnologies exploit natural processes and can be used for revegetating degraded lands (such as quarries, road sides), removal of excessive nutrient loads (phytoamelioration), and the cleanup of wastewater (e.g., road runoff, municipal and industrial effluents, surface and seepage water) using soil-plant filters, buffer strips, or constructed wetlands (phytoprevention). Phytotechnologies are beginning to offer efficient tools and environmentally friendly solutions for the cleanup of contaminated sites and water, the improvement of food chain safety, and the development of renewable energy sources, and thus contributing to a sustainable use of water and land. Phytoremediation is a phytotechnology, specifically dedicated to the removal or the destruction of contaminants, and has a strong potential as a natural, solar-energy-driven remediation approach (Siciliano and Germida 1998, Meagher 2000, Trapp and Karlson 2001, Van der Lelie et al. 2001, Schwitzguébel 2004).

Constructed wetlands, hydroponic systems, or nutrient-film techniques based on selected plant species can remove pollutants from wastewater (Biddlestone et al. 1991, Furukawa and Fujita 1993, Davies and Cottingham 1994, Dumont et al. 1999, Haberl et al. 2003). Such systems have great potential to treat a wide range of xenobiotics and industrial effluents containing recalcitrant organics such as priority pollutants and synthetic dyes (Davies and Cottingham 1994, Haberl et al. 2003). They can offer a low-cost, low-maintenance green approach to treat wastewater.

3 Anthraquinones as Secondary Metabolites

Anthraquinones (AQ) are an important group of secondary plant metabolites occurring in several genera like *Rheum, Cinchona, Rumex*, and *Rubia* (Trease and Evans 1983, Van den Berg and Labadie 1989, Van der Plaas et al. 1998, Han et al 2002). Even if the precise functions of anthraquinones in plants remain largely unknown, these chemicals have been reported to exhibit antioxidant, antimicrobial, antifungal, antiviral, hypotensive, analgesic, laxative, antimalarial, and antitumor activities (Demirezer et al. 2001, Matsuda et al. 2001, Morimoto et al. 2002, Onegi et al. 2002).

Knowledge of the biosynthetic pathway of AQ and its regulation in higher plants is not yet complete. However, two main biosynthetic pathways leading to AQ have been described: the polyketide pathway and the chorismate/o-succinylbenzoic acid pathway (Han et al. 2002). In the polyketide pathway, AQ is biosynthesized from acetyl-CoA and malonyl-CoA via an octaketide chain. These types of AQ often exhibit a characteristic substitution in both rings A and C. The other pathway occurs in the Rubiaceae: ring A and B are derived from shikimic acid and α-ketoglutarate via o-succinylbenzoate, whereas ring C is derived from isopentenyl diphosphate (IPP), a universal building block for all isoprenoids. IPP can be formed from the mevalonic acid pathway, or from the condensation of glyceraldehyde 3-phosphate and pyruvate, via the 2-C-methyl-D-erythritol 4-phosphate pathway (Han et al. 2002). In any case, precursors (pyruvate) or intermediates of the tricarboxylic acid cycle (acetyl-CoA, α-ketoglutarate) are involved, indicating close relationship with energy and carbon metabolism. Furthermore, it has been reported that peroxidases are able to catalyze the formation of dimeric AQ from hydroxy-9,10-anthraquinones (Arrieta-Baez et al. 2002). Finally, most of the natural AQs are glycosylated, whereas glycosyl-transferases are enzymes known to be involved in the conjugation of many xenobiotic compounds (Khouri and Ibrahim 1987, Pflugmacher and Sandermann 1998, Jones and Vogt 2001, Brazier et al. 2002).

These features raise the question of the relationship between the secondary metabolism and the detoxification of organic xenobiotics, in particular whether the metabolism of xenobiotics uses the same enzymes as secondary metabolism. Although the pathways are similar they are not identical, and the addition of xenobiotics probably induces specific enzymes needed for their modification. Nevertheless, detoxification seems to be integrated into that part of metabolism involved in the interactions of the plant with its environment.

4 Detoxification Enzymes and Redox Status within Plant Cells

Cytochrome P450 mono-oxygenases represent a multigenic family of enzymes involved in the detoxification process of many xenobiotic compounds (Khatisashvili et al. 1997, Werck-Reichhart et al. 2000, Morant et al. 2003, Isin and Guengerich 2007). They catalyze complex reactions, NAD(P)H reducing the iron in the heme center to the ferrous form, which then binds O_2. The oxygen molecule is cleaved: one atom is inserted into the xenobiotic molecule, whereas the second is released as water. Electrons from NAD(P)H are transferred to P450s via flavoproteins called cytochrome P450-reductases. Both plant P450s and their reductases are usually bound via their N-terminus to the cytoplasmic surface of the endoplasmic reticulum. The hydroxylation of an aromatic ring can generate a phenolic intermediate that may have the potential to dissipate proton gradients and uncouple oxidative phosphorylation. In addition to activating xenobiotics, cytochrome P450 plays an important role in the normal secondary metabolism of plants which produce compounds involved in cell signaling and defense mechanisms. Overloading a plant with high concentrations of xenobiotics requiring oxidation by cytochrome P450 may thus compete with the normal functions of these enzymes. There is evidence that the presence of xenobiotics can induce the synthesis of numerous cytochrome P450s. An overall increase in the activity of the mixed functional oxidases may impose a major demand on both the oxygen and the NAD(P)H pool within the cell. This could have significant effects on the redox balance of the cell, thus compromising the primary metabolic processes in the plant, making it more sensitive to environmental stresses (Gordeziani et al. 1999, Korte et al. 2000).

Conjugation occurs by nucleophilic or electrophilic interaction with an endogenous metabolite. Most of the natural metabolites involved in conjugation are important molecules or coenzymes that play a central role in the metabolism of plants. The tripeptide glutathione is known to play an important role in both primary metabolism and xenobiotic metabolism (May et al. 1998, Noctor et al. 1998, Cole and Edwards 2000, Foyer et al. 2001). In primary metabolism it maintains the redox balance of the NAD(P)/NAD(P)H pools and destroys hydrogen peroxide and organic peroxides. Glutathione is also known to form a conjugate with xenobiotics in plants. Glutathione-S-transferase enzymes catalyze the formation of these conjugates (Edwards et al. 2000, Pflugmacher et al. 2000). The presence of xenobiotic compounds like herbicides induces the biosynthesis of glutathione-S-transferases and thus an increased use of glutathione at the cellular level. The glutathione levels and redox state of the plants are thus affected under such stress conditions.

The addition of sugars, usually glucose, to endogenous and exogenous organic molecules is widespread among plants. Many natural plant products are glycosylated to oxygen (O), nitrogen (N), sulfur (S), and carbon (C) atoms by glycosyl-transferases using nucleotide-activated sugars (Jones and Vogt 2001, Lim et al. 2002, Messner et al. 2003). Xenobiotic compounds can also be glycosylated by plants when functional substituents such as hydroxyl, amino, sulfhydryl or carboxylic acid groups are present or introduced by enzymatic reactions (Pflugmacher and Sandermann 1998, Cole and Edwards 2000, Brazier et al. 2002, Messner et al. 2003). Plant glycosyltransferases comprise a large family of enzymes and, interestingly, certain species show particularly high activities with several specific xenobiotic chemicals (Pflugmacher and Sandermann 1998, Paquette et al. 2003). The use of combined approaches should provide additional insight into interactions between endogenous metabolic pathways and detoxification of xenobiotics, the developmental and spatial expression patterns of glycosyl-transferases, as well as the inducibility by xenobiotic substances (Mazel and Levine 2002, Taguchi et al. 2003).

Many of the enzymes involved in the early stages of the detoxification process are closely associated with the redox biochemistry of the cell. The activity of enzymes such as cytochrome P450, peroxidase, and glutathione transferase all have implications with respect to the maintenance of redox homeostasis (Gordeziani et al. 1999, Edwards et al. 2000, Stiborova et al. 2000, Werck-Reichhart et al. 2000, Foyer et al. 2001, Mackova et al. 2001). The regulation of the redox status is of central importance and is closely associated with mitochondrial respiratory processes, also involved in maintaining the energy status of the cell. Redox and energy balance are considered to be among the most important regulatory parameters in cellular metabolism in determining the relative flux of metabolites through the anabolic and catabolic pathways. Factors that disturb either the redox or energy balance of the cell will have serious ramifications throughout the entire metabolic network.

5 The Uniqueness of Plant Respiratory Chains

Plants have a complex respiratory system that provides a great deal of flexibility. All plants contain a "classical" respiratory chain, located in the inner membrane of the mitochondrion and responsible for the vectorial translocation of protons across this membrane during the process of oxidative phosphorylation. This process represents a major point of interaction between the redox and energy status of the cell.

However, plants are autotrophic and thus need to assimilate nitrogen for amino acid biosynthesis. The carbon skeletons required for this process are derived from tricarboxylic acid cycle intermediates. Consequently, this metabolic cycle has to turn independently of the energy needs of the cell. Plant mitochondrial respiratory activity has therefore to be tightly regulated to satisfy different metabolic demands in a flexible fashion. As a consequence, plant mitochondrial respiratory chains differ in several aspects from the mammalian one (Affourtit et al. 2001, Möller 2001, Rasmusson et al. 2004, Umbach et al. 2006).

In addition to Complex I, they contain four other dehydrogenases, nonprotonmotive and insensitive to rotenone, that enable the controlled oxidation of matrix and cytoplasmic NAD(P)H (Palmer et al. 1982, Schwitzguébel and Siegenthaler 1984, Schwitzguébel et al. 1985, Cook-Johnson et al. 1999, Möller 2001, Rasmusson et al. 2004). Furthermore, plant mitochondria are characterized by the presence of an alternative respiratory pathway, through which reducing equivalents can be transferred to molecular oxygen without the involvement of cytochromes. This pathway branches at the level of the ubiquinone pool and comprises a single enzyme, the alternative oxidase, not coupled to the synthesis of ATP and insensitive to cyanide (Vanlerberghe and McIntosh 1997). The activity of this oxidase can overcome any interaction between the energy level and the redox status of the cell (Ribas-Carbo et al. 1997, Millenaar et al. 2001, Affourtit et al. 2002). Consequently, plant mitochondria can oxidize both internal and cytoplasmic NAD(P)H in a manner that is not energy conserving and is thus beyond the control of the protonmotive force.

The enzyme responsible for the reaction with oxygen is a 32 kDa protein; it can exist as either a monomer, each containing an –SH group (reduced form), or a dimer, in which the monomers are linked by an S-S bond (oxidized form). The monomer is the most active form; and it seems that the enzyme is potentially regulated by the general redox status of the cell. Any major interference could thus have significant effects on the respiratory pathways.

Increase in the expression and/or activity of the alternative oxidase has been observed during temporal events such as seed conditioning, leaf development, elevation of salicylic acid levels, thermogenesis, fruit ripening, oxidative stress, physical wounding, and plant pathogenic attack (Geraldes-Laakso and Arrabaça 1997, Lennon et al. 1997, Noguchi et al. 2001, Nun et al. 2003, Umbach et al. 2006).

The alternative oxidase is inhibited by benzhydroxamate compounds, antioxidants such as propylgallate, and copper chelators such a disulfiram. It is able to oxidize polyphenolic substrates that resemble ubiquinone and could thus directly participate in the metabolism of xenobiotic compounds like sulfonated anthraquinones. Its direct or indirect involvement in detoxification process appears likely.

6 The Ability of Plants to Accumulate and Metabolize Sulfonated Anthraquinones

Anthraquinones occur naturally in several plant genera like *Rheum* (rhubarb), *Rumex, Cinchona*, and *Rubia*, and these plants might possess enzymes able to accept sulfonated anthraquinones as substrates. To test this hypothesis, cells have been isolated from *Rheum palmatum* and cultivated in shake flasks and bioreactors in the presence of up to 700-800 mg/l anthraquinones – a usual concentration in primary effluents from dye production lines – with sulfonated groups in different positions: 1; 2; 1,5; 1,8; 2,6 (Fig. 16.1). Samples have been taken up and cells separated from the growth medium by filtration. The filtrate has been analyzed by HPLC, capillary electrophoresis, and spectrophotometry for its content in anthraquinones and metabolites possibly derived from them (Schwitzguébel et al. 2002, Schwitzguébel and Vanek 2003). The most significant results indicate that: AQ-1-S totally disappears from the medium, phytotransformation occurs, but not desulfonation; AQ-2-S is partially but rapidly taken up by rhubarb cells, metabolized and desulfonated, as indicated by the stoichiometric release of sulfate into the medium; AQ-1,5-SS and AQ-1,8-SS rapidly disappear from the medium, but no intermediates are released.

The ability of rhubarb cells to accumulate, transform, and/or degrade other sulfonated aromatic compounds, all being byproducts and pollutants present in effluents from a dye-producing company, has also been investigated: 2-hydroxy-4-sulfo-1-naphthalenediazonium is completely removed and transformed; 2-hydroxy-4-sulfo-6-nitro-1-naphthalenediazonium is efficiently accumulated and desulfonated; 7-nitro-1,3-naphthalenedisulfonic acid, 7-amino-1,3-naphthalenedisulfonic acid, and 2-chloro-5-nitro benzene sulfonic acid are taken up and possibly transformed by rhubarb cells (Duc et al. 1999).

The large differences observed between the fate of several sulfonoaromatic compounds of the same family in plant cells look extremely interesting from a scientific standpoint and can be of great significance for future environmental applications. Although cell cultures have the advantage of giving rapid evidence of the behavior of a particular plant species in contact with certain compounds, they cannot be used for wastewater treatment at large scale due to the high cost and the fragility of such a system. The cultivation of whole plants is thus required for such a purpose.

Since rhubarb is a hardy plant species, it may prove a promising candidate in developing new biological processes to decontaminate effluents containing sulfonated aromatic compounds. It was precisely the purpose of a further investigation: whole rhubarb, rumex, and other plants have thus been grown under hydroponic conditions using artificial effluents. Five sulfonated anthraquinones have been chosen as model pollutants for this study, used at the same total concentration as with isolated cells. Results show that mono- as well as di-sulfonated anthraquinones are efficiently taken up by rhubarb and other anthraquinone-producing species and translocated to the shoot. The presence of putative metabolites, not yet identified, in leaves has also been observed, indicating the transformation of several sulfonated anthraquinones (Aubert and Schwitzguébel 2002, 2004, Haberl et al. 2003). The metabolites found in the different plant species tested are not similar among them, confirming that the transformation of sulfonated anthraquinones is plant specific.

On the other hand, the direct injection of sulfonated anthraquinones in cut stems of rhubarb, red sorrel, and maize has also been done. Under such experimental conditions, parent compounds and metabolites derived from them have been found in leaves of all investigated species. The metabolites formed depend on plant species, but are different from those obtained in whole plants. This indicates that the transformation of the sulfonated anthraquinones occurring in the root/rhizome is distinct from that taking place in the leaf, or that the presence of the root/rhizome is needed for such a metabolism.

Enzymatic investigations have been performed to determine if sulfonated anthraquinones might be transformed by enzymes of the classical detoxification pathways in plants. The results obtained with glutathione S-transferases show that this class of enzymes is apparently not significantly involved in the observed metabolism of sulfonated anthraquinones. In contrast, preliminary results suggest the possible involvement of cytochrome P450 and of glycosyl-transferases in the metabolism of synthetic sulfonated anthraquinones.

7 The Potential Role of Apoplast in the Accumulation and Metabolism of Sulfonated Anthraquinones

The apoplast is a continuous system of cell walls and intercellular air spaces present in all plant tissues. It is a complex and dynamic extracellular compartment surrounding the plant cells; it is challenging to study and until recently poorly understood. However, the extraction and characterization of the apoplastic fluid has taken more and more importance in recent years, and has increased insight into its role and physiology (Feito et al. 2001, Lohaus et al. 2001, Baker et al. 2005). It appears that apoplast plays a major role in many processes, including intercellular signaling, plant-microbe interactions, transport and storage of nutrients and contaminants. The activity of several antioxidant enzymes has also been reported (Lohaus et al. 1995, Haslam et al. 2001, Sattelmacher 2001, Kevresan et al. 2003, Tasgin et al. 2006). Its isolation and analysis should thus be useful to estimate the balance of xenobiotic compounds like sulfonated anthraquinones present inside and outside plant cells.

Three different and complementary approaches were used to estimate the partitioning of AQ-1,5-SS inside rhubarb leaves (Aubert 2003). This chemical was chosen as a model, since its metabolism appears to be limited inside the plant. First, protoplasts were isolated from plants grown in the absence of AQ-1,5-SS, and their ability to remove the xenobiotic from the medium was investigated. Second, vacuoles were isolated from leaves of plants grown in the presence of AQ-1,5-SS, and the different fractions obtained during the procedure were analyzed for their content in the xenobiotic. Finally, the apoplastic fraction was extracted from leaves of plants grown with AQ-1,5-SS and analyzed (Lohaus et al. 2001).

Rhubarb plants were cultivated in a greenhouse under hydroponic conditions (Aubert 2003). For isolation of vacuoles and apoplast extraction, adult plants, preadapted to hydroponic conditions, were cultivated in the presence of 5 mM AQ-1,5-SS for a minimum of 3 weeks and a maximum of 2.5 months. For experiments done with isolated protoplasts, plants were grown without any sulfonated anthraquinone.

7.1 Isolation and Incubation of Protoplasts

The lower side of freshly cut rhubarb leaves was peeled by hand, and leaf slices were laid in Petri dishes on a minimal volume of washing medium diluted three times and to which a small amount of bovine serum albumin (BSA) was added to increase the surface tension (Aubert 2003). When the surface of 8 Petri dishes was covered with leaf slices, the washing medium was sucked up, and 5 ml of digestion buffer was added per Petri dish (Aubert 2003). Leaf slices were then incubated at 30 °C for 2-3 hours to digest cell wall.

Protoplasts were extracted from the leaf slices by soft horizontal shaking of Petri dishes. The leaf's "skeleton" was removed with tips, and the buffer containing protoplasts was transferred into two 50 ml tubes, stored on ice. Two milliliters of a 35% Percoll® cushion (pH 6.0) was layered at the tube's bottom, and tubes were then centrifuged for 5 min at 1200 g and 4 °C with a soft start and stop. Protoplasts were concentrated on the Percoll® cushion, whereas cell wall fragments remained in the digestion buffer on top. The digestion buffer was removed and the protoplasts were then gently mixed to the Percoll® buffer layer. Protoplasts were used for incubation experiments or for isolation of vacuoles (Fig. 16.2).

A gradient was prepared in 400 μl tubes: 30 μl of 35% Percoll® (pH 6.0) on top of which 200 μl of silicone oil was layered. Isolated protoplasts (0.39 ml) were diluted in 1.56 ml incubation buffer (Aubert 2003). At the beginning of the experiment, sulfonated anthraquinones were added to the mixture at a final concentration of 130 μM (or 3 × 120 μM when 3 sulfonated anthraquinones were tested simultaneously). Samples were incubated at room temperature, whereas controls were incubated in parallel on ice to measure the passive absorption that might occur. The first measurement of each condition, after 2 min, was taken as a reference for possible adsorption of sulfonated anthraquinones on the protoplasts. After different incubation times, samples of the protoplasts' suspension were taken from the tubes (3 replicates of 100 μl) and layered on the top of 3 tubes containing the gradient

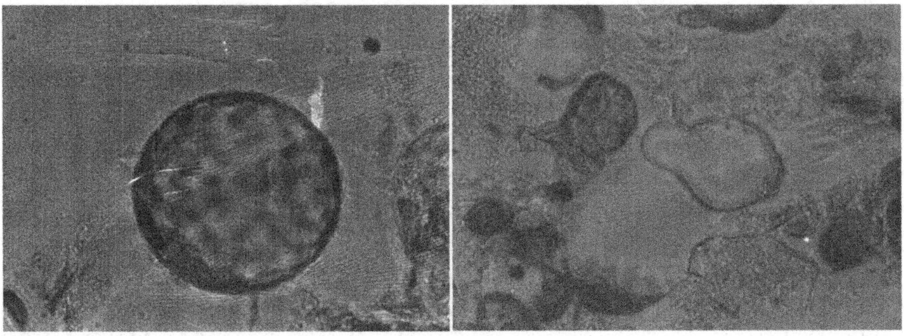

Fig. 16.2 Protoplast (left) and vacuole (right) isolated from rhubarb leaf, as seen under optic microscope (direct lighting)

previously prepared. Tubes were then centrifuged for 20 sec at 16,100 g. Supernatants of the replicates were mixed in 1.5 ml Eppendorf tubes to obtain a final volume of 240 μl sample. The disappearance of sulfonated anthraquinones from the incubation medium was measured by capillary electrophoresis after removal of proteins (Aubert 2003).

Protoplasts were incubated in the presence of either only one or three sulfonated anthraquinones simultaneously (Table 16.1). Only AQ-2-S rapidly and significantly disappeared from the medium, and three new peaks were observed in the electropherogram of samples taken after 180 min (not shown), indicating that protoplasts were able to metabolize this compound, as shown previously with cells isolated from rhubarb and cultivated in bioreactors (Schwitzguébel and Vanek 2003). However, in contrast to results obtained with isolated cells cultivated in vitro, protoplasts isolated from rhubarb leaves were able to accumulate only a small amount, approximately 10%, of AQ-1-S and AQ-1,5-SS. No metabolites were released into the medium.

7.2 Analysis of Isolated Vacuoles

To break cell membranes and isolate vacuoles, isolated protoplasts were transferred into a 22 ml glass tube, and four volumes of lysis buffer were added per volume of protoplasts (Aubert 2003). After very soft homogenization of the suspension, the tube was incubated for 4 min at 37 °C, homogenized softly again, and incubated for 3 more min at 37 °C. To help the release of vacuoles out of protoplasts, the liquid was transferred to a 50 ml tube using a 1 ml pipette whose tip had been cut to increase its pint hole. A second gradient was used to purify vacuoles: 8 ml of solution 1/1 (1 volume of lysis buffer added to 1 volume of vacuole buffer) was layered on top of the liquid containing the vacuoles and protoplasts fragments, and 3 ml of vacuole buffer was then layered on the solution 1/1 (Aubert 2003). The tubes were centrifuged at 1200 g and 4 °C, with soft start and stop, allowing the migration of

Table 16.1 Removal of sulfonated anthraquinones by isolated protoplasts from the incubation medium

Sulfonated	% Sulfonated anthraquinones removed after					
anthraquinone	2 min	20 min	40 min	60 min	120 min	180 min
AQ-1-S	0	0	0	0	0	0
AQ-1-S*	0	9	8	Nd	8	13
AQ-2-S	0	0	60	62	70	67
AQ-2-S*	0	Nd	18	Nd	58	Nd
AQ-1,5-SS	0	0	0	Nd	8	Nd
AQ-1,5-SS*	0	3	4	Nd	2	12

The different sulfonated anthraquinones under investigation were added separately at a final concentration of 130 μM or simultaneously* (3 × 120 μM) to the protoplasts to start the incubation. Nd means not determined.

vacuoles in between solution 1/1 and the vacuole buffer. Cell wall digestion buffer, protoplast fragments, and vacuoles were collected into 50 ml tubes and mixed separately on a Vortex to break all membranes (Fig. 16.2). The different fractions were filtered and analyzed by capillary electrophoresis (Aubert and Schwitzguébel 2002).

Fig. 16.3 shows that more than 60% of anthraquinone-1,5-disulfonate accumulated in rhubarb leaves was recovered from the cell wall digestion buffer. Approximately 35% of the amount was found in the protoplast fragments and only less than 5% in the vacuole fraction. This is in contrast to the vacuolar compartmentation previously described for the detoxification of Lucifer yellow, another sulfonated chemical (Klein et al. 1997).

To determine if the quantity of AQ-1,5-SS found in the different fractions could be due to an artifact of the experimentation, a calculation was made to evaluate the percentage which would be obtained from each fraction if the molecule would be uniformly distributed in the different fractions. Taking into account the respective volumes of the different fractions, the calculation indicated that 80% of the AQ-1,5-SS would be present in the digestion buffer, 18% in the protoplast's fragments, and only 2% in the vacuole fraction. The comparison of these figures from results experimentally obtained indicates that AQ-1,5-SS was not uniformly distributed in the fractions. Moreover, the concentration in the cell wall digestion buffer (up to 420 μM) and in protoplast fragments (up to 190 μM) was much higher than in the vacuole fraction (around 40 μM). The possibility that the recovery differences were due to the different volumes taken from each fraction could thus be excluded.

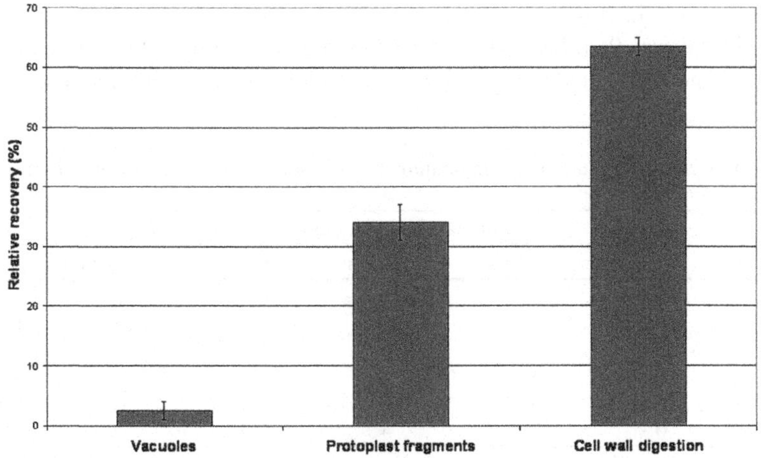

Fig. 16.3 Recovery of AQ-1,5-SS in each fraction (vacuoles, protoplast fragments, and cell wall digestion buffer). Mean values with SD

7.3 Analysis of the Apoplastic Fluid after Infiltration and Centrifugation

For each of the experiments performed, only one leaf of rhubarb was used, since the quantity of the accumulated xenobiotic has been shown to be similar on a fresh weight basis in different leaves from the same or from different plants. Leaves of rhubarb grown in the presence of AQ-1,5-SS were cut into square slices approximately 1.5 cm wide with a cutter. Leaf slices were then immersed into deionized water (upper face of leaf on top), and a vacuum was made. This allowed the extraction of the air present in the leaf slices through the cut sides and the stomata. After 4 min, the vacuum pump was stopped and the system gently opened to allow the entry of air. This induced the entry of water through the stomata and through the cut sides of the leaves. This step was repeated 5 times, until the leaves were completely infiltrated with water. The cut leaves were then gently dried on a surface with absorbing paper. They were divided into groups of 3 g each, delicately put into 20 ml syringes, which were disposed into centrifuge tubes. They were centrifuged for 5 min at different speeds, ranging from 270 to 7740 g, with one or two replicates per speed. The resulting apoplastic washing fluid was transferred to 1.5 ml tubes, kept on ice, diluted if necessary, and filtered before being analyzed (Aubert 2003).

Results of a typical experiment done with rhubarb leaf cut from plants cultivated in the presence of AQ-1,5-SS for 53 days are shown in Fig. 16.4. It has been reported that, below 1000 g, no cytoplasmic contamination occurs or should be

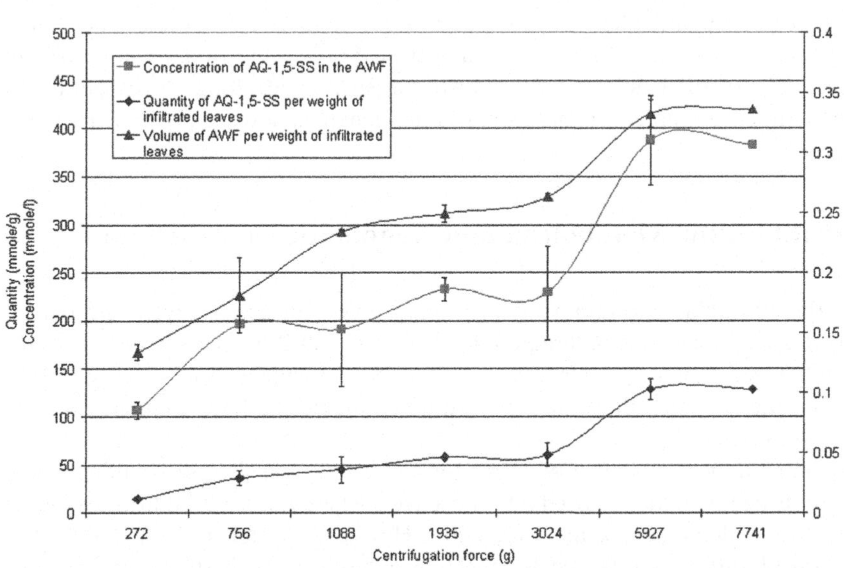

Fig. 16.4 Infiltration/centrifugation experiment with a leaf cut from a rhubarb plant cultivated for 53 days in the presence of 5 mM AQ-1,5-SS

negligible (Lohaus et al. 2001). However, even at very low centrifugation forces, which should avoid any cytoplasmic contamination, AQ-1,5-SS was recovered, indicating that the xenobiotic compound was present in the apoplast. The results show that the concentration and quantity of AQ-1,5-SS increased with the centrifugation force, suggesting that water infiltrated in the leaf was not completely homogenized with the apoplast liquid. At a low centrifugation force, the apoplastic washing fluid recovered was mainly the liquid which was added. In contrast, at high centrifugation speed, a higher proportion of the original apoplastic liquid was extracted in the fluid. The increase in AQ-1,5-SS concentration as a function of centrifugation speed might also be due partially to an increase of the cytoplasm leakage. Unfortunately, no clear-cut cytoplasmic marker is presently available for such estimation.

Furthermore, the overall quantity of AQ-1,5-SS per leaf biomass, as well as its concentration in the apoplastic washing fluid, increased markedly over time, from 53 to 75 days' (not shown) exposure of whole plants cultivated under hydroponic conditions. For both times of exposure, the quantity of AQ-1,5-SS recovered at high centrifugation force (> 6000 g) was always around two times higher than that recovered at low centrifugation force (< 3000 g), indicating that approximately half of the xenobiotic did not enter plant cells and remained in the apoplast.

Results obtained with three different and complementary approaches suggest that apoplastic storage plays an important role in the phytoaccumulation of AQ-1, 5-SS, and possibly of other xenobiotics. This is a very exciting and still largely unexplored phenomenon that deserves to be further investigated to understand the different pathways involved in phytoremediation processes, since antioxidant enzymes like peroxidase, catalase, and polyphenol oxidase have been found in the leaf apoplast of plants like winter wheat (Tasgin et al. 2006). On the other hand, peroxidases isolated from horseradish and from *Senna angustifolia* can transform natural hydroxy-9,10-anthraquinones (Arrieta-Baez et al. 2002); their ability to accept sulfonated anthraquinones as substrates remains, however, unknown.

8 Plant Sulfur Metabolism and Xenobiotic Detoxification

Rhubarb cells are able to remove the sulfonate group present in several sulfonated aromatic compounds, like anthraquinone 2-sulfonate or 2-hydroxy-4-sulfo-6-nitro-1-naphthalenediazonium (Duc et al. 1999, Schwitzguébel and Vanek 2003). However, the enzyme(s) involved in desulfonation and the fate of sulfate released are not yet known.

The most important source of sulfur for higher plants is precisely sulfate, taken up by roots and then translocated into the leaves, where it is reduced in chloroplasts (Ernst 1998, Mendoza-Cozatl et al. 2005, Hawkesford and de Kok 2006). The reduction of sulfate is required for the biosynthesis of methionine and cysteine, both sulfur-containing amino acids, as well as of glutathione and phytochelatins, important small peptides involved in the detoxification of xenobiotic chemicals and of heavy metals, respectively. When present in excess, sulfate can be accumulated in the vacuoles as is, or as flavonoid sulfate and phenol sulfate (Ernst 1998, Hawkesford and de Kok 2006). Flavonoid sulfate esters are common in plants, and the enzymatic sulfonation is catalyzed by sulfotransferases transferring sulfonate group from 3′-phosphoadenosine 5′-phosphosulfate to the hydroxyl group of a flavonoid (Varin et al. 1997, Gidda and Varin 2006). Even if its biological significance is still unclear, sulfonation seems to play an important role in signaling processes, and could be involved in the detoxification and excretion of hydrophobic molecules (Gidda and Varin 2006).

Glucosinolates present in many Brassicaceae are derived from amino acids and contain at least two sulfur atoms; these plants require more sulfur than other plant families and could be useful for phytoextraction of sulfur-enriched sites (Ernst 1998). Glucosinolates are preformed sulfur-containing defense compounds and play an important role in plant survival. Under stress conditions, thioglucosidases convert glucosinolates to different volatile compounds, which have been shown to have antimicrobial effects (Rausch and Wachter 2005).

It thus appears that sulfur metabolism and sulfur-containing compounds play a key role in the defense strategy of many plants against biotic and abiotic stress, including xenobiotics detoxification.

References

Affourtit C, Krab K, Moore AL (2001) Control of plant mitochondrial respiration. Biochim Biophys Acta 1504:58-69

Affourtit C, Albury MS, Crichton PG, Moore AL (2002) Exploring the molecular nature of alternative oxidase regulation and catalysis. FEBS Lett 510:121-6

Arrieta-Baez D, Roman R, Vazquez-Duhalt R, Jimenez-Estrada M (2002) Peroxidase-mediated transformation of hydroxy-9,10-anthraquinones. Phytochemistry 60:567-72

Aubert S (2003) Accumulation and transformation of sulphonated anthraquinones by higher plants: a first step towards the phytotreatment of wastewater from dye and textile industry. Thesis Nr 2809, Ecole Polytechnique Fédérale de Lausanne, Switzerland

Aubert S, Schwitzguébel JP (2002) Separation of sulphonated anthraquinones in various matrices by capillary electrophoresis. Chromatographia 56:693-7

Aubert S, Schwitzguébel JP (2004) Screening of plant species for the phytotreatment of waste water containing sulphonated anthraquinones. Water Res 38:3569-75

Baker CJ, Roberts DP, Mock NM, Whitaker BD, Deahl KL, Aver'yanov AA (2005) Apoplastic redox metabolism: Synergistic phenolic oxidation and a novel oxidative burst. Physiol Molec Plant Path 67:296-303

Banat I, Nigam P, Singh D, Marchant R (1996) Microbial decolorization of textile-dye-containing effluents: a review. Bioresource Technol 58:217-27

Biddlestone AJ, Gray KR, Thurairajan K (1991) A botanical approach to the treatment of wastewaters. J Biotechnol 17:209-20

Brazier M, Cole DJ, Edwards R (2002) O-glucosyltransferase activities toward phenolic natural products and xenobiotics in wheat and herbicide-resistant and herbicide-susceptible black grass (*Alopecurus myosuroides*). Phytochemistry 59:149-56

Claus H, Faber G, König H (2002) Redox-mediated decolorization of synthetic dyes by fungal laccases. Appl Microbiol Biotechnol 59:672-8

Cole DJ, Edwards R (2000) Secondary metabolism of agrochemicals in plants. In: Roberts T (ed) Metabolism of agrochemicals in plants, John Wiley and Sons, pp. 107-54

Cook AM, Laue H, Junker F (1999) Microbial desulfonation. FEMS Microbiol Rev 22:399-419

Cook-Johnson RJ, Zhang Q, Wiskich JT, Soole KL (1999) The nuclear origin of the non-phosphorylating NADH dehydrogenases of plant mitochondria. FEBS Lett 454:37-41

Davies TH, Cottingham PD (1994) The use of constructed wetlands for treating industrial effluents (textile dyes). Water Sci Technol 29:227-32

Demirezer LO, Kuruüzüm-Uz A, Bergere I, Schiewe HJ, Zeeck A (2001) The structures of antioxidant and cytotoxic agents from natural source: anthraquinones and tannins from roots of *Rumex patientia*. Phytochemistry 58:1213-17

Duc R, Vanek T, Soudek P, Schwitzguébel JP (1999) Accumulation and transformation of sulfonated aromatic compounds by rhubarb cells (*Rheum palmatum*). Int J Phytorem 1:255-71

Dumont P, Philipon P, Xanthoulis D (1999) Du céłery pour épurer les eaux. Biofutur 188:40-3

Edwards R, Dixon DP, Walbot V (2000) Plant glutathione S-transferases: enzymes with multiple functions in sickness and in health. Trends Plant Sci 5:193-8

Ernst WHO (1998) Sulfur metabolism in higher plants: potential for phytoremediation. Biodegradation 9:311-18

Feito I, Gonzalez A, Centeno ML, Fernandez B, Rodriguez A (2001) Transport and distribution of benzyladenine in *Actinidia deliciosa* explants cultured in liquid and solid media. Plant Physiol Biochem 39:909-16

Foyer CH, Theodoulou FL, Delrot S (2001) The functions of inter- and intracellular glutathione transport systems in plants. Trends Plant Sci 6:486-92

Furukawa K, Fujita M (1993) Advanced treatment and food production by hydroponic type wastewater treatment plant. Water Sci Technol 28:219-28

Geraldes-Laasko S, Arrabaça JD (1997) Respiration and heat production by soybean hypocotyls and cotyledon mitochondria. Plant Physiol Biochem 35:897-903

Gidda SK, Varin L (2006) Biochemical and molecular characterization of flavonoid 7-sulfotransferase from *Arabidopsis thaliana*. Plant Phys Biochem 44:628-36

Gordeziani M, Khatisashvili G, Ananiashvili T, Varazashvili T, Kurashvili M, Kvesitadze G, Tkhelidze P (1999) Energetic significance of plant monooxygenase individual components participating in xenobiotic degradation. Int Biodeter Biodegrad 44:49-54

Greim H, Ahlers J, Bias R, Broecker B, Hollander H, Gelbke H, Klimisch H, Mangelsdorf I, Paetz A, Schön N, Stropp G, Vogel R, Weber C, Ziegler-Skylakakis K, Bayer E (1994) Toxicity and ecotoxicity of sulfonic acids. Structure activity relationship. Chemosphere 28:2203-36

Haberl R, Grego S, Langergraber G, Kadlec RH, Cicalini AR, Martins Dias S, Novais JM, Aubert S, Gerth A, Hartmut T, Hebner A (2003) Constructed wetlands for the treatment of organic pollutants. J Soil Sediments 3:109-24

Han YS, van der Heijden R, Lefeber AWM, Erkelens C, Verpoorte R (2002) Biosynthesis of anthraquinones in cell cultures of Cinchona "Robusta" proceeds via the methylerythritol 4-phosphate pathway. Phytochemistry 59:45-55

Haslam R, Raveton M, Cole DJ, Pallett KE, Coleman JOD (2001) The identification and properties of apoplastic carboxylesterases from wheat that catalyse deesterification of herbicides. Pestic Biochem Physiol 71:178-89

Hawkesford MJ, De Kok LJ (2006) Managing sulphur metabolism in plants. Plant Cell Environ 29:382-95

Isin EM, Guengerich FP (2007) Complex reactions catalyzed by cytochrome P450 enzymes. Biochim Biophys Acta 1770:314-29

Jones P, Vogt T (2001) Glycosyltransferases in secondary plant metabolism: tranquilizers and stimulant controllers. Planta 213:164-74

Kevresan S, Kirsek S, Kandrac J, Petrovic N, Kelemen D (2003) Dynamics of cadmium distribution in the intercellular space and inside cells in soybean roots, stems and leaves. Biol Plant 46:85-8

Khatisashvili G, Gordeziani M, Kvesitadze G, Korte F (1997) Plant monooxygenases: participation in xenobiotic oxidation. Ecotoxicol Environ Safety 36:118-22

Khouri HE, Ibrahim RK (1987) Purification and some properties of five anthraquinone-specific glucosyltransferases from *Cinchona succirubra* cell suspension culture. Phytochemistry 26:2531-5

Klein M, Martinoia E, Weissenbock G (1997) Transport of lucifer yellow CH into plant vacuoles – evidence for direct energization of a sulphonated substance and implication for the design of new molecular probes. FEBS Lett 420:86-92

Kochhar S, Kochhar VK (2005) Expression of antioxidant enzymes and heat shock proteins in relation to combined stress of cadmium and heat in *Vigna mungo* seedlings. Plant Sci 168:921-9

Korte F, Kvesitadze G, Ugrekhelidze D, Gordeziani M, Khatisashvili G, Buadze O, Zaalishvili G, Coulston F (2000) Organic toxicants and plants. Ecotoxicol Environ Safety 47:1-26

Lennon AM, Neuenschwander UH, Ribas-Carbo M, Giles L, Ryals JA, Siedow JN (1997) The effects of salicylic acid and tobacco mosaic virus infection on the alternative oxidase of tobacco. Plant Physiol 115:783-91

Lim EK, Doucet CJ, Li Y, Elias L, Worrall, Spencer SP, Ross J, Bowles DJ (2002) The activity of Arabidopsis glycosyltransferases toward salicylic acid, 4-hydroxybenzoic acid, and other benzoates. J Biol Chem 277:586-92

Lohaus G, Winter H, Riens B, Heldt HW (1995) Further studies of the phloem loading process in leaves of barley and spinach. The comparison of metabolite concentrations in the apoplastic compartment with those in the cytosolic compartment and in the sieve tubes. Bot Acta 108:270-5

Lohaus G, Pennewiss K, Sattelmacher B, Hussmann M, Muehling KH (2001) Is the infiltration-centrifugation technique appropriate for the isolation of apoplastic fluid? A critical evaluation with different plant species. Physiol Plant 111:457-65

Macek T, Mackova M, Kas J (2000) Exploitation of plants for the removal of organics in environmental remediation. Biotechnol Adv 18:23-34

Mackova M, Chroma L, Kucerova P, Burkhard J, Demnerova K, Macek T. (2001) Some aspects of PCB metabolism by horseradish cells. Int J Phytorem 3:401-14

Matsuda H, Morikawa T, Toguchida I, Park JY, Harima S, Yoshikawa M (2001) Antioxidant constituents from rhubarb: structural requirements of stilbenes for the activity and structures of two new anthraquinone glucosides. Bioorganic Medicinal Chem 9:41-50

May MJ, Vernoux T, Leaver C, van Montagu M, Inzé D (1998) Glutathione homeostasis in plants: implications for environmental sensing and plant development. J Exp Bot 49:649-67

Mazel A, Levine A (2002) Induction of glucosyltransferase transcription and activity during superoxide-dependent cell death in Arabidopsis plants. Plant Physiol Biochem 40:133-40

McMullan G, Meehan C, Conneely A, Kirby N, Robinson T, Nigam P, Banat IM, Marchant R, Smyth WF (2001) Microbial decolourisation and degradation of textile dyes. Appl Microbiol Biotechnol 56:81-7

Meagher R (2000) Phytoremediation of toxic elemental and organic pollutants. Curr Opin Plant Biol 3:153-62

Mendoza-Cozatl D, Loza-Tavera H, Hernandez-Navarro A, Moreno-Sanchez R (2005) Sulfur assimilation and glutathione metabolism under cadmium stress in yeast, protists and plants. FEMS Microbiol Rev 29:653-71

Messner B, Thulke O, Schäffner AR (2003) Arabidopsis glucosyltransferases with activities toward both endogenous and xenobiotic substrates. Planta 217:138-46

Millenaar FF, Gonzalez-Meler MA, Fiorani F, Welschen R, Ribas-Carbo M, Siedow JN, Wagner AM, Lambers H (2001) Regulation of alternative oxidase activity in six wild monocotyledonous species. An in vivo study at the whole root level. Plant Physiol 126:376-87

Möller IM (2001) Plant mitochondria and oxidative stress: electron transport, NADPH turnover and metabolism of reactive oxygen species. Annu Rev Plant Physiol Plant Mol Biol 52:561-91

Morant M, Bak S, Möller BL, Werck-Reichhart D (2003) Plant cytochromes P450: tools for pharmacology, plant protection and phytoremediation. Curr Opin Biotech 14:151-62

Morimoto M, Tanimoto, Sakatani A, Komai K (2002) Antifeedant activity of an anthraquinone aldehyde in *Galium aparine* L. against *Spodoptera litura* F. Phytochemistry 60:63-166

Nigam P, Armour G, Banat IM, Singh D, Marchant R (2000) Physical removal of textile dyes from effluents and solid-state fermentation of dye-adsorbed agricultural residues. Bio Tech 72:219-26
Noctor G, Arisi ACM, Jouanin L, Kunert KJ, Rennenberg H, Foyer CH (1998) Glutathione: biosynthesis, metabolism and relationship to stress tolerance explored in transformed plants. J Exp Bot 49:623-47
Noguchi K, Go CS, Terashima I, Ueda S, Yoshinari T (2001) Activities of the cyanide-resistant respiratory pathway in leaves of sun and shade species. Aust J Plant Physiol 28:27-35
Nun NB, Plakhine D, Joel DM, Mayer AM (2003) Changes in the activity of the alternative oxidase in Orobanche seeds during conditioning and their possible physiological function. Phytochemistry 64:235-41
Nyanhongo GS, Gomes J, Gübitz GM, Zvauya R, Read J, Steiner W (2002) Decolorization of textile dyes by laccases from a newly isolated strain of *Trametes modesta*. Water Res 36:1449-56
Onegi B, Kraft C, Köhler I, Freund M, Jenett-Siems K, Siems K, Beyer G, Melzig MF, Bienzle U, Eich E (2002) Antiplasmodial activity of naphthoquinones and one anthraquinone from *Stereospermum kunthianum*. Phytochemistry 60:39-44
Palmer JM, Schwitzguébel JP, Möller IM (1982) Regulation of malate oxidation in plant mitochondria - Response to rotenone and exogenous NAD. Biochem J 208:703-11
Paquette S, Möller BL, Bak S (2003) On the origin of family 1 plant glycosyltransferases. Phytochemistry 62:399-13
Pflugmacher S, Sandermann H (1998) Taxonomic distribution of plant glucosyltransferases acting on xenobiotics. Phytochemistry 49:507-11
Pflugmacher S, Schröder P, Sandermann H (2000) Taxonomic distribution of plant glutathione S-transferases acting on xenobiotics. Phytochemistry 54:267-73
Rasmusson AG, Soole KL, Elthon TE (2004) Alternative NAD(P)H dehydrogenases of plant mitochondria. Annu Rev Plant Biol 55:23-39
Rausch T, Wachter A (2005) Sulfur metabolism: a versatile platform for launching defence operations. Trends Plant Sci 10:503-9
Ribas-Carbo M, Lennon AM, Robinson SA, Giles L, Berry JA, Siedow JN (1997) The regulation of electron partitioning between the cytochrome and alternative pathways in soybean cotyledon and root mitochondria. Plant Physiol 113:903-11
Riediker S, Suter MJF, Giger W (2000) Benzene and naphthalenesulfonates in leachates and plumes of landfills. Water Res 34:2069-79
Robinson T, McMullan G, Marchant R, Nigam P (2001) Remediation of dyes in textile effluent: a critical review on current treatment technologies with a proposed alternative. Bio Tech 77:247-55
Saidi Y, Domini M, Choy F, Zryd JP, Schwitzguébel JP, Goloubinoff P (2007) Activation of the heat shock response in plants by chlorophenols: transgenic *Physcomitrella patens* as a sensitive biosensor for organic pollutants. Plant Cell Environ 30:753-63
Sandermann H (2004) Molecular ecotoxicology of plants. Trends Plant Sci 9:406-13
Sattelmacher B (2001) The apoplast and its significance for plant mineral nutrition. New Phytol 149:167-92
Schwitzguébel JP (2004) Potential of phytoremediation, an emerging green technology: European trends and outlook. Proc Indian Natl Sci Acad B70:131-52
Schwitzguébel JP, Siegenthaler PA (1984) Purification of peroxisomes and mitochondria from spinach leaf by Percoll gradient centrifugation. Plant Physiol 75:670-74
Schwitzguébel JP, Vanek T (2003) Fundamental advances in phytoremediation for xenobiotic chemicals. In: McCutcheon SC, Schnoor JL (eds), Phytoremediation: transformation and control of contaminants, John Wiley and Sons, Hoboken, NJ, pp. 123-57
Schwitzguébel JP, Nguyen TD, Siegenthaler PA (1985) Calmodulin is not involved in the regulation of exogenous NADH oxidation by plant mitochondria. Physiol Plant 63:187-91
Schwitzguébel JP, Aubert S, Grosse W, Laturnus F (2002) Sulphonated aromatic pollutants – Limits of microbial degradability and potential of phytoremediation. Environ Sci Poll Res 9:62-72

Siciliano SD, Germida JJ (1998) Mechanisms of phytoremediation: biochemical and ecological interactions between plants and bacteria. Environ Review 6:65-79

Singer AC, Crowley DE, Thompson IP (2003) Secondary plant metabolites in phytoremediation and biotransformation. Trends Biotech 21:123-30

Stiborova M, Schmeiser HH, Frei E (2000) Oxidation of xenobiotics by plant microsomes, a reconstituted cytochrome P450 system and peroxidase: a comparative study. Phytochemistry 54:353-62

Taguchi G, Nakamura M, Hayashida N, Okazaki M (2003) Exogenously added naphthols induce three glucosyltransferases, and are accumulated as glucosides in tobacco cells. Plant Sci 164:231-40

Tasgin E, Atici O, Nalbantoglu B, Popova LP (2006) Effects of salicylic acid and cold treatments on protein levels and on the activities of antioxidant enzymes in the apoplast of winter wheat leaves. Phytochemistry 67:710-15

Trapp S, Karlson U (2001) Aspects of phytoremediation of organic pollutants. J Soils Sediments 1:1-7

Trease GE, Evans WC (1983) Pharmacognosy, 12th ed., Baillière Tindall, London.

Umbach AL, Ng VS, Siedow JN (2006) Regulation of plant alternative oxidase activity: A tale of two cysteines. Biochim Biophys Acta 1757:135-42

Van den Berg AJJ, Labadie RP (1989) Quinones. In: Dey PM, Harborne JB (eds), Methods in plant biochemistry, Vol. 1, Plant Phenolics. Academic Press, London, pp. 451-91

Van der Lelie D, Schwitzguébel JP, Glass DJ, Vangronsveld J, Baker A (2001) Assessing phytoremediation's progress in the United States and Europe. Environ Sci Technol 35: 446A-52A

Van der Plaas LHW, Hagendoorn MJM, Jamar DCL (1998) Anthraquinone glycosylation and hydrolysis in *Morinda citrifolia* cell suspensions: regulation and function. J Plant Physiol 152:235-41

Vandevivere PC, Bianchi R, Verstraete W (1998) Treatment and reuse of wastewater from the textile wet-processing industry: review of emerging technologies. J Chem Tech Biotech 72:289-302

Vanlerberghe GC, McIntosh L (1997) Alternative oxidase: from gene to function. Annu Rev Plant Physiol Plant Mol Biol 48:703-34

Varin L, Marsolais F, Richard M, Rouleau M (1997) Biochemistry and molecular biology of plant sulfotransferases. FASEB J 11:517-25

Wang W, Vinocur B, Shoseyov O and Altman A (2004) Role of plant heat-shock proteins and molecular chaperones in the abiotic stress response. Trends Plant Sci 9:244-52

Werck-Reichhart D, Hehn A, Didierjean L (2000) Cytochromes P450 for engineering herbicide resistance. Trends Plant Sci 5:116-23

Willetts J, Ashbolt N (2000) Understanding anaerobic decolourisation of textile dye wastewater: mechanism and kinetics. Water Sci Tech 42:409-15

Wink M (2003) Evolution of secondary metabolites from an ecological and molecular phylogenetic perspective. Phytochemistry 64:3-19

Wong Y, Yu J (1999) Laccase-catalyzed decolorization of synthetic dyes. Water Res 33:3512-20

Young L, Yu J (1997) Ligninase-catalysed decolorization of synthetic dyes. Water Res 31:1187-93

Zaalishvili G, Khatisashvili G, Ugrekhelidze D, Gordeziani M, Kvesitadze G (2000) Plant potential for detoxification (Review). Appl Biochem Microbiol 36:443-51

Chapter 17
Effects of Fertilization with Sulfur on Quality of Winter Wheat: A Case Study of Nitrogen Deprivation

Anna Podlesna(✉) and Grazyna Cacak-Pietrzak

Abstract Experiments with a bread type of winter wheat var. Kobra were performed at a very good rye complex for three years. The first experimental factor (I) was sulfur fertilization: control without sulfur (-S) and 50 kg S ha^{-1} (+S) and the second (II) – levels of nitrogen fertilization: 0, 30, 60, 90, 120, and 150 kg N ha^{-1}. Plant harvest was performed at full maturity of wheat, and flour and dough technological parameters were analyzed. All levels of nitrogen increased grain yield, while applied sulfur showed effects in the middle range of nitrogen doses. However, sulfur positively affected the quality of flour by decreasing ash content and increasing gluten and protein concentration. Furthermore, these conditions positively affected development, stability, softening, and quality of dough as well as bread volume. Wheat fertilization with S and N affected S content in grain. The N:S ratio ranged from 8 to 10.5:1, and the highest values indicate the best technological properties of winter wheat grain.

1 Introduction

Winter wheat is the main bread cereal in Poland. The most important criterion of wheat cultivation is stable yield and high technological quality. In field cultivation the weather conditions are the basis for proper plant growth and development for profitable production (Johansson and Svensson 1999). However, previous studies showed that agricultural practices played an important role in the development of grain quality. Achremowicz et al. (1993) and Luo et al. (2000) showed that genotype has a strong influence on quality parameters. Plant protection and applied fertilization are the other factors which influence cereal yield and quality (Podolska and Stypula 2002). Nitrogen is known as the most yield-creative factor, and many authors reported its positive effect (Fotyma 1999, Hřivna et al. 1999, Podlesna and Cacak-Pietrzak 2006). Apart from this, Luo et al. (2000) found that nitrogen application

Anna Podlesna
Plant Nutrition and Fertilisation Department, Institute of Soil Science and Plant Cultivation – National Research Institute, Czartoryskich 8, 24-100 Pulawy, Poland
ap@iung.pulawy.pl

N.A. Khan et al. (eds.), *Sulfur Assimilation and Abiotic Stress in Plants.*

significantly increased the wholemeal and white flour protein percentage, hardiness, SDS sedimentation, and midline peak value of the mixograph. According to Flæte et al. (2005), the increase of total N fertilization has been the major strategy recommended for intensive wheat production. However, at high N fertilization levels balanced S application is required to obtain an optimal N/S ratio and SDS sedimentation volume. There are many papers which confirm the important influence of sulfur on the breadmaking quality of flour, but there is little information concerning the need for combined nitrogen and sulfur fertilization for wheat in Poland (Podlesna and Cacak-Pietrzak 2006). According to Motowicka-Terelak and Terelak (1998) over 70% of Polish soils are characterized by low concentration of available sulfur. In addition, light, acid soils prevail in Poland, and for the two last decades, the considerable reduction of gaseous sulfur compounds emitted to the atmosphere has been observed (Podlesna 2002). A certain amount of sulfates are also found at drains and in drainage ditches water (Igras 2004).

The aim of this study was the estimation of winter wheat requirement for sulfur and nitrogen with respect to yield and baking parameters.

2 Methods

The winter wheat grain variety Kobra was grown on a very good rye complex location. Experimental plots were 20 m^2 in area, and the plant population was 5 million per hectare. The first experimental factor (I) was sulfur fertilization: control without sulfur (-S) and 50 kg S ha^{-1} (+S) and the second (II) – levels of nitrogen fertilization: 0, 30, 60, 90, 120, and 150 kg N ha^{-1}. The soil was characterized by very low sulfur content (0, 18 mg S-SO_4 100 g^{-1} soil) and about 60 kg N per ha (Fotyma and Fotyma 2000). Sulfur was given in the autumn together with phosphorus and potassium. Mineral nitrogen was applied in the spring as ammonium sulfate at the rate of 30 kg N ha^{-1} in two-week intervals from 18 (8 leaves stage) up to 64-65 (full flowering) according to Zadoks et al. (1974) scale. Winter wheat plants were protected against pests and disease as well as against competition by weeds. Harvest of plants was performed at full maturity. Nitrogen and sulfur concentrations were determined in dried and ground samples of grain. Nitrogen was analyzed by flow spectrophotometry after wet mineralization of samples, and sulfur was determined by the roentgen fluorescence method (Haneklaus and Schnug 1994). The grain N:S ratio was calculated from the N and S concentrations.

The evaluation of milling and baking parameters of wheat grain was performed at the Department of Food Technology, Agricultural University in Warszawa. Grain samples of about 1,5 kg were milled at Quadrumat Senior laboratory mill. On the basis of obtained results the balance of milling was done. The milling efficiency factor K and ash numbers (AN) were calculated according the formulas:

$$K = Wm/Pm$$

$$AN = Pm \cdot 100000/Wm$$

where:
Wm – yield of flour (%)
Pm – content of ash in the flour (%)

The nitrogen concentration in flour was estimated by the use of the Kjeldahl method, and then the total protein content was calculated by multiplying %N and 5.83 value was calculated. The amount and quality of wet gluten after washing on Glutomatic (Polish Norm 93-A-74042/02) and total ash content were determined in the flour (Jakubczyk and Haber 1983). Furthermore, dough density evaluation was done using Farinograph Brabendera. The interpretation of results was carried out according to Brabender/ICC (PN-ISO5530-1:1999) with computer readings on the following features: water absorption, dough development, dough stability, dough softening, and quality number. Laboratory baking was assessed according to the direct method (Jakubczyk and Haber 1983).

The presented data show the mean from 3-year studies. Analysis of variance and regression were made with Statgraphic v. 5.1.

3 Results

3.1 Grain Yield

The results showed that mineral sulfur had a positive effect on grain yield of winter wheat only in the range of 30-90 kg N per ha (Fig. 17.1). The increase of grain yield from 25% to 45%, which conforms to 130-220 kg of grain per ha, was observed. Sulfur applied in the two highest N doses did not show a positive interaction with nitrogen, so the grain yields were about 16% to 33% lower in comparison to yields obtained on plots without sulfur. Nitrogen was very profitable for winter wheat and caused the yield to increase up to the highest dose.

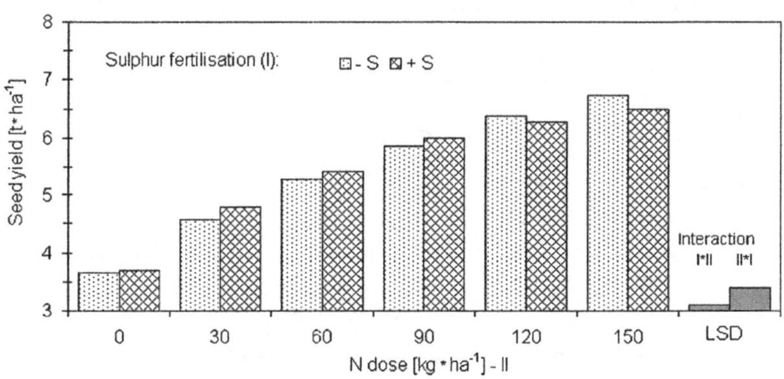

Fig. 17.1 Interactive effect of nitrogen and sulfur on seed yield of winter wheat

3.2 Quality Parameters

Although sulfur had no significant effect on grain yield increase, it positively affected gluten content in flour (Fig. 17.2). The highest increase of gluten content with sulfur was found with the control. Together with nitrogen the increase in gluten concentration was consistently higher. The positive effect of sulfur on gluten had a lowering tendency but was visible up to the highest rate of N. However, mean results of gluten softening indicated values in the range of 17 and 20 mm for −S and +S, respectively.

The protein content in flour of winter wheat grain was highly dependent on applied nitrogen (Fig. 17.3). Its concentration increased linearly together with N levels. Sulfur increased protein content with higher levels of nitrogen.

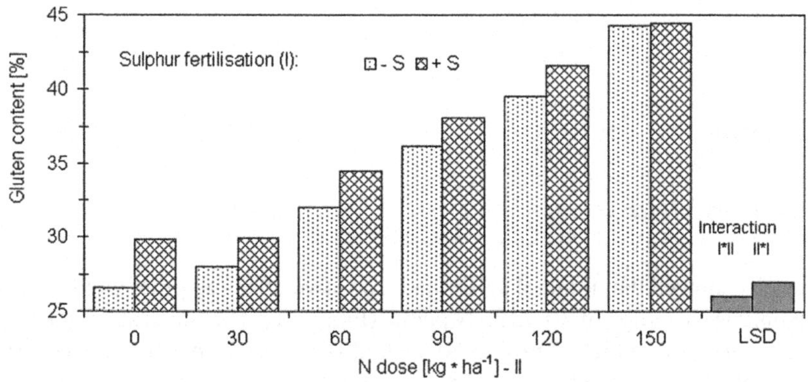

Fig. 17.2 Interactive effect of nitrogen and sulfur on gluten content of winter wheat grain

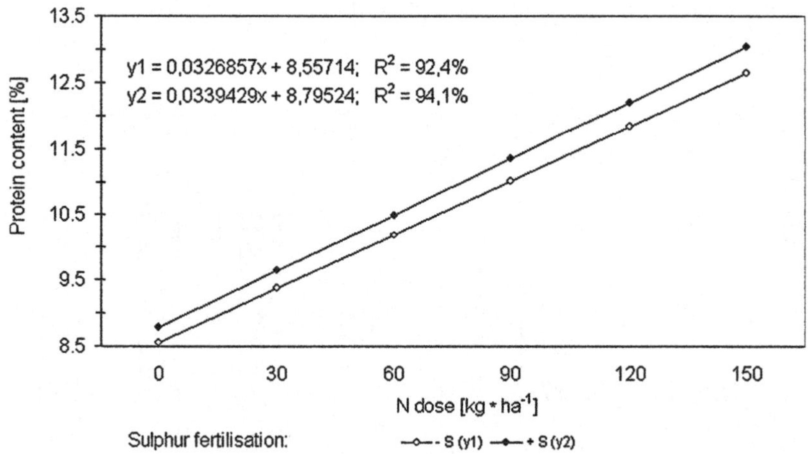

Fig. 17.3 Interactive effect of nitrogen and sulfur on protein content of grain flour of winter wheat

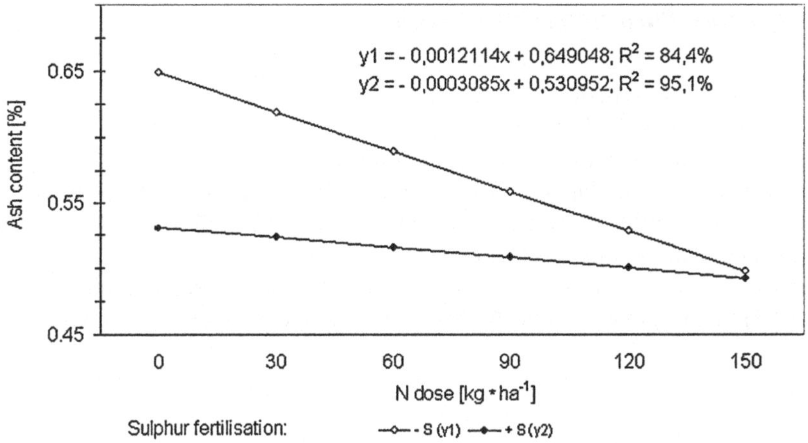

Fig. 17.4 Interactive effect of nitrogen and sulfur on ash content of flour of winter wheat grain

Table 17.1 The results of laboratory milling

Description	Yield of flour (%)	Bran (%)	Shorts (%)	Milling efficiency factor K	Ash number
N dose (kg · ha^{-1}):					
0	77.6a*	16.2b	5.2a	138a	730c
30	78.0ab	16.2b	5.3a	132a	770d
60	78.8bc	15.4a	5.4a	148bc	677b
90	78.0ab	15.4a	5.5a	153c	655a
120	79.5c	15.0a	5.2a	154c	652a
150	79.1c	15.0a	5.4a	149bc	674b
S fertilization:					
–S	78.7a	15.4a	5.2a	138a	732a
+S	78.3a	15.6a	5.5a	153b	654b

* The data in a column with same letters did not differ significantly.

The highest ash content in the flour was found with no nitrogen and decreased with the increase in nitrogen dose (Fig. 17.4). Considerably higher values of ash content were found in flour in the treatment –S than +S ones.

The results of laboratory milling show that some of these parameters are not dependent on fertilization (Table 17.1). Therefore, differentiated nitrogen doses did not significantly change the bran content in the flour. The applied sulfur did not influence the yield of flour as well as the percent of bran. A more evident effect of nitrogen and sulfur fertilization was found in the case of milling efficiency factor K and ash number.

3.3 *Quality Parameters of Dough*

Most of the tested rheological parameters of dough indicate their significant dependence on applied fertilizers (Table 17.2). All studied features had significantly greater values with increasing nitrogen levels, however, an exception was dough softening. A similar response was noted with sulfur application. In general, fertilization with sulfur caused a significant increase of dough development, dough stability, and quality number of flour.

After the bread was baked, bread volume was found to depend on nitrogen dose (Fig. 17.5). It was greatest in the highest N rate application; however, an additional

Table 17.2 Rheological features of dough

Description	Water absorption (%)	Dough development (min)	Dough stability (min)	Dough softening after 10 min (FU)	Quality number
N dose (kg · ha^{-1}):					
0	50.4a*	1.4b	1,8a	112e	27a
30	50.7a	1.1a	2,1b	102d	28a
60	51.6b	2.3c	3,0d	84c	38b
90	52.2b	2.8d	3,8e	86c	46c
120	54.0c	2.9d	2,7c	76b	53d
150	55.3d	3.6e	3,0d	64a	56d
S fertilization:					
–S	52.8a	2.2a	2,6a	90b	37a
+S	51.9a	2.5b	2,9b	84a	45b

* Explanations in Table 17.1.

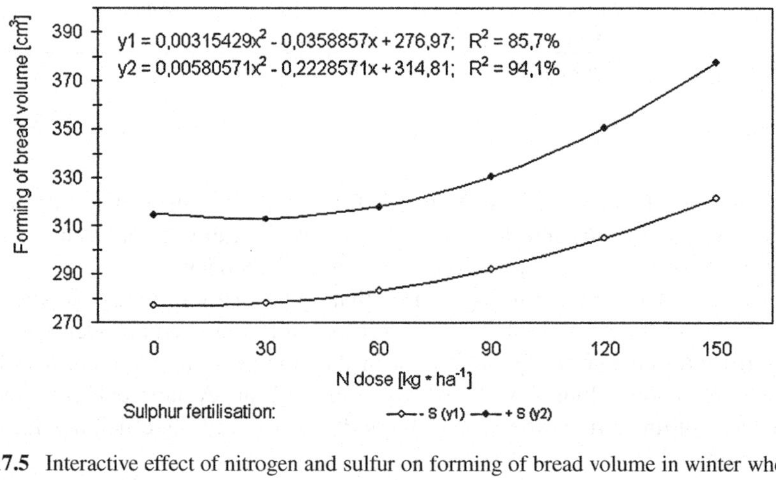

Fig. 17.5 Interactive effect of nitrogen and sulfur on forming of bread volume in winter wheat

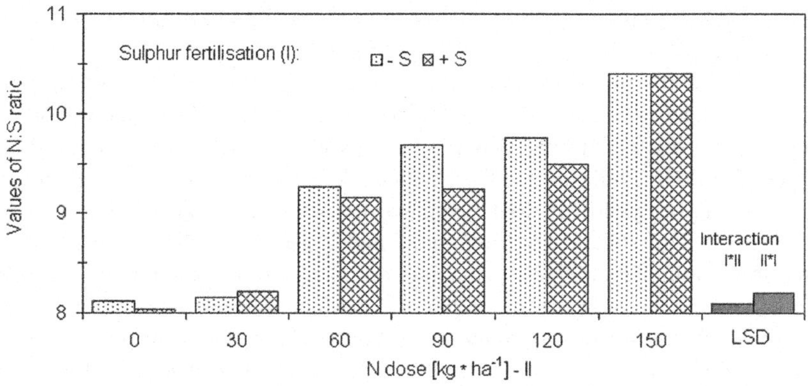

Fig. 17.6 Interactive effect of nitrogen and sulfur on values of N:S ratio in winter wheat

supply of plants with sulfur caused a considerable increase of bread volume at all nitrogen levels in comparison to bread volume without S treatment. The greatest increase was obtained with 150 kg N and 50 kg S per hectare, which was about 20% in comparison to the same N treatment but without sulfur.

3.4 Grain S and N Concentration and N:S Ratio

In general, the higher doses of nitrogen increased sulfur and nitrogen concentrations. Thus the grain N:S ratio was lower than 10:1 (Fig. 17.6). The greatest value of N:S ratio was achieved at the highest levels of nitrogen fertilization irrespective of sulfur dose. The application of sulfur decreased N:S ratio in the range of middle nitrogen levels.

4 Discussion

It is generally known that plants treated with mineral nutrients increase not only in yield but also in grain quality – in their chemical composition as well as nutritive and technological what means its good parameters to bread production (Cacak-Pietrzak et al. 1999, Ceglinska et al. 2005). The earlier reports of Fotyma (1999) and Sticksel et al. (2000) also showed the important role of nitrogen in yield formation. Our results showed that sulfur application also positively affected grain yields in comparison to control. Boreczek (2000) did not find any influence of sulfur-containing fertilizers on yields of winter and spring wheat, while Zhang et al. (1999) obtained wheat yield increase in the range from 3.5% to 14.5% with the use of sulfur. A field experiment performed by Hrivna et al. (1999) confirmed that sulfur has a positive effect on yields and on some quality indicators of winter wheat grain.

Flour yield is the fundamental parameter used for evaluating grain milling value, and it had a tendency to increase with the increase in nitrogen dose. According to the results of Cacak-Pietrzak (1999) differentiated nitrogen fertilization did not influence the amount of flour (Podlesna and Cacak-Pietrzak 2006). It is worth emphasizing that we obtained higher yields of flour compared to mean values for varieties of winter wheat cultivated in Poland (Achremowicz and Zajac 1993). This may be due to greater grain plumpness and less ash content in the grain. It is known that the higher values of factor K and lower ash number are considered as a better milling value for grain. These relationships were found in our study with sulfur application.

One of the factors that decides the baking value of wheat flour is the amount and quality of gluten. Its presence gives a gluey and plasticity which are very important features of flour. Gluten concentration in flour was highly dependent on nitrogen in our experiment, and this has also been shown in studies performed by Achremowicz and Zajac (1993) and Podlesna and Cacak-Pietrzak 2006). In their research a positive effect of sulfur on higher gluten concentration of wheat flour was found. However, the quality was poor with combined application of high doses of sulfur and nitrogen. The tested flour of spring wheat showed lower softening of gluten with these treatments (Podlesna and Cacak-Pietrzak 2006). In the opinion of Ewart (1978) the positive influence of sulfur is the disulphide bridges formed between groups of cysteine which are responsible for the connection of glutelin fractions.

Protein content in flour is an important parameter of wheat milling value. According to Bechtel et al. (1982) the importance of protein in baking value results from easiness of gluten net creation during the kneading. Gluten net makes easy to keeping of CO_2 produced during dough fermentation and this way porous crumb of bread is forming. Its concentration in our experiment depended on applied nitrogen dose and sulfur fertilization. It is known that nitrogen and sulfur are required in protein biosynthesis by their necessary presence in amino acids. Mortensen and Eriksen (1994) found the decrease of methionine and cysteine concentration in grain with sulfur deficit conditions. In consequence less than 25% of total nitrogen present in the plant was found in a form of protein, whereas with a normal supply of sulfur about 75% of nitrogen was found as protein. The change in baking quality of wheat grain due to sulfur deficiency has been partly attributed to the reduction in the proportion of proteins enriched in sulfur amino acids, which are considered as crucial in formation of rheological properties of bread dough (Castle and Randall 1987). The studies of Uziak and Szymanska (1969) showed that sulfur application positively affected glutamic acid concentration. However, Zhao et al. (1999) found that sulfur application did not affect grain protein concentration but directly increased the gel protein weight in flour as well as the proportion of polymeric proteins.

According to Czubaszek et al. (2005) protein content in flour influences flour water absorption. Water in dough plays important functions as solvent and creates an environment for some reactions of organic and inorganic substances. It also forms organoleptic properties and is a structure-forming factor (Piesiewicz 2004). Therefore, the greater the amount of water in dough, the softer the bread. The flour water absorption increased significantly with an increasing N rate in six of the seven experiments performed by Zhao et al. (1999). Similar results were obtained previously with spring wheat (Podlesna and Cacak-Pietrzak 2006). Czubaszek et al. (2005) found that dough made with barley flour, which is characterized by greater water absorption, took longer to develop and was more stable, which is in agreement with the results obtained in our experiments. Cacak-Pietrzak et al. (1999) found that flour water absorption and dough development and stability are dependent most of all on amount and quality of gluten contained in flour and its ability to join water. So dough made from flour which has weak gluten is characterized by shorter development and less stability than flours with strong gluten. These parameters are important because they create dough resistance to kneading, which is closely related to dough softening. Longer time of dough resistance increases softening, and in this way flour shows higher technological value. The quality number of flour is also a very positive parameter from the technological point of view, and greater values of this feature indicate better value of flour.

Volume is important in the evaluation of technological properties of wheat grain loaf. The significant increase in loaf volume with nitrogen levels was found in our experiment. The interaction of nitrogen and sulfur supply had a very positive effect on loaf volume. These results confirm the findings of Zhao et al. (1999). These authors observed the increase of loaf volume after application of sulfur. According to Moss et al. (1981) this positive effect of S can be explained by greater S content in the wheat grain, which induces higher baking volumes. Zhao et al. (1999) also reported that grain S status (S concentration and N:S ratio) is an important parameter that affects not only loaf volume but also breadmaking performance of wheat. However, the best N:S ratio ranged from 8 to 10.5:1 in our experiment and but was lower than 17:1 (Luo et al. 2000, Haneklaus et al. 1992). However, higher N fertilization gave significantly higher grain yield and grain N and S content, as well as N:S ratio, compared to results obtained at lower N fertilization as reported recently by Flæte et al. (2005).

5 Conclusions

1. Sulfur showed a positive effect on winter wheat grain yield with moderate doses of nitrogen.
2. The influence of sulfur on flour quality was shown as the increase in gluten content and decrease of ash content.
3. Fertilization of wheat with sulfur and higher doses of nitrogen was profitable for dough quality with respect to dough development, stability, dough softening, and quality of flour.
4. The best technological value of winter wheat is obtainable in Polish conditions when the N:S ratio is greater than 9.5-10:1 but lower than 17:1.

References

Achremowicz B, Zajac J, Styk B (1993) The effect of nitrogen fertilisation increase on technological quality of some spring and winter wheat. Roczniki Nauk Rolniczych Ser A110 (1-2): 149-57 (in Polish)

Bechtel DB, Gaines RL, Pomeranz Y (1982) Protein secretion in wheat endosperm formation of the matrix protein. Cereal Chem 59:336-43

Boreczek B (2000) Balance of sulphur at four field rotation. Fertilizers Fert 4(5): 173-84 (in Polish)

Cacak-Pietrzak G, Ceglinska A, Haber T (1999) The effect of differential nitrogen fertilisation on a technological value of selected varieties of winter wheat. Pamietnik Pulawski 118:45-56 (in Polish)

Castle SL, Randall PJ (1987) Effects of sulfur deficiency on the protein synthesis and accumulation of proteins in the developing wheat seeds. Aust J Plant Physiol 14:503-16

Ceglinska A, Samborski S, Rozbicki J, Cacak-Pietrzak G, Haber T (2005) Estimation of milling and baking value of winter triticale varieties depending on nitrogen fertilisation. Pamietnik Pulawski 139:39-45 (in Polish)

Czubaszek A, Subda H, Karolini-Skaradzinska Z (2005) Milling and baking value of several spring and winter barley cultivars. Acta Sci Pol Technol Aliment 4(1):53-62 (in Polish)

Ewart JAD (1978) Glutamin and dough tenacity. J Sci Food Agric 29:551-6

Flæte NES, Hollung K, Ruud L, Sogn T, Færgestad EM, Skarpied HJ, Magnus EM, Uhlen AK (2005) Combined nitrogen and sulphur fertilisation and its effect on wheat quality and protein composition measured by SE-FPLC and proteomics. J Cereal Sci 41:357-69

Fotyma E (1999) Nitrogen uptake and utilization by winter and spring wheat. Pamietnik Pulawski 118:143-52 (in Polish)

Fotyma E, Fotyma M (2000) Content of mineral nitrogen in soil as an indicator of fertiliser requirements of plants and state of soil environment. Fertilizers Fert 4:91-101 (in Polish)

Haneklaus S, Evans E, Schnug E (1992) Baking quality and sulphur content of wheat. I Influence of grain sulphur and protein concentration of loaf volume. Sulphur Agric 16:31-4

Haneklaus S, Schnug E (1994) Diagnosis of crop sulphate status and application of X-ray fluorescence spectroscopy for the sulphur determination in plant and soil materials. Sulphur Agric 18:31-40

Hřivna L, Richter R, Ryant P (1999) Possibilities of improving the technological quality of winter wheat after sulphur fertilisation. Zeszyty Naukowe Akademii Rolniczej w Krakowie 349:143-50

Igras J (2004) Mineral element concentrations in drainage water from agricultural area in Poland. Monography and Scientific Dissertations 13, Institute of Soil Sciences and Plant Cultivation, Pulawy (in Polish)

Jakubczyk T, Haber T (1983) Analysis of cereals and cereal products. Materials for students of Agricultural University in Warszawa (in Polish)

Johansson E, Svensson G (1999) Influence of yearly variation and fertilizer rate on bread-making quality in Swedish grown wheats containing HMW gluteninsubunits 2+12 or 5+10 cultivated during the period 1990-96. J Agric Sci Camb 132:13-22

Luo C, Branlard G, Griffin WB, McNeil DL (2000) The effect of nitrogen and sulphur fertilisation and their interaction with genotype on wheat glutenins and quality parameters. J Cereal Sci 31:185-94

Mortensen J, Eriksen J (1994) Effect of sulphur on amino acid composition. Norw J Agric Sci, suppl. 15:135-42

Motowicka – Terelak T, Terelak H (1998) Sulphur at Polish soils – state and menace. PIOS Biblioteka Monitoringu Srodowiska, Warszawa (in Polish)

Piesiewicz H (2004) Four types of wheat dough. Przeglad Piekarski i Cukierniczy 6:1-2 (in Polish)

Podlesna A (2002) Air pollution by sulfur dioxide in Poland – impact on agriculture. Phyton 42 (3):157-64

Podlesna A, Cacak-Pietrzak G (2006) Formation of spring wheat baking parameters by nitrogen and sulphur fertilisation. Pamietnik Pulawski 142:381-92 (in Polish)

Podolska G, Stypula G (2002) Yielding and technological value of winter wheat grain in dependence on protection method against diseases and weeds. Pamietnik Pulawski 130 (II):587-96 (in Polish)

Sticksel E, Maidl F-X, Retzer F, Dennert J, Fischbeck G (2000) Efficiency of grain production of winter wheat as affected by N fertilisation under particular consideration of single culm sink size. Eur J Agron 13:287-94

Uziak Z, Szymanska M (1969) The effect of sulphur on nitrogen use by oilseed rape plants. Annales UMCS, Sectio E, XXIV 13:187-201 (in Polish)

Zadoks JC, Chang TT, Konzak GF (1974) A decimal code for growth stages of cereals. Weed Res 14:415-21

Zhang Z, Sum K, Lu A, Zhang X (1999) Study of the effect of sulphur fertilizer application on crops and balance of sulphur in soil. J Henan Agric Sci 5:25-7

Zhao FJ, Salmon SE, Withers PJA, Monaghan JM, Evans EJ, Shewry PR, McGrath SP (1999) Variation in the breadmaking quality and rheological properties of wheat in relation to sulphur nutrition under field conditions. J Cereal Sci 30:19-31

Index

The manufacturer's authorised representative in the EU is Springer Nature Customer Service Centre GmbH, Europaplatz 3, 69115 Heidelberg, Germany. If you have any concerns regarding our products, please contact ProductSafety@springernature.com

Printed and bound by CPI Group (UK) Ltd, Croydon, CR0 4YY

07/01/2026

02030921-0009